W9-CPR-201

BUILDING CONSTRUCTION

for the Fire Service

Third Edition

Francis L. Brannigan, SFPE
Professor (retired)
Fire Science Curriculum
Montgomery College
Rockville, Maryland

National Fire Protection Association
Batterymarch Park, Quincy, Massachusetts

Third Edition

J2 2121035

Seventh Printing June 1999

Editor: Gene A. Moulton
Consulting Editor: John S. Petraglia
Graphic Artist: George Nichols
Cover Design: Joyce C. Weston
Interior Design: Susan Gerould
Index: Joan Croce
Text Processing: Louise Grant
Marilyn Lupo
Elizabeth Turner
Composition: Cathy Ray
Production: Stephen Dornbusch
Donald McGonagle
Debra Rose

Photographs not specifically credited are from the extensive collection of the author.

NFPA No. BCFS-3
ISBN 0-87765-381-x
Library of Congress Catalog Card No. 91-68102
Printed in the United States of America

Table of Contents

1 Initial Thoughts and Recommendations 1

Watch Your Language, 2 / **Overhauling and Safety,** 3 / **Prefire Planning,** 3 / **References,** 8

2 Principles of Construction 9

Why Study Building Construction? 9 / Cases in Point, **10** / Gravity, **10** / *Tactical Considerations,* **11** / The Limitations of Experience, **12** / Collapse Is Not the Only Hazard, **12** / Fire Fighter Fatalities Report, **13** / A Near Tragic Example, **15** / *Tactical Considerations,* **15**

Definitions of Loads, 16 / Dead Load, **16** / Added Dead Load, **17** / Live Load, **18** / Added Live Loads, **18** / *Tactical Considerations,* **19** / Recycled Buildings, **19** / *Tactical Considerations,* **20** / Impact Loads, **22** / Lateral Impact Loads, **23** / Static and Repeated Loads, **24** / Wind Loads, **25** /*Tactical Considerations,* **26** / Concentrated Loads, **26** / Imposition of Loads, **26** / Axial Loads, **27** / Eccentric Loads, **27** / Torsional Loads, **29** / *Tactical Considerations,* **29** / *Tactical Considerations,* **29**

Fire Load, 29 / *Tactical Considerations,* **31** / Building Classification Fallacies, **32**

The Characteristics of Materials, 32 / *Tactical Considerations,* **34** / *Tactical Considerations,* **34** / Single-Ply Roofing, **34** / Tactical Considerations, **35** / *Tactical Considerations,* **35** / *Tactical Considerations,* **36** / *Tactical Considerations,* **37** / Fire Characteristics of Plastics, **38** / *Tactical Considerations,* **39**

The Meanings of "Combustible," 40 / Non-combustible Buildings, **42** / Confusing Language, **42** / Effect of Energy Conservation, **42** / Fire Resistance Distinguished, **43** / *Tactical Considerations,* **45**

Applying Forces to Materials, 45 / Stress and Strain, **45** / Compressive and Tensile Strength, **46** / Effects of Shape, **47** / Suspended Loads, **48** / *Tactical Considerations,* **48** / Suspended High-Rise Buildings, **49** / Safety Factor, **49** / *Tactical Considerations,* **49**

Composites, 50 / Composite Materials, **50** / Composite Structural Elements, **50**

Structural Elements, 52 / Beams, 52 / Carry Capacity and Depth of Beams, **53** / Types of Beams, **56** / *Tactical Considerations,* **59** / Beam Loading, **60** / Reaction and Bending, **60**

Columns, 61 / Decrease in Load Carrying Capacity, **61** / Shapes of Columns, **61** / Tactical Considerations, **62** / Wood Columns, **62** / Tactical Considerations, **63** / I-Beams vs. H-Columns, **63**

Types of Columns, 64 / Euler's Law Columns, **64** / *Tactical Considerations,* **66** / Temporary Bracing, **66**

Walls, 67 / *Tactical Considerations,* **70** / Cantilever Wall, **70** / *Tactical Considerations,* **71** / Wall Bracing, **72** / Wall Breaching, **73** / *Tactical Considerations,* **74**

Roofs, 74 / **Arches,** 74 / *Tactical Considerations,* **76**

Rigid Frames, 77 / **Shells and Domes,** 78 / **Transmission of Loads,** 78

Foundations, 79 / *Tactical Considerations,* **80**

Connections, 80 / Failure of Connections, **81** / *Tactical Considerations,* **84** / Overhanging and Drop-In Beams, **85** / Spliced Beams, **85** / *Tactical Considerations,* **86**

A Note to the Reader, 86 / **References,** 86 / **Additional Readings,** 87

3 Wood Construction 89

Tactical Considerations, **89** / The California Bungalow, **90** / The Ranch House, **90** / Tile Roofs, **91** / *Tactical Considerations,* **91** / Definitions, **92** / Uses of Wood, **93**

Types of Wood Building Construction, 93 / Log Cabin, **93** / Post and Frame, **94** / *Tactical Considerations,* **96** / Deceptive Imitation Post and Frame, **96** / Balloon Frame, **96** / Interior Walls, **99** / Fires in Balloon-frame Buildings, **99** / *Tactical Considerations,* **100** / Platform Frame, **100** / *Tactical Considerations,* **101** / Plank and Beam, **101** / *Tactical Considerations,* **103** / Truss Frame, **103** / *Tactical Considerations,* **104**

Wood in Non-combustible and Fire-resistive Buildings, 104 / *Tactical Considerations,* **105**

Exposure to Fire-resistive Buildings, 105 / *Tactical Considerations,* **106**

Firestopping, 106 / Types of Firestopping, **107** / *Tactical Considerations,* **108** / Firestopping and Truss Floors, **109** / *Tactical Considerations,* **109**

Wood as a Building Material, 110 / **Protecting Wood from**

Ignition, 110 / Impregnation, **110** / Corrosion, **111** / Loss of Strength, **111** / Plywood Roofing Problems, **111** / *Tactical Considerations,* **112** / Surface Coating, **112**

Dangerous Treated Wood, 112 / *Tactical Considerations,* **113**

Engineered Wood, 114 / Plywood, **114** / *Tactical Considerations,* **115** / Spliced and Laminated Timbers, **115** / *Tactical Considerations,* **117** / Paper Wrapping, **117** / Planks, **117** / Finger Joints, **117** / Chipboard, **117** / *Tactical Considerations,* **118** / Flitch Plate Girders, **118** / Trusses and Wooden I-Beams, **118** / Wood and Plastic Roof Panels, **119**

Sheathing, 119 / **Siding, 120** / Brick Veneer, **122** / *Tactical Considerations,* **124** / Brick Veneer on Non-wooden Buildings, **124** / Stone Veneer, **124**

Wood Shingle Roofing, 124 / *Tactical Considerations,* **126**

Row Frame Buildings, 126 / *Tactical Considerations,* **127**

Wooden Cooling Towers, 129 / **Heavy Timber Construction, 129** / Slow Burning? **130**

Advantages and Disadvantages, 130 / **Use of Unprotected Steel, 131** / **Rehabilitation and Demolition Hazard, 132** / **Vacant Buildings, 133** / Tactical Considerations, **134**

Open Area Structures, 134 / *Tactical Considerations,* **134**

Imitation Timber, 135 / **Wooden Suspended Ceilings, 136** / **Strength of Wood, 136** / The Rising Truss, **137** / *Tactical Considerations,* **137** / Wood Dimensions, **137**

Making Wood Construction Safe, 137 / **References, 138** / **Additional Readings, 139**

4 Ordinary Construction 143

Tactical Consideration, **143** / **Construction Characteristics, 144** / Masonry Construction Terms, **145** / Hollow or Cavity Walls, **146** / Composite Walls, **147** / Cast Iron, **148** / Lintels, **148** / Imitation Brick, **148**

Characteristics of Ordinary Construction, 149 / Renovation and Restoration, **151** / *Tactical Considerations,* **151** / Fire Resistance, **152** / Preservation, **152** / *Tactical Considerations,* **152** / Recent Construction, **52**

Problems of Ordinary Construction, 153 / Discovery of Hazard, **154** / The Owner's Rights, **154**

Structural Stability of Masonry Walls, 155 / General Collapse Indicators, **155** / Signs of Potential Collapse, **156** / Bricks, **157** / Sand Lime Mortar, **157** / Wooden Beams, **158** / Poor Quality Bricks, **160** / Cracks, **160** / Arches, **160** / Tiles, **160** /

Unstable Walls, **161** / Holes in Walls, **161** / Steel Lintels, **162** / Braced Walls, **162** / Hanging Signs, **163** / Balconies, **163** / Canopies, **163** / Carbon Monoxide, **164** / Masonry Walls, **164** / Plane of Weakness, **164** / Effect of Interior Structure on the Wall, **165** / Appearances Are Deceiving, **166** / Steel Beams, **166** / Telephone Poles, **167** / Materials Soaked by Water, **167** / Effect of Fire Streams on Brick Walls, **167** / *Tactical Considerations,* **167**

Interior Structural Stability, **169** / Fire-Resistive Combustible Assemblies, **169** / Interior Support Systems, **169** / Deficiencies of Materials, **170** / *Tactical Considerations,* **172** / Connections, **173** / Beam to Girder Connections, **173** / Beam to Column Connections, **173** / Cast-Iron Columns, **174** / *Tactical Considerations,* **176** / Suspended Loads, **176** / Floors, **178** / Roofs, **180** / *Tactical Considerations,* **180** / Fire Characteristics of Conventional Wood Roofs, **180** / Roof Hazards, **181** / *Tactical Considerations,* **182** / *Tactical Considerations,* **183** / Treated Plywood Roofs, **183** / *Tactical Considerations,* **184**

Void Spaces, **184** / *Tactical Considerations,* **184** / Interior Sheathing, **185** / Light Smoke Showing, **185** / *Tactical Considerations,* **186** / Ceiling Spaces, **186** / Joist Spaces, **189** / *Tactical Considerations,* **189** / Combustible Gases, **189** / *Tactical Considerations,* **190** / Large Voids, **190** / *Tactical Considerations,* **192** / Fire Extension, **192** / *Tactical Considerations,* **192** / Interior Sructural Walls as Fire Barriers, **192** / *Tactical Considerations,* **193** / Voids in Mixed Construction, **193** / *Tactical Considerations,* **193** / Cornices and Canopies, **193** / *Tactical Considerations,* **194**

Masonry Bearing Walls as Fire Barriers, **195** / Fire Doors, **195** / *Tactical Considerations,* **195** / *Tactical Considerations,* **196** / Protection from Exposures, **197** / *Tactical Considerations,* **197** / Party Walls, **198** / *Tactical Considerations,* **199** / Partition Walls, **199**

Risk Analysis, **200** / *Tactical Considerations,* **201** / Fireground Safety, **203**

Mill Construction, **204** / Conversions, Modifications, Preservation, **205** / Code Classified Heavy Timber Construction, **206** / *Tactical Considerations,* **206**

Case Studies for Ordinary Construction and Heavy Timber Structures, **207** / Ordinary Construction, **208** / Heavy Timber Structures, **211**

References, 213

5 Garden Apartments and Other Protected Structures 215

Protected Structures, 215 / **Characteristics of Garden**

Apartments, 216 / *Tactical Considerations,* 216 / Educating the Management and Tenants, 219 / *Tactical Considerations,* 219 / General Advice to Tenants, 220

Fire Department Problems, 221 / Parking, 221 / Building Location, 221 / Gas Service, 221 / *Tactical Considerations,* 222 / Water Supply, 222 / *Tactical Considerations,* 223

Protected Combustible Construction, 223 / Effect of Fire of Gypsum Board, 223 / Fire Rating of Gypsum Board, 223 / Deficiencies of Gypsum Board, 223 / Fire-Resistance Testing, 224 / Installation, 227 / Protective Sheathing, 227 / Penetrations, 228 / Protected Combustible Is Not Fire Resistive, 231 / *Tactical Considerations,* 231

Fire Walls/Barriers - Draft Stops, 233 / Fire Walls, 233 / Fire Barriers and Draft Stops, 234 / *Tactical Considerations,* 235

A Word about Sprinklers, 236 / *Tactical Considerations,* 236

Serving the Citizens, 237 / **Case Studies,** 237 / *Tactical Considerations,* 239 / *Tactical Considerations,* 239 / *Tactical Considerations,* 240 / *Tactical Considerations,* 240

References, 240

6 Principles of Fire Resistance 243

Early Fire Tests, 243 / Floors, 243 / Columns, 244 / Standards for Fire Resistance, 244 / What Does It All Mean? 245

Fire-Resistance Testing, 245 / Caution, 246 / Minimum Requirements, 248 / The Legal Path, 248 / Another Caution, 249 / *Tactical Considerations,* 250 / Estimating Fire Severity, 250 / Combustibility and Fire Resistance, 251 / Containment, 252 / Fire Intensity and Duration, 252

Preplanning, 252 / **The Ultimate Test,** 253 / **References,** 253

7 Steel Construction 255

Fire Characteristics, 255 / Unwarranted Assumptions, 257 / Water on Hot Steel, 259 / *Tactical Considerations,* 260

Definitions, 260 / **Steel as a Construction Material,** 262 / **Steel in Non-Steel Buildings,** 262 / **Steel Buildings,** 265 / Huge Spans, 267 / Walls of Steel-Framed Buildings, 267 / High-Rise Framing, 270 / Steel-Framed Buildings Under Construction, 270 / *Tactical Considerations,* 271 / "Plastic" Design in Steel Construction, 271

More on the Fire Characteristics of Steel, 271 / Steel Conducts Heat, 271 / Steel Elongates, 273 / Steel Fails, 274

Overcoming the Negative Fire Characteristics of Steel, 275 / Ignoring the Problem, **275** / *Tactical Considerations,* **276** / Steel Highway Structures and Bridges, **276** / *Tactical Considerations,* **277** / Hazards of Concentrated Fire Loads, **278** / Excavation Bracing, **278** / *Tactical Considerations,* **281** / Buildings Under Construction, **281** / *Tactical Considerations,* **282** / Relying on Inadequate Codes, **282** / Steel High Above the Floor, **283** / The McCormick Place Fire, **283** / McCormick Place Fire Report, **284** / The New McCormick Place, **285** / Important Test Experience, **285** / *Tactical Considerations,* **286** / Taking Calculated Risks, **287** / Financial Calculation, **287** / *Tactical Considerations,* **287** / Engineering Calculation, **287** / Forgetting It, **288** / Fireproofing (Insulating) the Steel, **289** / Types of Fireproofing, **289** / Encasement Method, **290** / Sprayed-on Fireproofing, **291** / *Tactical Considerations,* **292** / Membrane Fireproofing, **292** / Testing and Rating, **293** / Hazards of Floor-Ceiling Assemblies, **294** / Potential Column Collapse, **298** / Deficient Column Protection, **298** / Smoke and Gas Movement, **298** / *Tactical Considerations,* **299** / Experience, **299** / Equivalent Fire Resistance? **299** / Protecting Steel with Automatic Sprinklers, **299** / Fireproofing with Water, **300** / Locating Steel Out of Range of Fire, **300** / Suspension Cables, **301**

Insulated Metal Deck Roof Fire Problem, 302 / Fires of Interest, **303** / An Unrecognized Problem, **305** / Fighting a Metal Deck Roof Fire, **306** / *Tactical Considerations,* **307** / A Different View, **308** / Metal Decks on Non-Metal Buildings, **308**

Fire Walls, 309 / Tied Fire Walls, **309** / Freestanding Walls, **310** / Partition Walls, **310** / Combination Walls, **310** / *Tactical Considerations,* **310**

Fire Doors, 311 / Tactical Considerations, **311**

Types of Protection of Steel Structures, 311 / Unprotected Steel, **312** / *Cautionary Note,* **313**

Water Damage, 313 / *Tactical Considerations,* **313**

Unprotected Steel in Wood and Ordinary Construction, 314 / **Dynamic Protection,** 316 / **Passive Fire Protection,** 316 / *Tactical Considerations,* **317** / *Tactical Considerations,* **318**

Passive/Dynamic Protection, 319 / **Code Problems,** 319 / **Code Variances,** 319 / **Preplanning Your "McCormick Place,"** 320 / *Tactical Considerations,* **321**

An Idea Is Adopted, 322 / **Be Proactive,** 322 / **References,** 323

8 Concrete Construction 325

Concrete Structures, 325 / **Steel vs. Concrete Framing,** 326 /

Fire Department Problems, 327 / **Types of Concrete Construction,** 327 / Definitions, **327** / Tactical Considerations, **329** / *Tactical Considerations,* **330**

Concrete Structural Elements, 332 / Columns, **332** / *Tactical Considerations,* **333** / Beams and Girders, **333** / T-Beams, **334** / Concrete Floors, **334** / *Tactical Considerations,* **337** / Prestressed Concrete, **338** / Prestressing Methods, **339** / *Tactical Considerations,* **340** / Reinforced Masonry, **340** / *Tactical Considerations,* **342**

Collapse During Construction, 342 / An Industry Warning, **343** / Problems of Falsework, **343** / *Tactical Considerations,* **345** / Reshoring, **345** / Collapse of Floors, **345** / *Tactical Considerations,* **346** / A Widely Believed Fallacy, **346** / Hazards of Post-Tensioning, **347** / Collapse of Reinforced Masonry, **348** / Collapse of Precast Concrete, **348** / Lift Slab Collapses, **348** / *Tactical Considerations,* **349**

Fire Problems of Concrete Buildings Under Construction, 350 / *Tactical Considerations,* **350** / Fire Potentials, **351** / Hazards of Post-Tensioned Concrete, **354** / Protection of Tendons, **355** / *Tactical Considerations,* **355** / A Total Collapse, **356** / *Tactical Considerations,* **356,** / Hazard Awareness, **356** / *Tactical Considerations,* **356** / Precast Buildings, **357** / Cantilevered Platforms, **357** / Tower Cranes, **358** / *Tactical Considerations* , **358**

Fire Problems in Finished Buildings, 359 / Fire Resistance, **359** / *Tactical Considerations,* **360** / Signs of Trouble, **360** / *Tactical Considerations,* **361** / Unprotected Steel, **361** / Heat Absorption and Fire Load, **362** / *Tactical Considerations,* **362** / Fire Load vs. Fire Resistance, **362** / Ceiling Finish and Voids, **363** / *Tactical Considerations,* **363** / *Tactical Considerations,* **364** / Combustible Trim, **364** / *Tactical Considerations,* **364** / The Integrity of Floors, **365** / *Tactical Considerations,* **366** / A Management Gap, **367** / Wall Panels, **367** / Exterior Extension, **367** / Tactical Considerations, **367** / Imitation Materials, **368** / *Tactical Considerations,* **369** / Partitions, **369** / Concrete's Behavior in Fire, **369** / *Tactical Considerations,* **369** / Cutting Tensioned Concrete, **369** / *Tactical Considerations,* **371** / Precast Concrete, **372** / Tilt Slab Hazards, **373** / Bridges, **373** / Fires in Concrete Buildings, **374** / *Tactical Considerations,* **376** / Know Your Buildings, **377**

References, 378

9 Fire Growth 381

Examples of Fire Growth, 382 / **Building or Contents Hazard?** 383 / Hidden Building Elements, **384** / Interior

Finish, **386** / Low-Density Fiberboard, **387** / Combustible Acoustical Tile, **387**

Industry Opposition, 388 / **Void Spaces,** 389 / *Tactical Considerations,* **389**

Remodeled Ceiling Hazards, 389 / *Tactical Considerations,* **391**

Eliminating Existing Hazards, 392 / **Efforts to Mitigate the Hazard Are Sometimes Inadequate,** 392 / Adhesive, **392** / High-Density Fiberboard, **393** / Paper, **393** / Fabrics, **393** / Cork, **394** / Rattan, **394** / Wood, **394** / Plywood, **395** / *Tactical Considerations,* **395** / Plastics, **395** / Aircraft Interiors, **398** / Acoustical Treatment, **398** / Open-Plan Offices, **398** / Carpeting, **399** / *Tactical Considerations,* **399**

Alterations, 399 / **Decorations and Contents,** 400 / **Today's Fire Loads,** 402 / First Interstate Bank Fire, **402** / Dupont Plaza Hotel Fire, **403**

Regulations, 403 / **Certification of Interior Designers,** 403 / **Residential Fire Tests,** 404 / **Films Available,** 404 / *Tactical Considerations,* **405** / Difficulties, **406** / *Tactical Considerations,* **406**

Control of Rapid Fire Growth, 406 / Eliminate High Flame-Spread Surfaces, **407** / Separate the Material from the Combustion Source, **407** / Cut off the Extension, **408** / Flameproof the Material, **408**

Testing and Rating Materials, 409 / The Tunnel Test, **410** / Caution About ASTM E84, **412** / Fire Rated, **413** / The Radiant Panel Test, **413** / Factory Mutual Corner Test, **413** / Carpet Tests, **413** /Radiant Flux Test, **413**

Faked Tests, 415 / **Don't Be Mousetrapped,** 415 / **Conclusion,** 415 / **References,** 416 / **Additional Readings,** 418

10 Smoke and Fire Containment 419

Smoke, 419 / **Gases,** 419 / *Tactical Considerations,* **420**

Flammability, 421 / **Smoke vs. Gases,** 424 / *Tactical Considerations,* **424**

Smoke Damage, 425 / **Contaminated Smoke,** 425 / *Tactical Considerations,* **427** / *Tactical Considerations,* **427**

Corrosion, 428 / **Smoke Movement,** 429 / **Containment of Fire,** 431 / *Tactical Considerations,* **431** / *Tactical Considerations,* **431** / Fire Door Closure Devices, **433** / *Tactical Considerations,* **434** / Smoke-Sensitive Releases, **434** / *Tactical Considerations,* **436** / Horizontal Barriers, **436** / *Tactical Considerations,* **437**

Escalators, 437 / **Public Education,** 438 / **Fire Department Education - "Light Smoke Showing,"** 440 / **Smoke Detectors,** 440 / **Unwarranted Alarms,** 441 / **Ventilation,** 442 / Fire Department Ventilation, **444** / Positive Pressure Ventilation, **445** / *Tactical Considerations,* **446**

Summary, 447 / **Case Studies,** 447 / **References,** 448

11 High-Rise Construction 451

Historical Notes, 452 / **General Classifications of High-Rise Buildings,** 454 / Early Fire-Resistive Buildings, 1870-1930, **454** / *Tactical Considerations,* **458** / Later High-Rise Building Construction 1920-1940, **458** / Exposure Problems, **459** / Flame Spread, **460** / *Tactical Considerations,* **461** / Modern High-Rise Buildings, **461** / Significant Construction Characteristics, **462** / Windows and Glass Exteriors, **466** / Atriums, **467**

Parking Garages, 469 / **A Unique Design,** 470 / *Tactical Considerations,* **470**

General Problems and Hazards with High Rises, 474 / Exits, **474** / Occupancy, **475** / Accommodation or Access Stairs, **475** / *Tactical Considerations,* **476** / Forcible Entry, **476** / *Tactical Considerations,* **477** / Elevators, **477** / *Tactical Considerations,* **478** / Some Additional Points, **478** / *Tactical Considerations,* **479**

Smoke Movement in High-Rise Buildings, 479 / Thermal Energy, **479** / Atmospheric Conditions, **480** / Wind, **480** / *Tactical Considerations,* **481** / Stack Effect, **481** / Winter Stack Effect, **483** / Summer Stack Effect, **484** / *Tactical Considerations,* **484**

Air Conditioning, 485 / Individual Room Units, **485** / Single-Floor Systems, **485** / Multi-Floor Systems, **486** / *Tactical Considerations,* **486**

Smoke Removal Systems, 487 / *Tactical Considerations,* **489**

Compartmentation, 491 / *Tactical Considerations,* **491**

Pressurized Stairways, 491 / *Tactical Considerations,* **492**

Installation of Special Equipment, 492 / **Fire Load and Flame Spread,** 492 / Contents, **494** / *Tactical Considerations,* **496** / Maintenance Operations, **496** / Rubbish, **496** / *Tactical Considerations,* **497**

Alterations to Occupied Buildings, 498 / *Tactical Considerations,* **498**

Partial Occupancy of Buildings Under Construction, 499 / **Automatic Sprinklers,** 501 / *For You to Do,* 502 / *Tactical Considerations,* 504

Checklist, 504 / Building Inventory Items, 505

Case Studies, 506 / *Tactical Considerations,* 508

References, 511 / **Additional Readings,** 514

12 Trusses 517

The Truss, 518 / **What Is a Truss?** 518 / The Triangle Principle, 518 / Components, 519 / Principal Types, 519 / Bowstring Truss, 519 / Parallel-Chord Truss, 520 / Triangular Truss, 520 / Vierendeel Truss, 520 / Exterior Trusses, 521 / Compression vs. Tension, 521

Truss Principles, 521 / **Problems with Trusses,** 522 / Unique Alterations, 525 / *Tactical Considerations,* 526 / The Bar Joist Gap, 526 / Excessive Loads, 527

Arena Collapses, 528 / **Truss Failure in Fires,** 528 / *Tactical Considerations,* 532

Heavy Trusses, 533 / **Steel Trusses,** 535 / **Insulated Metal Deck Roofs,** 536 / **Rated Steel Floor-Ceiling Assemblies,** 536 / Wood Truss Floors, 537

Fire-Resistance Ratings and the ASTM E119 Fire Endurance Test, 538 / **Non-Standard Tests of Wood Assemblies,** 541 / **The Truss Void,** 542 / **Catastrophe Potential,** 544 / *Tactical Considerations,* 546

Miscellaneous Steel, 546 / **Hazard Control,** 548 / Automatic Sprinklers, 548 / Fire-Retardant Treated Wood, 549 / Technical Studies, 549

Wooden I-Beams, 551 / Automatic Sprinkler Problems, 552 / Collapses, 552

Inconsistent Construction, 553 / Second Roofs, 554

Case Studies, 554 / *Tactical Considerations,* 557

References, 561 / **Author's Notes,** 562

13 Automatic Sprinklers 565

Types of Sprinkler Systems, 565 / Wet Pipe Systems, 566 / Dry Pipe Systems, 566 / Preaction Systems, 566 / Deluge Systems, 566 / Combined Dry Pipe and Preaction Systems, 566 / Special Types, 567

Sprinkler Installation Incentives, 567 / Code Incentives, 567 / Site Development Incentives, 568 / Tax or Insurance Incentives, 568

Opposition to Sprinklers, 569 / *Tactical Considerations,* 571

Popular Misconceptions, 572 / **Fire Service Activities to Correct Erroneous Opinions,** 577 / Sprinkler Demonstrations, **577** / Sprinklers and Flammable Liquids, **578**

Fire Service Misconceptions, 578 / **Fire Department and Sprinklers,** 581 / Why Were the Sprinklers Installed? **584** / Fire Department Policy, **585** / Fire Department Instruction, **585** / Company Level Inspections, **586** / *Tactical Considerations,* **587** / Outside the Building, **587** / Inside the Building, **589** / Sprinkler Fraud, **596** / Management, **597** / Special Situations, **597**

Protection of Glass Fire Barriers, 598 / **Early Suppression Fast Response Sprinklers (ESFR),** 598 / **Prewetting,** 599 / **Case Studies,** 600 / **References,** 601

14 Rack Storage 605

Warehouse Fire Problems, 605 / Pallets, **606** / Sprinklers, **606** / A Serious Risk, **606** / Shelving, **608** / Modern Rack Storage Warehouses, **608** / *Tactical Considerations,* **609** / Dry Storage of Boats, **609** / Library Stacks, **610**

Some Major Losses, 611 / The Problem, **611**

Fire Defenses, 612 / Standards, **612** / The Building, **612**

Passive Defenses, 614 / Fire Walls and Fire Doors, **614** / Water Spray, **614**

Active Defenses, 614 / Automatic Sprinklers, **614** / Foam System Protection, **616**

Attitudes, 616 / Management Attitude, **616** / Attitudes of Public Officials, **618** / Insurance Company Assistance, **618** / Employees, **620**

Fire Department Actions, 620 / Initial Planning and Plan Review, **621** / Consultants, **622** / Collapse Potential, **622** / Fire Doors, **623** / Clearance, **623** / Inspections During Construction, **623** / Routine and Special Inspections, **623** / Warehouse Loss Examples, **625** / Preplanning, **626** / *Tactical Considerations,* **626** / Intelligence Acquisition, **627** / Contents, **627** / *Tactical Considerations,* **628** / On the Fireground, **628** / *Tactical Considerations,* **628** / Solid Rack Shelves, **628** / Ventilation, **629** / *Tactical Considerations,* **630** / Handline Operations, **630** / *Tactical Considerations,* **630**

Protecting the Department and the City Treasury, 633 /

Personal Safety, 634 / **References,** 635 / **Additional Readings,** 636

Appendix A: Last Minute Updates 637

Appendix B: Addresses of References 645

Index 649

Foreword

Francis L. "Frank" Brannigan is a man with a mission. He believes that a "structural collapse" needn't be recorded as a cause of fire service deaths. He does not imply that buildings will not collapse under attack from fire or that design errors will not cause failures. Instead, Frank believes that even if those things are "givens," something can be done about them. That "something" is the purpose of this book, now in its third edition after more than 20 years of teaching the fire service about the dangers of building collapse.

Through years of fire ground experience and observations of how fire acts on structures, Frank realizes there is no mystery surrounding what appears to be sudden and inexplicable building failures that often, on first appearance, seem unrelated to the fire itself. Frank believes that if you just look close enough *before* fire occurs, and from that knowledge apply the right tactics during the fire, the risk to fire fighters from the building itself can virtually be eliminated.

In this book, Frank draws upon his specialized knowledge and long practical experience to explain the hazards of various types of building construction. Frank presents the fire fighter's view — not the designer's or construction engineer's. Frank discusses in detail the many obvious and not so obvious dangers inherent in fighting building fires. He traces the paths fire can take in spreading throughout buildings of various designs and materials. He shows many of the things a good observer should notice in anticipating what will happen to structural stability under fire conditions. His message is clear: "Know your buildings!" Frank is convinced that if every fire officer and fire fighter took the time to study the peculiarities of how buildings are constructed in their response areas, their chances of coming away alive would be immeasurably improved.

This is Frank's book and the opinions and recommendations expressed in it are, of course, the sole responsibility of the author and do not necessarily reflect the recommendations of the National Fire Protection Association. The text is being published by the NFPA as a valued addition to the literature of fire suppression in buildings and is dedicated to the preservation of life in the most hazardous of occupations.

Gene A. Moulton
Senior Editor

Los Angeles fire fighters barely escape from a brick wall brought down by a collapsing roof. When a roof collapse is expected, all should be clear of the walls. (Alan Simmons)

Author's Preface

Robert Louis Stevenson said that a book is an open letter to the author's friends. This book is my open letter to my friends in the fire service.

In 1966 David Gratz, then Chief of the Silver Spring, Maryland Fire Department* was developing the pioneer fire science program at Montgomery College in Rockville, Maryland. This author was scheduled to teach one subject in the new program when the instructor who was to teach building construction suddenly became unavailable. I accepted the assignment for that course, confident that a lifetime of the study of fire, and 25 years of varied professional experience, were adequate qualifications. I soon found out that not only was I not completely qualified, but that there was no adequate or formalized body of knowledge to teach about building construction for the fire service. There were a very few articles in the fire service press, and some "word of mouth" precepts had been handed down, many of which were inaccurate. There was also some relevant material in the *Fire Protection Handbook*, which was the assigned text.

I soon found two books by consulting engineers, Feld[1] and McKaig[2], both of whom wrote from the experience and perspective of professionals trying to discover the cause of building failures. I was struck by the fact that they were able to examine a building's wreckage and determine that the signs of potential failure had often been discernible prior to the disaster, but had been ignored or went unrecognized.

This realization motivated me to take the then novel approach of looking at existing structures to determine whether factors which might be significant in a fire could be identified beforehand, even though there was no specific experience to identify such hazards. In fact, many of these conditions had never been identified. For example, in the first edition of this text in 1971, the hazard of cutting a chase into a brick wall to accommodate a drain was presented. In 1973, the Broadway Central Hotel in New York City collapsed suddenly. After an extensive investigation it was determined that an unauthorized 8-inch chase had been cut into a 12-inch brick wall, and that water from a leak had washed out the existing sand lime mortar.

During the course, I prepared detailed notes for my students. David Gratz repeatedly urged me to publish these notes as a textbook. In 1970, I offered the concept to NFPA and it was accepted.

That first edition was adopted as the required text in many fire science courses. It is also specified as a text to be studied for promotion

---

* Chief Gratz later became President of the IAFC and Director of International Operations for the NFPA.

examinations. It is one of the texts used in courses for certified fire inspectors, and a number of insurance companies use it to train consultants. Fire Protection Engineers find it useful for the information it provides about why certain code provisions were written. Above all, to me, its value is indicated in this letter from a Hackensack, New Jersey fire fighter, Ron Pezillo:

> "Some time in the year 1981 I purchased your book. Through the years I have read it so often that sections and its cover are falling apart. Early in my career I was told by a Battalion Chief in my department to always read to gain knowledge as it will help to save my life one day. Well, your book did on July 1, 1988 when my platoon lost five men in a tragic auto dealership fire which had a bowstring truss roof structure that collapsed. Thank you for making your knowledge and life's work available to every fire fighter as I'm sure you have helped many fire fighters from an early grave."

This third edition is a natural progression from the earlier editions. It reflects ten more years of experience in examining buildings — under construction, in use, and after fires. It also reflects interaction with thousands of fire fighters and others, many of whom have attended building construction seminars and contributed their personal knowledge and experience.

Some time ago I announced the preparation of a third edition and asked for suggestions. Many complied. One persistent request was for more case studies of actual fires. Several score of these are provided together with references to enable the reader to study the more detailed original accounts. There were suggestions to be more attentive to the differences in construction between the newer cities of the west and the older cities of the east. This has been done.

Some asked that a list of pertinent questions be provided at the end of each chapter. This tends to direct readers to study only the answers to those questions, slighting the other material. One problem with this is that everything in the text is important to someone, somewhere. Consequently, it is better to leave the picking and choosing of questions to local instructors or readers familiar with local conditions.

I offer a word of caution to those who teach the course and must prepare test questions. In my opinion, the multiple choice question is dangerous. Its structure requires one answer which appears to be correct, but which can be proven wrong. The wrong answer is the one which may flash up on the student's memory at some critical point, with disastrous results.

It is not always possible to present this subject matter in a straightforward manner in which A is followed by B and C. At times the subject of fire and building construction is more like a merry-go-round, with no definitive beginning or end.

Nor is this book always read from beginning to end. It is often used as a reference for a particular subject or as a general sourcebook. For this reason the same fire may be cited and the same observation may be found in more than one chapter. For those who wish to browse, what follows is a description of the changes and the organization of this book and a brief summary of each of the chapters.

The following changes to the previous edition are noteworthy. Chapter 1, Initial Thoughts, has been entirely rewritten. Chapters 2 through 11 have been heavily revised and updated. Chapters 12, 13, and 14 are new.

The most significant overall change is the addition of *Tactical Considerations*. This recurrent element involves separating the Author's opinion, insights and suggested fire tactics from the information about buildings. Every effort is made to make the building information as accurate as possible. In the main it is not subject to serious question. My opinions, insights and suggested fire tactics are another matter. Despite the fact that I have been a fire protection professional for 50 years, my last experience as a fire ground commander was in 1949. Nevertheless, I still empathize strongly with the individual who has the ultimate responsibility for both suppressing the fire and attending to fire fighter's safety. The Spanish philosopher Ortega Gasset provides a good analogy.

Bullfight critics ranked in rows
Crowd the enormous plaza full
But there is only one who knows
He's the man who fights the bull.

Not everyone is able to devote the time needed for detailed study of each chapter of this book. Consequently here are some salient points which might escape the glancing reader's notice, and which may highlight the need to read certain chapters more thoroughly. Many of these points are unique to this text and can be found nowhere else in fire-related literature.

Chapter 1 presents some initial thoughts and recommendations that can be applied to the information and lessons contained in all the subsequent chapters. Language and prefire planning concerns must be addressed by all fire fighters to achieve the greatest possible levels of safety and effectiveness.

Chapter 2, Principles of Construction, looks at the principal terms used in structural engineering, without all the attendant mathematics.

On a professional level, it is important to use the proper terminology of any field or discipline. When a fire officer addresses building professionals and mistakenly refers to columns as "vertical I-beams," or speaks of "masonAry" construction, valid fire protection recommendations may be disregarded.

More important, fire fighter safety may be at risk — many fatalities have resulted because personnel have not been adequately informed or have not understood potential failures in structures.

Early in my career I was told point blank by a fire chief "I don't want my guys even hearing you speak. I want them to do what I want done, period!" This attitude is illustrated by two almost unbelievable cases of fire officers who took promotion exams. In both cases they were failed for indicating fire ground operations which took the safety of fire fighters into consideration. Some of these attitudes will change over time and some will be changed after civil or criminal legal action is taken. A northwest fire department was assessed a substantial fine by its State Labor Department for safety deficiencies.

Live fire fighter training may inadvertently be delivering the wrong message. The training often emphasizes taking the punishment and "putting the wet stuff on the red stuff." There is no need to be concerned about collapse or hidden fire in the training building. However, hidden fire and collapse are major hazards to fire fighters — despite the fact that they are too dangerous for live training. Classroom lectures and discussions should present these hazards so forcefully that they become equally important to the fire fighter as hitting the fire. For this reason the full text of a "Special Analysis: Deaths Due to Rapid Fire Progress in Structures" is provided from the NFPA 1989 report of fire fighter fatalities.

Finally, connections are discussed in the last part of Chapter 2. All loads of the building itself, and any superimposed loads, must be delivered to the ground. The path must stay intact. Fire fighters should learn to look at the connection system which transfers the loads from one element to another along the path. A failure of any connection may have catastrophic results.

This walking tour of Chapter 2 reflects the scope and details contained throughout the other chapters in this book. The main points in the other chapters are cited briefly.

Chapter 3, Wood Construction, details the many elements and hazards associated with this construction and material, and looks closely at balloon frame methods, paneling, decorations, and imitation wood.

Chapter 4, Ordinary Construction, covers buildings of masonry exterior construction and combustible interiors. The subject of collapse of masonry walls is dealt with extensively. Of particular note is the fact that walls which appear to be of solid masonry often contain wooden lintels or wooden leveling beams which can cause failure early in fires.

Chapter 5, Garden Apartments, investigates the extension of fire in what are often called "protected combustible structures." Deficiencies of "fire resistive combustible" structures are discussed.

Chapter 6, Principles of Fire Resistance, treats fire resistance rating, testing and performance issues, as well as attempts to clear up potentially fatal misconceptions that exist, even among many fire service personnel.

Chapter 7, Steel Construction, discusses in detail the properties, performance characteristics, and uses of this common building material which can be critical to fire fighter safety. Of particular interest are the necessity of cooling unprotected steel, the hazard of steel columns left unprotected in the plenum space of rated floor ceiling assemblies, and a fire attack method for combustible metal deck roof fires.

Chapter 8, Concrete Construction, looks at the many uses of concrete in buildings and relates extensive fire experience with this commonly-used material. Concrete is inherently non-combustible, but must be specially formulated to be fire resistive. From the earliest, concrete experts stressed that failure of the bond between the concrete and the tensile steel means the failure of the reinforced concrete.

Chapter 9, Fire growth, (or fire spread) presents many problems for unsuspecting or unprepared fire fighters. Some of the information on fire growth is not well known to fire protection professionals generally.

Chapter 10, Smoke and Fire Containment, presents a detailed discussion of the hazard of smoke and the containment of smoke and fire.

Chapter 11, High-Rise Construction, examines the construction features of high-rise buildings that vary widely and are most significant in the growth of the fire and its suppression. Successful high-rise operations start with a good knowledge of the building.

As noted, the following three chapters are new to this edition: 12, 13 and 14.

The great increase in the use of trusses and wooden I-beams indicated a need to draw together all the material on trusses into a new chapter, Chapter 12, Trusses.

There has been a trend to use more active fire protection such as sprinklers in buildings, often with trade offs such as reduction in traditional passive fire protection. This move has created a need for a new Chapter 13, Automatic Sprinklers. With most or all of the fire protection eggs being placed in the sprinkler basket, it is vital that the basket be carefully watched. It appears that the fire department is the only available watcher.

Modern high rack storage methods have created huge potential fire problems within buildings. There have been staggering losses in warehouses where it was assumed that the fire protection was adequate. The problems presented by these structures are dealt with in a new Chapter 14, Rack Storage.

Sections of some pages, such as page 40, show a vertical line in the left margin which indicates specific information to be stressed. Watch for these vertical lines.

Around the year 2002 the NFPA will probably want a new fourth edition of this book. It rests with the Good Lord whether I will be around to do it. In the meantime Maurine and I plan to continue our seminars, help you, and learn from you. The fire service has been a wonderful and satisfying career. I wish you much success.

References

1. Feld, J. *Construction Failure*, Wiley, New York, New York, 1968.
2. McKaig, T. *Building Failures*, McGraw Hill, New York, New York, 1962.

Acknowledgements

As the three years of work on this third edition draws to a close, the 50th anniversary of the attack on Pearl Harbor has already been commemorated. It seems appropriate to acknowledge the WWII Naval fire fighting officers who hammered an enthusiastic amateur into a professional fire officer: Deputy Chief Charles Bahme, Los Angeles Fire Dept.; Chief Harold Burke, New York Fire Department; Deputy Chief Edward Gaughan, Boston Fire Department; Battalion Chiefs Gerard Crowley and Rodney Cox, Seattle Fire Department; Battalion Chief Ignatius Bradecamp and Deputy Chief John Werheim, Washington DC Fire Department; Deputy Chief Joseph Morris, Charlotte NC Fire Department and Captain Orville Emory, Los Angeles Fire Department. Commander Felix Hargrett, later Vice President of the Home Insurance Company, taught me the art of report writing.

Over the years many have shared with me the art and science of fire fighting, particularly Chief John Cashman, Barrington R.I., who spent many years in the high hazard district of New York, Chief Michael Kelly of New Smyrna Beach, Florida, Deputy Chief Vincent Dunn, FDNY, Captain Curtis Varone of the Providence, RI Fire Dept., and Captain Dave Mager of Boston.

Many former students have assisted, notably Chief Jack McElfish of Clayton County, GA, Wayne Powell of the National Fire Academy, Jeffrey Shapiro, P.E. of the Western Fire Chiefs Association, and Battalion Chief Richard Freas, Howard County, MD.

There are too many others to name, who sent me stories and pictures and told me of their experiences when we met at seminars. I thank all of you. I trust that all pictures have been properly acknowledged, I would appreciate hearing of any errors.

Editors Tom Brennan and Bill Manning of *Fire Engineering* and Harvey Eisner and Tom Rahilly of *Firehouse* have shared a great deal of information with me.

Nora Jason, Harold (Bud) Nelson and Dan Gross of the National Institute for Standards and Technology (NIST, formerly NBS) have always cheerfully assisted with technical problems.

At NFPA, Tom Klem, Mike Isner, Rita Fahy and Martin Henry have filled in many gaps in the text.

Special thanks are due to Jack Bono and Tom Castino of Underwriters Laboratories for Reviewing Chapter 6 and to Larry Davis and Larry Wenzel of IRI for reviewing Chapter 14.

Battalion Chief John Mittendorf of Los Angeles and Deputy Chief (Ret) Elmer (Bud) Chapman, FDNY, have generously permitted me to draw substantially on their expertise for this text.

My associates in the Department of Energy Fire Loss Management program, Pat Phillips, Don Keigher and Walter Maybee, all Fellows of the Society of Fire Protection Engineers, and Hank Collins of UL were always ready with advice and assistance. Dick Gewain of Hughes Associates, formerly Chief Fire Protection Engineer of the American Iron and Steel Institute, and Hank Spaeth, formerly of ISO, both assisted on fire resistive construction.

A very special word of thanks must go to my editor, Gene Moulton, who caught a number of errors and made many valuable suggestions for improvement.

By custom the author thanks his family, usually for being quiet while he works. In this case they all played a part

Vincent Brannigan for legal advice, Eileen Longsworth and John Brannigan for research, Christoper Brannigan for art sketches, Anne Brannigan/Kelly for computer assistance, and Mary Ellen Schattman for information in Texas.

I must particularly thank my wife of 49 years, Maurine, who has been my professional associate since as a Naval Supply Officer she diverted all foam and other fire equipment from decommissioned vessels to good hands. She is an active participant in all my programs but her best service for the book has been to argue, "If I can't understand this, how can the students?"

1

Initial Thoughts and Recommendations

Clear communication is a critical factor for all fire departments. It follows that a wide range of fire professionals, in trying to do their job at a facility or building site, or when that site becomes the fire ground, must rely on effective communications to succeed — and sometimes survive. Knowing how to use and understand proper language and industry jargon, and communicate with different publics, is critical to all fire fighters.

Prefire planning is another key element for fire professionals. Its value cannot be overstated. Without it fire fighters are reduced to just reacting to a fire rather than being prepared for its many potential hazards.

Watch Your Language

In 1989, an Avianca airliner crashed on Long Island, New York. A major contributing factor was "language." The pilot kept telling the controller that he was low on fuel. The pilot should have used the proper term and declared an EMERGENCY. The controller would have then given him an immediate priority for landing and the catastrophe could have been avoided.

In one building collapse, the fire ground commander was heard ordering "Pull your line out." This order might simply mean, to some, to reposition the lines. On the other hand, in this case it meant "get out of there fast."

All fire departments should have a specific command word indicating immediate evacuation. This should mean leaving any tools or equipment behind in order to get out fast. Fire fighters have been injured and killed pulling lines out, when collapse was imminent. The radio order should be supplemented with the use of air horn blasts and siren signals.

By nature and training, fire fighters are aggressive and often resist orders to back out. The overall gravity of the situation may not be apparent to those fighting the fire. "We've got a good shot at the fire" is often an excuse for delay. This author knows of no fire department that drills on immediate evacuation. This is a critical oversight.

Some aggressive fire departments resent any intimation of retreat. The U.S. Navy is a very aggressive organization and it continually drills on "Abandon Ship." Fire departments should regularly drill on getting out of buildings. Fire fighters should respond immediately to such an order. They should not look back at the collapse while fleeing. They should be trained to look where they are going.

This author continues to urge for a specific change in dispatching language. Apparatus should be dispatched to a BUILDING FIRE. As

Fig. 1-1. U.S. Navy personnel await orders in an "Abandon Ship" drill. Such drills are a vital part of Navy life. Nobody is excused from participating in them. Every fire fighter should drill regularly for immediate evacuation. (National Archives)

soon as fire command determines that the fire involves the structure, as distinct from the contents, there is a radically different situation. The gravity resistance system of the building is under attack. All units should be advised — and they should acknowledge the emergency message — "This is now a STRUCTURAL FIRE." The word "structural" should be reserved for just such a situation. In the case of trusses or wooden I-beams, a department's SOPs should provide for evacuation of the affected area as soon as it is involved in fire.

Many years ago, the aviators of the world settled on the French words "m'aider" - or help me - (in English, May Day) as a universal distress signal. All fire departments should include this term in their radio language. A "May Day" should trigger immediate silence on the radio so that fire command can communicate with the caller in distress, without interference from others at the fire ground. One fire department now provides for a company to stand by at the command post for the sole purpose of acting on such May Day calls.

At times the term "withdraw in good order" is used with fire departments. This is a military term, and its meaning is not appropriate in fire scenarios. When a military retreat is necessary, it is important to deceive the enemy. Rear guard troops may keep up a fierce fight while all others slip out. Finally the rear guard moves. The operation may succeed because the enemy can be deceived. Fire can't be deceived or bluffed. When the time comes to get out, speed is of the essence. There is much truth in this author's humorous dictum, "Panic will save your life, provided you are the first to panic and are headed in the right direction."

Overhauling and Safety

A number of fire fighters have been killed in collapses during overhauling. After the fire is suppressed, it is no longer a menace to the community. Urgency should be replaced with a sense of caution. Tired personnel may not exercise good judgment. They may hurriedly try to complete the overhaul. In many cases it may be prudent to wait for daylight and fresh personnel to do the overhaul. At least one fire department, Champaign, Illinois, has a structural engineer on retainer. When necessary, the engineer makes an inspection and determines the condition of a building. A fresh look at overhauling and the innovative use of structural engineers may serve to improve fire fighting safety.

Prefire Planning

When this writer started his fire service career as a buff and auxiliary fire fighter prior to WWII, there was no organized prefire planning. A few young officers made some informal efforts. Like Notre Dame's

Fig. 1-2. Tragedy sometimes happens while overhauling. It can be a very confusing time as shown here. The whole end of the building is falling on the fire vehicles. Nine fire fighters died in this collapse.

legendary Knute Rockne, Captain Terry Conaty of Engine 61, FDNY, had a habit of firing off questions, "What would you do if. . . .?" He fired a tough one at me one evening. I gave the same answer that a fourth string quarterback gave to Knute Rockne, "Slide over on the bench to get a better look."

In the past, most chief officers relied solely on experience. Many gained heavy experience coming up under the old continuous duty system. This writer recalls asking an older buff what he thought Deputy Chief Ross (midtown NY) would do with a specific problem. His answer was "Young fellow, you never ask a Deputy Chief what he is going to do." The reason was that the chief felt supremely confident of his ability to deal with whatever came up.

This author's great interest in prefire planning was forced by necessity. All of a sudden, as a naval officer, I had responsibilities for lives and property for which I was totally unprepared. My first fire companies were comprised of "kids" I had trained myself.[1] I decided that to be effective and safe we needed to minimize surprises on the fire ground by learning everything possible about potential problems ahead of time.

Thus the inspiration for the first article in the fire service press on prefire planning was born — "Surveys Aid in Preparation For Handling Large Fires" in *Fire Engineering*, January 1948. In it this author described the prefire survey of a huge wooden Navy warehouse full of vital spare airplane propellers.

Many fire departments do preplanning today. There are many concepts and suggestions involved with these efforts which fire professionals may find as useful improvements. Sharing prefire planning knowledge is very beneficial to all concerned.

The word plan, though commonly used to describe this prefire effort, might be a misnomer. A more accurate term may be prefire analysis. The use of the word plan sometimes leads to overemphasis on the planning of *action*. Strong efforts should be placed on accumulating the necessary information on which sound judgments can be made at the time of the emergency, particularly if the emergency force is well trained. It the emergency force is less well trained, then more detailed planning against specific problems is necessary. If the plan is too specific, there is the potential problem that failure to follow the plan, for whatever reason, may be taken as evidence of negligence or incompetence.

Perhaps unbelievably, at some locations the concept of preplanning is sadly lacking. At one government facility, there are 20 to 30 places where the fire department was instructed to take no action until approval was obtained from a department head or representative. While this rule can be enforced against the fire department, it has *no effect whatsoever on the fire*. The fire does not recognize any authority. The fire chief is not in charge of the fire, but of the fire department's effort to contain the fire. The fire sets its own agenda.

One school board forbade any *unannounced* inspections by the fire department. No one pointed out that fire is an *unannounced* visitor.

At one facility this author once saw a notice "In the event of an alarm from Building 209, do not respond until called by Dr. X." The notice was signed by the fire chief. The fire chief was ill advised to take on this responsibility. Dr. X should have had the responsibility. He was sent a memo through administrative channels to sign and return with a superior's signature. The memo read like this. "There is no possible way the fire department can be instructed to deal with a fire in Facility XYZ. In the event of a fire in Facility XYZ, I wish it to be allowed to burn until I can arrive at the scene and direct the operations. I accept the full responsibility for the loss of property in the amount of $2 million.

Such a document can be very intimidating. If sent through the proper channels, it is very likely that such a document will help all concerned to find a better solution as happened in the case cited.

Consider separating prefire planning and the inspection function. An inspection is essentially a policing function. In an inspection, the property should be examined to see whether or not it complies with standards. If it does not comply, penalties may result. The inspector, therefore, is principally a police officer. As Gilbert and Sullivan told us, "a policeman's lot is not a happy one." The reason is simple. You can't win with a police officer. The best you can do is come out even.

It may not be wise to just walk in on a facility and announce that you are there to develop a prefire plan. It may be the worst possible day of the year for the management. This author suggests a letter from the fire chief indicating the mayor's appreciation of the valuable contribution the facility is making to the economic life of the community. Since a large percentage of businesses which suffer a serious fire never reopen, it is in everyone's interest that such a fire not happen. Any pertinent local experience should be cited in the letter. The letter can continue: the mayor has directed the fire chief to make every effort to keep local businesses safe from fire. To prevent or minimize fires or to operate effectively on a fire, there are a number of things which the fire department should know and record ahead of time. Since such surveys require mutual cooperation, they are best done by appointment. Please call the fire chief to arrange a mutually agreeable appointment time.

Wow!! A business owner may never have received such a letter in his or her life from any government agency. Usually they begin "You are hereby directed to. . . ." Imagine the talk at the country club, the business association, or the Rotary club. While such enthusiasm may be a little overstated, this type of letter may produce cooperation that may be "contagious."

Few people have any idea that firefighting is a planned operation. "Firefighting" is a commonly-used term for taking care of unplanned emergencies. For example: "We're so busy fighting fires around here, we can't get next year's budget organized." This author has never seen a firefighting movie in which fire fighters were discussing a particular hazard or studying a book — but it happens, and it needs to happen more.

Prefire planning surveys are important. Sometimes the perceived importance of such planning is enhanced by the stature of the fire officer performing the planning survey. Furthermore, owners or managers of a facility may be more likely to cooperate with, and provide more information to, an officer, rather than to a rank and file fire fighter.

It the client is uncooperative, then other measures may be legally available to help ensure cooperation.

Fire departments operate differently than other services or industries operating around the clock, such as police departments, prisons, hospitals, military bases or manufacturing plants. In such organizations the decision-makers work the day shift. Serious events which happen on the night shift are often referred to these supervisors at home. By contrast, fire departments deliver a senior officer with broad authority to make important decisions immediately to the scene of an emergency at any time of day. Many fire departments using three or four platoon systems operate almost as if each platoon were a separate fire department.

A company or entity which functions in a "normal" organizational manner may not be able to fully understand an organization which has three chiefs — all of whom regard themselves as equal. Since there are obviously many ways of handling a possible emergency situation (perhaps all equally good), it is very likely that if each of these three senior officers develop an emergency plan, that three different procedures might emerge. This may be the cause for a lack of confidence by an owner or facility manager in the ability of the fire department to handle such a fire situation. Consequently, all contact with an owner or manager should be through a single individual who has the lead responsibility for that particular risk. Differences on procedures should be ironed out within the fire department.

While the basic objective of prefire planning is to consider potential emergency situations and develop the best possible plans for coping with them, preplanning also serves another valuable purpose. If a credible disaster scenario is developed, and the best planning provides an inadequate solution, then there is a clear-cut problem. If the risk has been calculated and the answer is unsatisfactory, either a solution must be found or those responsible must face the distasteful fact of that risk. When planning develops a conclusion of such a disaster potential, the conclusion should be shared immediately with the client management and appropriate officials.

Let me offer a relevant, personal experience. The famous 1947 Texas City, Texas ammonium nitrate explosion killed 468 people and did an equivalent of $372 million in property damage. Ammonium nitrate, which is both an explosive and a fertilizer, was being shipped by the U.S. Army to a devastated Europe. Despite the disaster, a week later the Army made a similar shipment through Galveston, Texas.

A year later, similar shipments were being made through the Norfolk, Virginia U.S. Army base where my naval base fire department provided fire protection but not inspection functions. At the time the boss was away so the problem became mine. Comparison with Texas City showed that an explosion could cause total devastation over a wide area including the fashionable Larchmont area of the City of Norfolk.

This author prepared a prefire plan which called for towing the ship away from the pier if the fire wasn't put out immediately. Of the alternatives, the best place to beach and abandon the ship was at Craney Island, the Navy's east coast gasoline reserve depot area. In addition to storage tanks there were 75,000 drums of gasoline. The ship might not explode, it might simply burn out. In any case, many Monday morning quarterbacks would have a field day. I decided to forestall such a problem.

Every potential expert was called on a recorded telephone and their advice was solicited. I hit paydirt with Calvin Dalby, the Director of Public Safety of Norfolk. He had been Coast Guard Port Commander during the war and had a low opinion of military munitions safety procedures. He had strong political clout. In 48 hours the ammonium nitrate was moved to the special ammunition piers at Naval Ammunition Depot in Earle, New Jersey.

(Don't write and complain that I didn't solve the problem, I only unloaded it. I was paid to worry about and protect the Fifth Naval District. I didn't solve the fire problems of the whole world, but at least I did my own job!)

One hundred years ago Edward F. Croker, after 14 years in the New York Fire Department, was appointed Chief. At the time, his uncle was the head of Tammany Hall. Many saw the appointment as a stepping stone to the Mayorality. Croker denied any ambition with the following stirring words:

> *I have no ambition in this world but one, and that is to be a fireman. The position may, in the eyes of some, appear to be a lowly one; but we who know the work which a fireman has to do believe that his is a noble calling.*
>
> *We strive to preserve from destruction the wealth of the world, which is the product of the industry of men, necessary for the comfort of both the rich and the poor. We are the defenders from fire of the art which has beautified the world, the product of the genius of men and the means of the refinement of mankind.*
>
> *But above all, our proudest endeavor is to save the lives of men - the work of God Himself.*
>
> *Under the impulse of such thoughts, the nobility of the occupation thrills us and stimulates us to deeds of daring, even at the supreme sacrifice. Such considerations may not strike the average mind, but they are sufficient to fill the limit of our ambition in life and to make us serve the general purpose of human society.*

The words are a bit flowery for today's modern taste, but my advice to fire fighters is, if you don't feel at least some of the sentiments Chief Croker expressed, find another line of work. You will never be satisfied.

Reference

1. Brannigan, F. "Navy Yarns." *Firehouse*, December 1991.

2

Principles of Construction

This will be the most difficult chapter in this book for many readers to understand. For many instructors who use this book it is the toughest to teach. There is a whole new vocabulary to learn relating to building construction and fire. There are some accepted definitions and concepts to unlearn. Some readers may pass it over lightly or skip it. This is a great mistake.

The author seeks diligently to make complex concepts herein understandable. But the learning process for this type of material is sometimes difficult, and requires real mental effort. However, the alternative of not learning the basic principles can be terribly costly.

This is not a simplified text for laypeople, written by a graduate engineer. The author had no technical training in construction. The material was assembled from experience and by reading and questioning practicing engineers, and having it reviewed by competent authorities.

Why Study Building Construction?

All fire officers need an accurate knowledge of building terms. If fire officers make glaring errors in terminology, such as referring to "I-beam columns," it may be very difficult for building professionals or tradespersons to believe their fire protection recommendations.

However, the most important reason for knowing building terms is safety. There are too many instances of fatalities that have occurred because fire officers did not recognize construction elements and the conditions under which they are likely to fail. By far, most fires are fought in buildings — the study of fire tactics should properly start with the study of buildings.

Cases in Point

Recently, a five fatality, building collapse in Hackensack, N.J. has drawn increased attention to bowstring-truss roofs. The **bowstring-truss** is only one type of truss. The hazard — and the fire fighter's concern — should be on the fact that there is a truss present.[1]

A major midwestern city suffered three roof collapses. Three fire fighters were killed in each collapse. The first fatal collapse was of a bowstring-trussed garage roof. Subsequently, fire fighters were told to locate and note trussed roofs in their area. Truss roofs were described as having a hump, which is characteristic of the bowstring-truss roof. Later, at a fire in a restaurant, the fireground commander, seeing a long clear span area, asked the officer on the roof, "Do we have a truss?" Seeing no characteristic hump, the officer replied, "Negative." The roof collapsed "without warning" and there were three fatalities. The roof had parallel-chord trusses. All trusses present hazards to fire fighters. (See Chapter, 12, Trusses, for a complete discussion of this topic.)

In the third incident, fire fighters operating on a wood-joisted roof had to chop away ice to get to the roof before ventilating the fire. No one recognized the added live load of the ice on a structure already being attacked by fire. Three fire fighters died in the collapse. In the aftermath, the commissioner reported that a chief officer told him, "I didn't learn anything from this fire. If we fought the same fire again, I would have followed the same tactics."[2]

The days when a fire fighter's death was followed by only a funeral and a plaque are coming to an end. Accountability is the order of the day. It may be inevitable that fire officers will be held personally liable by civil courts for incompetence. Officers may well face criminal charges for deliberate gross disregard of safety.

Gravity

The force of gravity is the eternal enemy of every building. Twenty-four hours a day, seven days a week, gravity exerts a force on the building. Buildings may appear to be just quietly sitting on their foundations, but, in fact, they are under great stresses. While this may not be apparent as you look at the structure, the stresses are there nonetheless.By one means or another, in assembling a structure, the builder has defied gravity. Not all builders recognize this. A student once asked an old

Fig. 2-1. Only a small fraction of buildings such as these San Francisco high rises are designed to be fire resistive. Buildings that are not fire resistive are "entitled" to collapse in a fire. Note the diagonal wind bracing on the dark building.

carpenter, "How do you defeat the laws of gravity?" He threw down his hammer and replied, "They've got so many !@#$%% laws now, you can't get anything done."

The gravity resistance system in a building consists of all the structural elements and the connections which support and transfer the loads. Fire destroys the structural elements and/or the connections in a building, and places loads on some structural elements which cannot handle them. Any of these can cause a collapse. Once the collapse starts, the results are unpredictable.

Some buildings are designed to be fire resistive. This means that to some degree they will resist fire-caused collapse. The vast majority of buildings are non-fire resistive, and thus may easily collapse as a result of assault by fire on the gravity resistance system.

Tactical Considerations

Fire departments should use the term "building fire" when dispatching. The term "structural fire" should be announced to all,

Tactical Considerations, continued

as soon as it is determined that fire is consuming or affecting (as in unprotected steel) the gravity resistance of a building. This should put all on notice to carefully and continuously observe the structure and prepare to evacuate instantly, if indicated or ordered.

Instant evacuation from a building should be a drill subject for all fire fighters. This sort of thinking does not conform to the macho image many fire departments have of their operations. Consider, however, that the proudest ships in the U.S. Navy drill regularly on "Abandon Ship."

The Limitations of Experience

"Experience keeps a dear school, but fools will learn in no other!"

— Benjamin Franklin

From experience some fire officers have learned to make useful, but limited judgments regarding the loss of structural stability in a fire. Many of the commonly taught indicators, such as sagging floors, strange noises, etc., may be too little too late. In addition, experience with one type of building element, such as solid sawn wood joists, is not valid for trusses, or wooden I-beams. Relying on experience alone is not sufficient. Fire fighters must be aware of the theories and principles involved.

Collapse Is Not the Only Hazard

It is imperative to note that collapse is not the only hazard facing the fire fighter in a structure. Concealed fire which bursts out of a hidden void, and lightning-like spread of fire over combustible surfaces, are equally hazardous and may well account for as many fire fighter casualties as collapse.

Fire training schools conduct live training in fire-resistive structures where collapse and hidden fire hazards do not exist. Whether intended or not, the message delivered to the fire fighting students is simply to "Put the wet stuff on the red stuff."

It is not possible to train in the same way for collapse and hidden fire hazards. Special effort must be taken to cover these hazards in the classroom so that it is evident to the student that they are as important as putting out the visible fire.

Live fire training in abandoned buildings has resulted in casualties when the training supervisors did not recognize the hazards in the structure.

No fire fighter should be allowed to complete basic training without viewing "Fire: Countdown to Disaster," an NFPA film which

graphically shows flashover occurring two minutes after the start of a bedroom fire.[3]
See Chapter 9, Fire Growth, for a discussion of fire growth, and Chapter 4, Ordinary Construction, for a discussion of the hazards of void spaces.

Fire Fighter Fatalities Report

Annually, *NFPA Journal* publishes separate reports and analyses of fire fighter fatalities, and fire fighter injuries. The 1989 report of fatalities contains a most pertinent, "Special Analysis." Because of its significance, it is reprinted here in its entirety. The text is verbatim; some words have been printed in boldface for added emphasis.

> Special Analysis: Deaths Due to Rapid Fire Progress in Structures
>
> "One of the dangers fire fighters face while operating inside structures is being caught in the rapid development of a fire which can trap and fatally injure them. The types of rapid fire growth included in this analysis are **flashover, backdraft,** and **flameover**.
>
> Flashover is defined as the stage of a fire at which all surfaces and objects in a room or area are heated to their ignition temperature and flames develop on all contents and combustible surfaces at once.
>
> **Backdraft** is defined as the burning of heated gaseous products of combustion when oxygen is introduced into an environment whose oxygen supply has been depleted due to fire. This burning often occurs with explosive force.
>
> A **flameover** is the rapid spread of flame over one or more surfaces.
>
> Distinguishing among these specific phenomena can be difficult for observers outside the building or observers inside the building but removed from the involved area. For this reason, the following analysis will not deal with which phenomenon was involved, but with the circumstances that led to the fatalities in all fires that involved any of these phenomena. From 1980 through 1989, there were 455 fire fighters (36.7 percent of all line-of-duty deaths) who were killed while operating at structure fires. Of these deaths, 44 were the result of some sort of rapid fire development. Not included in this analysis are seven fire fighters killed under similar circumstances during live fire training, and 13 fire fighters caught by rapid fire spread at wildland fires.
>
> In analyzing the circumstances of these 44 fatalities, three possibilities were examined:
>
> 1) that the fire fighters did not recognize the signs of flashover or backdraft conditions developing;
> 2) that the suppression approach which was used made rapid fire development more likely; and
> 3) that the fire fighters were inadequately protected.

In general, no obvious signs of backdraft and flashover (e.g., the puffing of smoke out any opening in the building so that the building appears to be breathing; thick smoke and heat with no visible fire; rolling flames across the bottom of a heavy smoke layer) were reported. In several cases, fire fighters specifically reported that smoke was only light as they moved through the building.

It is important to be aware, then, that backdraft or flashover can occur with no obvious warning signals. Whenever a fire has been smoldering or burning in an oxygen-starved atmosphere for a long time, there is the potential for large quantities of carbon monoxide and other unburned products of combustion to be present. These may not involve evidence of thick smoke. In larger areas, these dense concentrations of gases and potential fuel could extend a considerable distance from the seat of the fire to positions where they would have been cooled by the atmosphere. Under the right conditions, these concentrations can be ignited and can flash across an area, creating tremendous radiant heat, generally from the ceiling downward. This type of phenomenon has led to a number of fire fighter deaths over the years.

The second hypothesis — that the suppression approach made rapid fire development more likely — may have been confirmed in a few incidents where there was inadequate ventilation. In one incident, two fire fighters were killed during the overhaul of a clothing store fire that had not been ventilated either horizontally or vertically. The roof had not been opened and the windows were left intact. It was about 45 minutes after the alarm, and the fire was thought to have been knocked down, when the area where the fire fighters were working suddenly became involved in an intense fire. This may have been due to a buildup of heat and unburned products of combustion above a false ceiling or to the sudden introduction of air into an area of the building where there was still some burning and an accumulation of unburned gases. In other cases, applying a hose stream into an unventilated basement or attic space appears to have introduced oxygen and provided the mixing needed to create a backdraft.

The third hypothesis — that fire fighters were not adequately protected — was not supported by the incident reports. In every incident but one, fire fighters reportedly were wearing protective clothing and using SCBA. In that one incident, a fire fighter was attempting to extinguish a fire in his own home. Specifics on the type of protective clothing worn and whether it met NFPA standards were unavailable. Equipment or clothing indeed may have been overwhelmed by these severe fire conditions in some cases, but failures of clothing or equipment within their design limits were not reported as factors in any deaths.

What can be done to prevent such fatalities? The ventilation of involved structures is essential in limiting risks to fire fighters operating inside, but *the ventilation process can expose fire fighters to the very dangers they are trying to reduce.* Fire scene managers must recognize conditions that could cause backdraft, flashover, or flameover and could endanger fire fighters operating inside the structure.

It is also possible that these fire development phenomena are not factored into local fire scene planning and decision-making as universally as the special hazard would warrant. A careful and thorough investigation of all abnormal situations should be initiated, whether or not there are deaths and injuries. Encountering unusual heat, smoke, or burning at locations remote from the main fire area could indicate problems. Fire fighters at the fire should use care when dealing with these conditions and, following the fire, attempts should be made to understand why the conditions existed. Additional fire research may be necessary to understand how some materials burn in actual building fires, how some of these unexpected situations occurred, and how fire command and attack procedures might be better designed to reduce this risk."[4]

A Near Tragic Example

Fire fighters in New England were operating at a fire in a three-story, wood-frame building with a **cockloft**, a void space between the roof and the ceiling of the top floor. Fire on the second floor was apparently extinguished. Units proceeded to the third floor leaving no unit to watch the second floor. Fire burst out of the second floor walls and almost instantly blocked the exits. Fifteen fire fighters were trapped. Only quick work by fire fighters outside the building, police and off-duty fire fighters watching the fire prevented a disaster. If the fire had occurred late at night there would have been fatalities.

Tactical Considerations

Hard charging, gung-ho engine companies must be restrained from moving too far too fast. Fire breaking out behind or below fire fighters is a constant threat, particularly in combustible buildings with characteristic hidden voids. Interior attack lines must be backed up. If the backup is not available, the attack must be slowed. The hazard of fire breaking out behind fire fighters is also very real in warehouse fires.

Definitions of Loads

Specific terms are used to describe the different loads placed on a building. It is important to understand and use them correctly. Dictionary definitions are not always identical to the definitions used in the construction field. The various terms, however, are not mutually exclusive. For example, a load may be a **live load** and an **impact load** at the same time.

Dead Load

Dead load is the weight of the building itself and any equipment permanently attached or built in. A more accurate term is **self-weight**.

Years ago deadload was often piled on the building without regard for the consequences, or even in the mistaken belief that strength was being added to the structure by use of the added weight.

The modern designer more clearly understands that deadweight breeds deadweight. Efforts are made to lighten all parts of a building. Removing a pound of deadweight at the top of a building enables the builder to reduce ounces at many points in the supporting structure. Advertisements for newly-developed building materials such as gypsum shaft liners emphasize weight saving over masonry enclosures. Today, buildings can be considered as being bought "by the pound." The late Chief John O'Hagan discussed these issues in his textbook.[5]

Fig. 2-2. The tank is a dead load. The water may have been in the tank for years but it is still a live load. If the tank is permitted to continue to overflow in freezing weather, the undesigned live load of ice will increase until the tank collapses.

Fire resistance is closely related to mass. All other things being equal, a heavier steel beam will take longer to fail due to heating than a lighter beam. A 4-inch by 10-inch wooden beam will support its load longer while burning than will a 2-inch by 10-inch beam.

This provides a valid general principle: Any substitute structural element which is of less mass than the element previously used to carry an equivalent load is inherently less fire resistant.

Added Dead Load

A structure's dead load is often increased during alteration. An example is the addition of air conditioners to the roof of a building previously without air conditioning. These loads are often added without any strengthening of the structure.

At a fire in a large city, fire fighters were overhauling a fire in a restaurant, seeking out hidden pockets of fire in the overhead. They attacked the fire-weakened supports of the building. Above them, added-on air-conditioning units fell, causing several fatalities.

Fig. 2-3. The additional dead load of the cooling tower has been added as close to the wall as practical. However, some unknown portion of the factor of safety designed into the wall has been consumed. Such a wall demands attention during a fire.

The fact that the structure is strengthened to cope with the added load may not be adequate in a fire. For instance, if light-weight trusses are strengthened by additional trusses, there is no improvement to resistance from fire collapse, since all the trusses will burn and lose strength at the same time.

A grocery supermarket was converted to a Japanese restaurant with many grills. Each grill required a heavy fume hood. These hoods were hung from the roof. When the building was a supermarket, the roof did not have to support such a dead load. Now the same roof must carry the hoods -- safety has been greatly reduced.

Live Load

Live loads are any loads other than dead loads.

An elevated water tank is dead load. The water in the tank may be there for years but it is live load. A concrete vault is dead load. A movable safe, no matter how heavy, is live load. These are not just picky distinctions.

- Dead loads can be accurately calculated.
- Live loads are indeterminate. The live load must be estimated based on the projected use of the building and such variables as snow, wind or rain.

Buildings are designed with a given use in mind. The building code specifies the minimum live-floor load design for specific types of buildings.

Typical building code minimum design load requirements are spelled out in the Building Officials Conference of America (BOCA) Code.

Proposed Use	Design Load (lb/sq ft)
Classroom (fixed seats)	40
Classroom (movable seats)	100
Classroom corridors	100
Libraries (reading room)	60
Libraries (stacks)	150
Dwelling units	40
Stores (upper floors)	125

Added Live Loads

The water trapped in a building or on a roof can be a significant added live load. Flat-roof buildings often are built with little reserve strength.

An elevated water tank is designed for a certain live load -- the water in the tank. If the tank continues to overflow during freezing weather, the water will become an ice coating on the structure and could overload it to the point of collapse.

Tactical Considerations

Water from rain, snow or fire fighting may pond on a flat roof. In the case of such ponding the fire department may be asked to assist in water removal. Be careful -- a flooded roof may be near collapse. Don't add to the weight. Attempt to clear the debris from the drain, operating from a ladder without stepping on the roof. Merely stepping on the roof may precipitate collapse. In these litigious times, a lawsuit might ensue. Another possible solution is to build a salvage chute in a suitable location and puncture the roof from below.

Also, beware of the loading, and collapse possibilities, of such roofs from overspray when using master streams on a fire in an adjacent building, particularly in freezing weather.

The author, while wintering in Florida, received a call from a fire department up north. "We have this five and dime store with a big snow load. We got the snow off with hose streams, but we can see some cracked trusses. How should we brace them?" The advice was emphatic: "Get out of there and tell the owner to hire somebody competent, like a wrecking company."

Even staying outside such a building is not always enough. A collapse of the roof could set up an air ram which could blow the windows out. Often the best fire department tactic is to set up barriers well away from the building.

Recycled Buildings

When the use of a building changes, the design of the building should be reviewed to determine whether the structure should be strengthened for its new use. Unfortunately, this is often ignored, sometimes from pressure to eliminate "bureaucratic obstructions" to saving a historic building from wreckers.

A typical case is the conversion of a residence into a commercial structure. In a well known southern furniture-making city, I counted 14 sofas and associated furniture, in what had been the living room of a large private home.

The author observed a fire at a YMCA which was in a converted dwelling. An employee arrived on the scene, "Tell the fire fighters there is a big safe on the top floor." The safe had already collapsed into the

Fig. 2-4. This Florida residence was converted into a bank. It is probable that there are excessive floor loads which have reduced the factor of safety. Note also that because of its age, the building is probably of balloon-frame construction with unfirestopped connecting voids.

basement. Fortunately, no fire fighters were hit. Note that the untrained employee was aware of the hazard.

Tactical Considerations

In preplanning, make note of all concentrated loads, particularly in non-fire-resistive buildings. Old dwellings converted to offices should be examined with special care. Filing cabinets filed with paper are a very heavy concentrated load. The more heavily a structural unit is loaded, the sooner it will collapse.

Buildings which appear to be overloaded, or which have loads in excess of the original design, or structures which are internally braced such as with adjustable posts, should all be noted on prefire plans.

Fig. 2-5. Los Angeles fire fighters are making a concentrated aerial attack on this fire. Fire streams can add a substantial live load to the structure at a time when its structural strength is being diminished by fire. (Alan Simmons)

Tactical Considerations, continued

The existing overload situation in a structure could be so serious that the weight of hose lines, ladders and fire fighters might be too much additional load and cause a collapse.

The fire and fire fighting efforts will change or increase the loading of a structure. At the same time, the ability of the structure to resist gravity is being attacked by the fire. Unless a building is designed and built to be fire resistive, there is no provision in the design of the structure to prevent failure due to fire. The vast majority of structures built are not fire resistive by design. Whatever resistance such structures have to fire-caused structural collapse is purely accidental.

Bear in mind that a building's tendency to collapse is being increased by fire, while the fire suppression forces place additional demands on the structure. When the structure, or part of it, is no longer able to resist the loads, collapse occurs. Gravity acts instantly.

When heavy-caliber streams are used, fire fighters must be conscious of the tremendous added weight. A 600-gpm master stream will add 25 tons to a structure in 10 minutes. Some of this water will be absorbed into the contents. Paper and fabrics of any type will absorb huge quantities of water. Five fire fighters died when piles of scrap from the manufacture of tissues, soaked by water from sprinklers, collapsed on them.[6] Some fire departments have set

Tactical Considerations, continued

specific drain times to permit water to drain from structures subjected to heavy master streams before personnel are permitted to enter a building.

In one example, water thrown on a theater marquee in freezing weather turned to ice and collapsed the marquee, trapping several fire fighters. Fortunately, one fire fighter had access to a construction crane and the marquee was lifted.

Heavy water loads were once a problem only to departments with strong water supply. Today, with large hose and well-organized tanker and dam operations, heavy water loads can be delivered in many suburban and rural locations.

Impact Loads

Impact Loads are loads which are delivered in a short time. A load which the structure might resist, if delivered as a static load over time, may cause collapse if delivered as an impact load.

Buildings have a habit of standing up as long as they are left alone. However, fire is a violent insult to a building. All sorts of interrelationships among building elements, which have mutually supported one another and provided strength not even calculated by the designer, are disturbed. A fire can cause innumerable *undesigned* changes in loading, many of which can cause collapse. An explosion, the overturning of heavy live loads, such as a big safe, the collapse of heavy, nonstructural, ornamental masonry, or even the weight of a fire fighter jumping onto a roof, may be enough of an impact load to cause collapse.

There is a formula for tensile impact in the fourth edition of the *Civil Engineering Handbook.*[7] If a tension bar is supported at the upper end and a weight is allowed to drop from a height, the stress produced in the bar caused by the drop of the weight is as follows:

$$S_1 = S + S\sqrt{1 + \frac{2h}{e}}$$

★ *S_1 is the unit stress in pounds per square inch of the impact:*

★ *s* represents the unit stress in the bar if the weight were quiescent (just sitting there);

★ *e* denotes the total deformation in the bar in inches when the weight is quiescent;

★ *h* is the distance in inches that the weight falls.

★ The test bar is a piece of steel rod hanging freely.

★ The weight dropped is like a doughnut around the rod.

★ The rod is enlarged at the bottom so the “doughnut” can’t fall off.

Any structural member under load must change shape. Compressive (pushing) loads shorten a member; tensile (pulling) loads lengthen a member. If there is no change of shape, there is no load. The dropping weight exerts a tensile (pulling) load on the bar. The same weight quietly in place on the bottom of the bar also exerts a tensile load on the bar. In both cases, the bar is elongated. The value e in the formula represents the elongation when the weight is quiescent. The elongation is greater at the moment the weight hits.

The whole point of looking at this equation is to solve it using the value of zero for "h" (the height through which the weight drops), thus simulating what we might think of as "no impact."

Solution. Impact stress s_1 = quiescent stress s plus s times the square root of 1 plus *zero* (since h is *zero* the whole term is *zero*) or s plus $1s$ which equals $2s$.

Thus impact stress equals twice the quiescent stress even when the height through which the object drops is zero, which most of us would believe would cause no impact.

The point, of course, is that there is no such thing as a "no impact" load. No matter how gingerly personnel and equipment are placed on a structure, there is a substantial increase in the stress, at least momentarily. In any event, assaults on a structure already under attack by fire are very dangerous.

Lateral Impact Loads

Lateral impact loads can produce disastrously high stresses. A lateral impact load, such as from an explosion, can be delivered from a direction which has little or no stress resistance.

Walls are designed to accept vertical compressive loads. However, even impressive masonry walls are not necessarily resistant to undesigned lateral loads. The lateral thrust of collapsing floors has brought down many brick walls.

The sudden ignition of trapped carbon monoxide or other fire-generated combustible gases has sometimes caused detonations which have destroyed substantial masonry walls.

A cooking-gas explosion caused the multi-story collapse of a 24-story concrete-panel apartment house in England. The undesigned lateral impact load blew out one bearing wall. The floor it was supporting fell and became an impact load on the floor below, and the successive collapse of the underlying floors followed. This was the so called "Ronan Point Collapse."[8] Gas service was replaced with electricity, and repairs were made to other units. In 1984, tenants were moved out of eight other buildings in the same group because of a fear of collapse. The repairs after the first collapse were apparently inadequate.

The safety factors built into ordinary buildings are rarely large enough to assure that there will not be progressive collapse in the wake of the first excessive impact load. Progressive collapse is a particular hazard in the construction of concrete buildings. (See Chapter 8, Concrete Construction.) Backdraft explosions can occur when the carbon monoxide and oxygen mixture is exactly right, and can blow the building apart.

Building codes typically require that explosives be handled in buildings of "4-hour fire-resistive construction," almost guaranteeing the restraint which increases explosive destruction.

In industry, where codes may not apply, an opposite tack is often taken. Buildings housing hazardous processes are isolated and built of **friable** (easily disintegrated) construction elements. One such method employs a steel frame covered with an easy-to-replace material such as cement-asbestos board. If an explosion occurs, the structure vents, and the board becomes dust-like particles, not missiles. Much of the equipment survives the explosion, though pictures of the scene appear to show total devastation. After such an explosion, however, the asbestos particles represent a health hazard and all necessary precautions should be taken.

Special-purpose buildings may be designed to channel the force of an internal explosion in a desired direction. Transformer stations are often built with heavy walls to protect one transformer from an explosion in an adjacent transformer and to direct the blast upward or outward.

A national laboratory which is proud of its achievements in the physical sciences, decided to build a room of three sturdy walls and one "blowout wall" for handling hydrogen. The blowout wall, built of wood studs which would make jagged missiles, faced out on the lawn where many of the staff ate lunch on pleasant days!This author has often noticed that competence in one field is sometimes offset by ignorance in another, particularly fire. Persons who could not qualify in court as experts in fire protection do not hesitate to make disastrous fire protection decisions.

Static and Repeated Loads

Static loads are loads which are applied slowly and remain constant. A heavy safe is an example of a live, static load. It is not a dead load.

Repeated loads are loads which are applied intermittently. A rolling bridge crane in an industrial plant applies repeated loads to the columns as it passes over them.

Wind Loads

Wind load is the force applied to a building by the wind. The designer can counter this force in a variety of ways.

Small wood-frame buildings were formerly built with only a gravity connection to the foundation. The 1938 surprise hurricane, which devastated the northeast coast, destroyed thousands of homes by simply pushing them off their foundations. Since then, codes have required that wood-frame buildings be bolted down to the foundation.

Masonry, low-rise bearing wall buildings usually have enough mass in the walls that special consideration of wind load is unnecessary.

The framing of lightweight, unprotected-steel buildings, however, is tied together to resist wind forces. Fire-caused failure of one part may cause the collapse of other sections because of undesigned torsional loads transmitted through the ties.

Most floors simply carry loads to the support system of columns and floors. Diaphragm floors are designed to stiffen the building against wind and other lateral loads such as earthquakes.

One of the best examples of a need to counter wind load is found in high-rise buildings. The high-rise building must be braced against the shear force of wind. This is accomplished in one or more ways.

- Diagonal braces which connect certain columns are concealed in partition walls of such buildings. An arrangement of braces between columns which resemble the letter "K" is called K-bracing.
- Heavy riveting of girders to columns from top to bottom of the frame is called portal bracing.
- Masonry walls, needed to enclose vertical shafts, can do double duty as shear walls, resisting overturning. This method is satisfactory only for low- or medium-rise buildings. If used in a megastructure, it would require shear walls, with no other purpose than wind resistance to be erected above the floors where some elevators terminated.
- Very tall high-rise buildings are built to take the wind load on the exterior walls rather than on the interior core. Sometimes diagonal columns are visible on the exterior.
- Externally braced buildings are known as **tube construction** as contrasted with **core construction**.
- Unique giant **vierendeel trusses** (rectangular trusses with very rigid corner bracing), formed by exterior box columns and spandrels, provide the wind bracing of New York's World Trade Center.

The wind load can be particularly hazardous to fire fighters when operating in or near buildings under construction. This is because the full benefit of interconnecting all the parts of the completed building has not yet been realized. Temporary bracing may be inadequate.

In addition to its effect on structures, the wind is a force to be reckoned with in the movement of smoke in buildings. This force is often ignored by some whose solution to the high-rise fire and life safety problem rests first on "sophisticated, state of the art" smoke-handling equipment. This author's cynical definition of *sophisticated equipment* is "you can't get the parts when it breaks down," and similarly, *state of the art* is "we're trying this out on you."

Tactical Considerations

Fire forces should demand specific answers to the question, "How will this smoke system function when faced with a 30 mile-per-hour wind and all the windows in a substantial area are out?"

Concentrated Loads

Concentrated loads are heavy loads located at one point in a building. A steel beam resting on a masonry wall is an example of a concentrated dead load. A safe is a concentrated live load.

In a typical concrete block wall, the building designer may insert solid block, brick, reinforced concrete or a **wall column** to stiffen the wall and carry the weight of the concentrated load. If a wall is being breached and the structure is found to be stronger than normal, choose another location. You are probably right under a concentrated load.

Early fireproof floors were made of brick segmental arches. Many still exist. The wooden floor was leveled in some cases by supporting it on little **piers**. At some locations there was a gap of several inches between the floor and the arch. When such a floor burned, a live load such as a safe would fall the few inches. The impact was too much for the arch which had carried the safe for years, and the safe wound up in the basement. To this day the testing and rating of safes involves dropping the heated safe a considerable distance.

Someplace in the U.S. and Canada there are existing examples of almost every feature or practice of building construction. While you may not have every feature or problem in your area, you should be aware of them, even though you concentrate on those common to your locale.

Imposition of Loads

Loads are also classified in the manner in which they are imposed on the structure.

Fig. 2-6. This wall column of brick was built into a concrete block wall to sustain the concentrated load of a main girder. It is not visible from the exterior. If you are breaching a wall and find stronger material, stop and breach someplace else.

Axial Loads: An axial load is a force that passes through the **centroid** of the section under consideration. An axial load is perpendicular to the plane of the section. In simpler words, an axial load is straight and true; the load is evenly applied to the bearing structure. All other conditions being equal, a structure will sustain its greatest load when the load is axial.

Structural elements are not always loaded in the most efficient manner. Other considerations may be more important. For instance, a ladder is strongest when absolutely vertical, so that a person represents an axial load. But human nature and mechanics works against axial loading, and requires slanting of the ladder.

Eccentric Loads: An eccentric load is a force that is perpendicular to the plane of the section but does not pass through the center of the section, thus bending the supporting members. In simpler terms, the load is straight and true but is concentrated to one side of the center of the supporting wall or column.

Fig. 2-7. Note that the wooden girders are attached to the wooden columns equally on each side. The load therefore is balanced. The column is axially loaded.

A key factor in the infamous Washington, D.C. Knickerbocker Theater collapse in 1922 which caused the deaths of 97 persons, was the fact that the main roof trusses bore eccentrically, rather than axially, on the hollow tile walls. The ends of the trusses did not rest on the full width of the walls. The wall was not reinforced to carry this eccentric load. The roof survived for several years until an unusually heavy snowstorm deposited two feet of snow on it.

Fig. 2-8. This column is eccentrically loaded. It is probable that future expansion was planned. The column is under compression on the loaded side and in tension on the other. Although not visible, there must be reinforcing rods on the tensile side to hold the column together, otherwise the blocks would pull apart.

For an example of such a load place a stack of books on the floor. Sit on the stack squarely. Note how evenly applied weight stabilizes the stack. Shift your weight until it bears on only part of the stack. Note the tendency of the stack to bend outward. One side of the stack is being compressed, while the other is being pulled apart. These forces are called **compression** and **tension**, and are defined later in this chapter.

Torsional Loads: Torsional loads are forces that are offset from the shear center of the section under consideration. They are inclined toward, or lie in, the plane of the section, thus twisting the supporting member. In simpler terms, torsional loads are twisting loads. One sign that destruction was caused by a tornado, rather than by ordinary high winds, is trees twisted off by torsional loads.

The failure of one part of a steel frame building or steel truss will often place undesigned torsional or twisting stresses on other parts of the building, thus extending the area of collapse.

Tactical Considerations

Cooling of unprotected structural steel may well be the most important operation of a fire department at certain fires, where unprotected steel structural members are being heated. (See Chapter 7, Steel Construction.)

Within the limits of material available, any type of load can be dealt with successfully. Axial loads are more easily dealt with by the building designer than are eccentric and torsional loads. The **undesigned shifting of the loading** from axial to eccentric, or from axial or eccentric to **torsional**, may cause collapse.

Tactical Considerations

A structural fire upsets the delicate balance of forces in a building, and can result in disastrous undesigned changes in loading. Prefire plans should examine these possible changes in loading which may occur during a fire.

Fire Load

Some engineers do not recognize the term **fire load**. It represents the potential fuel available to a fire. The research work done on fire load

Fig. 2-9. These fiber-reinforced plastic concrete forms look harmless enough but they are a substantial fire load. The glass is non-combustible but the plastic is not.

was directed to the **contents** of buildings. When the building is combustible, the building itself is part of the fire load.

Fire load represents the total amount of potential heat in the fuel, while the term **Rate of Heat Release (RHR)** indicates how fast the potential heat in the fuel is released.

This distinction is important. All wood can generate approximately the same amount of heat per pound, but a pound of plywood will burn many times faster than a pound of heavy timber.

The basic measurement of **caloric value** is the **Btu** (British thermal unit), the amount of heat required to raise a pound of water one degree Fahrenheit. The metric equivalent is the **kilojoule**. One Btu is approximately equal to one kilojoule.

Each combustible material has its own caloric value. Two estimates are commonly used. Wood paper and similar materials are estimated at 8000 Btu/lb. For plastics and combustible liquids 16,000 Btu/lb is a common estimate, though some such fuels are much higher.

The weight of the fuel is multiplied by the caloric value and divided by the floor area, to arrive at Btu/sq ft, the measure of the fire load. The metric statement is kJ/m^2. You may find fire load expressed in pounds per square foot — a practice which dates back to when there was only one basic fuel, i.e., wood or paper, estimated at 8000 Btu per pound. Plastics are converted into "equivalent pounds" on the basis that one pound of plastics equals two pounds of wood.

A fire load of 80,000 Btu/sq ft or 10 lb of ordinary combustibles per sq ft is approximately the equivalent of a one-hour exposure to the stan-

Fig. 2-10. Plywood has the same potential heat per pound as other wood but it delaminates (layers come apart), increasing the surface-to-mass ratio so that it has a high rate of heat release.

dard fire endurance test; 160,000 Btu per sq ft is equivalent to a two-hour exposure.[9]

Rate of heat release refers to the rate at which the fuel burns. A five-pound solid chunk of wood will burn more slowly than five pounds of wood chips. Five pounds of plywood will burn very rapidly. The number of Btu remains the same in these fires, but the heat is released more rapidly from certain materials.

Tactical Considerations

In the case of an apparent heavy fire load in a fire resistive building, a calculation should be made of the estimated fire load. If this is substantially beyond the rating of the building, corrective measures should be sought. Interior store rooms in office buildings often have very heavy fire loads. Manual fire suppression would be difficult and dangerous because of the inability to ventilate properly.

Extensive data is not available on which to base estimates of rate of heat release (RHR). However, some estimates are necessary because RHR is an important criterion in the development of prefire plans. "How big and fast a fire can we anticipate?"

Only in recent years has the concept of calculating the fire load in advance gained acceptance, even in fire protection circles. Yet, the absorption of heat by water is the essence of fire suppression by fire departments.

Tactical Considerations, continued

Fire department engine companies could be called "heat removal units." This might then raise the appropriate question, "How much heat will we have to remove?" From this answer, water flow requirements could be accurately developed.

Many prefire plans note that a building is sprinklered. Not many contain an analysis of whether the sprinkler system is capable of delivering the amount of water needed to absorb the heat that must be absorbed if the fire is to be stopped. (See Chapter 13, Automatic sprinklers.)

Calculations of fire load and rate of heat release should be part of a truly adequate prefire plan.

The well known senior fire protection engineer, Harold (Bud) Nelson, SFPE (Fellow), has recently developed a computer program "FPETOOL" which facilitates fire calculations from known or estimated data. Contact the HIST Center for Fire Research Computer Bulletin Board for more information.[10]

Building Classification Fallacies

Building codes make a limited effort to attack the fire load problem by classifying buildings by their intended use. A warehouse will usually contain a greater fire load per square foot than a department store. A warehouse may have 4-hour fire resistance requirements, while a store may be required to have only 2-hour fire resistance. However, the warehouse may store sheet metal and the store may display merchandise made of foam rubber. It is difficult, if not impossible, to cope with this problem by code.

The Characteristics of Materials

Buildings are built of various natural and synthetic materials. Each material--and even different forms of the same material--has certain physical and chemical characteristics which make it more or less desirable for its intended function.

Some materials are good against compressive forces, others against tensile forces. Some are heat insulators, others absorb sound. Some resist weather; others lack structural strength but provide a good smooth surface that takes a nice coat of paint. Some materials are merely cheap.

There are those who consider fire solely from the point of view of chemistry. If a material contributes fuel to the combustion process, it is considered combustible and taken into account for fire formulas. If it does not burn, it is safe. This is untrue. All materials can be damaged by fire even if they do not burn, and the structural damage may be significant.

The fire characteristics of materials are often misunderstood or ignored unless the law directs attention to this vital facet. As each material is covered in the later chapters, its fire characteristics will be discussed. A brief summary of some important characteristics of common materials that can influence their behavior in fire is provided here.

Wood burns and thus loses structural strength and integrity. The vast majority of structural fires are fought by fire fighters standing on or under wooden structures.

Bricks are smooth-sided pieces of clay cured in an oven. They are quite fire resistant, but can spall (lose surface material) when subjected to fire. Like other masonry materials, brick cannot be effectively considered apart from the mortar used with it. Prior to 1880, when portland cement mortar became available, the only mortar used was sand-lime which is water soluble. Even today, regulations require the use of sand-lime mortar in the restoration of old buildings built with soft bricks. Ordinary water leaks or fire hose streams have dissolved sand lime mortar and caused collapse.

Natural stone buildings will spall when exposed to fire. Granite particularly is subject to spalling.

In older buildings, slate and marble were used for stairways, and sometimes for unsupported flooring in library stacks. These should be examined for support. If only self-supported, they should be avoided after exposure to heat. Marble can look perfect yet actually may have turned to chalk. It will collapse under the weight of a fire fighter.

Cast iron can have good fire characteristics. The casting method determines whether it is good or bad, and this cannot be determined by examination. Poor connections of floor beams to cast-iron columns are probably the chief cause of failure. This is discussed in Chapter 4, Ordinary Construction.

Structural steel has three negative characteristics in fires.

- It conducts heat.
- When heated it elongates, and may push through any barrier.
- It fails at about 1000 - 1100° Fahrenheit.

Cold-drawn steel, such as cables which are sometimes used to brace failing buildings or as tendons in tensioned concrete, fails at 800° Fahrenheit.

Tactical Considerations

In many fires, the most important heat to absorb with water streams is that which is being absorbed by structural materials, particularly steel. Heat absorption may change the structural characteristics which the designer relied on to carry the loads.

The use of water to cool steel doesn't cause failure; it simply cools the steel.

Reinforced concrete is a composite material. Failure of the bond between concrete, which provides compressive strength, and steel, which provides tensile strength, causes failures of reinforced members.

Inherently, concrete is only non-combustible. It may be further formulated to be fire resistive. Because most concrete buildings are merely non-combustible, they are subject to possible early failure in a fire.

Gypsum is useful material. It absorbs heat from the fire as it breaks down under fire exposure. It can be used as a barrier to fire and to protect combustible structures. Its value is often mistakenly exaggerated, as discussed later in Chapter 5.

Tactical Considerations

After being heated by fire, gypsum starts to break down. This can't be stopped. You are doing the building owner a favor to remove gypsum board which has been heated. If it is just painted over, its breakdown will continue. It is wise to let the owner know why you are removing it, to avoid any problems.

Many fire-resistive partitions are made of gypsum on studs. Sometimes the quickest access into an occupancy is to bypass a locked door and go right through the gypsum wall. Montgomery County, Maryland fire fighters made a fine rescue of an infant trapped on the second floor of a townhouse by going through the wall of the adjacent unit.

Single-Ply Roofing

Special thanks are due to fire fighters Tony Papoutsis and Scott Reichenbach. Single-ply roofs are made up of a layer or layers of insulation and a single sheet of membrane material used for waterproofing. The membranes are in sheets from 30 to 160 mils in thickness, and as large as 50 x 200 feet. The membrane may be glued or fastened to the insulation. If not fastened, gravel or concrete pavers are used to hold the membrance in place. (Single Ply Roofs, *Fire Engineering*, Feb., 1992)

There are a number of different systems used. The type of system used on a building should be part of the pre-fire plan.

The insulation may be a flammable plastic such as polystyrene, an easily fire-consumed plastic such as polyisocyanate, or fire-inhibited wood fibers.

The membrane roof may be installed over an existing roof.

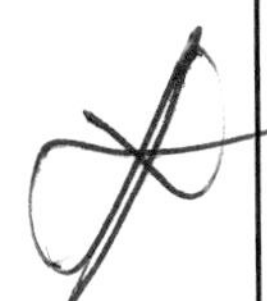

Tactical Considerations

To ventilate a ballasted roof, remove the ballast from an area several feet larger in all directions than the desired hole. Cut the membrane with a knife or scissors, not with an axe or power saw. Peel back the membrane and open the roof.

If the membrane is glued to the roof, peeling is difficult.

A mechanically fastened membrane is loose between fasteners. In high winds it can blow like a sail and throw fire fighters or equipment off the roof. White membranes make roof edges difficult to see in blinding glare or ice and snow.

There are a number of different compounds used for the membrane. All should be expected to emit toxic gases so SCBA should be worn, even through a roof fire might seem to permit otherwise.

Membranes are slippery when wet. Check for fire traveling under the membrane.

Insulation may mask the possibility of a structural collapse.

Installation is a hazardous operation due to the use of flammable compounds which emit heavy vapors. There may be a large quantity of hazardous flammable materials on the job site.

Aluminum (known as aluminium in Canada) melts at very ordinary fire temperatures. This is sometimes beneficial, such as when an aluminum roof melts and vents a fire. Working on an aluminum roof, on the other hand, is extremely hazardous.

Tactical Considerations

Some houses have been built with aluminum beams. These should be expected to collapse early in a fire. One magazine article described aluminum trusses to be used for homes.

Some house trailers have aluminum sides and steel roofs. Since the roof won't melt and vent, there is greater potential for extension laterally to any adjacent units.

Aluminum is widely used for railings. In a high-rise building, aluminum railings were used to separate the stairway from the smoke

Tactical Considerations, continued

shaft. They failed in a multi-alarm fire and presented a serious hazard to fire fighters working in the stairway. (The basic fallacy of "smoke shafts" is that the heat and fire cannot be separated from the smoke).

Aluminum trusses and beams are sometimes used for concrete falsework. They melt and lose strength rapidly when heated. Such a failure occurred in a fire in Fairfax County, Virginia. The glass fiber formwork often used with the aluminum can provide sufficient fuel for a raging fire. May other fuels are likely to be present.

Aluminum ductwork will melt and admit fire to the duct.

Glass is used in several ways, most commonly in windows and doors. Glass has little resistance to heat or pressures built up by the high temperature of a fire. Often the first sign of a serious fire in a closed structure is the violent pressure-failure of the windows, often described by witnesses as an explosion. Some glass windows are double or triple glazed for energy conservation. They are difficult, if not impossible, to break. Some codes require that some windows in high-rise buildings be of breakable glass. These are sometimes marked with a maltese cross.[11]

Fixed sashes of double-pane glass were a factor in a motel fire in which several persons died. Such windows cannot be readily broken by what is available in a motel room. They should be equipped with emergency opening devices.

Wired glass, properly installed, has greater resistance to fire than ordinary glass, but it passes radiant heat as readily as regular glass.

Tactical Considerations

Many high-rise buildings have large glass areas. Units operating inside should be notified immediately when the failure of such windows is seen from the outside. The inrush of oxygen will accelerate the fire, seriously endangering units on or above the fire floor.

Falling glass may make outside operations at a high-rise fire dangerous or even impossible. Falling glass has cut hose lines feeding standpipes. Be aware of any protected way by which the building can be entered, such as an underground garage.

Glass fiber insulation is a combination of non-combustible glass fibers and combustible binders. Glass fiber insulation is often installed above metal grid suspended ceilings. If the fire is above the ceiling, this will tend to conceal the situation from fire fighters below. If the fire is below the ceiling, the insulation will retain heat in the steel grids. If the glass fiber is not included in the listing of a rated floor or roof and ceiling assembly, the retained heat may cause the ceiling to fail earlier than expected.

Glass fiber-reinforced panels are rigid flat or corrugated plastic panels. These are sometimes assumed to be noncombustible, because of the glass. They are best described by their accurate generic name, **glass fiber-reinforced polyester resin plastic**. The plastic burns readily, leaving a mat of glass fibers which should not be disturbed. If freed, the fibers can cause severe skin irritation, and if inhaled could cause severe lung damage.

Asbestos was long used by itself or in combination with other materials as a fireproofing agent for steel. Inhaled asbestos fibers are a known carcinogen. Concern for this hazard has led to the removal of asbestos from buildings. If asbestos was used for fireproofing and removed, there is the serious question of whether any action was taken to restore the protection to the steel.

Another problem with the past use of asbestos is its discovery in the ruins of a fire during the investigation. This was the case in the fatal fire at the Du Pont Plaza hotel in San Juan, Puerto Rico in 1986. Three weeks after the investigation started, asbestos was discovered in the debris. It was learned also that the baggage of guests who had already gone home was contaminated.[12]

Much of downtown New York is heated by steam supplied at high pressure. An asbestos-insulated main blew up and heavily contaminated an apartment house. Fire fighters on the scene had no knowledge of the hazard, which was not recognized for several days. A number of pieces of apparatus were out of service until they could be decontaminated.

In August 1991, the American Medical Association's Council on Scientific Affairs stated that the asbestos should be removed only when damaged or crumbling or disturbed by renovation or demolition, and minimized the hazard. Experts at Mt. Sinai School of Medicine argued strongly that asbestos remains a significant danger to several groups including fire fighters. The AMA agreed.

Tactical Considerations

Buildings where asbestos might present a hazard during or after fire suppression should be identified in advance so that technical assistance can be summoned as soon as there is a working fire.[13]

Plastics are defined by ASTM as "A material that contains . . . one or more organic polymeric substances of large molecular weight, is sold in its finished state, and . . . at manufacturing or processing can be shaped by flow." There are thousands of plastics, including many which are used in construction, which are impossible to discuss in detail. However, the fire characteristics of this material are important to learn.

Fire Characteristics of Plastics

Many plastics can be ignited readily with small flames. Polyurethane imitation wood beams, a commonly used plastic, can be ignited with a match. Plastic imitations of other materials look very realistic and can be quite deceptive. It was reported that the multi-fatality disaster at the Six Flags Haunted House in New Jersey was started when a guest ignited polyurethane with a cigarette lighter.

Some plastics emit dense black smoke when burning. The smoke is sooty, sticky, and costly to clean up. It can cause acute distress to those breathing it.

Highly toxic gases may be evolved from the burning of plastics.

Thermoplastics can produce flaming, dripping plastic, which will produce secondary fires. Such fires can burn down through the wooden floor into the floor void of even one hour-rated combustible floor and ceiling assemblies. A **dripping-pool fire** can be seen in the NFPA film "Fire Power."

Fig. 2-11. The cast plastic in this passenger tunnel holds back tons of water and is an example of the structural use of plastic.

Small-scale laboratory tests may produce results which are more favorable to the plastic than real life situations warrant.

Foam core panels are foamed plastic panels in which foamed urethane or expanded polystyrene is sandwiched between two panels of **oriented strand board,** (see Chap. 3 for description) or oriented strand board on the exterior and gypsum board on the interior surface are being used for roof and wall panels. A Public Broadcasting System show, "This Old House," showed such panels being used on a timber framed house. The typical panel appeared to be 2 feet by 8 feet. Several panels were shown assembled together to make a large self-supporting roof panel. It appears that this might be a dangerous roof to depend on for ventilation or fire extinguishment. Some plastics melt, others might be consumed by the fire. Some might have a gypsum board thermal barrier, others might not.

Resin plastics reinforced with fiberglass are used for corrugated and flat plastic panels and some automobile panels. The plastic will burn and leave a matted sheet of glass fibers.

Fire-retardant panels made by a similar process, and which have been approved by Factory Mutual for installation without automatic sprinkler protection, are available (where sprinklers would otherwise be required if ordinary panels were used).

Structural members of plastic reinforced by continuous glass fibers oriented in the direction of the load are used for special buildings, such as those where no metal is permitted due to radio frequency interference, or where corrosion is a serious problem. Bridges with a clear span of 90 feet have been built to be used over tanks emitting corrosive fumes. Composite Technology, Inc. has information available on such composite elements.[14]

Even when combustion-inhibited, plastics may deform and cause costly damage under fire conditions. The NFPA *Fire Protection Handbook* discusses plastics in detail.[15]

Tactical Considerations

In fires involving plastics, anticipate severe disabilities among the occupants, possibly even mass casualties. Expect heavy toxic smoke conditions and fast-spreading fire. The toxicity of some plastics makes it imperative that SCBA be worn even through overhauling and outside operations such a plastic roofs.

Any fire fighting on foam core roofs should be done by personnel operating from an aerial apparatus. A building enclosed by foam core panels and fitted with the energy efficient windows discussed earlier might well present some flashover and backdraft conditions.

Fig. 2-12. This is a test setup for fiber-reinforced plastic beams. The heavy masonry units are resting on the bottom flange of the beams (light-colored horizontal units) to provide the weight necessary for the test.

Combustible fiberboard has been used for wall paneling, ceiling tiles, and sheathing. It burns persistently. Even when codes require new ceilings to meet flame-spreading requirements, the dangerous old tile is left above the new ceiling. This was a significant factor in the loss of 16 lives in a Tennessee retirement home fire in 1989.[16]

Failure to recognize the hazard of combustible fiberboard in a building to be burned for a drill was a significant factor in a multi-death training disaster. (See Chapter 9, Fire Growth, for a further discussion of this hazardous material.)

The Meanings of "Combustible"

The word combustible and its antonym "non-combustible" are frequently used in fire protection. Many of the usages are in legal contexts. In a legal context, a dictionary definition of the word may be misleading.

In the regulation of flammable liquids, for example, one code used the word **flammable** to describe liquids which have a flash point below 100°F; and the word **combustible** for those with a flash point above 100°F. Other codes have different thresholds.

With respect to building materials in the past, combustible was defined indirectly by defining non-combustible. The 11th edition of the NFPA *Fire Protection Handbook*, published in 1954, said:

> "Non-combustible as applied to a building material or combination of materials means that which will not burn and ignite when subject to fire, such as steel, iron, brick, tile, concrete, slate, asbestos, glass, and plaster."

This was a nice comfortable definition of non-combustible and by antithesis, of combustible. However, such a definition was not found to be precise enough nor did it take into account that some materials are more combustible than others.

NFPA 220, *Standard Types of Building Construction*, 1985, defines non-combustible material as:

> "a material which, in the form in which it is used and under the conditions anticipated, will not ignite, burn, support combustion, or release flammable vapors, when subjected to fire or heat. Materials which are reported as passing ASTM E 136 *Standard Test Method for Behavior of Materials in a Vertical Tube Furnace at 750°C*, shall be considered non-combustible materials."

The distinction between true non-combustible materials and materials that are to some degree combustible is provided in the definition of **limited combustibility**, also from NFPA 220.

> "a material, not complying with the definition of noncombustible material, which, in the form in which it is used, has a potential heat value not exceeding 3500 Btu per pound (8141 kJ/kg), and complies with one of the following paragraphs (a) or (b)."
>
> (a) Materials having a structural base of noncombustible material, with a surfacing not exceeding a thickness of 1/8 of an inch (3.2 mm) which has a flame spread rating not greater than 50.
>
> (b) Materials, in the form and thickness used, other than as described in (a), having neither a flame-spread rating greater than 25, nor evidence of continued progressive combustion and of such composition that surfaces that would be exposed by cutting through the material on any plane would have neither a flame-spread rating greater than 25 nor evidence of continued progressive combustion.

Paragraph (a) above enables building codes to more properly classify gypsum board. At one time, the definition of non-combustible was written to include gypsum board so that it would not be classified as combustible because of the paper facing.

Non-combustible Buildings

Some buildings are described in codes as non-combustible. These descriptions are by no means technically accurate. The local fire officer should know just what this means. Since the non-combustibility requirement is limited to certain designated components, non-combustible buildings can have very significant combustible components, including specifically, wooden roofs, cornices, wooden interior balconies, sheathing, and metal deck roofs.

Confusing Language

Many terms are used to describe materials in advertising or building codes. Some examples are **fire-rated, fireproof, flameproof, self-extinguishing, slow-burning, flammable, non-flammable, fire retardant, non-burning, flame retardant, fire resistant,** and **non-combustible.**

It is uncertain what these terms actually mean, and what testing has resulted in the properties described by them. A sample examined in one fire test may be "self-extinguishing," while it may rate as "combustible" if another test is used. Get full information on the tests and the standards used before accepting any claim of superior fire performance.

The term **inflammable** may still be found. For many years, the NFPA has striven to eliminate the term from fire protection literature because it is confusing — some people may think it means "not flammable." If it disappears from use, it will soon become archaic and its usage may be totally dropped and the confusion avoided.

Effect of Energy Conservation

The efficiency of heating and cooling buildings has received great attention recently due to the high cost of fuel.

Insulation added to a building changes its fire characteristics. In one test, a wood frame and steel building was subjected to a fire from a 140-pound wooden crib. The exposed corner of the structure was damaged and the fire went out. Various types of insulation were then applied to the building. In each case, the fire characteristics were altered. In some cases, the building was heavily damage by the same-sized fire.

Insulation can cause heat to be retained in light fixtures designed to disperse heat. Paper vapor seals on insulation placed directly on fixtures can ignite.

The hazards presented by double and triple pane glass insulating windows were discussed earlier.

Fig. 2-13. Note the Maltese cross on the hotel window. This indicates a window which can be broken easily by the fire department.

Fire Resistance Distinguished

You must distinguish between two uses of the term fire resistance.

- **Rated fire resistance** is a quality ascribed to a wall, floor, or column assembly that has been tested in a standard manner to determine the length of time it remains structurally stable (or resists the passage of fire) when attacked by a standard fire. Rated fire resistance is discussed more fully in Chapter 6, Principles of Fire Resistance.
- **Inherent fire resistance** is a structural member's resistance to collapse by fire because of the nature of its material or assembly. A heavy wood beam takes longer to fail from a fire than does a 2 X 4. A light steel beam will absorb heat faster and thus fail sooner than will a heavier beam. A masonry wall, itself quite fire resistant, may fail because of the lateral thrust of an elongating steel beam. The inherent fire resistance of a structure has never been formally rated or required by law.

By experience and tradition, the fire service has developed some criteria for determining the inherent fire resistance of a particular type of construction. Unfortunately the criteria are imprecise. Many times fire fighters risk life and limb without really knowing the degree of risk and without any risk/benefit calculation. The generally accepted

Fig. 2-14. Los Angeles fire fighters battled a raging fire in the massive concrete Main Public Library. The almost windowless building resembled an underground building. Like many other libraries, it was not sprinklered. This fire is discussed in detail in Chapter 8. (Alan Simmons)

indications of imminent collapse such as floors or roofs "softening," water flowing through bricks, smoke pushing out of mortar joints, "strange noises," are, at best, grossly inadequate, even for the buildings built many years ago. If these warning signs are relied on solely to alert fire fighters in today's lighter buildings, bloody disasters will be the certain result.

Some codes have a "protected combustible" classification. Combustible structures are "protected" with gypsum board and wood assemblies which have passed the standard fire resistance test. The many fire deficiencies of these structures are discussed in Chapter 5, Garden Apartments.

When a combustible structure is involved in fire, no code provision, however well written and however well meaning, provides real personal safety for the fire fighter. The building is the enemy, and you must know the enemy.

Underground buildings present special fire protection problems, similar to those posed by a fire in a submarine. Tightly sealed, unsprinklered structures present the potentially hazardous buildup of a heavy carbon monoxide concentration just waiting to blow when the unwary fire fighter admits oxygen.

Many years ago eight New York fire fighters died as they opened up a workroom in the basement of the Ritz Tower hotel.

Some years ago, this writer lectured at a major underground defense installation. It was several stories deep, loaded with combustibles — with no water available. Furthermore, there was no provision for breathing equipment!

Tactical Considerations

The hazards and fire suppression problems of such structures should be thoroughly discussed with the occupants. Every effort should be made to impress them with the absolute necessity of immediate evacuation. One visit might not be enough.

Applying Forces to Materials

Stress and Strain

The external force which acts on a structure is called a load. The internal forces which resist the load are called stress and strain. **Load** and **stress** are often used synonymously.

Stress is usually measured in pounds per square inch (psi). Occasional references to pounds per square foot (psf) are also found. The **unit area** measurement is at the discretion of the person making the calculations. Always be careful to note the unit area when examining any calculations.

KIP, a term meaning 1000 psi, is used in engineering calculations where the number of psi would be so large as to be unwieldy.

Strain is the percent of elongation which occurs when a material is stressed. Strain is measured in fractions of an inch of deformation per inch of original length of the material. In the vernacular, stress and strain are considered synonyms. A clear understanding of the technical definitions of these terms is critical.

In compression, the **deformation** takes the form of shortening. In tension, the deformation takes the form of elongation. There is a basic principal at work: If there is no deformation, there is no load.

Forces are applied to structures in various ways. Compressive forces push the mass of the material together. Tensile forces tend to pull the material apart. Shear forces tend to cause materials to slide past one another.

Observe a suspension bridge. The cables are in tension. The cables deliver the load to the towers which carry it to the ground in compression. The ends of the cables are buried in concrete. The shear resistance of the cable anchors in the concrete keep the cables from pulling out.

Fig. 2-15. It is clear in this view of San Francisco's Golden Gate Bridge that the cables are in tension and the towers are in compression. The cable ends are trying to shear out of the massive concrete anchors.

Compressive and Tensile Strength

In compression tests, crushing strains are exerted on the material until it buckles or crumbles. Concrete is tested in compression since it has little tensile strength. In tensile tests, the material is pulled apart until failure occurs.

Building materials are tested to determine their ability to resist compression, tension and shear. Then values for common materials are published.

Some materials are strong in tension and weak or unsuitable in compression. Manila hemp rope is a good example of this. Other materials are strong in compression and weak in tension, such as natural stone and concrete. The stone arch works because all members of the arch are under compression, and little or no tensile load is placed on a material that is weak in tensile strength.

Reinforced concrete is a **composite material**. That is, two elements act together under the load. Concrete provides the compressive strength; steel provides the tensile strength.

Steel is almost equally strong in compression and tension. Usually steel is tested only in tension. If the sample passes, it is assumed the steel also has the required compressive strength.

Wood varies greatly in its compressive strength, depending on the direction in which the load is applied. Wet wood has less strength than dry wood. **Plywood** is manufactured with the grain of alternate plies laid at right angles to develop approximately equal strength in either direction.

Effects of Shape: The shape of a material affects its ability to resist a compressive load or a deflective one. **Deflection** is bending which combines both compression and tension. Shape is not a consideration in tensile loads.

An interesting example concerning shapes and forms of materials involves paper. Set up two stacks of books of the same height, about eight inches apart. Lay two sheets of writing paper across the gap. The paper, in the form of thin sheets, is barely able to support itself and, in fact, may sag so badly that it falls.

Roll the two sheets of paper lengthwise into tubes about an inch in diameter. Prevent the tubes from unrolling with rubber bands or clips. Place the tubes across the gap. The tubes act as beams which could carry a sizable book without collapsing.

Another form of beam can be created from a sheet of paper by folding it into pleats. Such a **folded plate** is used at times for concrete roofs. Similarly, **corrugations** give steel a greater ability to span a gap, without unacceptable deflection or bending than the same steel would have as a flat plate.

Perhaps the most dramatic example of a material's strength also involves paper columns. Roll four paper tubes and stand them on end as columns. Carefully set a large book, such as the NFPA *Fire Protection Handbook*, in place atop the columns. It is possible to stack at least two of these books on the columns. This heavy weight is supported on less than a half ounce of paper arranged in the ideal shape of a column. Note that the load must be perfectly axial and that any eccentricity in the loading will cause the structure to fail.

Resistance to tensile loads, on the other hand, is directly related to the cross-section makeup of a material. Regardless of the shape, the paper would tear if tested for tensile strength.

T. H. McKaig's book, *Building Failures*, is useful for additional study on this topic, especially his discussion of the elements of structural design:[17]

> "A standard 3-inch pipe (outside diameter 3½ inches) weighs 7.58 pounds per foot; a 1 11/16-inch round bar weighs 7.6 pounds per foot, practically the same. However, according to standard design theory, the pipe, used as a column 7 feet high, will safely carry a superimposed load of 36,050 pounds, whereas the rod will only support a load of 13,900 pounds (when used as a column)."

Conversely, the pipe and the rod, made of the same quality steel and of equal cross-sectional area of metal (because they are of equal weight per foot of length), will have the same tensile strength (when the load is suspended).

Fig. 2-16. Instead of a column the designer of this library used a slender steel rod (arrow) to carry this beam which supports an end of the mezzanine. If fire causes the connection to the beam to fail, the mezzanine will collapse. Know your buildings!

Suspended Loads

Slender tensile members can carry a load which would require a compressive member of much greater size. This feature currently is being used in the design of many buildings.

Designers are eliminating interior columns at selected locations by hanging the ends of beams from the overhead structure. A slender rod can replace a much larger column. When this is done, however, several fire problems develop. The tensile member, having less mass, has less fire resistance. In addition, the load cannot be delivered to the ground in tension; it must be converted into a compressive load. This requires one or more **connections** in the overhead where fire temperatures are often greater than on the floor. These connections may be hidden in the structure. In such a case, an attic fire might cause the collapse of the first floor.

In old timber buildings, overloaded beams are sometimes restored by inserting a tie rod which goes up through the building generally to a truss or beam extending from wall to wall in the cockloft. This may be well hidden. Look for the telltale end of the tie rod about mid point of the beam.

Tactical Considerations

The connection which is weakest when subject to fire is the most significant connection to the fire fighter.

Suspended structures should be noted and evaluated as part of the prefire plan.

Suspended High-Rise Buildings

The concept of suspended beams has been applied in the construction of some high-rise buildings. Columns are replaced with cables suspended from beams **cantilevered** out from the top of the central reinforced concrete core. This provides some economy of construction, unobstructed floor space, and an open plaza at the entry level to the building. Possible fire problems of this type of structure are discussed further in Chapter 7, Steel Construction.

Safety Factor

The safety factor represents the ratio of the strength of the material just before failure, to the **safe working stress**.

The term safety factor is sometimes misunderstood. It is not practical to use a material in a structure so that it will be loaded to its **ultimate strength**, as shown in tests. The material used in a structure may not be as good as the sample tested; the workmanship may be inferior; the material may deteriorate over the years. For these reasons the design load is only a fraction of the tested strength of the material. If the design load is only a tenth of the tested strength, the safety factor is ten. If the design load is half the tested strength, the factor of safety is two.

Steel, which is made under controlled conditions, has a safety factor of two; masonry constructed in place might have a safety factor of ten. The less that is known about the characteristics of a material and its role in a building assembly, the greater the factor of safety required.

The factor of safety represents a measure of what is *not* known about the material. Lower safety factors made on the basis of knowledge of the material and its loading are not reductions in safety. Improper reductions in the safety factor which might survive normal loading may be significant under fire conditions.

Tactical Considerations

A brick masonry wall was designed for a certain load. During the life of the building a substantial additional load was added to the wall. Some of the safety factor has been absorbed. It is not known how much or if it is significant. All that is known is that the wall is more heavily loaded than the design called for, and thus bears watching.

The safety factor of steel used in excavation bracing is half what would be permitted in a structure because it is temporary. In other words, the permitted load is doubled. If a fire occurs, the steel will

Tactical Considerations, continued

fail much sooner because the temperature at which heated steel fails, decreases as the load increases.

A residential structure will usually be designed for a floor load of about 40 pounds per square foot. This includes a factor of safety which has been found adequate for residences. If the building is converted to mercantile or office use, the load will probably be much greater than the residential load. Often, nothing is done to improve the building. The factor of safety has been compromised, possibly to the point where some slight increase in the load or loss of structural strength might cause collapse.

Composites

Composite Materials

At times, two materials are combined to take advantage of the best characteristics of each. For instance, concrete is a relatively inexpensive material which is strong in compression but weak in tension. Steel is strong either way but is more costly than concrete. By providing steel at the locations where tensile stresses develop, a composite material, called reinforced concrete, is developed.

All elements of a composite material must react together if there is to be no failure. If the materials separate, the composite no longer exists, and the two materials separately may be unable to carry the load successfully.

Composite Structural Elements

Two different materials may be combined in a structural element.

Steel and concrete are combined in composite floors. In some cases, studs are welded to steel beams and then imbedded in the concrete of the floor. This is done to produce a diaphragm floor, which stiffens the structure. In a serious high-rise building fire, heat caused shear connectors, which joined the beams to columns, to fail. The beams did not fall because the imbedded studs kept the beam attached to the concrete.

Composite concrete-steel floors can be constructed with bar-joist trusses. The top chord of the truss is set below the tops of the web, allowing triangles of steel to project upwards. These are imbedded in the concrete.

A **flitch plate girder** is made by sandwiching a piece of steel between two wooden beams. The girder is much stronger than a piece of wood

Fig. 2-17. In this brick and block composite wall, the brick and block must react together to the load. This is not a veneer although there are block walls with a brick veneer.

of the same dimensions yet it can be installed by ordinary carpentry techniques. A sheet of plywood may also be used as the "meat" of this "sandwich."

A **brick- and block-composite wall**, in which cheaper concrete block substitutes for brick where it will not be seen, is another common example of composite structural elements. Do not confuse this with a brick veneered concrete block wall, in which the brick and block are not structurally united. In older construction, **hollow tile** is found instead of concrete block.

The term **composite construction** is sometimes used to describe buildings in which two different materials carry structural loads. Some concrete buildings, for instance, have a top floor or penthouse of lightweight steel.

Fig. 2-18. This high rise might be called composite construction because concrete supports the basic structure. The top floor is light-weight steel. Know your buildings!

Structural Elements

Buildings are made up of structural elements, such as beams, columns, arches, and walls, which differ in how they carry the load and transfer it to the next element. The principles of each element are the same regardless of the material.

Beams

The beam is probably the oldest structural member. It is not hard to imagine primitive man dropping a tree across a stream to form the first bridge.

A beam transmits forces in a direction perpendicular to such forces to the reaction points (points of support). Consider a load placed on a floor beam. The beam receives the load, turns it laterally, divides it, and delivers it to the reaction points.

The definition of a beam does not consider its attitude, that is, its vertical or horizontal orientation. While beams are ordinarily thought of as horizontal members, this is not always the case. A vertical or diagonal member that performs the functions of a beam, although it may have another name, such as a **rafter**, is structurally a beam.

When a beam is loaded, it deflects or bends downward. The initial load is its own deadweight. The load placed on it is a **superimposed** load. Some beams are built with a slight camber or upward rise so that when the design load is superimposed, the beam will be more nearly horizontal. A carpenter selects wood floor beams accordingly.

Deflection causes the top of a beam to shorten so that the top is in compression. The bottom of the beam elongates and thus is in ten-

Fig. 2-19. Beams change dimension when loaded. The triangles are the engineering symbol for the reaction points.

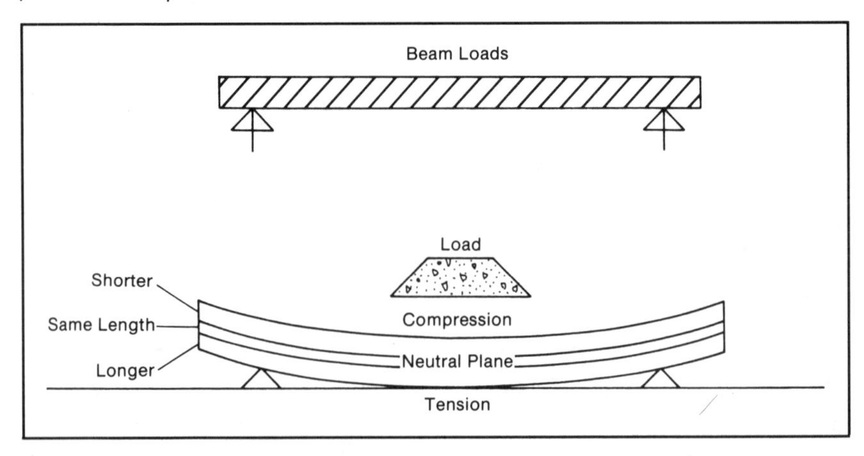

sion. The line along which the length of the beam does not change is the **neutral axis or plane**. It is along this line that the material in the beam is doing the least work and material can be most safely removed.

Along the byways of West Virginia, the simple suspension bridge spanning a creek is a common sight. This simple, adequate, but shaky bridge illustrates the designer's dilemma.

The cable is an ideal beam. Fully in tension, it makes the most economical use of the available material. Cable-supported roofs are being used for some large open area structures, but for a beam to be safe and economical is not always enough. It must deflect so little that the deflection will not be noticed. This is accomplished by using additional material, or by rearranging the shape of the material. In other words, stiffness or reduced deflection can be achieved by material **mass** or by **geometry**. In recent years, the economics of using geometry (e.g., truss shapes) over mass has had a tremendous effect on structures.

Carrying Capacity and Depth of Beams

The load-carrying capacity of a beam increases by the square of its depth. Consider a 2-inch by 4-inch wood beam carrying a certain load. If another 2-inch by 4-inch beam is laid alongside the first beam, the carrying capacity is increased by a factor of two. If the same amount of wood is sawn into a 2-inch by 8-inch beam, thus multiplying the depth by two, the carrying capacity is increased by the square of two, or four.

Truss loft

Fig. 2-20. Increasing the depth of a beam four times increases the load capacity sixteen times. Note that when the beam is a truss, the greater depth results in a greater truss void or trussloft in which toxic, explosive carbon monoxide can accumulate.

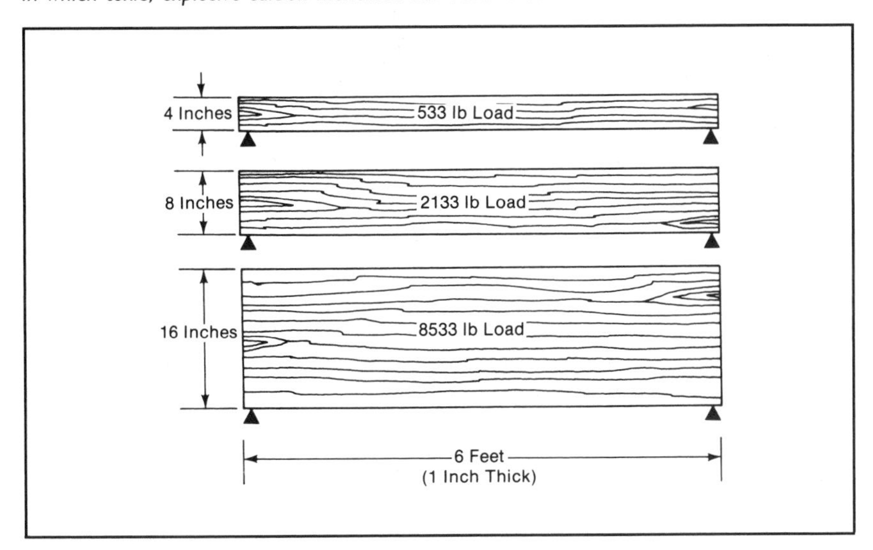

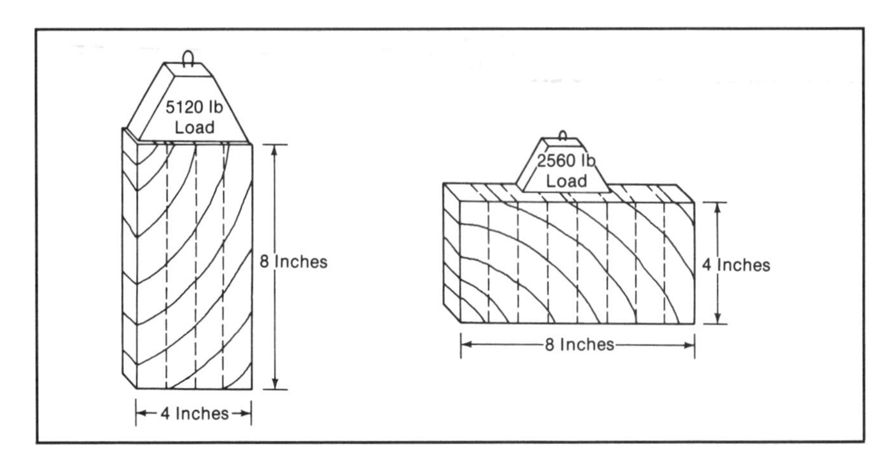

Fig. 2-21. Any beam can carry a greater load when set so that its depth is greater than its width.

The principles governing beams are the same for all beams, regardless of the material. For examples, we will look at wooden beams.

Standard carpentry manuals provide a table which details Safe Long Time Loads, Uniformly Distributed, for 1800-pound Structural Grade Lumber. For convenience in determining loads for beams of different thicknesses, the information is arranged by sizes from 1 inch by 4 inch to 1 inch by 16 inch, even though no beams are as thin as 1 inch. To find the strength of a 3-inch thick beam, the figure is multiplied by three.

A 1-inch by 4-inch beam spanning six feet can carry 533 pounds, a 1-inch by 8-inch beam spanning six feet can carry 2133 pounds (four times as much), and a 1-inch by 16-inch beam spanning six feet can carry 8533 pounds. This latter beam is four times the depth of a 1-inch by 4-inch beam.

To repeat, the load-carrying capacity or strength of a beam increases as the square of the depth, but increases only in direct proportion to increases in width. Look at this in another way. A 4-inch by 8-inch beam spanning 10 feet can carry 5120 pounds. In standard terminology, the width is given first. Suppose the 4-inch by 8-inch beam is used as an 8-inch by 4-inch beam. The 8-inch by 4-inch beam can carry only half the load.

The greater efficiency of a deeper beam must be balanced against other considerations, such as the desired thickness of the floor and the deflection of flooring between widely spaced beams. Floor boards are themselves beams. For a variety of reasons, then, the almost universal spacing for sawn wooden beams in ordinary construction is 16 inches and the depth of the beam is determined by the design load and the span.

TRUSS loft

The depth of the beam required is particularly significant in the case of **trusses**, because of the greater void in which combustible gases can be generated or accumulated.

The length of a span, or the distance between supports, is a determinant of the safe load of a beam. As the length of the span increases, the safe load capacity decreases in direct proportion.

The following example demonstrates the principle.

Span Distance (Feet)	Load Capacity (Pounds)
6	3333
12	1666
18	1111
24	833
30	667
36	555

The figures also consider the beam's weight, which must be deducted to get the permissible **superimposed** or added load. The figures assume a **uniformly distributed load**. If the load is **concentrated** at the center of a beam, the permitted load is half the distributed load.

The principles involved are the same for all beams, regardless of the material used.

Fig. 2-22. The strength of a beam decreases as its length increases.

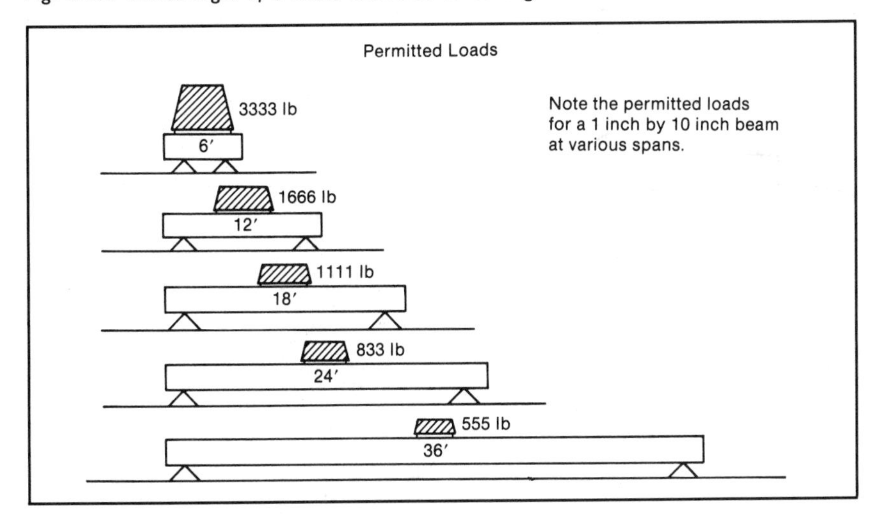

Types of Beams

Different terms are used for various beams.

- A **simple beam** is supported at two points near its ends. In simple beam construction, the load is delivered to the two reaction points and the rest of the structure renders no assistance in an overload.
- A **continuous beam** is supported at three or more points. Continuous construction is structurally advantageous because if the span between two supports is overloaded, the rest of the beam assists in carrying the load.
- A **fixed beam** is supported at two points and is rigidly held in position at both points. This rigidity may cause collapse of a wall if the beam collapses and the rigid connection does not yield properly.
- An **overhanging beam** projects beyond its support, but not far enough to be a cantilever.
- A **bracket** is a diagonal member which supports what would otherwise be a cantilever.
- A **joist** is a wooden beam.
- A **steel joist** or **bar joist** is a lightweight steel truss joist.
- A **girder** is any beam, of any material (not just steel), which supports other beams.
- A **built-up girder** is made of steel plates and angles riveted together, as distinguished from one rolled from one piece of steel.

- A **spandrel girder** is a beam which carries the load on the exterior of a framed building between the top of one window and the bottom of the window above.
- A **lintel** is a beam which spans an opening in a masonry wall. Stone lacks tensile strength so it can be used only for quite short lintels. Many precase concrete lintels have the word TOP cast into the top to be sure they are erected with the reinforcing rods which provide the tensile strength in the bottom.
- A **grillage** is a series of closely-spaced beams designed to carry a particularly heavy load.
- A **cantilever beam** is supported at only one end, but it is rigidly held in position at that end. It projects out over a support point. Beyond the support point, the tension is in the top and the compression is in the bottom. Cantilever structures are very likely to be unstable in a serious fire because the fire may destroy the method by which the beam is held in place. The cantilever beam resembles a playground seesaw. Many of us can recall being lifted up in the air by a playmate who then ran off and dropped us to the ground. Failure of the interior section of the cantilever can collapse the projecting section.

Fig. 2-23. Stone lacks tensile strength and thus is suitable for only short lintels. This historic Italian stone lintel cracked so it was patched. In a fire the patch would probably fall, dropping the load above.

As a cantilevered structure such as a balcony collapses, it is likely to pull down a wall. This happened in New Orleans and killed three fire fighters.

Cantilevers are used widely for both architectural and design economy considerations. Wooden cantilevered balconies are common.

Temporary construction platforms, for example, on which a crane delivers material, are cantilevered out from a building. They are supported by wooden compression members delivering the load to the floor above. A fire could destroy the wood and cause the platform to fall.

Fig. 2-24. This Colorado balcony provides a spectacular view of the Rocky Mountains. Fire in the area where the upper end of the suspension rod (arrow) is attached would release the rod, probably causing the balcony to collapse. It is said that this rod was added after the structure was built because the fully cantilevered balcony developed problems.

Fig. 2-25. Fire in the interior of the floors of this Alaska apartment could cause the interior part of the wooden I-beams to fail and thus drop the balcony, possibly with a load of fire fighters.

Exterior fire escapes are usually cantilevered structures. Exposed to the weather for many years, they can be hazardous to fire fighters.[18]

Fig. 2-26. Cantilevered platforms like these are common on construction jobs. Note that the interior support consists of columns which carry the load up to the floor above. If there is a fire on the platform, the columns would burn and allow the platform to drop. The entire area below should be cleared and the attack made from another side of the building.

Fig. 2-27. The needle beams support the wall while the foundation is being repaired in this building along the "Romantic Road" from Frankfurt to Munich. When the Washington, D.C. subway was being built, engineers tried to use needle beams to support some old government buildings. There were problems because the walls were of rubble masonry, so loose stones kept falling out. Rubble masonry walls would be vulnerable to collapse in a fire.

- **Needle Beam**. When any change is to be made in the foundation of an existing wall, the wall must be supported. Often holes are cut through the wall, and so-called needle beams are inserted and supported on both sides. They pick up the load of the walls.
- A **suspended beam** is a simple beam, with one or both ends suspended on a tension member such as a chain, cable or rod. The typical theater marquee is a suspended beam. Fire may destroy the anchoring connection. The beam becomes an undesigned cantilever. The ends of the chains were connected to the roof trusses in a Washington, D.C. theater. An attic fire destroyed the connections. Alert junior officers, who were fire science students, noticed that the front brick wall had been pulled out almost to the point of collapse.

Another old theater was converted into a furniture store. In a fire the suspended marquee collapsed, killing six fire fighters. The collapsing marquee pulled down the masonry along the length of the building. One victim was about a hundred feet from the initial collapse.

Tactical Considerations

Fire fighters, equipment and all other people should stay clear of the collapse area.

Tactical Considerations, continued

In the case of a cantilever, ask what the fire is doing to the other end.

In the case of a suspended beam, ask what is happening to the connection of the tension member inside the building. In short, ask what is holding the rope.

- A **transfer beam** moves loads laterally when it is not convenient to arrange columns one above the other — the ideal arrangement. If it is necessary to change the vertical alignment, a transfer beam must be designed to receive the concentrated load of the column and deliver it laterally to supports. Improper alterations of buildings may produce **undesigned transfer beams** which are points of weakness.

The Kansas City Hyatt Regency Hotel suffered a disastrous collapse of a walkway. Over 100 people were killed. A construction change caused a short section of the upper walkway beam to act as an undesigned transfer beam. It failed.[19]

Beam Loading

Beam loading refers to the distribution of loads along a beam. Assume a given simple beam which can carry eight units of distributed load. If the load is concentrated at the center, it can carry only four units. If the beam is cantilevered and the load is distributed, two units can be carried. If the load is at the unsupported end of the cantilever, only one unit can be carried.

Particularly in buildings under construction or under demolition by fire or otherwise, excessive loads may be concentrated in one area.

This may have been the case recently in New Orleans when scaffolding with stacks of bricks ready for use, collapsed suddenly at a building being reconstructed.[20] Do not increase the overload by the weight of fire fighters and equipment.

Reaction and Bending

The reaction of a beam is the result of force exerted by a beam on a support. The total of the reactions of all the supports of a beam must equal the weight of the beam and its load.

The **bending moment** of a beam can be simply described as that load which will bend or break the beam. The amount of bending moment depends not only on the weight of the load, but also on its position. The farther a load is removed from the point of support, the greater the moment. Heavy loads should be placed directly over or very close the the point of support.

(Note: Many beam loads are carried on trusses. The hazard of trusses is so important to fire fighters that the subject of trusses of all types and materials is discussed separate in Chapter 12, Trusses.)

Columns

A column is a structural member which transmits a compressive force along a straight path in the direction of the member. Columns are usually thought of as being vertical, but *any structural member which is compressively loaded is governed by the laws of columns, despite its attitude.*

Columns by themselves are often used for monuments. Non-vertical columns are often called by other names, such as **struts**, or **rakers** which are diagonal columns which brace foundation piling.

A **bent** is a line of columns in any direction. If a line of columns is specially braced to resist wind, it is called a **wind bent**. A **bay** is the floor area between any two bents.

A **pillar** is a free standing masonry load-carrying column, as in a cathedral.

Decrease in Load Carrying Capacity

Beams decrease in load-carrying capacity proportionately -- a twelve-foot beam can carry half the load of a similar six-foot beam.

Columns, however, lose strength by the square of the change in length. Thus a twelve-foot column can carry only one fourth the load of a six-foot column of the same material and cross section.

Shapes of Columns

The most efficient shape for a column is one which distributes the material equally around the axis as far as possible from the center of the cylinder. The thin-wall tubing used for the legs of a child's swing set is an example of an efficient use of steel in forming a column. In theory, column material could be paper thin, but it would be subject to local damage as by a puncture or dent. Most codes provide for minimum wall thickness for columns to prevent local damage.

To obtain a free floor space 63 feet x 357 feet in a 48-story Tokyo high-rise the designer provided super columns. They consist of four 39-inch square steel box columns spaced so they form a super column 21 feet on a side. The box-column steel walls are as thick as 3⅛ inches at the base and get narrower further up the height of the column.

Note that the principle of handling a compressive load — place the column material as far from the center as possible — was followed. The space inside the square is used for stairs and elevators.

It is more difficult to attach beams to round columns than to rectangular columns so less efficient rectangular columns are often used. In cast-iron buildings, interior columns are usually circular, while wall columns (within exterior walls) are rectangular. Rectangular columns often fit better into floor plans. Structural design is often a compromise between competing needs.

When beams were discussed, it was learned that a 2-inch by 8-inch joist set on the narrow edge would be a much more efficient use of wood than the same amount of wood in the 4-inch by 4-inch shape. On the other hand, the 4-inch by 4-inch shape would be the most efficient use of materials under a compressive load, because it is most nearly circular in cross section.

Masonry walls under construction are often braced against high wind. Available scaffold planks, usually 2 inches by 8 inches or 2 inches by 10 inches, are most often used. The load on the braces is a compressive load, so the braces should be shaped like columns, but are not. One builder described the 2-inch by 8-inch planks as "snapping like match sticks" at a construction site before the walls fell in a windstorm. If four 2-inch by 8-inch planks had been spiked together to make a hollow column and set diagonally against the wall, with 2-inch by 8-inch planks laid flat against the walls as beams, the wind load would have been much better resisted with the available material.

Tactical Considerations

If available lumber must be used to brace an excavation to rescue trapped workers, the horizontal members are under compression and should resemble columns.

During high winds fire fighters should keep clear of buildings under construction.

Wood Columns

Most wooden columns are simply smoothed off tree trunks. Large wooden columns, almost always ornamental as well as structural, are hollow. The column consists of curved, usually tongue-and-grooved, sections glued together.

Tactical Considerations

Fire moves easily through hollow columns.

Fire has been transmitted from under porch space to the attic through hollow columns and vice versa.

Cast-iron columns were sometimes used in part of the heating system as air ducts. Fire has extended vertically through cast-iron columns as well.

I-Beams vs. H-Columns

While steel beams are I-shaped, steel columns are H-shaped, box-shaped or cylindrical. Beams are shaped like the letter "I" because the depth determines strength. Columns are shaped like the letter "H," and of a dimension that permits a circle to be inscribed through the four points of the "H." A builder may use an available I-shaped steel section as a column but it is wrong to speak of I-beam columns.

Fig. 2-28. Wooden columns often are hollow. They can transmit fire from below the porch to the attic and vice versa.

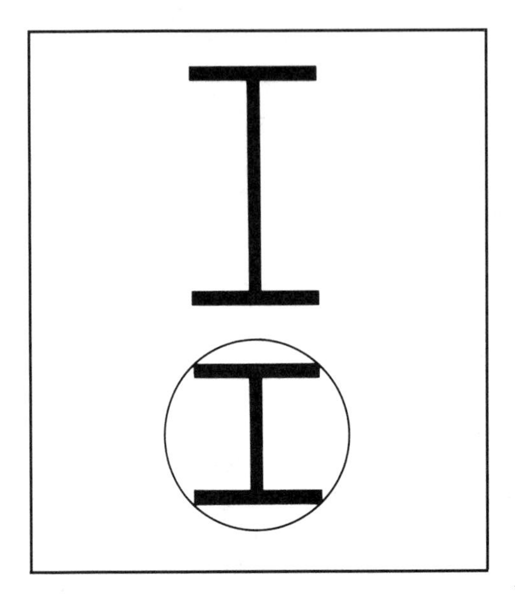

Fig. 2-29. On the top is a cross-sectional view of an I-beam. On the bottom is a column. Don't use the incorrect term, "I-beam columns."

Types of Columns

There are three types of columns which can be differentiated by the manner in which they generally fail: **Piers** are short, squat columns, which fail by crushing. **Long, slender columns** fail by buckling. In buckling, the column assumes an S shape. **Intermediate columns** can fail either way.

Understanding characteristics of long, slender columns yields information and principles applicable to columns and other structures under compressive loads, notably the top chord of trusses.

Euler's Law Columns

Very long, thin columns are known as Euler's law columns. A straight column, axially loaded, may suddenly collapse. Euler discovered that there is a critical load for a column and that the addition of even a single atom over the critical load can cause sudden buckling and collapse.

Examine the formula for Euler's law for the understanding it will give of the three vital variables in column stability. The formula is:

$$P_c = \frac{\pi 2EI}{L^2} \text{ in which}$$

P_c is the critical load, the absolute maximum load.

π^2 is 3.1415 squared.

E is the modulus of elasticity of the material in the column. (The modulus of elasticity is a measure of the ability of the material to yield and return to its original shape.)

I is the moment of inertia, a mathematical value for the geometric cross-sectional shape of the column.

L^2 is the length of the column squared.

Once the column starts to yield, there is generally very little reserve strength left and the column is on the verge of total collapse. By contrast, in beam action there is generally considerable reserve strength beyond initial yielding.

Consider a long, slender column. The load on the column tends to cause it to buckle. If the column is braced rigidly at the midpoint, the effective length of the column is cut in half. Since the square of the length is the divisor in the formula for column loading, cutting the length of the column in two increases the carrying capacity to four times what it was initially. The critical load is four times higher than it was initially. Shortening the effective length of a column by intermediate bracing pays dividends in load-carrying capacity.

The loss of bracing is a cause of column failure. Consider a high scaffold. Several long, narrow columns are securely braced at several points along its height, making several eight-foot segments. The effective length of the column, for this purpose, can be considered to be that of one of the segments. If a bracing connection fails, the effective length of that portion of the column becomes 16 feet. The decrease in load carrying capacity is geometric. The critical load is reduced to 25 percent

Fig. 2-30. In basic column theory, the load-carrying capacity of a column decreases geometrically as it gets longer. The two columns are the same width. The left column appears thicker due to an optical illusion.

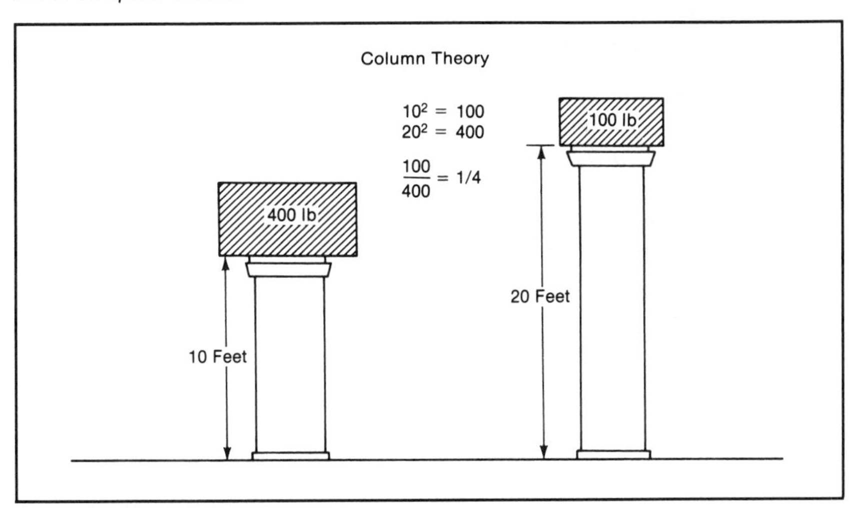

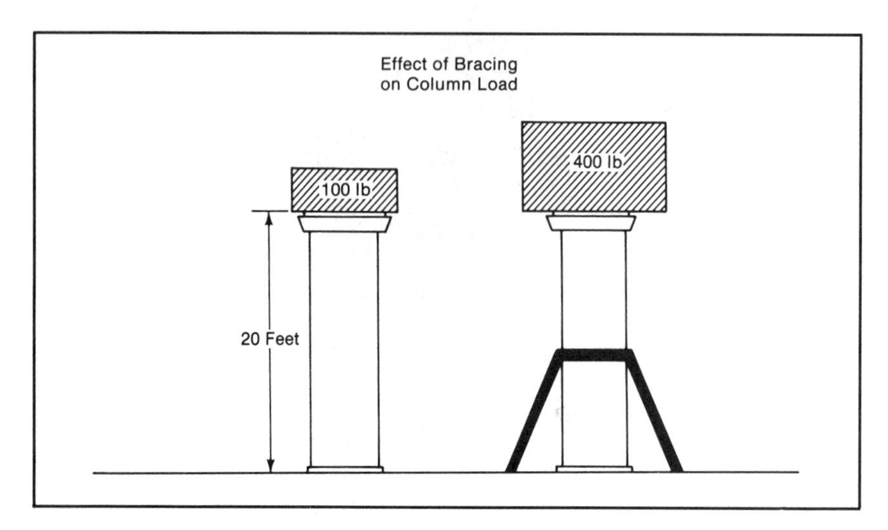

Fig. 2-31. Note the effect of bracing a column. Bracing it adequately at midpoint turns the 20-foot column into two 10-foot columns, one on top of the other, giving it enough strength to carry 400 pounds. Similarly, the top chord of a truss is under compression and acts like a column. If a connection fails, two short lengths of the chord become one length twice as long. The load-carrying capacity suddenly decreases to one fourth as much and the top chord buckles. Trusses are fully discussed in Chapter 12.

of what it was. Few, if any structures have the reserve strength to resist such a loss, so collapse results.In Chapter 12, Trusses, this knowledge is used to show how the top chord of a truss fails in compression.

Tactical Considerations

The slightest indication of column failure should cause the building to be cleared immediately. Once yield stresses are reached in column action, there is little reserve strength left and the column is on the verge of total collapse. Concepts of Structures *provides a good discussion of this phenomenon.*[21]

The failure of a column is likely to be much more sudden than the failure of a beam. The failure of a column may precipitate the collapse of the entire portion of the building dependent upon that column.

Temporary Bracing

When a building is under construction, there are many elements not in position or permanently connected. It is possible that columns might not have the full benefit of the bracing which will be provided by the

Fig. 2-32. The builder was concerned about the stability of this tilt slab wall. It is "braced" with a 2 × 10. The brace is under a compressive load so that it acts like a column. The overloaded column is buckled. Give a building like this a wide berth!

completed building. Guy or temporary bracing is common. This bracing may not be adequate to resist high winds or other unexpected loads, or it may be vulnerable to fire.

Walls

Walls transmit to the ground the compressive forces applied along the top or received at any point on the wall. A wall resembles a wide slender column. The wall may also be required to resist flexural or bending forces, as does a beam. Wind load is an example of flexural force.

Walls are classified in two main divisions: load-bearing walls; and nonload-bearing walls.

Load-bearing walls carry a load of some part of the structure in addition to the weight of the wall itself. **Non-load-bearing walls** support only their own weight. **Veneer walls, curtain walls, panel walls,** and **partition walls** are some examples of non-load-bearing walls.

A load-bearing wall is more stable than a non-load-bearing wall of identical construction because the weight of the superimposed load. The load carried by any component in addition to its own deadweight provides stability. Stack up a wall of books or blocks. The wall is easily pushed over. Superimpose your weight on this wall; it is much more difficult to push it over.

The structural consequences of the collapse of a load-bearing wall may be much more serious than those following the collapse of a non-load-bearing wall, because of the collapse of building elements supported by the bearing wall. To a fire fighter caught in a collapse, however, it makes little difference whether a masonry wall is load-bearing or non-load-bearing.

Within these two broad classes of walls, other descriptors can be applied to walls.

- A **cross wall** is any wall at right angles to any other wall. The walls should brace one another. Sometimes the bond is so poor that there is no benefit from the cross wall.
- A **veneer wall** is made of a single vertical thickness of masonry designed to improve the exterior appearance of the building. Decorative masonry such as stone, brick or marble may be veneered over common stone masonry, concrete block, reinforced concrete, or steel. By far the greatest use of veneered walls is brick veneer on wood frame.

The veneer wall totally depends for stability on the underlying wall. It should be tied to the wall with metal ties embedded in the mortar. It can be catastrophically unsafe if left to itself.

Brick veneer was applied to the visible parapetted portion of a masonry block fire wall in an apartment development near Charlotte, North Carolina to improve its appearance. The brick was applied without ties, and was resting on the wood roof. In 1989, the brick collapsed through the roof. Fortunately no one was in the area so injuries were avoided.

- A **composite wall** is composed of two or more masonry materials which react together under the load.

In order to save costly brick used in masonry walls, brick and concrete block (in the past, terra cotta tile) are used to form a composite wall. When walls were all brick, some bricks were "headers" which were turned on end to cross the wall and tie it together. This was unsatisfactory when concrete block was used in brick and block composite walls. A steel-wire masonry truss is now laid in the wall. As a result the bricks are all laid lengthwise as "stretchers." Thus, it is no longer possible to differentiate a brick veneer wall from a brick masonry wall

Fig. 2-33. This party wall is supporting the truss roofs on each side. It should be checked to see if the trusses are in a common opening. If so, fire can extend through the opening. Know your buildings!

at a glance, since now both walls can be laid as all stretchers. This is also known as a running bond.

- The terms **panel walls** and **curtain walls** are often used interchangeably to describe non-load-bearing enclosing walls on framed buildings. Technically, panel walls are one story in height and curtain walls are more than one story.
- A **party wall** is a bearing wall common to two structures. If structural members of both buildings are located in common sockets, fire can extend through the opening.
- **Fire walls** should be able to stop the fire with little or no assistance from fire fighting forces. All penetrations of the fire wall should be equal to the fire wall in fire resistance. Openings should be protected with properly rated fire doors. Fire walls in steel-framed buildings are probably unstable if there is fire on both sides of the wall. See Chapter 7, Steel Construction, for additional information.
- **Partition walls** are non-load-bearing walls which subdivide areas of a floor. They may be required to have some fire or smoke resistance. They may be required to extend to the underside of the floor above or only to the ceiling line, leaving a void above.
- A code term, **demising walls,** indicates walls bounding a tenant space.
- A **shear wall** in a framed building is designed to help resist the force of the wind. It is usually incorporated in some required enclosure such as an elevator shaft or stair shaft.

Fig. 2-34. An excellent example of a fire wall. In addition, the utilities were brought into each living unit from the street through the trenches in front of the building. There are no penetrations through the walls. (Howard Summers, Jr., Virginia State Fire Marshal)

- In a **rubble masonry wall**, the outer and inner wythes (a single vertical thickness of masonry) are laid in courses and finished. The interior consists of stones dumped in at random. Sometimes the stones are loose. In other cases mortar is dumped in around them.

Tactical Considerations

If you are breaching a stairway or an elevator enclosure and find reinforced concrete, it may be doing double duty as a shear wall. Examine the other walls. They may be required only to be fire resistive and may be of block, or even gypsum on studs, or gypsum plank, and thus much easier to breach.

Cantilever Wall

When a masonry wall is under construction, it acts as a cantilever beam with respect to wind loads received on the face of the wall. When the

Fig. 2-35. This concrete tilt slab wall collapsed when the steel bar joist roof failed. (Chief Bill Ennis)

roof is in place, the wall, with respect to wind loads, becomes a simple beam supported at both ends. High, free standing walls are common in the construction of large one-high-story buildings, such as shopping centers, churches, and industrial buildings.

Severe winds may topple a free standing wall. Earlier, in the section on columns, the inadequate design of construction wall bracing was discussed. In addition, stakes so short that they move when the rain turns the ground to mud are sometimes used and fail.

Eccentric loading of the wall by hanging work platforms on one side is another cause of failure. If the wall is braced at one end by another cross wall, the unbraced end will be more vulnerable to collapse. During high winds, there are often gusts which greatly exceed an already high wind speed. In several cases, failure of what had been regarded as well-braced walls has been attributed to gusts.

Precast concrete tilt slab walls are vertical cantilevers when being erected and are braced by tormentors or temporary bracing poles. The roof of the building provides the permanent bracing. If the roof fails in a fire, the walls revert to undesigned vertical cantilevers and may collapse.

Tactical Considerations

During high winds beware of all free-standing walls. Avoid both the direct collapse zone and the area which might be hit by missiles.

Wall Bracing

A wall can be compared to a column extended along a line. Most walls have a high height-to-thickness ratio which is comparable to a tall slender column. They are therefore similarly unstable. President Thomas Jefferson once tried to build a one-brick thin garden wall. It failed. He reasoned that it was buckling like a column so he built a buckle into it to brace it internally. The result is the famous **serpentine wall** at the University of Virginia.

Walls can be braced or stiffened by several means.

- **Buttresses** are masonry structures built on the outside surface of the wall. The flying buttresses of Gothic cathedrals, which cope with the outthrust of the roof on the high slender walls, are built away from the wall. A prominent English politician once said "I am not a pillar of the church, but a buttress since I support it from without rather than from within."
- **Pilasters** are masonry columns built on the inside surface of the wall.
- **Wall columns** are columns of steel, reinforced concrete, or solid masonry (such as brick or solid block) in a block wall. Concentrated loads such as main girders are applied to the wall directly above the wall column.

Fig. 2-36. For almost a thousand years these buttresses have resisted the weight of the roof on the Canterbury Cathedral in England.

- **Cavity or hollow walls** are built of two wythes (a single vertical thickness of masonry) separated by a space for rain drainage or insulation. Such a wall might be built of two rows of four-inch brick, separated by four inches of space and tied together with steel ties. It is more stable than an eight-inch solid brick wall, just as a column is more stable the further the material is from the center. Such hollow or cavity walls were originally used to control rain. Rain penetrating the outer wythe drains down the cavity to weep holes at the bottom. Currently, the void is often filled with foamed plastic, either rigid board or foamed in place. Some of what the author has examined has not been fire inhibited. The exact fire problem with these walls is not known, but one can postulate degradation, toxic fumes, and if oxygen becomes available, possibly violent combustion.

Wall Breaching

If a wall is homogeneous (i.e., good bond exists between bricks, blocks or mortar), it acts as a unit. A load coming down the wall to a window or doorway does not follow a vertical path. It splits and passes on both sides of the opening. Draw a line vertically above an opening, its height at least half the width of the opening. Draw two lines connecting the top of the vertical to the ends of the opening. The masonry in this triangle is the only masonry resting on the lintel.

Fig. 2-37. It is dangerous to breach a wall in this manner. If it must be done this way, the opening should be supported immediately.

Tactical Considerations

The proper way to breach a masonry wall is to open a triangular hole. If you cut the top of the hole as a line parallel to the ground, you are depending on the mortar above holding the masonry units in place, certainly an undesigned feature. If a parallel cut is made, it should be supported immediately. If wall elements fall apart easily, indicating that the wall is in poor condition, breaching should cease.

If you are breaching a wall and find any evidence that the wall is strengthened at that point, stop and start elsewhere. You are likely to be under a concentrated load.

Collapses of masonry walls have killed hundreds of fire fighters. Some signs of potential for collapse can be detected by inspection before a fire. (See Chapter 4, Ordinary Construction.)

Roofs

Roofs are very important to the fire service.

In some cases the roof is necessary only to keep the rain out. In other cases, such as tilt-slab concrete buildings, the roof is vital to the stability of the structure, and roof damage can cause wall collapse.

Conventional wood roofs are discussed in Chapter 4. The recent special attention to truss roofs should not cause fire fighters to ignore the hazard of the sawn joist roof.

Combustible metal deck roofs are discussed in Chapter 7. Trusses are discussed in detail in Chapter 12. The hazard of roofs of tilt-slab concrete buildings is discussed in Chapters 8 and 14.

Single-ply roofing is discussed in this chapter under "materials."

Arches

The **arch** combines the function of the beam and the column. The classic appearance of the Greek temple with its closely spaced columns was dictated by the fact that the Greeks did not know of the arch. Stone beams have little tensile strength, thus spans must be short. The Romans developed the arch. It is under compression along its entire length. If the arch is properly built, connections are not important to its strength. Many Roman arches used no mortar; the stones were cut and placed perfectly.

Fig. 2-38. This tied arch is on the front of a restaurant. The tie bar (arrow) keeps the two ends together.

Arches tend to push outward at the base and therefore must be either **braced** or **tied**. Arches usually are braced by heavy masonry buttresses. Failing buttressed arches often are tied together with steel rods. Some steel arches are tied together across the bottom of the arch, thus eliminating the need for a buttress. In some 19th century buildings, the brick or concrete floor arch tie rods are visible across the ceiling.

Bowstring-truss roofs are sometimes described as arches or arched trusses. This is incorrect. The truss is a beam and its thrust is straight down the wall or column. The thrust of an arch is outward. When the arch is tied the thrust is downward, but it lacks the triangles which would make it a truss. This is not the only case where one construction element slides into another. Steel arches as in heavy railroad bridges may themselves be trusses.

The arch is not complete until all its elements are in place. The keystone is the last stone set in place in an arch, but once in place, it is no more important than other stones (**voussoirs**).

An arch is said to spring from its point of support.

The word arch brings to mind the graceful Roman segmental arches. A segmental arch is one which describes a portion of the circumference of a circle. This is not the only arch form. Arches can have one, two or three hinges, points at which an arch changes direction. Gothic and trefoil arches are found in churches and similar buildings.

Fig. 2-39. Residents of Lake of the Ozarks, Missouri should recognize this perfect flat arch. Note the keystone. It is placed last and makes the arch, but then is no more important than any other stone.

Flat arches are constructed by tapering masonry units. Hollow-tile flat floor arches are common in 19th-century and early 20th-century buildings.

The availability of laminated wood has made wood arches possible. Many are used in churches.

Many exterior masonry arches over windows or doors are false. The exterior wythe is an arch. The interior wythes are carried on wood lintels. Though this author has never seen it cited, this must be the initiator of many masonry wall collapses. If the lintel is steel, elongation may displace the masonry.

Tactical Considerations

The removal of any part of an arch can cause the collapse of the entire arch. Cutting through tile arch floors, such as for a distributor or cellar pipe, is dangerous because the removal of one tile can cause the entire arch to fail.

A severely unbalanced load or a load outside the center third of an arch may cause it to collapse.

Fig. 2-40. A heavy railroad bridge that is an arch, but the arch is a truss. In this truss, all members are in compression.

Rigid Frames

The rigid frame is derived from the arch. Steel rigid frames are widely used for industrial and commercial buildings where clear space is required. Wooden rigid frames are often used for churches. Precast concrete rigid frames also have been used.

Such frames must be tied together at the bottom to resist the characteristic outward thrust of the arch. The ties often are made of steel reinforcing rods. The rods may connect the bases of the frame, or they

Fig. 2-41. A rigid frame is very suitable for wide-span, one-story buildings such as this bowling alley under construction. Such frames must be tied together under the floor. If the floor is combustible, failure of the ties may cause the frame to collapse.

may be "buttonhooked" into the concrete floor. In this case, the floor is structurally necessary to the building. It is possible that in the future a person cutting into such a floor may consider the tie just another rod and cut it with surprising results. In one huge hanger, the ties were cast into concrete beams to prevent this.

In one church it was observed that the ties, of a laminated wood rigid frame, apparently passed through an ordinary combustible floor-ceiling void. A fire in the basement could cause the ties to fail and perhaps collapse the frame.

This hidden steel might cause the same sort of tragedy as occurred in a Montreal church. In that case unprotected steel columns abutting the crawl space were heated by the fire and failed. This brought down the masonry wall and killed two fire fighters. This fire is discussed in Chapter 7, Steel Construction, page 314.

Shells and Domes

A shell is a thin plate that is curved. Shells are built of concrete. Wide areas can be spanned with extraordinarily thin shells. Shells can be less than two inches thick. The shell transmits loads along the curved surface to the supports. An eggshell provides a good example of the strength of a shell in relationship to the weight of material.

The dome is a shell. It can also be considered a three-dimensional arch. There are other forms of shells. These shapes produce roofs of varying architectural designs.

Geodesic domes are formed from a large number of triangles of equal size. They provide structures with very high volume-to-weight ratios. The sheath may be any desired material.

Transmission of Loads

It is important to understand how loads are transmitted from the point of application to the ground. Consider an ordinary brick and wood-joisted building with interior columns. A load is placed on a wood floor. The floor boards are beams and deliver the load to the joists. The joists deliver the load to a masonry wall at one end, and to a girder at the other. One end of the girder rests on a masonry wall. The other end rests on a column.

When a load is placed on the floor boards, they transmit the load to the joists on either side of the load point. The amount of the load transmitted to each joist depends upon the distance of the load point from each of the joists. The nearer joists, of course, receive the greater load.

The load transmitted by the floor boards to the joists is transmitted by the joists to the wall on one end and to the girder on the other. The proportion of load delivered to each support point again depends on the relative distances from the point at which the load is applied to the ends. The load received by the wall is delivered to the foundation, and thus to the ground. The load delivered to the girder is divided among both walls and the column, again in proportion to the distance. The structural engineer must calculate the distribution of the loads. The building does it automatically, according to the laws of physics.

This example makes use of a simple masonry and wood-joisted building. Regardless of the size of the building or the construction materials, the principles are the same. Any weight on the roof of a giant high-rise building is transmitted to the ground by the structure of the building.

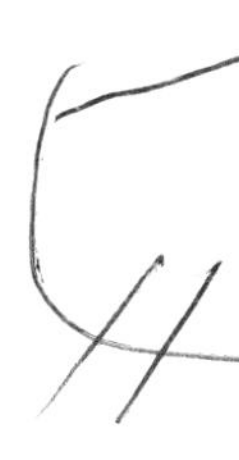

All loads must be transmitted continuously to the ground from the point at which they are applied. Any failure of continuity will lead to partial or total collapse. Having accurate knowledge of the ground on which a structure is built is vital to its stability.

There is a tendency among those who are concerned about building stability to make light of partial collapse and consider it unimportant. A partial collapse is very important to at least two groups -- those under it and those on top of it. If you are caught in a collapse, it is of little importance whether the building collapsed entirely or partially. The fatal collapse of the walkways in the Kansas City Hyatt Regency Hotel, which claimed more than a hundred lives, was only a partial collapse.

Foundations

Ultimately, all loads are delivered to the ground through the foundation. The nature of the ground and the weight of the structure determines the foundation. Foundations can range from simple footings to grade beams or a foundation under the entire wall to foundations which literally float the building on poor soil.

In some cases, wood or concrete piles are driven either to bedrock or until the accumulated friction stops the pile. Almost all foundations today are of concrete. In some locations, decay-treated wood is used for small houses.

Tactical Considerations

The fumes from burning decay-treated wood are toxic.

Any number of foundation problems can affect fire suppression in several ways. Masonry walls above foundations may develop severe cracks which make the wall vulnerable to collapse. Fire doors may not close properly. Openings may develop in fire walls, or in floor-wall connections permitting passage of fire. Dry-pipe sprinkler systems may not drain properly after "going wet."

Connections

Except for the very simplest structures, connections, which transfer the load from one structural element to another, are a vital part of a structure's gravity resistance system. The connection system must be absolutely complete. It is only as strong as its weakest link. A single failure can be disastrous.

Proponents of building materials trumpet the virtues of their products. They may have interesting information about their materials but it is important to know about the connections of these materials. For example, heavy wood beams are slow to ignite, but what is also significant is the connections of wooden girders to cast-iron columns, or laminated beams supported on unprotected steel columns. In this discussion of connections there is some unavoidable overlap with later chapters.

There are two general types of connections.

A building is said to be **pinned** when the elements are connected by simple connectors such as bolts, rivets or welded joints. These are usually not strong enough to reroute forces if a member is removed.

In a **rigid-framed** building, the connections are strong enough to reroute forces if a member is removed.

A **monolithic concrete** structure is one in which the successive poured castings are joined together so that the completed building is like one piece of stone. A monolithic concrete building is rigid-framed. If a column is removed, it is quite possible for the building to redistribute the load to the remaining members. If no member is thereby overloaded, the building will not collapse.

Precast concrete buildings may be pinned, or may be made monolithic by the use of **wet joints** in which cast-in-place concrete unites rods which project from precast sections. Some concrete buildings may have both types of connections.

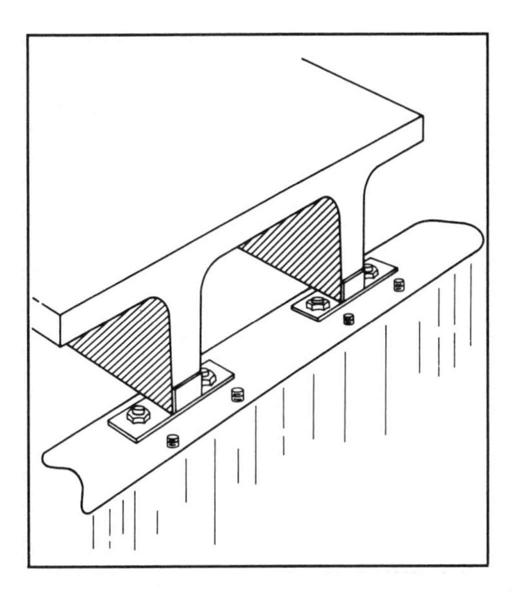

Fig. 2-42. The precast concrete double-tee floor unit is pinned to the foundation with nuts and bolts.

In an ordinary riveted steel-framed building, the collapse of a column will cause the collapse of all elements supported by that column. Some steel buildings have connections which redirect overloads to other sections of the building. This is called **plastic design.**

Failure of Connections

The great majority of buildings stand up. Up to a point a building can cope with undesigned loads, either vertical or lateral. Occasionally, a building not under construction or demolition collapses. A building on fire can be considered a building under demolition. Often structural failure is due to a failure of the connections.

There are numerous ways in which connections can fail.

- Masonry walls shift outward, dropping joists.
- Temporary field bolting of steel gives way in a high wind.
- Steel connectors rust.
- Concrete disintegrates.
- Of particular concern are connection systems which have lower fire resistance than the structural elements that they are connecting.

In many buildings, particularly older buildings, the connections may be adequate as long as the building is axially loaded. However, an eccentric or lateral load, or shifting wall, floor or column alignment may cause collapse even after the building has been standing for many years.

Be especially wary of buildings in areas where there was, or is, no building code supervision, such as rural areas that suddenly become urbanized.

Fig. 2-43. This Denver, Colorado building was "rehabilitated." Note that the cast iron column was short, so a block of iron was placed on top of it for the wooden girder to rest on. The beam is not connected to the column. Any lateral thrust could displace the column, and it would collapse.

A Los Angeles City fire fighter died in a roof collapse where once there had been a vacant lot between two brick buildings. To provide a roof over the vacant area, mortar had been removed from between bricks. Pieces of shingle were hammered into the gaps. A **ledger board** nailed to the shingle pieces supported a wood joist roof. The makeshift assembly survived for a number of years until a fire destroyed it.

Connections are concealed in finished areas of buildings. Do not pass up the opportunity to examine basements and attics. What you see there is probably typical of the building. Buildings of the same age were probably built by the same methods. Additional problems associated with connections are discussed in other chapters. However, here are examples of some important connection defects.

Sand-lime mortar was used exclusively in masonry work until about 1880. Sand-lime mortar is water soluble. A fire fighter, operating with his unit in the basement of a building, noticed that a hose stream had washed the mortar out from the bricks. He alerted the officer. The building was evacuated and shortly thereafter collapsed.

Gravity connections in buildings simply depend on the weight of the building element to hold them in place. This is especially true for cast-iron columns. A gravity fit, cast-iron column was one of the causes of the collapse of the Vendome Hotel in Boston in which nine fire fighters died.

Houses formerly simply sat on their foundations. Since the 1940s many codes have required that a structure be anchored to the foundation. The need for such anchoring is often disregarded by builders who know nothing of the reason for the requirement — the 1938 hurricane which pushed thousands of houses off their foundations.

Steel connections enter into the construction of almost every building. Light-weight wood trusses are held together with steel gusset plates or gang nails. Destruction of the fibers holding the gusset plate releases the plate and the truss fails.

Steel is often intermixed with other combustible construction elements. If the building is not required to be fire resistive, the steel is unprotected. Steel heated to 1000°F elongates nine inches per 100 feet of length. If the girder is restrained and cannot elongate, it will overturn and drop its load of wood joists. At higher temperatures, the steel fails.

Unprotected steel connections can often be found in precast reinforced concrete buildings.Unprotected steel rods and cables (which fail at 800°F) are often used to tie failing buildings together, or to provide some additional resistance to earthquake movement.

It was noted earlier that a load can be suspended on a thin rod as contrasted with the bulky column required to support it in compression. This advantage has not been lost on designers. The load cannot stay in tension, however. It must be changed to a compressive load and delivered to the ground. This requires a series of connections. The vulnerable point is the connection most susceptible to fire. This may be floors away from the suspended structure. The steel tension rod connection to a wooden beam may be hidden in the cockloft or the space between the top floor ceiling and roof. The burning of the beam will cause the connection to fail, dropping the rod and its load. Thus, a cockloft fire might cause an interior collapse.

Many codes require that wood joists in masonry walls be **fire-cut**. The end of the joist is cut off at an angle to permit the joist to fall out of the wall without damaging the wall. The removal of wood lessens the inherent resistance of the joist to fire and can precipitate floor collapse.

Heavy timber buildings are often built with **self-releasing floors**. Floor girders are set on brackets attached to columns. A wood cleat or steel dog-iron similar to a big staple is used to provide minimal stability. Such a floor can be expected to release sooner than if it were tightly connected. Some designers, recognizing this, required tight connections.

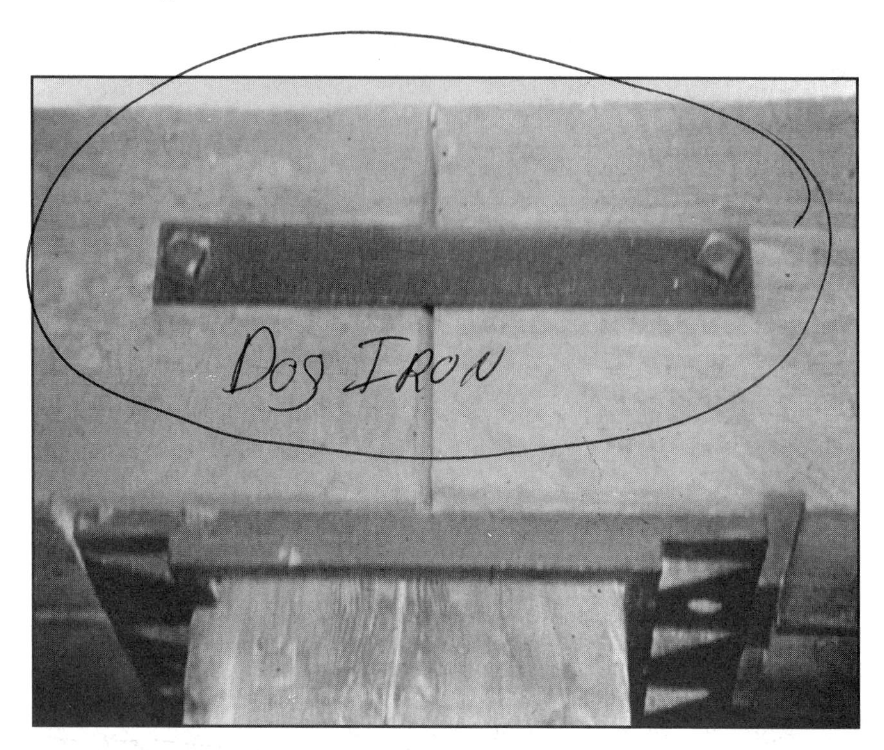

Fig. 2-44. This dog iron is a large staple. The concept is that it provides minimal stability. In a fire, however, one half of the floor could fall, while the other half might stay intact. This could be called "preferential collapse." Fire fighters should not be there when it carries out its design function. Know your buildings!

Tactical Considerations

A building with fire-cut or self-releasing floors is designed to collapse. It is the duty of a fire officer to see to it that fire fighters are not under the structure when the designed collapse occurs.

Heavy timber girders, as big as two feet by two feet, are often connected to cast-iron columns in a particularly dangerous manner. A shelf is cast onto the column. The girders are cut away in a half moon in order to fit the two opposite girders snugly around the column. Only the two "ears" of wood are resting on the shelf. Failure of the shelf or burning of the wood can cause collapse. An old building handbook points out this hazard and recommends that the big cast-iron column be reduced to a solid pintle of much smaller diameter to pass through the wood. This leaves very much more wood in the connection. This author has never seen such a pintle used in a cast-iron columned building. I think this is far more likely the cause of collapse than the poetic "red hot cast iron, struck by cold fire stream."

Overhanging and Drop-In Beams

Structural design is often intended to be as economical as possible. The economy may be in material or in the work of erection. Consider a space three bays wide. There are two masonry walls and two lines of columns supporting girders. Sometimes it is more economical to let the two outermost beams overhang the girders by two or three feet. The gap is then closed by a beam dropped in and nailed to the overhanging ends of the outer beams. As a result, the drop-in beams are connected only by the nailing. They have no support underneath.

Spliced Beams

Long wooden beams are not always available for building needs. Shorter lengths are often spliced together with metal connectors to produce the desired length. The resultant beam will carry its design load, but the connectors may fall out when heated sufficiently, causing collapse. In some buildings these connectors may have been made to look decorative. Take a second look!

Some years ago the sports arena in Daytona Beach, Florida was destroyed by fire. The roof was supported on laminated wood arches. The owners of the building sued the supplier of the foam plastic insulation on the roof, alleging that the flammable plastic had destroyed

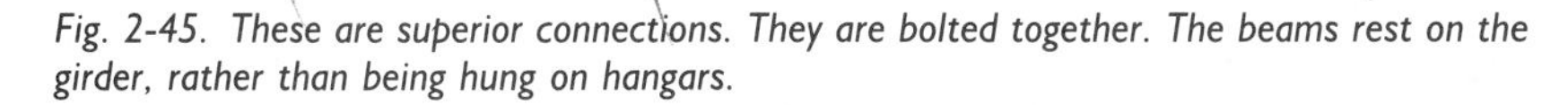

Fig. 2-45. These are superior connections. They are bolted together. The beams rest on the girder, rather than being hung on hangars.

their sturdy heavy timber building. Pictures of the fire clearly showed that the arches had fallen apart at the connections. When the plaintiffs learned that the defendants had these convincing pictures, they withdrew their suit.

These examples are but a few of the many connection failures which can occur, possibly catastrophically in a fire.

Tactical Considerations

When you preplan a building, take a hard look at the ***connections*** *so that you will not be in the collapse path if they fail.*

A Note to the Reader

This chapter provides a basic introduction to the most important terms in structural engineering, and relates structural principles to the behavior of buildings on fire. The serious student should read further. The publications cited as references for this and other chapters, and similar books on construction, contain a wealth of useful information. After acquiring a vocabulary and some understanding of structural principles, visit construction job sites and ask questions. There are no stupid questions, but there are many dangerous actions taken on the basis of ignorance.

References

The addresses of all referenced publications are contained in the Appendix.

1. Eisner, Harvey. "Hackensack, NJ: One Year Later." *Firehouse*, August 1989, p. 58.
2. Eisner, Harvey. "Three Fire Fighters Are Killed in a Collapse." *Firehouse*, April 1985, p. 58.
3. "Fire: Countdown to Disaster." Film or Video. National Fire Protection Association, Quincy, Massachusetts.
4. Washburn, A., *et al.* "1989 Fire Fighter Fatality Report - Special Analysis: Deaths Due to Rapid Fire Progress in Structures." *Fire Command*, June 1990, pp. 35.
5. O'Hagan, J. *High Rise--Fire and Life Safety*. Saddle Brook, N.J.: Fire Engineering, 1977.
6. Best, Richard. "Storage Collapse Kills Five." *Fire Command*, November 1980, p. 20.
7. Urquhart, Leonard C., ed. *Civil Engineering Handbook, Fourth Edition*. New York, New York: McGraw-Hill, 1959.

8. *Engineering News Record*, May 23, 1968 and October 16, 1984. New York, New York: McGraw-Hill.

9. Cote, A., *ed., Fire Protection Handbook*, 17th ed., National Fire Protection Association, Quincy, Mass., 1991, p. 6-55.

10. National Institute for Standards and Technology (NIST). Fire Research Computer Bulletin Board. Dial (301) 921-6302. Uses 300, 1200 or 2400 baud, 8 data bits, 1 stop bit, no parity.

11. Kennedy, T., "Energy Efficient Windows in Multiple Dwellings." *Training Notebook, Fire Engineering*, January 1991, p. 14.

12. Hays, Steve. "Asbestos at a Fire Scene." *Fire Journal*, Vol. 84, No. 2, March/April 1990, p. 45.

13. Earhart, G., "The Silent Hazardous Material." *Speaking of Fire*, IFSTA, Spring 1991.

14. Composite Technology, Inc., 1005 Blue Mound Road, Ft. Worth, Texas 78131.

15. Cote, A., *ed., Fire Protection Handbook*, 17th ed,. National Fire Protection Association, Quincy, Mass., 1991, p. 3-116.

16. *NFPA: Alert Bulletin*, No. 90-1, Feb. 1990. Research & Fire Information Serices, National Fire Protection Association, Quincy, Massachusetts.

17. McKaig, Thomas H. *Building Failures*. New York, New York: McGraw-Hill, 1962.

18. Dunn, V. "The Danger of Fire Escapes." *Fire Engineering*, May 1990 p. 38.

19. *Engineering News Record*, Mar 4, 1982. New York, New York: McGraw-Hill.

20. *Engineering News Record*, Aug 31, 1989. New York, New York: McGraw-Hill.

21. Zuk, W. *Concepts of Structures*. Litton Educational Publishing, Inc. (by permission of Reinhold Publishing Co.), 1963, p. 37.

Additional Readings

Many of these publications might not be available in the local library. Check for an inter-library loan or a nearby technical library, such as one at an engineering school. Three books cited (Condit, McKaig, and Feld) and one magazine (ENR) were invaluable in the self education of this writer.

Allen, E. *How Buildings Work*. New York, New York: Oxford University Press, 1980.

Condit, C. W. *American Building*. Chicago: University of Chicago Press, 1968.

Cote, A., *ed. Fire Protection Handbook*, 17th ed. National Fire Protection Association, Quincy, Mass., 1991, pp. 6-130 through 6-139.

Degenkolb, John. "Will Energy Conservation Have an Effect on the Fire Protection of Buildings?" *Building Standards*. September-October, 1976.

Ellison, Huntington, Mickadeit. *Building Construction, Materials and Types of Construction*, Sixth Edition. New York, New York: Wiley, 1988. (Previous editions were called Huntington, W. C., Building Construction, and are excellent reading if the Sixth Edition is not available.)

Engineering News Record (ENR). New York, New York: McGraw-Hill (a weekly publication).

Feld, J. *Construction Failure*. New York, New York: Wiley, 1968.

NFPA 220 *Standard on Types of Building Construction* (1985). Quincy, Massachusetts: National Fire Protection Association.

Gordon J. *Structures, Or Why Things Don't Fall Down*. New York, New York: De Capo Press, 1981.

McKaig, T. H. *Building Failures*. New York, New York: McGraw-Hill, 1962.

Urquhart, Leonard C., Ed. *Civil Engineering Handbook*, Fourth Edition. New York, New York: McGraw-Hill, 1959.

Zuk, W. *Concepts of Structures*. New York, New York: Litton Educational Publishing, 1963.

3

Wood Construction

In the United States and Canada most fires are fought by fire fighters standing on, in, or under wooden structures.

- Wood is combustible. As it burns, its structural strength is lost. Unless the fire is checked, at some point gravity will act and a collapse will result.
- Wood construction creates combustible void spaces in which fire can hide and burst out on the unwary.
- Wood, in thin sections can have a very rapid flame spread.

This chapter considers buildings in which wood elements carry the structural loads. The next chapter considers Ordinary Construction, in which exterior loads are carried by masonry, and interior loads are carried on wood, masonry, or unprotected steel. Much of the material in this chapter applies also to ordinary construction. Combustible multiple dwellings such as garden apartments, town or row houses, or any dwelling structure in which a unit is directly exposed to a fire in another unit are a different problem. They are discussed in Chapter 5.

Tactical Considerations

Fire fighters operating on, in, or under burning wooden structures are in a hazardous situation.

I have long urged a change in fire department terminology. Units should be dispatched to "A building fire." When fire command has determined that the structure is involved, and thus its gravity resistance is being attacked, all should be notified, "This is now a structural fire." This warning will alert all that structural collapse is a serious threat, and prepare all to respond quickly to an evacuation order.

Tactical Considerations, continued

There should be a clear signal when evacuation is ordered. Some fire departments use a series of short air-horn blasts, supplemented by radio communications. The radio alone is insufficient.

All fire fighters should be trained to get out of the structure immediately upon hearing the signal, leaving equipment behind. No length of hose or tool is worth a life. Fire departments should drill on "Clear the building." Even the most macho ships in the U.S. Navy drill regularly on "Abandon ship."

Wooden structures under attack from fire are subject to collapse. Always be prepared for immediate evacuation.

The California Bungalow

In older parts of cities across the country, multi-story wood residences, with small rooms, are the most common residences. A one-story residence was often nothing more than a shanty with a shed roof.

California has developed a reputation as a style setter in many areas. This also is true in some aspects of construction. The California bungalow became popular all across the country. This dignified one-story house has a peaked roof. The rafters can be as small as 2″ × 4″ or 2″ × 6″ rafters. Often there is no ridge beam. The attic usually contains a high fire load of stored materials and is reached by a small scuttle hole. The house can have one or more dormer windows.

The Ranch House

In the Sunbelt, which includes states from California, Arizona, New Mexico and Texas to Florida, one-story ranch houses are very popular. These houses have open interiors, large attics, extended overhangs, light-weight roof supports due to no snow loads, and wood shingle roofs. This California-style dwelling is now found in all parts of the country. In a Washington, D.C. suburb, one owner of such a house substituted fire-resistive shingles for the usual wooden shingles. The neighbors were up in arms because he had destroyed the "California Look."

These houses are most often of wood exterior. In some areas they are spaced so close they create a serious exposure hazard. Fast-spreading fires in these types of houses, heavily involved on arrival of the fire department, are not uncommon. Untreated wood shingles present a conflagration hazard.

Properly installed and maintained smoke detectors can provide a high degree of occupant life safety in these buildings since escape from them is relatively easy. Many bedrooms have direct routes to the outside. One common deficiency in these dwellings is failure to place a detector in the attic. Heating and air conditioning units are one potential cause of attic fires. There are known cases of attic fires in which smoke detectors in the living spaces did not activate until the attic was well involved and collapse was imminent.

Nine residents died in a fire in a Colorado Springs personal care home in March 1991. The fire started in an attic ceiling fan. About a week before the fire, smoke detectors had been installed in the dining room and hallways. There was no smoke detector in the attic space. The code required sprinklers in such facilities but it did not apply to existing buildings.

Tile Roofs

The California-style home may feature a heavy tile roof instead of wood shingles. The Apple Valley, California fire department recently had a narrow escape in the collapse of thousands of pounds of tile when making a fog attack on an attic fire. Typically, the failing roof showed no visible signs of weakness. The interior residence was clear. The fire fighters were not accustomed to thinking of roof collapse.[1]

This points up the need for better exchange of information among fire departments. Some years ago fire fighters in Clearwater, Florida told me that an attic fire in a tile-roofed structure indicated an immediate defensive attack, based on close calls they had experienced. In January 1989, two Orange County, Florida fire fighters died when the truss-supported tile roof of a store, in which the attic was involved, caved in on them.[2] Recently, three members of the Phoenix, Arizona fire department fell through a tile roof supported on light trusses, after ventilating the roof. Review of the fire on videotape revealed that the acceleration of the fire, immediately after ventilation, seemed to provide the final weakening of the roof supports.

Tactical Considerations

Fire departments in areas where such wood construction is popular are often doubly handicapped. As with many areas, response distances can be great and units are often understaffed. Dry weather and high winds can accelerate these fires. Tactics should include a so-called "blitz attack" — a master stream, initially from the booster tank, backed by a supply line — to make the heaviest possible immediate hit.[3]

Fire department policy should be established calling for a shift to defensive operations when certain pre-defined conditions are confronted.

In many fire departments much of the fire experience is obtained in one-family dwellings. This can lead to the use of tactics suitable for such fires in other situations where they are unsuitable and probably dangerous. There is an excellent discussion of this subject by Captain Robert Obermayer.[4]

Definitions

Some basic definitions are necessary for a better understanding of the elements of wood construction. Building terms are often local in origin, and very often the construction trades use different words for components than do engineers or laypersons. The term wood-framed is typical of the imprecise terminology common in the building field. A framed building is usually understood to be a building with a skeleton of beams and columns, in which the walls are just curtain walls. The opposite term is wall bearing, indicating a structure where the load is carried on the walls. The wooden-walled building, which is usually referred to as a frame building, is actually a wall-bearing building. It can be very confusing.

Boards: Lumber less than two inches thick.

Chamfer: To cut off the corners of a timber to retard ignition.

Dimension lumber: Lumber between board and timber sizes. 2X4

End matched: Lumber with tongues and grooves at the ends.

Engineered wood: In the construction trade, this typically refers to laminated timbers. In this text, it refers to wood modified from its natural state.

Glued laminated timbers: Planks glued together to form a solid timber.

8" 10"

Heavy timber: Lumber eight inches or more in its smallest dimension. (Note: some building code definitions of heavy timber specify smaller dimensions.)

Joists: Floor beams.

Lumber: Wood that has been sawn and planed.

Matched lumber: Tongue and grooved lumber (usually lengthwise).

Oriented Strand Board: (OSB) is made of layers of strands of wood cut from logs, with a fairly constant width-to-length ratio. OSB is said to have less expansion from moisture. It is available in sizes larger than 4 feet × 8 feet. There is also an oriented flakeboard.

Plywood: Layers of wood veneer laid in different directions, glued together under pressure.

Rough lumber: Lumber which is left as sawn on all four sides.

Splines: Wooden strips which fit into grooves in two adjacent planks to make a tight floor.

Studs: Columns in frame buildings, usually two inches by four inches (two by four) or two inches by six inches (two by six).

Timber: Lumber five inches or more in the smallest dimension. (Note: dimensions of wooden structural members are nominal or sawn size. Planing reduces each dimension of a sawn timber.)

Wood: A hard fibrous material forming the major part of trees. It is usually milled or otherwise processed for use in construction.

Wood lath: Narrow, rough strips of wood nailed to studs. Plaster is spread on wood laths. Practically no longer used, wood lath is present in many existing buildings.

Uses of Wood

Wood is used in many ways. Its significance relating to fire varies with its uses: structural, non load-bearing, decorative.

- Wood is used structurally to carry the building loads.
- Non-load bearing uses of wood include wood-lath carrying plaster, and wood wainscoting, thin grooved board used for ceilings and wall panels.
- Wood also is commonly used for roofs of buildings which may be code-classified as, or appear to be, of non-combustible construction.
- Wood also is used for interior trim of otherwise fire-resistive buildings.

Types of Wood Building Construction

Wood is used to carry the major structural loads in many types of building construction. These include:

- log cabin
- post and frame
- balloon frame
- platform frame (sometimes called western framing)
- plank and beam (also sometimes called western framing)
- truss-frame (a new development)

Wood also often carries major interior loads in masonry wall-bearing buildings.

Log Cabin

The log cabin was not invented by Daniel Boone or Davy Crockett. It came to the east coast of the United States from Sweden, and to the west coast from Russia. Prefabricated log cabins were sold in Moscow in the 17th century.

Fig. 3-1. Look carefully at the exposed side of this brick building. Note that the original building was built of 15" × 15" white oak logs.

It was easier for early settlers to use whole trees or slightly dressed lumber for home building than saw the trees into boards. Later, the humble log cabin would be concealed under siding or even brick veneer.

Many older buildings, even multi-story buildings, are, in fact, concealed log cabins. The walls of such a building carry unexpectedly heavy loads and therefore have potential for serious collapse.

Log cabins are coming back into fashion. A number of companies offer dressed logs, sometimes these are prepared for assembly like the popular toy "Lincoln Logs." The interior surfaces of many of these cabins are usually wood, either boards or plywood, which would increase the intensity of a fire, as contrasted to the typical residential gypsum board interior. Fire departments in areas where such buildings exist should be aware of this hazard.

Post and Frame

Post and frame construction came to America from England, Germany, and France. Such a building has an identifiable frame or skeleton of timber fitted together. Joints are constructed by mortise and tenon (socket and tongue), fitted together to transfer loads properly. They are pinned with wooden pegs called trunnels, (a New England term for a tree nail).

Fig. 3-2. A 400 year-old post and frame building in Cambridge, England.

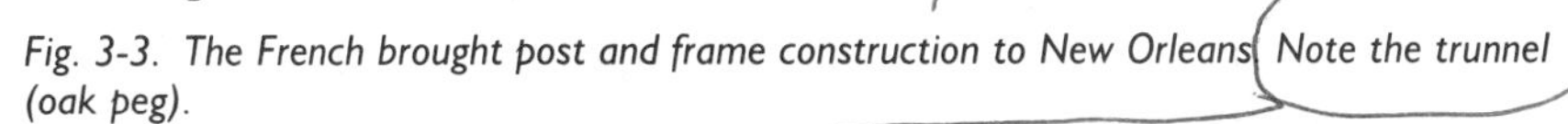

Fig. 3-3. The French brought post and frame construction to New Orleans. Note the trunnel (oak peg).

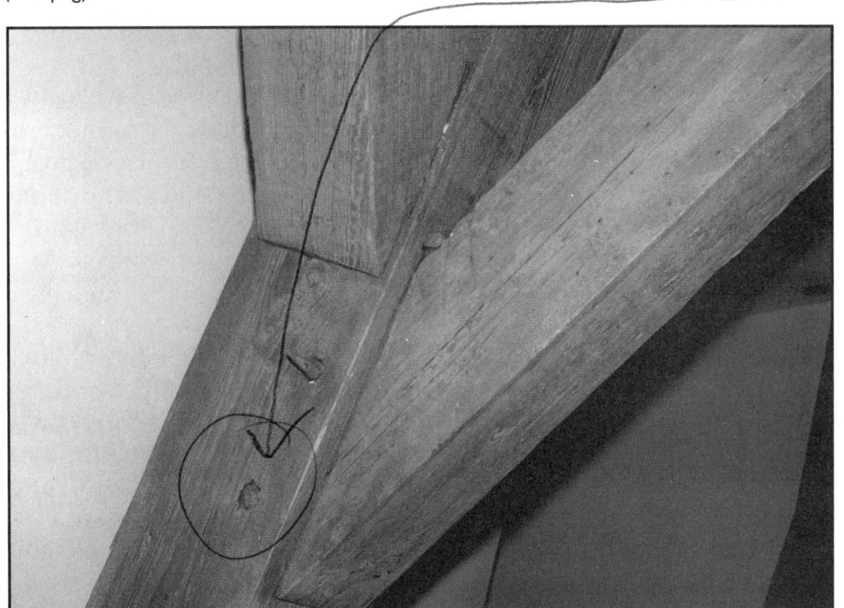

Post and frame buildings are the ancestors of the high-rise, for they are framed, not wall-bearing, buildings. The walls are not structural or load-bearing. Original post and frame buildings are part of our heritage and every effort should be made to preserve them. Recently, new post and frame buildings have been constructed as homes and commercial buildings. Like log cabins, the interior finish often consists of wood.

Many post and frame buildings are enclosed with foam core panels. Foam core roofs present the possibility of failure. See the discussion of these panels under Plastics in Chapter 2.

Tactical Considerations

Do not be misled by the massive size of the timbers. Their size and strength is greatly reduced at the tenon. The failure of a tenon could precipitate a collapse.

Trinity Church in Newport, Rhode Island was built in 1726. Over the course of 250 years, winds pushed it out of plumb. Steel tubular truss frames were built into the structure of the church to stabilize it. The effect of a fire on the unprotected steel frame would be crucial to the stability of the church.[5]

Deceptive Imitation Post and Frame: In the 1920s, many balloon frame houses were built with the exterior finished to resemble post and frame construction. They were called "English Tudor" style homes. Recently, this style has had a renaissance. Buildings of several other types of construction have been decorated to resemble the post and frame style. Some ordinary brick wall-bearing buildings have been dressed up to give the appearance of post and frame. In one case, the fake post and frame trim was removed two decades later to modernize the building. At Disneyworld and other amusement parks, buildings of steel construction are decorated to resemble post and frame appearance.

Balloon Frame

Balloon-frame construction started in 1833. The architect of St. Mary's Church in Chicago hit upon the idea of fabricating a wall of ordinary studs, nailing it together. Inexpensive machine-made nails had become available. This made it possible to throw up the entire wall at once without the skilled labor necessary to do the cutting and framing required for post and frame construction. The type of construction was derided by many who said it was so light that it was like a balloon.

Fig. 3-4. This ordinary brick and wood joist building is being decorated to look like post and frame.

Unfortunately, as will be seen, it can be compared to a balloon in another way. In any event, balloon-frame construction became an almost universal construction method for multi-story wooden buildings until the middle of the 20th century.

In a balloon-frame building, the studs run two or more stories high from the foundation to the eave line. At the floor line, a horizontal board called a ribbon board is nailed to the studs. The joists rest on the ribbon board. The channels between the studs may be open from the cellar to the attic, and the joist channels (space between the joists) are open to the stud channels. Thus, fire can spread through all the interconnected spaces from cellar to attic and across the ceiling. Present-day methods may use firestopping, but earlier construction methods, when many of these buildings were built, did not.

The interconnected voids can be considered to be one big balloon. The author saw this confirmed at a fire in an old balloon-frame house. There was a heavy fire in the porch ceiling. Each time it was hit with a fog stream, the fire ballooned out at the rear of the building.

When balloon-frame buildings are remodeled, fire stopping may or may not be installed. Installing fire stopping in an old building is very costly. In one case an historic balloon-frame college building was fully rehabilitated. Even though sprinklers were installed, fire could have

Fig. 3-5. This Montana mining town dance hall shows the vertical voids open to the floor voids, a characteristic of balloon-frame construction. The interior is "one big balloon."

Fig. 3-6. Pennsylvania fire fighters worked from the outside to cut off this balloon-frame building fire, thus minimizing interior damage.

raced through the voids if they were not firestopped. The final cost was about two-thirds of the cost of a new building.[6]

Until recently it was reasonable to assume that any wooden building built before 1940 and more than one-story high was probably of balloon-frame construction. Later buildings used platform-frame construction. Recently, however, the Victorian-style construction has come back into favor, and expensive multi-story platform-frame suburban homes are being built to look like one hundred year-old farmhouses.

The book *Building Construction*[7] recommends balloon-frame construction as superior to platform frame for masonry-veneer houses (see discussion of brick-veneer later in this chapter). Shrinkage of the wooden building could damage the brick veneer. The smaller vertical shrinkage potential associated with a balloon-framed structure contrasts with greater horizontal shrinkage potential of platform-frame construction.

Two six-story brick-veneer apartment houses recently built in an Ohio suburb were of balloon-frame construction. Studs were spliced to achieve the required height. The apartment houses had many deficiencies and presented a major fire hazard. The fire chief fought successfully to have them torn down before they were ever occupied.

> 25

Interior Walls: Masonry buildings with spans greater than 25 feet must have interior bearing walls. In older buildings these walls are usually of balloon-frame construction unless made of masonry. Interior masonry walls can be distinguished from wooden interior walls by their greater thickness. Large wooden buildings also must have interior bearing walls, which usually are of balloon-frame construction in older buildings.

Fires in Balloon-frame Buildings: A chief officer, who was a student in a fire science course, described a fire which the first-due company reported as "outside rubbish." The officer ordered some of the siding removed and found fire in the walls. Shortly thereafter, another officer reported that an upstairs bedroom on the opposite side of the house was untenable, even though no fire was showing. The fire had quickly crossed over through the joist channels and had literally surrounded the bedroom. A good fight saved the building, but the damage was extensive.

Los Angeles City fire department Battalion Chief John Mittendorf provides an excellent discussion of the hazards of fire in a large balloon-frame "Queen Anne" building in which the roof collapsed and a fire fighter escaped with only minor injuries.[8]

Another California fire demonstrates that sometimes even sprinklers cannot always protect balloon-frame structures. An incendiary trash fire, against an exterior wall of the Victorian building, penetrated into void spaces. Responding fire fighters checked for extension but found nothing. Fire broke out four hours later. Much structural damage occurred before the sprinklers activated on fire which broke out from the voids.[9]

Tactical Considerations

When a fire has entered the inner structure of a balloon-frame building, it can spread to every part of the building in all directions. Investigate all parts of the building immediately. Be aware of the potential for intense fire buildup in void spaces. Don't wait for the smoke, heat and fire to make conditions untenable before you decide to examine the attic. Fire moving up or down through exterior stud channels can often best be stopped by removing siding from the outside at the second floor line, and eave and foundation lines.

The author was recently lecturing in Pennsylvania, where fire officers described a fire in a balloon-frame wall. They had followed the suggestion given above, to open the wall from the exterior, rather than the interior, and stopped the fire. If they had made a conventional inside attack, they might have lost the building.

The problem with an attack limited to inside operations is that fire moves up through many stud channels. As soon as the first channel is opened, the situation becomes obscured with smoke and steam. It is easy to miss a channel, and the fire may get away. Many fire departments today lack the bent swivel tip which was used on the old controlling nozzle. With this swivel tip, the fire fighter could quickly give it a dash up and down the channel. This is not as easily done with today's equipment.

Fire burning in balloon frame walls destroys the structural integrity of the building. Collapse is a serious threat. Fires in balloon-frame buildings should be observed from the exterior, by an officer specifically assigned to that function, and positioned far enough away to see the entire building. This officer should watch for signs of fire spread — heavy volumes of smoke pushing out from voids in the building, as well as intense heat buildup and structural failure, which may not be apparent from close range. Chief Donald Loeb discusses fire fighting in the most common type — the balloon-frame 2½-story dwelling.[10]

Platform Frame

In platform-frame construction, the first floor is built as a platform. This means the subflooring is laid on the joists, and the frame for the first-floor walls is erected on the first floor. The second-story joists are then placed, the second-story subfloor is laid, and the second-floor walls are erected on the second-story subfloor. The third floor, if any, is built in the same way. Three stories are typically the limit for platform-frame construction.

In a balloon-frame building, there is structural continuity from top to bottom. In the platform-frame building there is no continuity from top to bottom. Thus, the structure is vulnerable to the wind.

If construction methods are as described above, there are inherent barriers to limit the spread of fire through the walls. There is a good possibility of confining a fire to one portion of the structure, even if the fire gains access to a concealed space.

Unfortunately, only rarely is the construction as described. One common construction feature provides a bypass for fire, and from the fire protection point of view, converts a platform-frame building into a balloon-frame building.

Soffits (or false spaces above built-in cabinets, usually in a kitchen, or in the undersides of stairways and projecting eaves) provide a connection, generally without a firestop, between wall and joist spaces. Furthermore, only a thin sheet of wood or composition board makes up the top of the built-in kitchen cabinets . A kitchen fire that extends into such a cabinet will enter the soffit space quickly. Fire can then extend to the joist spaces in a multi-story building or to the attic spaces in a one-story building. Such soffits are often built back to back in multiple dwellings so that fire may extend immediately to the adjacent occupancy. Interconnection of the soffit void with pipe openings may cause fire to spread vertically throughout the building.

Tactical Considerations

The split-level house, in which portions of the building are one-half story above other portions, is a variation of platform-frame construction. There are so many variations that it is impossible to categorize them, but a thorough study by fire fighters of locally-built housing is warranted. Expect interconnections between void spaces, and don't rely on firestopping.

Note the wide open interior of this type of building. A fire can deliver deadly combustion products to any part of the house. The inherent barrier to the extension of a basement fire provided by a basement door, is often absent.

Plank and Beam

In recent years, architects and interior designers, looking for more attractive, economical, or at least different construction methods, have adapted industrial plank and beam construction techniques to residential, assembly, and commercial structures.

Fig. 3-7. In this plank and beam roof the laminated beams are formed into a catenary curve (the opposite of an arch). Void spaces are eliminated but fire can spread quickly over the surface.

Instead of using multiple floor beams 16 inches apart, heavier beams are used which are spaced much farther apart. Instead of thin, rough subflooring or plywood, thick, finished tongue and groove planks are used. The planks are thick enough to span the wider gap between the beams without deflecting. The finished plank surface is the ceiling.

From a fire fighting point of view, this type of construction has the positive benefit of reducing the volume of concealed space in which fire can burn shielded from hose streams. The interior finishes used, however, often have high flame spread and smoke-developing characteristics. Surface damage may be serious, even in an easily controlled fire in which collapse or hidden fire are not factors.

In some residences, to provide an open effect, partitions are only head high, except for bedrooms and bathrooms. This increases the life hazard for nighttime fires because the kitchen, dining, living and family rooms are all one open area in which fire can easily build up.

Many churches and similar buildings with high open spaces are built of rigid laminated wood frames or arches and heavy plank roofs. Any fire that reaches the surface of the wood will spread rapidly and develop into a huge volume. Any intermediate structure, such as a sacristy or robing room, should be sprinklered or of non-combustible construction. Eliminating the intermediate fire load greatly reduces the potential for ignition of the roof.

Tactical Considerations

Where a plank and beam ceiling is already — or about to be — involved in a fire, prepare for a heavy stream attack. A solid stream tip should be used to reach the surface of the wood with maximum water. Fog would be eaten up by the heavy fire. When the surface is wet the fire is extinguished.

In preplanning, select locations for the streams so that all areas are covered. Recently, this attack worked successfully at a fire in a Florida church.

Truss Frame

The United States Forest Products Laboratory has developed a method called truss-framed construction. This is different from conventional construction in which truss roofs, and perhaps truss floors, are substituted for sawn beams. However, the structure is still held together by nails.

Truss frame is engineered construction in which the roof and floor trusses and studs are tied into a unitized frame. The studs are an integral part of both the roof and floor trusses. There must be meticulous attention to detail in the manufacture and erection of these structures. A chief advantage of this type of construction is speed. One manufacturer reported that an experienced crew of four erected the truss frame in 90 minutes and had the house under lock and key the same day.

The hazards of trusses are discussed in detail in Chapter 12, Trusses. It is sufficient to note here, however, that the small dimension lumber (two by fours or less) will burn faster than larger solid lumber. In

Fig. 3-8. This sketch of "truss-frame" construction shows how the wall studs are an integral part of the truss.

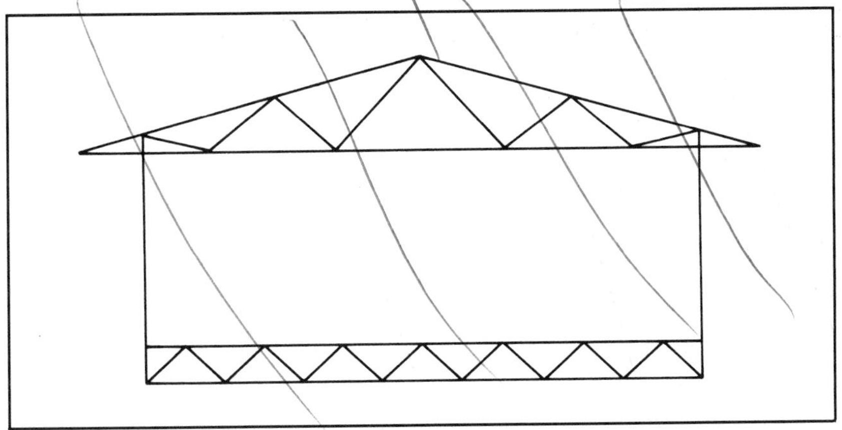

addition, unsupported spans are in trussed structures subject to total collapse, and the loss of a stud due to fire could precipitate the collapse of the integral roof or floor truss. The floor truss void (trussloft) in which hidden fire can spread, provides a hazard to fire fighters which affects fire fighting tactics. From the fire point of view these are truly disposable buildings.

"Truss-frame structures are plane structural components. Their design assumes that every truss member will remain in its assigned position, under load. Permanent bracing must provide adequate support to hold every truss in its design position and to resist lateral forces due to wind or seismic load."[11]

Tactical Considerations

The fact that every load member must be in its assigned position under load supports the contention that the failure of even one member of a truss "entitles" the truss to fail. In a particular instance, it may not fail due to some undesigned advantage, such as a solid nailed floor. However, this does not negate the concept that trusses are subject to one point failure.

Wood in Non-combustible and Fire-resistive Buildings

Some buildings which may appear to be steel or concrete construction, are, in fact, ordinary construction and have a wooden roof.

Many so-called non-combustible buildings have substantial wooden components, particularly roofs and roof adaptations such as mansards which in effect provide a combustible top floor. A "non-combustible" steel-framed Louisiana church has a large wooden balcony.

For example, a California apartment house was built of fire-resistive construction, but portions of the fifth floor and the entire roof were of wood. An undivided cockloft extended the length of the building. A mansard roof was covered with wood shingles. A loss of over $1 million was incurred when an electrical fire involved the entire cockloft.

In another case, the owners of an otherwise fire-resistive motel decided to use the space at the top of a stairway for a wooden storage room. Heavily involved in a fire, it came close to collapsing on unsuspecting fire fighters. A post-fire critique revealed two important points. The attack should have been made via the aerial ladder to the roof to avoid the hazard of the room falling on fire fighters advancing up the stairway. The pile of lumber used for building the storage room had lain on the sidewalk for several days —in the path of fire fighters reporting for duty at the nearby station.

Fire fighters should be ever curious about building alterations. Any alteration of a building, except the proper installation of sprinklers, potentially could be detrimental to fire suppression efforts and may create hazardous surprises.

Fire-resistive buildings can have substantial wooden interior finish. Plywood or heavy wood paneling is common in executive offices, club, dining, court, and ceremonial rooms. Often installed after the building is finished, sometimes it is done without a building permit.

Such paneling usually is installed with air space on the unexposed side where fire can burn unseen and unchecked. Surface flame retardants, applied after installation, are ineffective because the rear surface is not affected. Some codes require flame resistant impregnated wood to be used. Four women jumped to their deaths, and two fire fighters died in a rescue attempt in a fire that took place in a wood-paneled restaurant on the 15th floor of a New Orleans high rise. Inadequate $1\frac{1}{2}$-inch hose streams were a factor in this tragedy as fire fighters were unable to push the fire back into the restaurant.[12]

2 1/2" LINE

Tactical Considerations

Wooden structures within otherwise fire-resistive buildings can cause serious problems. Many fire departments use only small diameter hose lines on high-rise fires. All should be prepared to shift to heavier caliber streams with large diameter hose when warranted.

The Wilmington, Delaware fire department has provided all its companies with light-weight portable deluge guns to provide heavier fire streams when necessary. Tests show that the guns can exceed 300 gpm with two 100-foot, $1\frac{3}{4}$-inch lines.

Exposure to Fire-resistive Buildings

In December 1989, the largest commitment of fire fighters and apparatus to a structural fire in the history of the Los Angeles City fire department was made for the Wilshire fire. The fire started in a wooden apartment house under construction, and extended to an adjacent 14-story, fire-resistive residence building before the arrival of fire apparatus. Flying brands and direct exposure extended the fire to other nearby buildings.[13] Some high-rise structures, similarly exposed, are protected by installed exterior water spray systems designed to protect windows.

Tactical Considerations

Potential fire exposure to high-rise structures should be considered in pre-fire planning. If a water curtain is provided, units should be aware of how to operate it. It should be tested periodically. While the concept of water curtains may be outdated, some structures still have them.

The best and fastest tactics for covering the exposure should be studied and developed before the fire gains a serious foothold. A heavy-caliber stream from the street directed at a specific spot on the exposure might fan out to provide effective coverage. In another case, it might be necessary to use the slower procedure of getting lines to each floor. If fire extension protection is planned for, the defense will more likely be successful. Adequate assistance should be summoned and staged immediately. Traffic control also can become a serious problem at such spectacular fires, and should be included in planning as well.

Firestopping

One of the major problems of combustible construction is fire spread through hidden voids. Firestopping is often required by code to be installed to prevent the spread of fire.

To the best of this author's knowledge, there are no tested standards for fire stopping or draftstopping.

The National Forest Products Association (NFoPA) distinguishes between firestopping and draftstopping. Firestopping and draftstopping limit the spread of fire by preventing the movement of flame, hot gases and smoke to other areas of the building.[14]

Firestops limit movement through relatively small concealed passages such as under stairs and inside walls. Firestopping material may consist of at least 2-inch nominal lumber, two thicknesses of 1-inch nominal lumber with broken up lap joints, or ¾-inch plywood or other approved materials.

Draftstops limit movement through large concealed passages such as open web floor trusses or attics. Draftstopping may consist of at least ½-inch gypsumboard, ⅜-inch plywood, or other approved materials usually applied parallel to the main framing members.

Types of Firestopping

This book defines two types of firestopping.

Inherent firestopping comes as a result of the normal building construction. This kind of firestopping is incidental to necessary structural purposes. An example is the floor-to-wall seal when a masonry panel wall is built directly on a concrete floor. It is reasonable to assume that such inherent firestopping is in place and effective.

Another type of firestopping which this author calls "legal firestopping" is installed with no other purpose than to meet the requirements of a code. Such firestopping may provide a barrier to the spread of fire in the interior voids of the building. However, this type of firestopping is often ineffective and questions regarding "legal firestopping" may always remain.

"Was the firestopping properly and completely installed?" "Was the building inspector competent and vigilant?" "Was it tampered with after installation?" In many of the cases this author has seen and photographed, unfortunately the respective answers were: No, No, and Yes.

All firestopping must be in place to be effective. The lack of firestopping in one stud channel (the space formed by the sides of two studs and the exterior and interior walls) is sufficient to transmit fire from the cellar to the attic. In older houses, both the exterior sheathing and the lathing on the interior walls are made of wood, so all four sides of the chimney-like stud channels are combustible.

Lack of firestopping is particularly critical in balloon-frame buildings. It is important to note that firestopping was unheard of by many builders when many of these structures were being built over 40 years ago.

Not all building inspectors are familiar with the basic gas law: *If the temperature rises and the volume remains the same, the pressure rises.* They do not know that leaving an "inconsequential" opening in the firestopping will, in effect, create a nozzle. Nor do they realize that if firestopping isn't perfect, it might as well be omitted. Unfortunately, it is rarely perfect.

Wood firestopping is often made from the cut-off ends of joists. Unbelievably, the cut-off ends of wooden I-beams have been used as firestopping. There is, of course, no seal because the wood "cut out" to make an I-beam creates a space. A few remarks that have served to rationalize less-than-perfect efforts with firestopping include: "Well, that's all we had," "We always did it that way," and "It's just a little opening."

Fig. 3-9. The fire stopping was removed to install the fire main. Now the fire can follow the fire main through the building.

There are those who believe building practices are different on the west and east coasts. Regarding firestopping, however, I have observed uniformly unsatisfactory practices. Recent firestopping or draftstopping practices consist of a sheet of gypsum board "buttered" into place, or a piece of thin plywood or flake board poorly fitted and often penetrated by utilities, and wiring. These are about as effective as a posted notice from the building department forbidding fire to pass.

Firestopping, once installed, often is removed for the installation of such items as heat ducts, electrical cables, sprinkler systems, and central vacuum cleaner systems. In one ironic case, it was removed to install the fire main requested by the fire department and was not replaced. Since the firestopping is often not "necessary" to an installer making an improvement or fixing an existing item, it is unlikely that it will be replaced. The author has read many articles on the restoration of old houses; firestopping is rarely mentioned.[15]

Tactical Considerations

In a fire in a combustible structure, it is best to assume that no firestopping was installed. Then, any surprise will be pleasant rather than unpleasant.

Fig. 3-10. Fire travels vertically through unfirestopped pipe chases.

Firestopping and Truss Floors

Trusses are discussed in detail in Chapter 12. However, trusses create a new hazard, the truss void or trussloft in each floor. When this hazard is discussed, truss proponents sometimes argue that firestopping will mitigate the problem.[16]

If draftstopping or firestopping is installed by persons who understand that its purpose is to stop a gas under pressure, and not to satisfy an inspector, it may limit the spread of fire. Note, however, that even if the draftstopping installed around the perimeter of the affected space is successful, this will not prevent the collapse of the affected area — at least 500 square feet in a single-family residential building and 1000 square feet in other buildings. This is a large enough area to develop a severe backdraft explosion or to provide a significant collapse. In multiple dwellings, the recommended location for firestopping is along the tenant separations so that the entire floor-ceiling area, above and or below the unit on fire, is a collapse area. Firestopping will not make such construction materially safer.

Tactical Considerations

The solution is a change in tactics. Fire fighters can no longer rush pell-mell into burning structures. Progressive fire departments will

Tactical Considerations, continued

make the changes after analyzing the problem. Others will learn as a result of disasters and lawsuits.

Wood as a Building Material

As a building material, wood has both advantages and disadvantages. Its basic defect is its combustibility, producing about 8000 British thermal units (Btu) per pound. Wood also varies greatly in strength. Its chief advantages are its relatively low cost, and how easy it is to work with using hand tools.

Wood construction is subject to insect infestation, wet and dry rot, and other forms of decay, which may cause weakening without being apparent to the casual observer. Deterioration of this nature in structural members should be carefully noted on prefire plans. Following a fire in New York City, a heavy-timber building collapsed. Investigation disclosed that dry rot had decayed the timbers or turned them into punk, which continued to smolder after the visible fire had been extinguished.

Protecting Wood from Ignition

It is sobering to realize that most fires are fought by fire fighters standing on, in, or under a combustible structure. Eliminating this basic undesirable combustibility characteristic of wood as a building material has long been a goal of those interested in fire protection. Early fire prevention books extolled the virtues of whitewash in fireproofing wood. In the early 19th century, ads for products like "Blake's Patent Fireproof Paint" were common.

Attempts to protect wood by encasing it in cement-like products were found to be dangerous, since the wood, not being exposed to the air, tended to decay. In a San Antonio, Texas building, the author has seen wood columns that had been "fireproofed" with sheet metal cladding. Serious wood decay can occur unobserved with this practice.

Impregnation

Wood cannot be made fireproof or non-combustible. It can, however, be made fire retardant by impregnation with mineral salts which slow its rate of burning. Some codes accept impregnated wood as non-combustible. It is not. "The mayor could also call a giraffe a camel, but this doesn't make it one."

Impregnation is accomplished by placing the wood in a vacuum chamber, drawing out moisture from its cells, and forcing mineral salts into the wood. Sometimes the wood is pricked to increase penetration. Mineral salts do not penetrate deeply, and the removal of surface wood may destroy fire retardant treatment. Such wood is sometimes called pressure-treated, but should not be confused with wood pressure-treated to resist decay.

Pressure treatment can significantly reduce wood's flame spread, as measured in the ASTM-E 84 Tunnel Test (See Chapter 9, Fire Growth). Underwriters Laboratories Inc. (UL) subjects pressure-treated wood to a test for 30 minutes rather than the usual 10 minutes. Pressure treatment can significantly reduce the hazard of wood construction where there is insufficient other fuel to provide a strong exposure fire. Given enough exposure to fire, treated wood will burn, although at a slower rate.

UL provides various classifications of treated lumber. For detailed information consult "Lumber, Treated (BPVV)" in the latest issue of the *UL Building Materials Directory*. Such wood often is called Fire-Retardent Treated wood (FRT).

Corrosion

Some chemicals previously used to impregnate the wood have leached out and corroded metal connectors. Some advertising literature for impregnated wood claims materials now used are non-corrosive.

Loss of Strength

Pressure treatment of wood decreases its allowed load by 25 percent. In the construction of apartments with truss floors in Reedy Creek, Florida, this strength penalty was accepted.

Plywood Roofing Problems

Some years ago, to avoid parapetting fire walls through the roof, builders adopted a practice of placing FRT (fire retardent treated) plywood, one sheet wide, on both sides of a firewall, which did not penetrate the roof. At the time, I set forth potential deficiencies of this method of stopping the extension of the fire over the top of the firewall since the plywood delaminates (layers separate) and fire passes over the top of the wall.

Recently it also has been found that plywood treated with certain chemicals decays from heat and is subject to failure if walked on.[17, 18]

The problem is compounded if treated plywood is used not only along the line of the fire wall, but randomly throughout many roof areas. In some cases, the entire roof is treated plywood.

In an attempt to provide some sort of barrier to the spread of fire over an unparapetted fire wall, some jurisdictions require a sheet of gypsum board to be attached to the underside of the roof, on each side of the fire wall. Presently, there is no evaluation of the adequacy of this concept.

Tactical Considerations

Fire fighters should suspect all such roofs. Some jurisdictions have banned certain products, but at the time of this writing there is no consensus as to what should be done about existing roofs.[19]

Surface Coating

The fire hazard or flame spread of wood can also be reduced by the application of intumescent coatings which swell up when heated. As a substitute for pressure impregnation, one manufacturer suggests that all surfaces be painted with the proper coating at the job site and touched up as necessary. This way eliminates the objection by carpenters, that treated wood dulls cutting tools.

One of the major problems with a surface coating is the tendency for it to be spread beyond its recommended thickness. A gallon of surface coating must not be applied to more than the area specified for it. One inspector requires that surface coating container labels be delivered to show that the required amount of coating was purchased. This author has always recommended that the owner purchase the material and hire a painter to apply it, in an attempt to achieve the recommended coverage.

Applying surface coatings to existing combustible installations such as plywood does not completely solve the problem because the unexposed surface of the paneling remains untreated. However, in one fire, intumescent-coated plywood wainscotting did not burn significantly, while an untreated chair rail molding burned completely.

Dangerous Treated Wood

Plywood and lumber treated to resist insects and moisture (pressure treated in the trade), has been used for basement walls. It is also

Fig. 3-11. Millions of board feet of toxic treated wood, such as the wood in this deck, are used in many wildland, urban interface settings.

widely used for exterior structures such as decks. It is, of course, combustible and might cause a collapse. The fumes from this pressure-treated wood are toxic. Do not use scraps of this wood for kindling.

A fire investigator became ill and required hospitalization after spending some time investigating the ashes of a fire which had involved pressure-treated (for retarding decay) wood.

Tactical Considerations

All exposed personnel must wear SCBA during the entire exposure. Particular attention should be given when wildland fires extend to structures. SCBA is rarely worn on such fires and might not even be available to forest fire fighters.

Wood should not be handled with bare hands. Fire fighters should wash up after exposure.

Engineered Wood

For many years, wood was used as it was found with all its limitations. For instance: many of the famous California Mission churches were long and narrow because their width was limited by the longest tree trunks which could be found for use as roof beams. Some of the other limitations of wood are:

Wood splits along the grain.
Long straight trees of large diameter are scarce.

However, wood can be engineered or modified to overcome these limitations. The term engineered wood is commonly used in the trade to refer to glued laminated heavy timbers. In this text it is used to denote any wood material which has been modified from its natural state.

Some engineered wood products are:

plywood
laminated timbers
chipboard
wooden I-beams
wood trusses
flitch plate girders
wood and plastic composite roof panels

Each of these products presents the fire service with specific problems.

Plywood

About one hundred years ago, it occurred to someone that one of wood's limitations — the lack of shear strength along the grain — could be overcome by slicing the wood into thin layers, placing the layers at right angles to one another, and gluing the entire mass together. We know the product today as plywood, which is just about equally strong in all directions. This was the first of many ways in which natural wood was engineered into a useful product.

A basic problem of plywood exposed to fire is that it delaminates. This increases the surface area and its rate of heat release.

Plywood can be impregnated to render it fire retardant. The problem of some fire-retardant plywood roofs has been noted earlier in this chapter.

Plywood can be used as an interior finish, a building sheathing without structural value, and as a structural material in floors, roofs, or walls. Such construction is often described as stressed skin or diaphragm, in which the plywood provides some of the structural strength of the building, particularly in providing resistance to shear stresses.

Fig. 3-12. Plywood delaminates readily, increasing the surface area. Thus plywood burns fiercely.

Oriented strand board is used to replace plywood in many applications. It consists of layers of wood strands cut from logs.

Tactical Considerations

Expect a hot, fast fire from plywood, with high flame spread and a high rate of heat release.

At a fire involving treated plywood, do not be surprised at dense smoke.

A Fire Technology article noted that "the two most effective fire retardant chemicals, monoammonium phosphate and zinc chloride, greatly increased smoke density index values for plywood."20

The cutting or burning of structural plywood can have more serious consequences than the destruction of plywood which was used for simple sheathing.

Spliced and Laminated Timbers

The shortage of big trees from which solid timbers could be sawn led to the development of spliced timbers. Various metal connections are used to transfer loads so that spliced timber acts like a single timber.

Fig. 3-13. Not even Hitler's engineers could get timbers long enough for the roof of his Eagle's Nest mountaintop retreat at Berchtesgaden. These heavy beams are spliced with steel connectors. In a fire the heated connectors could fall out of the beam.

In a fire, the heated metal connections can destroy the wood and the timber may fail.

More recently laminated timbers have been developed. Plank-like sections of nominal 2-inch (or thinner) boards are glued together under pressure to produce large arches, beams, girders, and columns. Such timbers are also known as Glulam, a trade name. Sometimes bolts are used to supplement the glue. When highly finished, these timbers are very attractive. Combined with wood plank they can provide the necessary structural strength together with an aesthetically pleasing interior finish. Laminated timbers apparently burn like solid heavy timbers, and do not delaminate like plywood.

In *Fire Command*, a writer mistakenly equated laminated wooden beams (which do not laminate) with plywood as used in wooden I-beams (which does delaminate).[21]

Laminated wood sections were spliced together to form arches for a Daytona Beach, Florida sports arena. In a fire, the wood was only charred but the arches fell apart at the metal connections.

In the days of wooden ships, shipwrights sought out trees that had grown in the special shape required for certain structural members. Today, almost any shape can be fabricated.

We usually think of arches as having a characteristic segmental arch shape. Two hinge arches of laminated wood are available, combining in one member both column and girder. It provides a straight-walled structure with a flat roof and a clear floor area. The Forest Products Laboratory in Madison, Wisconsin is constructed of arches that provide a floor area the equivalent of five stories in height with a 60-foot span.

Tactical Considerations

Beams spliced with steel connectors should be noted on prefire plans. It may be possible to protect the spliced area with hose streams. A collapse is a serious threat.

Paper Wrapping: Laminated timbers and other finely finished wood are shipped in a protective paper wrapper. This cover is kept on as long as possible during the construction period. This paper is hemp-reinforced and coated with a bituminous moisture repellant. It ignites readily, has a high flame spread, and could contribute to a severe loss in a building under construction. In one fire, this paper was responsible for the extension of a grass fire to piles of packaged lumber.

In another case, an addition to a hospital was under construction. A wooden snow roof was erected over the excavation. Hemp-reinforced treated paper was selected to waterproof the roof. A potbellied stove in a change room spewed out sparks. A fire could have easily sent a sheet of flame up the face of the hospital.

Planks: Tongue and groove roof planks, for plank and beam construction, were fashioned in the past by wasteful methods. Such planks are now often fabricated without waste by gluing three boards together with the center board protruding on one side and indented on the other. The author wonders whether during a fire these would separate like plywood, causing the boards to fall from the overhead. A sample, ignited by this author, burned like a solid piece of wood. It might be wise for a fire department to run its own test on materials used locally.

Finger Joints: Light lumber such as two by fours are sometimes glued together. Finger joints are made by cutting a series of long points into the end of each piece. The joints are glued together. Some open web trusses with finger joints manufactured in Oregon between 1974-1980 reportedly came apart at the joint in normal use.[22]

Chipboard

Often wood chips are glued together to make flat sheets. These chipboards are sometimes used for the floors of mobile homes. Some chipboard is water soluble and has dissolved in fires.

Tactical Considerations

Be cautious at mobile home fires. Walk on the beams.

Flitch Plate Girders

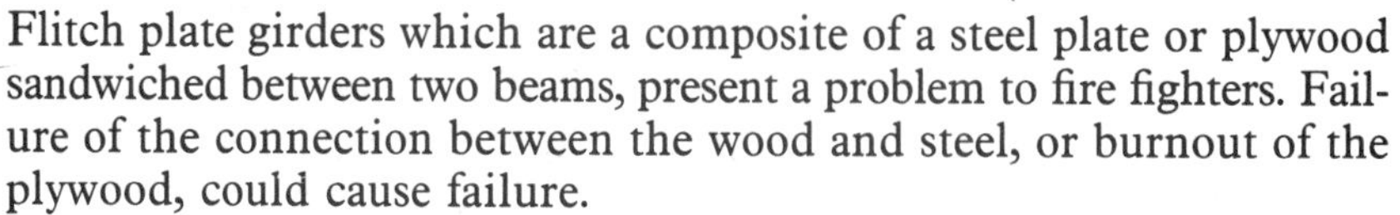

Flitch plate girders which are a composite of a steel plate or plywood sandwiched between two beams, present a problem to fire fighters. Failure of the connection between the wood and steel, or burnout of the plywood, could cause failure.

Trusses and Wooden I-Beams

Sawn wooden beams contain excess wood which is unnecessary dead weight. A beam requires strength in the top to resist compression and in the bottom to resist tension. All that is needed in between is a web to separate the top from the bottom. In recent years wooden I-beams have been developed which save weight, and have other desirable construction characteristics.

Another method of lessening the weight of wood beams is to use parallel chord wood trusses.

The serious hazards of trusses and wooden I-beams are fully discussed in Chapter 12, Trusses.

Fig. 3-14. A flitch plate girder. The long black line is a piece of one-half inch steel. In some cases plywood replaces the steel.

Wood and Plastic Roof Panels

In a television show, large plywood or chipboard roof panels with a gypsum board interior surface sandwiched a thick plastic foam core. The panels were installed between widely spaced roof beams to provide a roof that was otherwise unsupported.

It could be dangerous to vent such a roof. Even away from the fire area, heat from the fire might melt some foamed plastics and the roof may fail. If the foam is uninhibited, a fierce fire with dripping plastic is possible.

Sheathing

Sheathing is the covering which is applied to the studs or framing of a structure. The exterior surface covers the sheathing.

Tongue and groove boards laid diagonally to provide shear strength were used in older houses. Many old houses were built without sheathing; fire spreading though walls can come out through joints in the siding.

In recent years, low-density black fiberboard which has been moisture-and vermin-proofed has been used in residential construction because it can be installed quickly and has relatively high insulation value. (Celotex® is often misused as a generic name for any low-density fiberboard. In fact, the Celotex company manufactures many building products, and many other companies manufacture low-density fiberboard.)

This low-density fiberboard material carries the warning "Combustible. May burn or smolder if ignited." A common method of ignition of this material is the plumber's torch. A similar material is often used for sound conditioning. It was applied directly over wood studs and covered with a gypsum board in a school. A propane torch was used to sweat copper pipe in the same school. The flame ran along the pipe and found the fiberboard. The fire extended vertically to the metal deck roof. The brand new school was destroyed. It was uninsured because it was believed to be fireproof.

Plywood is also used for sheathing. In some cases, a building is totally sheathed in plywood. In other cases, plywood is used at the corners to provide shear strength.

Gypsum sheathing is found in some localities, particularly where combustible sheathing is not permitted. It is similar to the gypsum wallboard used in interiors. Combined with brick veneer surfacing and gypsum interior surface, it is used to provide rated fire-resistive exterior walls.

Foamed plastic is now widely used for sheathing. It may or may not be flame-inhibited. In any case, if exposed to fire, it may degrade and give off noxious fumes.

Siding

The siding on a building is the outer weather surface installed over the sheathing. Many different sidings are used on frame buildings. All combustible siding readily extends fire vertically, and is easily ignited by an exterior fire such as trash or burning foliage.

Wood siding is usually **novelty siding**, laid on horizontally. Novelty siding is often called **clapboard**, but true clapboards, cut from a log in a wedge-shape, are rare.

Board and batten siding consists of boards laid on vertically, butt to butt, with strips nailed over the joints. Sometimes plywood is used, and battens are nailed on one-foot centers to simulate board and batten. Battens have been used on painted concrete block walls to simulate such siding.

Plywood siding is delivered in four-foot wide sheets, side matched. It gives the appearance of four-inch wide strips about a half inch apart.

Shingles and shakes (longer, thicker shingles are called shakes) are also used for siding. As siding, they do not present the extreme conflagration hazard of wood-shingled roofs. However, they can become ignited when burning shingles or brands fly about and set fire to rubbish or dry foliage.

Asbestos cement shingles have been used both in new construction and as a replacement over deteriorated wood siding. They are noncombustible, but the presence of wood trim, and often the old wood siding, makes such a building as vulnerable to a grass or rubbish fire as a wooden building. Asbestos cement shingles can explode when heated, and flying particles can cause eye injuries, and possible inhalation hazards. One fire officer got into serious trouble with the Environmental Protection Agency for holding a live fire drill in a building with asbestos shingles.

Asphalt felt siding, usually made up to look like brick or stone, is often used as replacement siding over wood. It burns readily, and produces dense black smoke. Asphalt siding is very similar to asphalt roofing which meets certain roofing standards. This has led people to believe that it has some fire resistance value. It has none.

Vinyl siding is made to look like wood siding. It will burn as well.

Metal siding, when used on residences, is usually made up to resemble some other material. Currently, aluminum siding is made to look

Fig. 3-15. Asphalt felt siding burns readily, producing great amounts of smoke.

like clapboards. Embossed sheet metal has been made to look like imitation stone.

Corrugated metal siding is used on industrial buildings of wood or steel framing. Metal siding can present severe electrical hazards, both from stray electrical currents and from lightning. Thorough bonding and grounding of the siding and metal roof will do much to eliminate the hazard. This is rarely done, and the possibility that metal siding is energized should be a factor in fire fighting.[23] (Corrugated metal siding is discussed in Chapter 7, Steel Construction.)

Stucco is a thin concrete surface. Stucco can be used on any structure such as brick, block, hollow tile or wood. It is not correct to describe a structure as a stucco building. The correct description is "stucco over something else." In the case of wood, stucco is laid on metal lath. The lath makes it difficult to make openings in the wall. Stucco is used both in initial construction and in rehabilitating old buildings. Some assert that the stucco makes the building safer from fire. This is untrue as far as life safety is concerned. However, the use of stucco does improve exposure protection over that of comparable wood-surfaced buildings.

In some cases, a coat of gray stucco is applied. A coat of red stucco is laid on over the gray. Red stucco is then removed in lines leaving the gray stucco visible. The effect is that of brick masonry. Some concrete block buildings are stuccoed so as to resemble clapboard. The houses are then painted and appear to be wood.

Fig. 3-16. Stucco is concrete put on steel lath. Sometimes there is no wood siding. It is very difficult to remove steel lath.

Brick Veneer

Brick structures have so-called "good press." Brick veneer siding is very popular for wood frame residences, garden apartments, and smaller commercial buildings in areas where brick is economical. This brick is not structural. It carries no load except itself. While initially more costly than wood siding, the savings on painting over the years may make brick veneer economically attractive.

Fig. 3-17. Brick veneer is being laid over foamed plastic sheathing.

Fig. 3-18. These bricks will appear as all stretchers (laid lengthwise) but this is not a brick veneer wall. The bricks and blocks are tied together into a composite by the masonry truss, visible at the top of the wall.

Brick veneer walls are laid up from the foundation in one wythe (one thickness of brick in a wall). Such a wall, if not attached to something, is very unstable because it is thin. In a veneered wall, galvanized steel anchors are nailed to the studs. (Nailing anchors to the sheathing is a poor work practice.) The anchors are bent at a right angle and embedded in the mortar between two brick courses (or layers). If there is a fire in the wall, the nails may detach from the studs due to pyrolytic decomposition. The resultant free standing wall is unstable. It is not stabilized by the compressive load of the structure as is a bearing wall. On the other hand, its collapse will not affect the stability of the structure. The collapse of a veneered wall of an upper story may be an impact load to a first-floor extension or an adjacent structure.

Until a few years ago, it was easy to identify a brick-veneered wall compared to a solid masonry wall. All the bricks were laid as stretchers (laid lengthwise). There were no headers (laid across the wall with the ends showing). In a bearing wall, headers can be seen, usually every seventh row. More recently, the use of metal truss masonry reinforcement has made it possible to omit the headers from a solid masonry wall. If the brick wall is all stretchers, suspect brick veneer, but you cannot be certain this is the case.

Many buildings combine brick veneer and solid brick masonry. Bearing walls may be solid masonry, non-bearing walls may be veneer. Lower floors may be solid and upper floors veneer. This cannot be detected after the building is finished, and points to the necessity for recording information about buildings as they are built.

An old wooden house in New York had been rehabilitated by the addition of a brick veneer wall. It was heavily damaged by fire. During overhaul, an interior collapse knocked down the brick wall. A lieutenant was killed while pushing an injured probationary fire fighter out of harm's way.

Tactical Considerations

As noted, an interior collapse can bring down a veneer wall with fatal results.

If fire involves the sheathing and studs behind the brick veneer, there is no point in extraordinary and dangerous exertions to save the veneer. It must be removed to resheath the house and to reattach the veneer.

Whether the wall is a veneer or a structural wall is immaterial to the fire fighter struck with falling bricks. Some place too much reliance in newer helmets. No matter how strong the helmet, it is worn on a human body of limited impact resistance. The wearer of the super helmet may find that the frontpiece becomes a belt buckle!

Brick Veneer on Non-wooden Buildings

Brick veneer is also applied to buildings other than wood. Some prefabricated metal buildings use brick veneer to improve their appearance. Brick can also be veneered onto concrete block. This is not the same as a composite brick and block wall in which both elements react to the load together.

Stone Veneer

Natural or artificial stone and cast concrete — Permastone® is one trade name — are also used as veneer. Stone can be freely substituted for brick in the discussions above.

Wood Shingle Roofing

Wood shingles or shakes are split pieces of wood used for roofing or siding. Shakes are larger than shingles. Some of the greatest fire disasters in history have been due to the spread of fire by wood shingle roofs.

If consideration of a fire is limited to the problem of the building in which the fire originated, the wood shingle roof might be a more desirable material because by burning through, it vents the fire. However, the conflagration hazard presented by wood shingles should be the dominant consideration.

In recent years, wood shingles and shakes have made a strong comeback. It seems as if each generation must learn anew. In 1959, the NFPA warned that a major conflagration risk was present where large areas of wood shingle roofs existed. The Los Angeles conflagration of 1961 proved the prediction to be accurate. There are many areas now where the majority of houses have wood-shingled roofs. In some areas, they are permitted wherever frame buildings are permitted.

Fast fire department response, one-story buildings, wider spacing between buildings than in bygone years, and the fact that we have no extensive amount of shingles on closely spaced 2½-story buildings which existed 50 years ago have combined to keep the conflagration rate low. But given the coincidence of hot dry weather, brush fires that engage a large portion of available fire fighting forces, high winds, and a hot fire in a wood-shingled structure, the threat of conflagration is still very great.

On the afternoon of July 31, 1979, the Houston City Council was considering an ordinance with minor restrictions on the use of wood shingles. The ordinance was tabled. The day was hot and dry. At about the same time, the fire department responded to an alarm for a fire at the Woodway Square Apartments. By that evening, 30 apartment buildings were destroyed, hundreds of people were left homeless, and an estimated $44 million in damage was incurred. The fire spread because of the wood shingle roofs. Ironically, the owner of the apartments was in the process of replacing the shingle roofs. The next day, the shingle ordinance passed.

In the early evening of March 14, 1988, residents of apartments in Davis, California also were the victims of a fire spread by wood shingles. Nine of 18 buildings in the complex were damaged. Over 100 University of California students lost personal property. Damage was in excess of $1.5 million.[24, 25]

Wood shingle proponents stress that brush fires are more responsible for conflagrations than shingles. The Davis and Woodway Square fires, and many other fires, did not involve a brush fire. They were purely structural fires.

Testing laboratories rate wood shingles in accordance to NFPA 256, *Standard Methods of Fire Test of Roof Coverings*. Elements considered include flame exposure, spread of flame, and resistance to burning brands or flying pieces of burning wood. Roofing materials are classified "A," "B," and "C." The least fire resistant is "C" rated. In addition, for wood shingles, a flying brand test is required to determine if the roofing will produce brands. Wood shingles that meet Class "A" "B" or "C" standards are available.

For many years some asphalt shingles have carried Class C ratings. Asphalt shingles burn and generate dense black smoke, but conflagration hazard, not the combustibility of an individual roof, is the important consideration. One of the complaints against Class C asphalt shingles is their appearance. Recently, several suppliers started offering Class A glass fiber reinforced shingles, which are very attractive.

Know the building code provisions governing wood shingles in your area. If rated shingles are required, bear in mind that only bundles of shingles are labeled, not individual shingles. You can't tell treated from untreated shingles by looking at them.

In July, 1989 the Los Angeles City Council passed, by an 11-1 vote, an ordinance banning new wood shingle roofs. Wood shingle roofs can be repaired if not more than 10 percent is damaged.[26]

Tactical Considerations

In wood shingle areas prepare for a blitz attack to knock down the original fire quickly. Call additional units for downwind patrol immediately. Don't wait for confirmed extension.

Row Frame Buildings

In older parts of many cities, frame buildings were often erected in rows. These structures are contiguous and often have a common attic or cockloft, and may even have party walls which provide support to both buildings. In older construction, there may be no fire barrier between buildings. Crude attempts at making a fire barrier by using brick nogging (brick and mortar filling between studs) have been made in row houses. The brick nogging does not cut the floor voids or the cockloft. Brick or stone nogging is also found in old individual houses. It served as a heat sink for winter warmth. This presents an additional hazard in collapse.

More recently, row houses have been given a dignified new name, town house. In any case, unless there is an adequate masonry fire wall between the separate buildings all the way through the roof, the entire structure is all one building and should be so described in stating the size for planning or fire report purposes. It doesn't matter to the fire that these houses are separately described and owned on the land records. As long as fire can move from unit to unit, totally or partially unimpeded, the structure is one building.

Fig. 3-19. Brick noggin is shown in this farm house. It served as a heat sink.

In November 1990, Los Angeles County fire fighters fought a serious fire in closely spaced wooden buildings on a movie set. The structures were the wooden false fronts of many buildings "acting as one building."

Tactical Considerations

The change in description may be of real value. Consider a row of ten frame houses, each 20 feet by 40 feet. The row of houses is actually one building 200 feet by 40 feet. A fire involving one house might require a greater alarm, and it is much easier for the fireground commander to justify sending a greater alarm for a building 200 feet long rather than one described as 20 feet wide.

In any area where there are closely spaced wooden buildings, the prime tactical consideration must be to confine the fire to the building of origin. This makes it mandatory that units be prepared to place heavy-caliber streams in operation immediately upon arrival. Preconnected small lines serve very well in the majority of situations, but the really competent fire officer is one who can recognize a big fire and respond to it immediately with big fire tactics, not by working upward through small fire tactics because of bad habits.

Since many fire departments lack the personnel to handle big handlines, there is a clear need for equipping apparatus with

Fig. 3-20. Closely spaced frame buildings demand fast work to prevent extension. (W. Noonan, Boston Fire Dept.)

Tactical Considerations, continued

preconnected deluge guns, so as to combine heavy-caliber streams with minimal personnel requirements. Today all pumpers can deliver far more water than responding personnel can deliver through hand lines.

The fire department in Henderson, Nevada reported the use of the "bombline," a portable monitor with an automatic nozzle,

Tactical Considerations, continued

connected to a rear discharge with 200 feet of lightweight 3-inch hose. It is deployed in 90 seconds by two fire fighters and can be operated by one, and be left unattended if necessary. It can deliver from 150 to 1000 gpm.[27]

The use of lightweight deckguns by the Wilmington, Delaware fire department in high-rise situations is noted in Chapter 11, High-Rise Construction. These would also be useful for a fast-moving fire in wooden buildings.

Wooden Cooling Towers

Cooling towers are used to disperse the heat from air-conditioning systems. When the tower is functioning, cascades of water pour down the inside. Despite this, much of the wood in the tower remains dry. If the tower is not operating, all the wood is dry. Electrical short circuits, fan-bearing friction, or even birds carrying lighted cigarettes to their nests in the towers are few of the possible fire causes. Such a fire can be spectacular and disastrous.

Many cooling towers are equipped with specially designed sprinklers, particularly where interruption of cooling would be costly. Other towers are built of noncombustible materials. When erected as part of a modern high-rise building, the tower is often concealed by masonry walls or constructed in a well, accessible only from the roof. Wooden water tanks have also been known to burn. One such fire was reported as a "fire in the fence around the water on the roof."

A fire in Toronto, Canada illustrates the vulnerability of wooden cooling towers. A fire in a trash disposal container was fanned by high winds. Embers were scattered for several blocks. A third of a mile away, a wooden cooling tower atop a 26-story building was ignited and virtually destroyed.

In March 1991, a spectacular fire consumed a wooden cooling tower atop a 43-story highrise in midtown New York at the Olympia York Building.

Heavy Timber Construction

Much ingenuity has gone into the development of heavy timber construction capable of sustaining massive loads or spanning great distances. Structures of wooden interior construction, with masonry walls, are classified as ordinary construction. A special type of heavy timber and masonry construction, which is designed to control ease of ignition and eliminate void spaces, is called mill construction. Much heavy timber construction does not meet the standards of true mill construction (See Chapter 4, *Ordinary Construction*).

A typical description of the advantages of heavy wooden construction is: ". . . large sections of wood, over six inches thick, can actually prove to be more resistant to fire than exposed steel. Although steel does not burn, fire temperatures in excess of approximately 1000°F cause steel to soften and bend like a pretzel, whereas the surface of the heavy wood merely chars, limiting further burning."[28]

It is important to note that the fact that a particular material may burn slowly is much less important than how the structure is connected. All loads must be delivered to the ground; the entire path must be examined. Study the connections.

Earlier in this chapter it was noted that heavy laminated arches fell apart at steel connections in a sports arena fire in Florida. The insurance company for the owner sued the firm which applied the foam plastic roof, on the grounds that the roof had destroyed the heavy laminated timber building. The suit was dropped when the plaintiff learned that the defense had pictures showing how the building had fallen apart.

Don't fall for propaganda; look at the connections.

Slow Burning?

Proponents of heavy timber construction often advance the term "slow-burning" as an unqualified advantage because of collapse-resistance. In fact, the slow-burning characteristic is an advantage only as long as the fire department can maintain interior offensive operations. Once the fire department must operate defensively, slow burning delays extinguishment and prolongs air pollution.

Advantages and Disadvantages

Each building material has advantages and disadvantages. An advantage which might limit loss might make fire fighting operations more hazardous. Proponents of one material, such as trade associations, advance their own interests.

Wood construction has many merits. Properly protected with automatic sprinklers, or subdivided into manageable fire areas by fire walls,

Fig. 3-21. This fully involved heavy timber building threatened midtown New York. It collapsed onto the adjacent quarters of Rescue Company One. The sprinklers had been turned off. (Harvey Eisner)

wood construction can hold its own among various materials. An article in *The Forest Products Journal* includes a good discussion of the merits of timber as opposed to other building materials; the author points out the necessity for sprinkler protection, since no building material can detect or extinguish a fire.[29]

It is a bit discouraging, therefore, to find in a recent article describing the newly-built laminated timber warehouse of a lumber company that "in order to comply with local fire regulations, the building will be equipped with a sprinkler system." Thus, again, the myth is perpetrated that fire protection requirements are a burden imposed by local authorities, and by inference, not really necessary.

Use of Unprotected Steel

The substitution of unprotected steel for heavy timber construction, or the use of steel to reinforce aging or overloaded timber structures, should be noted. Such steel can fail early and trap fire fighters who believe they are operating in a heavy timber building. Fire departments should study alteration permits and investigate when building materials are delivered to a completed building.

Beams are often penetrated for electrical conduit, and in older buildings for gas pipes. Long bolts and nuts are used to attach timbers to other members. Any of this metal can provide a path for heat to reach the interior. Destruction of wood by pyrolytic decomposition may cause failure. Long solid timbers are hard to obtain. Watch for spliced timbers with overlapping joints and metal connectors. These would be subject to early failure at the joint.

The Methodist Church of Gibson City, Ilinois is a building of masonry exterior walls and interior timber framing. It became necessary to remove a timber column carrying an 18-ton load. A one-story-high steel truss of steel tubes and solid steel tie bars was inserted in the building to permit the removal of the column. A description of this engineering feat was silent as to any fire protection provided to the steel.[30] Unless the steel was protected, the truss could fail in a fire prior to the time the wooden column might have failed.

Many wooden buildings are tied together with steel tension rods or cold-drawn steel cables (which fail at approximately 800°F). Some were built with the ties. Other times the ties were installed because the structure was failing. A Paterson, New Jersey fire officer was killed when steel ties failed in a church fire, causing the wall to fall. Many wooden buildings built in recent years have unprotected steel beams located within the structure. The failure of such a beam could cause a collapse.

Rehabilitation and Demolition Hazard

Full sprinkler protection, adequately maintained, is the only fire protection measure which can reasonably be expected to prevent a disaster in a heavy timber building.

Problems arise when buildings are abandoned, or converted to multiple occupancy, or are being prepared for rehabilitation into apartment or office occupancies.

The fire department should anticipate these problems and plan accordingly. The political authorities are often inclined to "go easy" on a developer who promises to provide employment or housing or whatever is perceived as good. The cost of sprinklers may be perceived as a stumbling block to the project.

Vacant Buildings

Baltimore, Philadelphia, Lynn (Massachusetts), Minneapolis, Indianapolis and Montreal are some of the cities which have suffered massive downtown fires in old combustible interior masonry buildings which were being remodeled. In these types of fires, the buildings were usually wide open with fire barriers removed or inoperable. Sprinkler systems had been removed or disabled and the fire had possession of the building upon fire department arrival. The only fuel was the timber interior structure of the building. Multi-million dollar damages also had been sustained by nearby structures. Many valuable assets and plans for downtown rehabilitation have gone up in smoke.

In the Minneapolis fire, floors six through sixteen of an adjacent fire-resistive high-rise office building were gutted when exposed to a fire in a department store that was being demolished. The loss was almost $100 million.[31]

Similar problems are presented by huge structures, not planned for rehabilitation but simply vacant because of economic changes. Sprinklers are often turned off to prevent freezing, and the structures, often occupied by vagrants who kindle cooking and warming fires, are an open invitation to arson. Three Detroit fire fighters died in a spectacular fire in a vacant building. Detroit had appropriated over $300,000 only days before the fire to tear down the building.[32]

Full automatic sprinkler protection is absolutely necessary to prevent the heavy timber building from becoming a spectacular fire. Some such buildings have been recycled for other uses and the sprinklers eliminated. Spectacular total loss fires are a credible risk.

Tactical Considerations

No fire department would leave a leaking gasoline tank truck on a downtown street. Vacant buildings may be far more hazardous than the tank truck. They demand more than routine attention. A control system should assure that a disaster will not occur.[33]

Fire departments should be well prepared. Articles and videotapes of fires like these and others should be on hand. It is obvious from the pictures that such fires are beyond the suppression capability of even the strongest fire department. Existing sprinkler protection should be maintained floor by floor as the building is demolished. If the loss of sprinkler protection is unavoidable, arrangements should be made for a continuous fire protection presence, possibly including a pumper with lines laid. The cost of this may be chargable to the owner.

Open Area Structures

Many large open-area structures, such as churches and arenas, are built of wood, or have exposed wood plank roof which also comprises the ceiling. Such a valuable, building should be fully sprinklered.

The objection to sprinkler piping in an open decorative wood structure is understandable. The exposed piping destroys the decor. However, in some cases, the sprinkler piping is run close to beams, and sidewall sprinklers are used. A competent painter, or a grainer charged with making the pipes less apparent, can work wonders.

If a building is not to be fully sprinklered, try to keep out the kindling—the minor light combustible structure or elements within the building which can ignite the whole building. If a structure, such as a robing room, sacristy, ticket office, hot dog stand, etc., is to be built within such a building, urge that it be built using noncombustible material.

This author has recommended sprinkler protection, even if only a couple of sprinklers taken off the domestic system are installed to protect small enclosures in heavy-timbered buildings. Don't call it partial sprinkler protection, call it a super extinguisher. If there is no "kindling" to set it on fire, the open-area timber structure should be a good bet to stand a long time.

The structure at Busch Gardens at Tampa, Florida, in which trained primates perform, is a wooden dome. It is not sprinklered but the dressing room within, which might otherwise serve as kindling, is sprinklered.

Tactical Considerations

Fire fighting plans should provide for heavy caliber SOLID streams with adequate reach to cover the whole inside surface of such structures. The objective is to wet the entire wooden surface. Fog streams would be eaten up by the fire before the surface is wetted.

Operating points should be determined in advance, and dry-run drills should be undertaken. A deluge gun set inside the building (pickup truck-mounted if access permits) with a relatively small solid-stream nozzle, to guarantee reach of the stream to all points as fast as possible, might be useful. Speed in getting all the interior surfaces wet might be the difference between a refinishing or a rebuilding job.

To sum up, heavy timber structural elements may provide a fair measure of structural stability during a fire if the connectors and any

Tactical Considerations, continued

supporting elements are equal to the timber in fire resistance. If not, as is usually the case, anticipate collapse due to failure of the weakest element. This is particularly true of timber trusses.

In a Florida country club, huge wooden parallel-chord trusses are featured in the dining room. The fire department has noticed that the trusses are supported on an unprotected steel girder. They are aware that the girder may rotate when elongating and drop the trusses, or simply collapse. Any serious fire in such a structure should be planned as a defensive operation—as the building is "designed to collapse."

Expect a massive fire if there are concealed spaces that cannot be reached with hose streams; there are other obstructions to manual fire fighting efforts; if the contents' fire load has a fast, high-heat-release potential; and the building is unsprinklered. Given this massive fire, heavy water supplies will be required.

A key point of the prefire plan is to determine how the water is to be applied in sufficient quantity to stop the production of heat.

Imitation Timber

Styles in interior decor seem to run in cycles. In contrast to the modern interiors popular a few years back, many current styles feature wood. Very often wood is made to appear structural when it is not. Often it is not even wood. Few if any of these practices are effectively regulated by code. Some of these deceptions should be noted:

- Unprotected steel beams or columns boxed in wood to look like wooden structural members
- Unprotected steel encased in plaster surfaced to look like wood
- False wood beams which are actually hollow wooden boxes. (The support system should be examined. Decorative beams have been found suspended from aluminum wire. In one case, they were just glued to the ceiling. The temperature up at the ceiling can be hundreds of degrees higher than it is at the floor level.)
- Polyurethane imitation wood beams and fittings such as brackets. (These are readily ignitable and burn furiously. Such imitations are often found in conjunction with plywood veneers. The hazards include the sometimes massive fire load overhead, and the potential for collapse.)

It is becoming fashionable to make a masonry building appear to be of wood construction. The Teton Lodge, in Teton National Park,

Wyoming is made of concrete. The wood formwork marks were left evident and the entire building stained brown. Much of the chief interior finish is plywood.

Brick or block buildings have been covered with a wood veneer which is mounted on furring strips, making it easy for the wood to burn on both sides. In one case, a concrete block store was covered with shingles stapled to plywood sheets and mounted on vertical wood trusses. The shingles were said to be treated but apparently the plywood and wood trusses were not.

Wooden Suspended Ceilings

The popularity of wood for interior decor has led to the development of suspended wooden ceilings, often used in otherwise fire-resistant buildings. These can range from a few timbers, suspended from the ceiling of a sporting goods department, to sturdy assemblies providing a large suspended load. In some cases, such false ceilings interfere with the coverage of automatic sprinklers. This situation can be remedied by open work ceiling or by installing sprinklers above and below the ceiling.

The auditorium of a major national association located in an unsprinklered fire-resistive building in Washington, D.C., is decorated with 2-inch by 10-inch planks hanging from the ceiling. This could give fire fighters a fatal surprise.

The fire resistance of the supporting system of an extensive, heavy suspended ceiling is worth investigating. It would not be surprising to find it hung from aluminum wire, or supported in some other equally fire-vulnerable way. It might be important to inquire about the local code provisions governing the fire protection of supports for suspended structures. Compliance with flame-spread requirements is not sufficient. The supports are up high where heat concentrates. The fact that the suspended ceiling may be about to fall may not be apparent to fire fighters operating below.

Strength of Wood

Lumber is a natural product, and two pieces of lumber of similar size can vary greatly in strength. This may be apparent to the skilled eye because of defects such as a large or loose knot or other weaknesses, or it may be determined only by testing.

In February, 1967, the trussed roof of St. Rose of Lima Church in Baltimore collapsed and injured 48 worshippers. The *Washington Post* reported that a consulting structural engineer found that a single two by four of inferior grade had failed, triggering collapse of the entire roof. The newspaper purchased samples of wood and had them tested for strength. It then reported that the wood should have had a modulus of elasticity of 1,760,000 pounds per square inch. Instead, the samples had values of 410,000 and 1,240,000 pounds per square inch. One half of the samples tested graded above the required figure and one half below.[34] In some cases, averaging values may be satisfactory, but not where failure of any single unit can precipitate a general collapse. Bear in mind the sad fate of the man who drowned in a river, the average depth of which was only six inches.

The Rising Truss

The housing industry has seen hundreds of cases of the rising truss. It has been reported that metal-plate truss roof systems are bowing upward, causing separation of ceilings from walls. It appears that the cause is the use of juvenile wood, cut from trees harvested from tree plantations where fast growth is the objective.[35]

> ***Tactical Considerations***
>
> *Fire department personnel should make it their business to be aware of what is going on in their community with respect to buildings. When there is evidence of deficiencies, the effect on firefighing operations and fire fighter safety should be analyzed, and where appropriate worked into pre-fire plans.*

Wood Dimensions

As explained earlier, actual lumber sizes are less than nominal due to planing. In recent years, the lumber industry has reduced the actual sizes another seven percent (2-inch lumber, formerly 1⅝ inches thick, is now 1½ inches thick). This, of course, reduces the fire fighter's margin of safety further, as there is less wood to burn away before the fire reaches the wood that is providing the designed structural strength.

Making Wood Construction Safe

Wood is a uniquely renewable resource. It has served humanity for ages and will continue to do so. Unfortunately it is combustible. The problem of its combustibility can be dealt with only by *complete* automatic sprinkler protection, properly designed with adequate water supplies, and competently maintained. *Structures with unsprinklered floor or attic voids do not have complete sprinkler protection.*

Given this protection, a wooden structure is more superior from a life safety point of view than many unprotected fire-resistive structures. A case in point — visitors to Teton and Yellowstone Parks may see two hotels. One is of concrete but it was lined with plywood when last seen. There were no sprinklers. Another is an old wooden hotel but is fully sprinklered. This author would much rather sleep in the sprinklered wooden hotel.

References

The addresses of all referenced publications are contained in the Appendix.

1. Watkins, T. "Spanish Tiles Weigh Heavily in Roof Collapse." *Fire Command*, July 1985, p. 33.
2. "Tragedy Knows No Boundary." *Firehouse*, May 1989. Also "Orange County Fatal Fire." *Fire Engineering*, July 1989.
3. Curtis, Don. "Blitzing." *Fire Engineering*, July 1985, p. 19.
4. Obermayer, R. "The Dwelling Fire Mindset." *Fire Engineering*, March 1991, p. 154.
5. de Boer, C. "Trinity Church Good for another 250 years." *Modern Steel Construction*, Vol. XXIX, No. 1, Jan-Feb 1989.
6. Porter, W., and Schofield, B. "Making a Landmark Victorian College Safe." *Fire Journal*, Vol. 70, No. 1, July 1985, p. 32.
7. Ellison D. C., Huntington, W. C., Mickadeit, R. E. "*Building Construction*, Sixth Edition," John Wiley & Sons, N.Y., 1978, p. 264.
8. Mittendorf, J. "Victorians in the '90s." *Fire Engineering*, April 1990, p. 41.
9. Courtney, N. "Fire Record - General Office Building," *Fire Journal*, Vol. 84, No. 4, July/August 1990, p. 20.
10. Loeb, Donald. *Fire Chief*, Vol. 24, No. 2, February 1980 (et seg.) An excellent series of articles starting in the February 1980 issue.
11. "Truss Framed Construction." NAHB Research Foundation Inc., (P.O. Box 1627, Rockville, MD 20850). (No date given)
12. Watrous, L.D. "High-Rise Fire in New Orleans." *Fire Journal*, Vol. 67, No. 3, May 1973.
13. Meradows, M. "One the Job, California." *Firehouse*, June 1990, p. 29.
14. "Improved Fire Safety: Design of Firestopping and Draftstopping for Concealed Spaces." National Forest Products Association.
15. NFPA, *Fire Protection Handbook*, 17th Edition, July 1990, Section 6, Chapter 11, pp. 6-132 and 6-133.
16. Sylvia, R. "New Type of Construction." *Fire Engineering*, March 1981.

17. "Razing the Roofs." *Washington Post*, 9/16/90 Business Section, p. H1.

18. Parkin, L. "FRT Plywood, Not As Safe As It Sounds." *Fire Engineering*, May 1990, p. 27.

19. "Treated wood Linked to Deterioration in Roofs." *Fire Command*, August 1989, p. 12.

20. Eickner, H. W., and Schaeffer, E. L. "Fire Retardant Effects of Individual Chemicals on Douglas Fir Plywood." *Fire Technology*, Vol. 3, No. 2, May 1967, pp. 90-104.

21. Clark, A. "Wooden I-Beams and Laminated Beams." *Fire Command*, July 1984, p. 34. In the October 1984 issue, *Fire Command* published a number of letters about Clark's article.

22. "Trusses Suspect in Collapse." *Engineering News Record*, Sept. 10, 1987.

23. Stone, W. "The National Electrical Code and You." *Fire Journal*, Vol. 64, No. 3, May 1970, p. 33.

24. Eisner, M. "Untreated Wood-Shingle Group Fire, Davis, California," *NFPA Investigation Report*, March 18, 1988.

25. Isner, M. "Untreated Wood Shingles Spread Apartment Fire." *Fire Journal*, March/April 1989, Vol. 83, No. 2, p. 83.

26. See "An NFPA Perspective on Untreated Wood Shingle Roofs." *Fire Journal*, April/May 1989, p. 83.

27. Shapiro, Paul. "Bombline," *American Fire Journal*, November 1969, p. 34.

28. Zuk, W. "Concepts of Structures." Litton Educational Publishing, Inc. By permission of Reinhold Publishing Co., 1963, p. 12.

29. Thompson, H. "The Fire Performance of Timber." *Forest Products Journal*, Vol. VIII, No. 4, pp 31a-34a.

30. Gurfunkel, G. "Built in Place Truss Aids Remodeling Job." *Engineering News Record*, Feb. 23, 1967, p. 25.

31. Best, J. "Demolition Exposure Fire." *Fire Journal*, Vol. 77, No. 4, July 1983.

32. Shinske, R., Reardon, J. "Detroit's Fatal Warehouse Fire." *Fire Engineering*, June 1987. See also Eisner, H. "Trapped." *Firehouse*, June 87.

33. NFPA 241, *Standard for Safeguarding Construction, Alteration, and Demolition Operations* (1989)

34. "Lumber Misgrading, Builders Beware." *Washington Post*, November 19, 1968, p. 1.

35. Senft, J., Bendtsen, B. A., Galligan W. L. "Weak Wood." *Journal of Forestry*, Vol. 83, No. 8, August 1985.

Additional Readings

Case Histories "Fire Fighter Casualties as a Result of Roof or Floor Collapses in Wood Frame Buildings" contains the stories of 38 such fires. Contact NFPA's One Stop Data Shop for information.

The following articles from the Fire Service press contain additional valuable information. The citations given below sometimes include this author's comments.

Best, R. "Fire in Community Home Causes Seven Deaths." *Fire Journal*, Vol. 78, No. 2, March 1984, p.19. Combustible interior finish and an open stairway were principal factors in the loss of life. Also "Recent Important Fires," p.91. Wood shingle roofs caused the loss of 15 homes in Oklahoma.

Carlson, G. "Residential Attic Fires." *Fire Engineering*, February 1987, p.8. A good discussion of tactics on attic fires. Unfortunately, the author does not discuss the hazards of attempting to ventilate the common light-weight trussed roof.

Curtis, Don. "Blitzing." *Fire Engineering*, June 1985, p. 19. An excellent explanation of the blitz attack. The blitz attack direct from the tank is an excellent tool for controlling fast-spreading fires in closely spaced, wooden structures.

Dunn, Vincent. "Floor Collapse in Residential Structures." *Firehouse*, October 1984, p. 40. An excellent discussion of the hazards of interior collapse, especially around kitchens and bathrooms. In discussing the handling of a building after heavy stream attack, Dunn limits the discussion to fireground tactics.

Dunn, Vincent. "Hazards of Ceiling Collapse." *Firehouse*, December 1984, p. 26. Chief Dunn makes the good point that the death of a fire fighter caught in a collapse is often listed by statisticians as suffocation or burns, whereas collapse of part of the structure is the real cause.

Eisner, H. "Boston Firefighter Killed by Collapse." *Firehouse*, June 1986, p. 90. On a bitter cold windy morning, a three-story house was found fully involved. The fire rapidly spread to adjoining and neighboring houses. Twenty minutes into the alarm the original structure collapsed. It knocked down a tree which killed a fire fighter.

When a wooden building is fully involved, the structure which resists gravity is fast dissolving. Also note the problem of closely spaced wooden buildings. Lots of help is needed fast in this type of scenario. The sixth alarm was ordered only 11 minutes after the initial box alarm. The small diameter hand line is fine for the majority of fires, but 25 or 250 small fires provide no data base for handling a potential mass fire. Realistic preplanning must fill the gaps in the limited experience of many officers.

Featherstone, J.J. III. "The Problems With Converted Buildings." *Fire Engineering*, February 1984, p.14. An old school is converted into flats for elderly people. A severe potential life hazard is created, as the apartments converted from classrooms are located on the second floor. An open stairway leads from the first floor where the kitchen and recreation room, both likely fire origination spots, are located.

"Granny Flats" are permitted in existing one-family homes. A code provides that there must be no change in the exterior appearance of the house, thus there will be no clue as to the occupancy. Know your buildings!

Fire Engineering, August 1984. The cover photograph of a Syracuse, New York house fire shows excellent coverage of exposures.

Gardner, Jack. "Fatal Flash." *Firehouse*, September 1985, p 127. Three fire fighters were injured in the second of two flashovers discussed. Fire fighters were on the second floor of a house with fire in the attic when the flashover occurred.

It would be useful to have more specifics about the structure and interior trim. My best surmise is that the fire on the first floor reached the attic through wall voids, bypassing the second floor. It burst downward when a portion of the ceiling collapsed. The pictures show no evidence of ladders to the second floor to provide a secondary exit for the search crew. It cannot be repeated too often, "Heavy fire can be above your head with little or no smoke showing below." On page 130 there is a spectacular picture of two fire fighters jumping from the building.

Harvey, H. "Working Fire." *Firefighting in Canada*, June 1991, p. 15. A Canadian fire director forcefully argues the case for a thorough job of opening up the interior of the fire building to discover and cut off the fire.

Hoffman, J. "Innovative Firefighting Procedure Restricts Danger and Damage." *Firehouse*, Jan. 1991, p. 56. Lee's Summit, Missouri fire fighters have developed a new approach to shingle roof fires. Exposures are covered, but no attack is made on the burning roof. Contents are protected with salvage covers. The roof is allowed to burn off. The concept is that the contents are more valuable than the roof. I would caution that this procedure should be restricted to fires determined to involve only the shingles. If the attic structure is on fire, the potential for collapse of the roof, trussed or not, should govern interior operations.

Howard, F. "Vertical Foam Insulation." *Fire Engineering*, May 1987, p.10. A good description of the serious effects of adding foamed plastic to the sides of wooden buildings before installing new siding. Such siding is often laid over the end of fire walls which have been built to inadequate code provisions . . . "fire wall extends to exterior wall." The fire can pass right around the end of the wall.

Kyte, G. "Tragedy Strikes Biloxi." *Fire Command*, May 1987, p. 18. Two fire fighters died in an old house converted into seven small apartments. Fire was in concealed, interconnected voids. "No smoke showing" was the first report. The fire was reported at 2:55 A.M; it spread into the habitable spaces at about 3:38 A.M. The victims were not immediately missed.

Fire lurking in void spaces is like a tiger perched in a tree above a trail. It is hard for many fire fighters to realize the amount of fire which may be concealed, or the speed with which the fire can grow when a void is opened up, by either the fire or the fire department. "What you see ain't necessarily all you've got." Many fire departments need improved discipline on the fireground. The British Fire College teaches procedures for keeping track of every fire fighter who enters a fire structure.

Lewis, Bart. "A $70,000 Car Fire." *Firehouse*, August 1985, p. 46. Finding a house fully involved on arrival, the fire department successfully protected the exposures. The author justifies the strategy used.

It is a sad fact that an article like this is necessary. Some fire departments still concentrate on where the fire was, rather than where it is going. The first fire suppression principle is to confine the fire to the area involved on arrival. Every conflagration starts with extension to the next structure. Heroic efforts to save half-burned (and thus all destroyed) structural elements may give satisfaction to fire fighters, but cannot withstand rational analysis. Every fire department should be prepared to go to heavy caliber cutoff tactics on arrival. One mark of a good fire officer is the ability to immediately recognize a big fire.

The hard dollar costs of fire fighters killed in fires is so high that consideration of this problem should be policy rather than tactics. Champaign, Illinois has a structural engineer on a retainer. When called, the engineer responds to the fire and examines the safety of the building. Could not a fire department place this cost on the owner of the building? "The city is safe. You hire an expert. We will give him a lift in a bucket to give us an opinion as to the stability of the structure."

McFadden, P. "Residence Fires Starting in the Cellar." *Fire Engineering*, May 1987, p.8. The author describes the difficulty of cellar fires. In most codes a cellar is defined as having half its height below the grade of the street, a basement has at least half its height above the grade of the street. Usually basements but not cellars can be used for living space. From the suppression point of view the distinction may be academic. Ask the owner where the gasoline for the lawnmower is kept. If it is in the cellar, react as if it were "Dynamite."

High-expansion foam is sometimes very effective on a cellar fire — if it can be confined. Fumes from some burning plastics will destroy the foam structure. The fire may be pushed up into the joist spaces and continue to burn. Be aware of the potential for fire extending up into the voids in the house above. This is almost a certainty in older homes of multi-story balloon-frame construction.

Meade, J. "Fire Rips Through California Complex." *Firehouse*, February 1984, p.51. Several closely-spaced wood homes with wood shingle roofs were damaged or destroyed in a fire started by a plumber's torch. Note the use of a deck gun immediately to stop the fire's extension.

Pennachi, G. "The Fire at Oceanview." *Fire Command*, April 1985, p. 30. The story of a fire in a huge wooden l9th-century mansion that was stopped before a total loss occurred.

Perry, D. "Wildland-Urban Fires." *American Fire Journal*, April 1986, p. 14. Discusses the hazard of wood shakes (shingles) and wood construction generally in a "keep it wild" environment.

Phelps, B., and McDonald, E. "Model Incident Command." *Fire Engineering*, June 1985, p. 25. Note the emphasis on safety in overhauling fire-weakened buildings.

Taafe, M. "Firefighter Beware." *WNYF*, (First Issue) 1987, p. 12. This article covers the hazards of operations in residential buildings with burglar-proof barred windows.

Wilson, Charles. "Fire Rips Through Unsprinklered Building." *Firehouse*, March 1985, p. 63. Typical fire in a large old-frame building. The sprinkler system in this old school was removed when it was converted into apartments. Void spaces, balloon-frame construction, and a fire wall which failed when part of the structure collapsed, contributed to the loss.

4

Ordinary Construction

There have been some comments that this chapter is concerned with hazards which exist primarily "back east." It is true that many of the examples cited have occurred in older eastern cities, but many of the best pictures which illustrate these examples were taken west of the continental divide. Those who built the structures of the east migrated westward and exercised their talents from Seattle to San Diego. The fact that a fire department has only a few of the buildings described in this chapter is one reason to learn from the bad experiences of others, so that they may avoid these problems themselves.

In addition, lateral transfer is becoming more common among the highest positions in the fire service. An ambitious junior fire officer cannot afford to neglect any aspect of this subject simply because that problem doesn't exist in his or her current department.

This chapter should be studied in conjunction with the chapters on every other type of construction. A wide range of construction characteristics—void spaces, balloon frames, tile arch floors, marble stairs, timbers, unprotected steel, plain and reinforced concrete—may be found in ordinary and other types of construction. Some elements may masquerade as being fire-resistive.

Tactical Considerations

More fire fighters have been killed in fires in ordinary construction buildings than in any other type. Fire officers should become more analytical of the hazards of this type of structure and rely less on experience which may not be adequate.

Tactical Considerations, continued

Under pledges of anonymity, this author often hears stories which reflect on the competence of senior officers, particularly those who have not kept up with new tactics but still respond to serious fires and take command. The stories of several disasters have a common thread. The senior officer, arriving after the fire was already in progress, insisted on tactics appropriate for first-arriving units at a fire which have not had a chance to get a grip on the building.

When a fire chief's duties as administrator allow no time to study, it is time to hang up the helmet, and become "fire administrator," and let competent tactically up-to-date subordinates fight the fires.

Construction Characteristics

The term "ordinary construction" describes an almost infinite variety of buildings. In simpler days it was generally known as brick- and wood-joisted construction. Today this construction uses a variety of walls and interiors.[1]

The chief common characteristic of ordinary construction is that the exterior walls are made of masonry. However, a few cast-iron-front buildings still stand.

25' ABOUT MAX

Fig. 4-1. A 25-foot span is about the maximum possible with wood joists. Shown here is the basic brick and wood joisted building.

The incorrect spelling or pronunciation of words such as masonry — (not masonARY), lintel (not LENtil), and spalling (not spallDING) — reflects on the competence of fire professionals, and underscores the need for accuracy in usage and pronunciation. Proper use of technical words or trade jargon help to bolster the perception of technical knowledge and authority. A fire service magazine ran an article on this subject.[2]

Ordinary construction is "Main Street, USA." When the Disney organization wanted to build their version of Main Street they chose the appearance of typical 19th century brick-front commercial buildings. The actual structures in the Disney parks are imitation brick on steel frame, fully sprinklered.

Masonry walls may consist of brick, stone, concrete block, terra cotta tile, adobe, precast or cast-in-place concrete. The wall may be all of one material, different materials may be used in different areas, or different materials may be combined to form composite construction.

Most building codes have a provision for so-called **fire limits**. Within the fire limits a structure may not be built unless the outer walls are of masonry to limit fire extension.

Codes and standards attempt to divide types of buildings into various exclusive classes. Unfortunately, many of the buildings a fire fighter must cope with were built by people who used building materials which seemed best suited to their purpose or were readily available. Often, little or no thought was given to distinctions between types of construction as classified by building codes, the NFPA, or insurance underwriters.

Masonry Construction Terms

Some of the principal terms used in masonry construction are:

Adobe - Bricks made of clay, water and straw dried in the sun. The building is then parged with a slurry of the same material.

Ashlar masonry - Stone cut in rectangular units.

Cantilever wall - A free standing wall unsecured at the top. This wall acts like a cantilever beam with respect to lateral loads, such as wind or a hose stream. Precast concrete walls are very dependent on the roof for stability. If the roof is affected by the fire, the walls are likely to fall. (For a discussion of the hazards of cantilever walls, see Chapter 2, Construction Principles.)

Cavity walls - Hollow walls in which wythes are tied together with steel ties or masonry trusses.

Composite wall - Two different masonry materials, such as brick and concrete block, used in a wall and designed to react as one under load.

Concrete masonry unit - A precast hollow or solid structural block made of cement, water, and aggregates.

Course - A horizontal line of masonry.

Cross wall - Any bracing wall set at a right angle to the wall in question.

Flying buttress - A masonry pier at a distance from the wall and connected to it. Such buttresses resist the outward thrust of the roof. They are mostly used in Gothic architecture.

Header or bond course - Bricks laid so that the end is visible.

Hollow masonry walls - Two wythes of masonry with an air space in between. The wythes are tied together or bonded with masonry.

Masonry columns - Masonry bracing incorporated into unstable masonry walls are often called piers, buttresses, pilasters, or columns. These may be built inside or outside the building. Where visible, they indicate where the wall is strongest, often where the concentrated loads are applied, and where not to attempt to breach the wall.

Parging (or pargetting) - The process of covering a masonry wall with a thin coat of concrete.

Rubble masonry - Masonry composed of random stones.

Rubble masonry wall - A wall composed of an inner and outer wythe of coursed masonry. The space between is filled with random masonry sometimes mixed with mortar. Such walls are unstable to a lateral thrust.

Solid masonry walls - Masonry units (either solid or hollow) laid contiguously with the joints filled with mortar. CONCRETE Blocks

Stretcher course - Bricks laid so that the long side is visible.

Terra cotta tiles - Made of clay and fine sand, terra cotta is fired in a kiln. Terra cotta is both structural and decorative and is used in ornamental facings. Structural terra cotta has been replaced to a large extent by concrete block.

Unreinforced masonry - Ordinary masonry walls are not reinforced, so they have no resistance to lateral movement. They are therefore very vulnerable to earthquake collapse. **Reinforced masonry** is discussed in Chapter 8, Concrete Construction.

Veneer wall - A wythe of masonry attached to a bearing wall of any material but not carrying any load but its own weight.

Wythe - Vertical section of a wall, one brick or masonry unit thick.

Hollow or Cavity Walls

A wall collapse caused five fire fighter deaths. The roof beams rested on only the inner wythe of a cavity wall.[3]

Hollow and cavity walls limit penetration by rain. This author does not know of any recorded cases, but it is possible that carbon monoxide from a fire could accumulate in the hollow space or cavity and explode disastrously.

Gas from a leaking main outside a building found its way into the hollow of a cinder block wall in a Florida supermarket. Filtered by seeping through the ground, natural gas can lose its artificial odorant. Unaware of the gas, an employee flicked his lighter and the gas exploded.

It is an accepted practice to place sheet or foamed-in-place plastic insulation in hollow walls. These plastics have various ignition characteristics. Burning plastic produces large quantities of smoke. If the source of smoke cannot be found, it would be wise to check for plastic insulation in the walls.

Hollow walls built of hollow terra cotta tile present a special hazard. Originally the exterior and interior wythes were connected with steel ties. The tile industry developed special tiles which connected the wythes. If either wythe starts to move, a tensile load is placed on a clay tile which has no tensile strength.

Composite Walls

Masonry walls were, for centuries, built only of bricks. In recent years, the composite wall, once incorporating hollow tile, now using concrete block, has been developed. Recall from Chapter 2 that both composite wall elements must react together under a load. When composite walls were first developed, bonding the wall together was accomplished by inserting brick headers and stretchers according to various design practices.

Uneven settlement between the brick and the block has caused header bricks to crack. Consequently, the **masonry wire truss** was developed. This wire truss is bedded into the mortar in specified courses. As a result, the header course is no longer necessary, and the appearance of a masonry bearing wall may be no different than that of a veneer wall of all stretchers. In some veneer walls, bats or half bricks were inserted to give the appearance of a bonded wall. Thus, it is often impossible to tell a bearing wall from a veneer wall by external appearance alone. Brick veneer walls are discussed in Chapter 3, Wood Construction.

Not all brick-and-block walls are composite. Brick may be veneered onto concrete block or cast concrete using the same ties that are used in brick veneer-on-wood construction.

Cast Iron

An intermediate step in the development of cast-iron-front buildings was the use of cast-iron columns and wrought or cast-iron arches or lintels at the street floor level carrying masonry above the first floor. This enabled the builder to provide larger show windows or entranceways than would have been possible with masonry.

The small area of a masonry wall available for windows led to the acceptance of walls made of prefabricated cast-iron sections. Usually a unit was cast integrally in two sections. The sections were bolted together at the crown to form the arch. There are few buildings with cast-iron fronts left, and the demolition of one today is regarded as an act of barbarism in some circles.

The cast-iron front may separate from the masonry side walls. Some walls are tacked onto the masonry with steel straps. The columns, like interior cast-iron columns, may transmit fire vertically.

Lintels

Arches and beams are the two ways to carry the wall above an opening. Such a beam is called a lintel. The lintel commonly used today is a steel "L" or channel section. When such sections were first used, the practice was to arrange the bricks above the lintel as a false flat arch. Later, the pretense of an arch was abandoned and the bricks above the window were set vertically. Today, the common practice is to carry the brickwork horizontally across the lintel. Steel lintels are tied tightly into the masonry wall. When heated, they elongate and the masonry can fail.

Imitation Brick

Some imitation brick is made by spreading a coat of grey concrete on lath. A coat of red concrete is then applied. The red concrete is cut away to expose the grey concrete in horizontal and vertical lines to simulate mortar joints.

Another method is to cement thin slices of brick onto panels of gypsum board. These are mounted on a steel frame and joined together to give the appearance of a brick wall. Several bank branches have been built using this method.

Imitation masonry is also created by applying a thin coat of concrete over blocks of cast foamed plastic. The exteriors of many of the buildings at Universal Studios in Orlando, Florida are constructed this way.

Characteristics of Ordinary Construction

The simplest ordinary construction building consists of masonry bearing walls, with wood joists used as simple beams spanning from wall to wall. The joists usually are parallel to the street frontage of the building. The roof may be similar to the floor in construction, or it may be peaked by using rafters or trusses. In most cases there will be a cockloft between the top floor ceiling and the roof. Ventilators for the cockloft are often seen in the side walls.

Bearing and non-bearing walls use similar construction materials and are often identical in appearance. In the typical downtown business or commercial building, the side walls are the bearing walls, while the front and back are non-bearing.

The simple wood beam floor is satisfactory for buildings up to a practical limit of about 25 feet in width. For a wider building, or one with an irregular floor plan, interior masonry walls or a column, girder, and beam system must be provided. There is no limit to the ingenuity of builders. Many possible combinations of building materials are used. Columns may be wood, brick, stone, concrete block, steel or cast iron. Different materials may be used for columns in the same building. Balloon frame stud walls may provide intermediate support. Girders may

Fig. 4-2. This "column" and others were located in a Virginia mercantile building visited by many tourists. A letter to the governor brought about some structural improvements.

Fig. 4-3. Note the crack in the column of this Portland, Oregon building. It may only be in the veneer of the finished brick, but it is a problem that should be investigated.

be wood or unprotected steel. The connection systems which attach the beams to girders, and the girders to the columns, are of almost infinite variety. It is in the weakness of these connections that the principal interior collapse potential is often found.

As in wood construction, void spaces are an inherent part of ordinary construction. Some fire protection measures such as embossed metal or tin ceilings, which were intended to prevent the extension of fire from the usable space to the void space, also prove to be barriers to the fire department's efforts to reach the fire, once the fire penetrates the void space.

As a general rule, there is no effective fire separation within the ordinary construction building, either from floor to floor or within floors. Even where fire separations exist up through the regular floors of the building, they often are imperfect or non-existent in attic spaces.

There is an inherent limit to the height of masonry buildings. This is due to the necessity for increasing the thickness of the wall as the height of the building increases.

The tallest old style masonry bearing wall building in the United States is the Monadnock Building in Chicago. It is 15 stories high, and the masonry walls at the ground level are several feet thick. About the time it was built, steel frame construction was developed, and masonry soon was being used only for fire resistant panel walls, supported on the steel frame, and was not load-bearing.

In recent years, high-rise brick or concrete block buildings with no wall thicker than 12 inches, and medium-rise brick buildings with no wall thicker than 8 inches have been developed, supplanting the traditional practice of ever-increasing wall thickness. Since these reinforced masonry structures depend in great measure on integration with reinforced concrete, they are discussed in Chapter 8, Concrete Construction.

Renovation and Restoration

Over the years, most old buildings have undergone extensive modifications. Usually, such modifications have had a detrimental effect on the structure from the fire suppression point of view, creating collapse potential, or interconnected voids from which fierce fire can burst out on the unwary.

This is not a new problem. Recently a fire service magazine reviewed an 1890 fire in which 13 Indianapolis fire fighters were killed.[4]

The article relates:

> "... The Brown Merrill Building was essentially a front and back with the stores on the other two sides providing the side walls. Over the years some interior walls on the first floor had been removed and iron poles placed above the joists to support the upper floors. Additionally an arch in the basement had been removed, further weakening the structure."

"Old Stuff" you say! Within the past few years I have seen these conditions and many others in existing buildings.

It is also important to note that alterations over the years, including reconstruction from fires that may have occurred many years ago, may cause one part of a building to be quite different from another. What you see on first observation may not be representative of the entire building. Furthermore, interior alterations and finish may make it difficult or impossible to determine the true nature of the building.

A case in point is the dining hall at the National Fire Academy which was used during the Civil War as a temporary hospital for victims of the battle of Gettysburg. When the building was first taken over from St. Joseph's College, it was not too difficult to trace the growth of the building over the years by examining the basement and the attic. The interior improvements have made it much more difficult. Cast-iron columns mark the boundaries of the original structure.

Sometimes automatic sprinklers are installed as part of a modernization program. In some cases, these sprinkler systems do not cover the structure completely. Chapter 13 discusses possible deficiencies of sprinkler systems.

Tactical Considerations

Fire departments should make every effort to become involved in restoration or modernization projects. Some of their beneficial recommendations or suggestions may be accepted. In any event, the department will be well aware of structural deficiencies.

Fire Resistance

The boundary line between ordinary construction and early "fireproof" construction is not at all clear-cut because building development is evolutionary. Many of the hazards of early fireproof buildings can be found in ordinary construction. Unsupported marble stairs is a good example.

Portions of an ordinary construction building may have some degree of fire resistance, either initially or as a result of legal action. Some examples include installing a properly enclosed, fire-resistive stairway in an old school, or providing a rated fire-resistive barrier around a special hazard such as a boiler room. Rarely does this piecemeal provision of fire-resistive features alter the fundamental nature of the building.

In some codes, concrete topping was required over first floor wood floors for fire resistance or to provide a sanitary floor. This represents additional dead weight and may confuse fire fighters as to the nature of the floor. It may also conceal the heat below. Concrete topping on the floor was a factor in at least two multi-fire fighter fatality collapses.

Some additions or modifications may be truly fire resistive, particularly in public buildings. It is a principle of most building codes that the fire resistive and non-fire resistive sections should be adequately separated. Unfortunately, this author has noted a number of cases where this principle was not carried out.

Preservation

Old buildings may be preserved for a number of reasons. They may have undeniably historic value such as Independence Hall, or they may be a part of our architectural heritage. They may be preserved to rehabilitate a specific neighborhood, or a developer may fix up an old building into attractive apartment, shop or office rental space.

In such projects it sometimes seems as if the firesafety of the occupants has a lower priority than the preservation of the original structure. In any case, it is important to remember that fire fighter safety is left up to the fire department — in short, you are on your own.

Tactical Considerations

Buildings are constructed to last for a long time. Be very wary of the building which has been altered.

Recent Construction

Many buildings constructed in recent years depart from simple ordinary construction. This is not necessarily an improvement.

In modern construction, a non-combustible void can accumulate explosive carbon monoxide gas as readily as a combustible void. A non-combustible void can contain combustible wiring and thermal insulation.

Metal deck roofs can have self-sustaining fires because of bituminous vapor seals. A wall pushed out of line by an expanding heated steel truss can be just as lethal as a wall pushed out of line by collapsing wood joists.

In early 1991, a Los Angeles County, California fire fighter suffered fatal injuries when a wooden facade collapsed. The use of wood-truss roofs, wooden cornices, canopies, "colonial" belfries, combustible interior wall and ceiling finish, and even wood veneer over masonry leaves few truly noncombustible commercial or institutional buildings.

The desire for wider spans and the availability of construction cranes has led to the widespread use of unprotected steel for roof framing. One office building is a case in point. The floors are of bar-joist construction with left-in-place corrugated metal forms and concrete topping. The ceilings are of lay-in tile. A casual inspection might indicate that the building is of fire-resistive construction, using floor and ceiling bar-joist assemblies. In fact, the building is of ordinary construction. It did not need to be fire resistive because the code does not require a fire-resistive building. The steel framing of the roof is unprotected steel. The roof is plywood. The ridge beam runs the length of the units. Given a fire in the wooden roof, enough heat would be generated to elongate the beam and push out the brick gable ends of the building.

In another building steel beams run end to end through two openings in fire walls, on which they are supported. The roof is plywood on wood trusses. The elongating steel would push down the end gables and might pull down the fire walls. The use of wood trusses and wood I beams has become almost universal. These are discussed in Chapter 12.

Problems of Ordinary Construction

The problems presented to the fire department by ordinary construction can be divided into the following areas:

- The structural stability of the masonry wall.
- The stability of the interior column, girder and beam system.
- Void spaces.
- The masonry wall as a barrier to fire extension.

These are not hard and fast divisions - they are all interrelated. An exterior collapse will usually cause an interior collapse. Conversely, the elongation of steel beams or an interior collapse may cause the walls to come down. A carbon monoxide explosion in a void space can demolish a building. A minor failure of a wall may permit fire to extend to another building.

Discovery of Hazard

Many fire texts cite indications of building failure which may be observed on the fire ground. Typical indicators include smoke or water through walls, soft floors, small partial collapse, walls out of plumb, and time since arrival on the scene. While these are all good indicators, sometimes they are insufficient, or, tragically, too late.

Early in this author's study of this subject, two books were noted as key texts: *Construction Failures*[5] and *Building Failures*[6]. Both of these excellent texts cover similar ground. The authors were consulting engineers who were often retained after a collapse to fix the cause. If an investigation after the fact could determine that the clues to impending disaster were evident before the disaster, perhaps the fire service could examine buildings beforehand and determine whether a particular building was likely to fail in a fire. These authors helped to move the consideration of fire-caused building collapse from anecdotal experience to analysis.

Much of the ordinary constructed buildings in a typical city were built before present members of fire departments were born. There is ample opportunity to study buildings ahead of time and establish preplanned tactics to minimize the risk from collapse. Unfortunately, too many fire reports indicate the fire fighters were completely unaware of situations which presented serious fire suppression and collapse problems.

In this text, many clues to potentials for disaster are given. The list could, of course, be almost endless. Use these examples as a guide to help you estimate hazards in buildings where your department must fight fire.

Some of the clues are evident from the street; others require detailed examination, which may not always be possible. In any event, be aware that an ordinary construction building, however sturdy and well maintained, was built without any thought as to what would happen to the building in the event of fire. By code definition, such buildings are non-fire resistive; they have no designed resistance to collapse in a fire.

The Owner's Rights

The right to own a building and to maintain it as one pleases is a fundamental right anchored in our legal system. Some laws have whit-

Fig. 4-4. Note the many indicators of potential collapse in this building. There is an added dead load on the roof, vertical bracing channels, many alterations, and anchors for tie rods.

tled away at this right, rarely without legal challenge, however. Often only in the presence of clear evidence of existing public danger will the courts order an owner to repair or demolish the property.

It is certainly not a commonly held belief, that the potential collapse of a building in the event of a fire is adequate grounds to infringe on the owners' rights. The fire department is left, then, with the duty of examining buildings for possible collapse and adjusting fire suppression tactics accordingly.

Structural Stability of Masonry Walls

Traditional fire service training on the subject of falling walls provides little real guidance. While some trainers speak wisely of walls falling a fraction of their height, bricks can fly in all directions and great length. If a single brick gets you, the fate of the rest of the wall is immaterial.

General Collapse Indicators

Is it probable that a masonry wall will collapse during a fire? Some do; many do not. No study has been conducted to determine if there is a common pattern in fire collapses. There are, however, certain gen-

Fig. 4-5. Falling walls have cost many fire fighters their lives. (Bill Noonan, Boston Fire Department)

eral indicators of probable collapse, some of which can be observed beforehand and noted in a prefire plan. Other indicators may be observed during a fire.

Collapse may be due to a variety of causes.

- Inherent structural instability, aggravated by the fire.
- Failure of a non-masonry supporting element upon which some portion of the masonry depends.
- Increase in the live load due to fire fighting operations, specifically retained water.
- Collapse of a floor or roof with consequent impact load to the masonry wall.
- Impact load of an explosion.
- Collapse of a masonry unit due to overheating.
- Collapse of another building onto the building in question.

Signs of Potential Collapse

Some of the signs of potential collapse are derived from building elements and materials. These can be observed prior to a fire and should be noted in the prefire plan.

PREFIRE PLAN

Fig. 4-6. This old Oklahoma building contains water soluble sand lime mortar and deteriorating bricks. The parging will not fix the problem; it merely hides it. The parging has fallen off the building and shows fire fighters its true condition. (Harold Cook, Shawnee, Oklahoma Fire Department)

Bricks: Bricks have a reputation for toughness and permanence which can be misleading. Properly made, they can be a tough and long lasting building material. However, poorly made bricks can deteriorate readily. Poorly made bricks absorb moisture and can fail due to freezing. Poor quality brick work is not uncommon.

Look for signs of poor brick in buildings. At times, brick defects are covered with parging (sometimes pargetting). This parging may be scored to look like brick. The parging itself may be falling off. This often indicates poor brick.

Mortar which reacts chemically may "grow," thus forcing the masonry out of alignment. Parapet walls, which project above the roof line, are particularly subject to weather deterioration. Many such walls, as well as some chimney tops, consist of bricks held in place by gravity, with non-adhering mortar simply acting as a shim which is used to wedge a component into position.

Sand Lime Mortar: As has been noted, the use of sand lime mortar creates a potential failure hazard because it is water-soluble. Suspect any building built in the last century or early in this century. To get rehabilitation tax credits, sand lime mortar must be used to repair old sand lime mortar buildings. Consequently, even currently restored

Fig. 4-7. Elongated bar joists collapsed this wall in a Texas community, and the roof fell in.

buildings should be suspect. If a wall is torn down, look at the bricks. If they are clean with little mortar clinging, sand lime mortar was used.

A number of non-fire collapses have occurred when water leaks washed out sand lime mortar. The old Broadway Central Hotel in New York, and the Empire Apartments in Washington D.C. collapsed because of a water leak which washed out sand lime mortar. A probationary fire fighter noted and reported the washout of sand-lime mortar at a fire. The building was cleared — it collapsed shortly thereafter.

Wooden Beams: Wooden beams carry an amazing load. The destruction of the wood in a fire causes the failure of the otherwise impressive masonry structure. This deadly construction is not apparent from the exterior. It can only be detected by competent prefire evaluations, focusing on collapse potential. The practice of using wooden lintels in the interior portions of masonry walls, while showing substantial masonry arches on the exterior, is centuries old.

In a number of European buildings, heavy masonry walls are carried over openings on wooden beams. This is cheaper and faster than building full arches. The craftsmen who knew this technique emigrated to the U.S. and taught it widely. It has been seen across the country. Look at masonry buildings. Check the basement and the attic where the walls are exposed. Determine whether the wall over the openings

Fig. 4-8. Four hundred years ago, this house was built at Carrick on Suir, Ireland. Note the fine flat arch over the windows. On the interior, however, the masonry was carried on wooden lintels, which would burn out and cause collapse. The Irish Government is replacing the "historically authentic" wood with reinforced concrete. See Figure 4-9.

Fig. 4-9. Skilled artisans emigrated to the United States and taught the same dangerous practice as shown in Figure 4-8. The wooden lintel supporting the interior of a brick segmental arch at the 200 year-old Paca house in Annapolis, Maryland is still in place. Also note the wooden leveling beam in the wall. If this burns out the wall will fall.

Fig. 4-10. Many stone or brick barns have wooden lintels.

for doors and windows is carried on masonry arches, or on wood. If on wood, expect early failure. Stone barns have had fatal collapses resulting from fires attacking wooden lintels.

Poor Quality Bricks: Poor quality brick work is not uncommon. Poorly made bricks absorb moisture and fail due to freezing. An expert on brick wrote a tongue-in-cheek article—"How to Build Leaky Brick Walls with Good Materials."[8] This article might also be called "How to Build a Wall that Will Collapse in a Fire." The time and money-saving techniques described, which produce a fine-looking but inadequate wall, are widely used.

Cracks: Cracks, even if patched, indicate weakness in the wall which may be due to inadequate foundations. If there is a thrust against one part of the wall, the rest of the wall may not resist the thrust.

A horizontal crack may indicate that the wall is being pushed out by steel roof beams, elongating in summer heat. When they contract in winter, the wall may be left out of plumb and the beams may have little bearing. This happened in the Knickerbocker Theatre in Washington, D.C. in 1922. Steel trusses had elongated and contracted repeatedly over the years, pushing the hollow tile walls out of plumb. On a cold winter night the trusses, then at minimum length, were bearing on very little of the wall. A 22-inch snowfall brought down the wall. Almost one hundred people died.

Arches: A brick or stone may fall out of an arch. If any arch unit (vousoir) is out, there is no arch. The window or door frame becomes an undesigned structural post and beam. If it burns out the masonry will collapse.

Tiles: Hollow clay tiles were substituted for brick before concrete blocks were available. Many hollow tile walls are poorly done. Earlier the hazardous use of clay tile connectors, instead of steel, was noted in cavity hollow tile walls.

Unstable Walls: A wall is inherently unstable. One stabilizing method uses intersecting or cross walls. Typically, the front and side walls are designed to brace one another.

Often the walls are made of different materials, such as brick and stone or different types of brick, which expand and contract at different rates. Look at wall joints. In many cases the separation is readily apparent.

Other ways to stabilize a wall include the use of:

- buttresses or masses of masonry outside the wall,
- pilasters or columns built inside the wall,
- wall columns, built within the wall.

In a concrete block wall, the wall columns may be solid concrete blocks, concrete bricks, regular clay bricks, cast-in-place reinforced concrete, or made of steel.

Holes In Walls: Often sizable holes are cut through walls. Until a proper lintel is installed, this is a serious weakness in the wall.

Fig. 4-11. This hole, cut horizontally, leaves concrete blocks unsupported so that the integrity depends on the shear strength of the mortar. Such an opening should be supported immediately.

Steel Lintels: Steel lintels have been used for years without any protection for the steel. In recent years there has been a great increase in fire loads and the rate of heat release of contents, due principally to plastics. Steel lintels have deformed and thrown bricks off the wall.

Reinforced-concrete lintels are commonly used in masonry walls. Since the building is not required to be fire-resistive, the concrete will not be required to be fire-resistive. The bottom concrete may spall off, exposing the reinforcing rods, which provide the tensile strength to the composite structure, causing the lintel to fail. It is a common myth that concrete is inherently fire resistive. Concrete is inherently noncombustible. It can be formulated to be fire resistive. (See Chapter 8 for a discussion of noncombustible versus fire resistive concrete.)

Braced Walls: Braced walls are another basic sign that a wall is in distress. (This is signalled by stars, plates, channel sections, or other spreaders, or straps tying the front wall to the side wall.) The walls may be tied together across the building or tied to floor beams. The braces of unprotected structural steel, or worse yet, of steel cable (complete failure at 800°F), will fail at fire temperatures. Wooden walls are tied together in a similar manner. Failure of a steel tie caused the collapse of a wooden wall which cost a Paterson, New Jersey Fire Captain his life.

Braces are not always an indication of instability. Some buildings had walls which were tied to the floors to make the structure more rigid. Generally these can be identified by spreaders, usually stars, in a regular pattern.

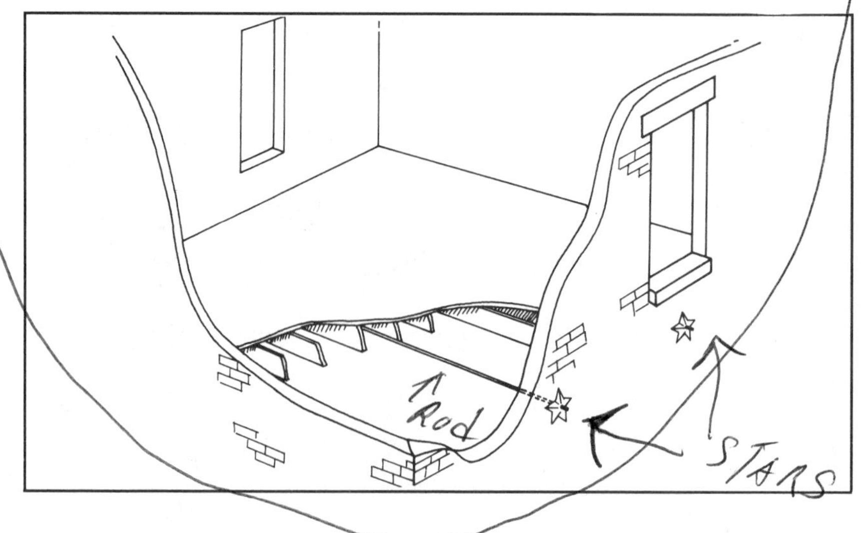

Fig. 4-12. One of the ways defective masonry walls are tied together. This may also be done to increase earthquake resistance of unreinforced masonry buildings.

In California and other earthquake areas, bracing is used initially or as a retrofit, to upgrade the earthquake resistance of unreinforced masonry walls. However, the building may adapt to the bracing so that it is necessary at all times. In this case fire caused failure of the steel may cause failure of the structure.

Hanging Signs: Some walls carry an eccentric load, as from a projecting sign. This must be counterbalanced, usually by tying the wall to the interior structure. A large abandoned garage with masonry walls and a wood-truss roof had a projecting sign. The wall was tied to the roof trusses. The trusses and the roof burned. There were no contents. When the trusses weakened, the sign pulled the wall down, injuring 17 fire fighters. There was no anticipation of collapse.

A peaked-roof restaurant has a brick front parapet. The projecting sign is counter balanced by two metal rods, lag screwed into roof rafters. If the lag screws burn out, the wall will be seriously overbalanced, and will likely come down.

Balconies: Cantilevered beams project from the wall to form balconies. In the cantilever, the question is "What is happening to the other end?" Three New Orleans fire fighters died when a balcony collapsed and pulled down the wall. Cantilevered canopies and balconies are very common. Those supported on trusses or wooden **I**-beams are particularly dangerous.

Canopies: Suspended beams which form canopies are quite common. They are secured by tie rods connected to the interior structure. If the connection fails, the canopy becomes an undesigned cantilever. As it falls it will bring the wall down with it. Pictures of such a collapse in Long Beach, New York show the parapet wall being pulled down for about 50 feet. The collapse zone is not merely at the canopy, but the entire front of the structure.

Fire fighters were battling a blaze in a furniture store for some time when a canopy fell. It pulled down the coping for over 75 feet. Six fire fighters died. With suspended beams the question is "What are the suspension rods tied to?"

During freezing weather, fire fighters continued to pour water onto a theatre marquee which was on fire. The water froze on the marquee. It collapsed, trapping several fire fighters. Fortunately a member of the department had access to a construction crane.

The canopy support rods of a Washington, D.C. theatre were tied to the roof trusses, and a multi-alarm fire was in the roof void. Alert officers noted that the top of the front wall had been pulled out just short of the collapse point and cleared the area. The insurance company engaged a wrecking company to erect posts to convert the canopy back into a simple beam.

Carbon Monoxide: Carbon monoxide gas trapped in unvented voids can detonate violently and blow down walls. Fire fighters were battling a blaze in an abandoned icehouse. Carbon monoxide accumulated in a nine-foot cockloft and exploded, bringing down the rear wall. Thirty-nine persons watching the fire from an apartment house died when the wall fell on the apartment house. Ironically, a small boy who was saved from the rubble, grew up to become a fire fighter.

Carbon monoxide generation in a void with restricted air flow can be as much as 50 times what would be evolved if burning in the open. This estimate is based on experiments performed at NIST.[9, 10, 11]

Masonry Walls: Masonry walls are not designed to resist lateral impact loads. Explosions have blown down many masonry walls. In one fire an explosion of coal dust blew down the five story masonry curtain wall of a steel framed building.

Brick or masonry walls which separate one electrical transformer from another are specially reinforced to resist lateral impact.

Plane of Weakness: One of the problems in masonry buildings is to get the floor beams level, due to masonry irregularities. A simple method is used to solve this problem across the country — in buildings as much as 200 years old. A wooden beam is laid in the brick wall at the line of the bottom of the floor beams. The floor beams are then set on the wooden beam. This produces a level floor without shimming. The hidden danger occurs when the wood burns out and a plane of weakness (like the scoring to cut glass or gypsum board) is created in the wall. A modest lateral thrust may be enough to break the wall along the plane of weakness. The historic Paca House in Annapolis, Maryland was restored some years ago. Yet these weaknesses were ignored during the restoration. (See Fig. 4-9.)

This author has never seen such a plane of weakness described as being responsible for the fatal collapse in any fire account.

A horizontal plane of weakness was noted in a two-story wall under construction. A slot was left in the wall to accommodate a concrete floor yet to be poured. When the floor is poured, if the concrete is not worked well into the gap, a hidden weakness would be left in the wall. Such a weakness may cause a collapse. This weakness will give no warning during a fire. The warning is in knowledge and prefire planning, not in anything you can hear or see on the fire ground.

It is possible to put a vertical plane of weakness into a building by cutting a chase into a wall. A chase is a passageway, usually vertical, cut into a masonry wall for a pipe or conduit. Close examination of the rear wall of an old building in Toronto disclosed an unusually weakened wall. The roof downspouts were often damaged by trucks moving through the alley. A chase had been cut vertically down each end

Fig. 4-13. This Salt Lake City building shows two defects. A vertical chase was cut into the wall, and a wooden beam was inserted into the wall to level the floor beams.

of the face of the wall, from the ceiling to grade. The downspouts were set into the chases, free from exposure to damage. When chases are built into a wall initially, they represent a weakness, but at least the joints can be properly laid to forestall disintegration. The chase cut into an existing wall presents a serious weakness.

In 1973, the Broadway Central Hotel in New York City collapsed.[7] The investigation disclosed that 30 years earlier an unauthorized chase had been cut into the wall for a drain. The drain leaked water which washed out the sand-lime mortar. Without being able to cite any such experience, this author warned of this potential in the first edition of this book - two years before the collapse. This is just one example of how fire officers can identify a hazard by analysis, not experience.

Downspouts that need protection from vehicles should not be recessed into a wall, but rather should be protected by masonry or cast iron.

Effect of Interior Structure on the Wall

The interior structure may push down the wall in several ways. Wooden floor beams fitted tightly into the wall may act as a series of levers on the wall, and when they collapse, pull the wall down.

The usual solution is the so-called "fire cut." The end of the beam is cut at an angle so the beam can fall out of the wall. This may save

Fig. 4-14. The "fire cut" on the ends of joists is designed to let them pull out of the wall easily.

the wall but weakens the beam since there is less wood to burn. Floor beams are placed with an upward camber or rise. Over the years they may sag. To restore the upward camber, the fire-cut beams are sometimes turned upside-down. This often results in a very small bearing surface of the beam on the wall.

In some areas the beams are not let into the wall, but sit on a masonry ledge, corbelled out from the wall.

In heavy timber construction, a cast-iron beam box may be inserted into the wall. The beam end is inserted into the box. If it falls out there is no problem with the wall.

Appearances Are Deceiving: The heavy masonry walls of a Montreal church were supported on an unprotected steel grillage of three I-beams bolted together. The grillage was supported by unprotected steel columns which extended down into the crawl space. Steel columns inserted into the wall rested on the grillage and supported the roof trusses. The wall therefore was not a bearing wall, but a curtain wall. Nothing of this construction was evident. The wall appeared to be a typical solid masonry wall.[12] See Figs. 7-23 A, B, C on page 315.

Fire in the crawl space cause the steel columns to fail and dropped the wall. Two fire fighters died. Be careful of potentially fatal assumptions made on the fire ground.

Steel Beams: Steel elongates when heated. A 100-foot steel beam reaches 100 feet, 9 inches at 1000°F. Elongating I-beams and trusses have pushed down many walls.

Six Dallas, Texas fire fighters were on a metal deck roof, some distance from the visible fire. Heat from the burning combustible metal deck roof fire, rolling through the truss void, elongated bar joists under their feet. The bar joists pushed out the concrete block wall of the building. The six fire fighters were injured in the collapse.

Several Orlando, Florida fire fighters narrowly escaped injury when a steel wind brace between the block gable and a roof truss elongated and pushed the gable down. If the wall is strong enough to resist the push, then the steel will buckle. Beware of tilt slab concrete and steel truss or I-beam roofs. Elongating steel will push the wall panels.

The cause of elongating steel collapse is rarely reported after a fire because all that is evident at that time is failed steel. The elongation temperature range is narrow.

Telephone Poles: Telephone poles being creosoted were stacked on wooden racks. Two fire fighters, attempting to breach a wall in a fire, died when the wall collapsed. The fire caused the failure of the racks, and the poles rolled into the wall. The report noted at some length the inadequacies of the helmets.

Materials Soaked by Water: Material which absorbs water, such as rags, waste paper, and fibrous material, should be stored at least a foot away from masonry walls to allow for expansion. Expanding water-soaked material can push the wall out.

Effect of Fire Streams on Brick Walls

Some fire texts advance the concept that wall collapse is caused by cold water hitting a hot wall, causing unequal expansion. This is suspect. Too many brick walls have been heated to high temperatures, struck by heavy cold streams and survived. I believe that the factors mentioned above, particularly the never-cited wooden components, are probably far more significant.

However, fire streams can damage brick walls. Older brick buildings have an exterior veneer like the surface of finish brick. A heavy stream can penetrate the veneer and strip it off the wall, sending deadly missiles in all directions.

Heavy streams can rip loosened bricks, as from a deteriorated parapet. Very heavy streams, such as from fireboats, can smash through a brick wall.

Tactical Considerations

Prefire planning and training are absolutely essential. Information must be widely disseminated, not confined to command officers. The size-up of an acting officer on the first due unit may be vital.

Consider marking particularly dangerous buildings on the exterior. In Boston, a red X is placed in several locations on dangerous

Tactical Considerations, continued

structures. This signifies that a defensive attack is required. A red diagonal line indicates that extreme caution should be exercised.

Possible wall collapse must be uppermost in the constantly changing evaluation of the fire problem assessed by the Fire Ground Commander.

If good prefire planning has been done, information on possible weak points in the walls will be available. In any event, a sector officer should be assigned to wall stability. Apparatus should be located outside the danger zone, which extends further than most people think.

In Spokane, Washington the collapse of a six-story wall was anticipated. Two fire fighters were in a bucket 12 feet from the wall. It was hit by bricks and part of the cornice. One fire fighter died. Apparatus is expensive but, entirely apart from the human suffering, killing fire fighters is even more expensive.

Walls may come down in several ways. The worst or most dangerous collapse is when the wall breaks at ground level and falls straight out along the full height. Bricks and other debris will fly even further. In one case bricks were measured as far as 200 feet away from the wall.

Examine a wall before breaching. If it is deteriorated, discontinue breaching because the entire wall may come apart. If you are breaching a wall and you find materials different from the rest of the wall, such as a brick, steel, solid concrete blocks or reinforced concrete, you are probably into a wall column, and under a major load. Start again someplace else.

When breaching a wall make a triangular cut, with the apex at the top. Do not cut a straight line across the top of the opening. Masonry units mortared together are not designed to hang upside down. You may very likely cause a collapse. Available material should be used as soon as possible to support the opening in the wall.

There should be strict control of the number of personnel in or close to the collapse zone. It doesn't take four or five people to operate a portable deluge set. In fact, it can be left unattended, but this "one spot" delivery is rarely effective. A heavy stream nozzle can be directed with special equipment, by means of radio signals, thus providing effective delivery without much risk.

The department should have a recognizable signal, such as air horn blasts in addition to radio orders, for all to evacuate IMMEDIATELY. Drills should be held on evacuation. I have a number of pictures of escaping fire fighters looking back. Train personnel, "DON'T LOOK BACK." Look where you are going, there are many tripping hazards.

Interior Structural Stability

Interior stability is critical to fire fighters. This section should be studied in conjunction with Chapter 12, Trusses.

In this text, interior and exterior collapse are discussed separately, although they are interrelated. An interior collapse of an overloaded floor can cause the walls to collapse. The collapse of an exterior wall into the structure or onto the adjacent structure may cause an interior collapse.

The interior structure consists of floors, any necessary interior support for the floors, any suspended loads, and the roof.

Historically consideration of interior structures assumed collapse in case of fire. Questions arose as to whether to design connections to permit easy collapse of girders (with self releasing floors) to prevent collapse of columns, or to design easy collapse of floors to prevent collapse of walls with fire cut joists. The opposite point of view called for tying the building together tightly, and disregarding the concept of releasing the floors in case of fire.

In a series of fire-resistance tests carried out many years ago, interior wooden columns were tested. Those tests showed that wooden columns were unable to survive a 1-hour exposure to the standard fire test.[13]

Fire-Resistive Combustible Assemblies

In recent years, combustible floor and wall assemblies of wood and gypsum board have been developed which can achieve fire-resistance ratings in standard fire tests conducted in accordance with NFPA 251, *Standard Method of Fire Tests of Building Construction and Materials*. They are used to produce code-classified protected combustible structures. These structures have many deficiencies which are described in Chapter 5, Garden Apartments.

Interior Support Systems

The list of inherent defects of interior supports is endless. Those that are most important to fire fighters, however, are found in structures for which you are responsible. Observe, study, ask questions.

In older masonry buildings, the interior walls, if not of masonry, are of balloon frame construction. Balloon-frame wall-carrying interior loads may fail very early due to the small cross-sectional area of the studs. Masonry walls are usually much thicker than wooden walls.

Adjustable steel jack posts or simple steel posts may be used to support overloaded beams. They are vulnerable to early failure.

Interior merchandise racks are often of very lightweight steel construction, and subject to early collapse.

Amateur builders often use water pipe for columns and beams, porches and canopies. This is very dangerous. Water pipe is designed for internal pressure. The metal removed to cut threads is replaced by the heavy collar around the fitting. As a structure the connection is weak because metal has been removed where the shear load is greatest.

Interior masonry bearing walls may stand in the way of projected improvements, such as combining several rooms. If an opening is made in the wall, or the wall is entirely removed, this arrangement will have much less inherent fire collapse resistance than the original brick wall. Such alterations were major factors in the Washington D.C., Empire Apartment collapse, and the Vendome Hotel collapse in Boston which killed nine fire fighters.

A beam and cast iron column unconnected to the beam were often used. In the remodeling of the Vendome Hotel in Boston, an air duct opening was cut through the wall directly below a cast iron column which supported a brick wall. Nine fire fighters were killed in the resultant collapse in June, 1972.[14]

In the Empire Apartments the wall above an opening was supported on a beam cut into the old sand lime masonry wall. Leaking water washed out the mortar and the structure collapsed with several fatalities.

The bearing walls of masonry buildings adjacent to one another, such as stores on "Main Street, U.S.A.," have served very successfully as fire barriers, but permit a space usually only 25 feet wide. It has long been a practice to unite two or more buildings into one store, at ground floor level. Column and beam arrangement support the remaining heavy masonry walls of the upper floors. The columns are often unprotected steel, or cast iron possibly unconnected to the beams. Some beams are wood. I call this the hidden wall trick because it is often not suspected. Occasionally, true masonry arches are formed in the wall openings. Suspect any old building with a ground floor area more than 25 feet wide.

Brick and stone interior walls can also have all the defects of exterior walls except perhaps weathering. Some very deteriorated mortar joints have been noted in brickwork which, while not exposed to weather, was inaccessible. It was apparent that visible and accessible deterioration had been rehabilitated while the inaccessible deterioration had been neglected. Stone walls and interior walls also may lose their strength due to spalling during a fire.

Deficiencies of Materials

Structures originally built with heavy timber columns, girders and beams are found reinforced with added steel in a variety of ways. In

Fig. 4-15. This cut-away view of the Vendome Hotel shows a common arrangment by which a brick wall is removed to enlarge an area. It cost nine Boston fire fighters their lives. (Boston Sparks Assn.)

such cases, the failure of the steel may precipitate collapse. Assume that any structural support which was added after the building was built was added because it was absolutely necessary.

The addition of unprotected steel to a building with a wooden interior structure is sometimes apparent, but is often fully concealed. Steel bar joists had been installed between each pair of wood joists to strengthen a building. This was not apparent until demolition.

Many newly built ordinary construction buildings have some unprotected steel structural members.

In structures, wood beams may have been trussed initially or at a later date. This is usually accomplished by erecting a strut downward from the center of the beam and stretching tension rods from one end of the beam to the other, over the strut. The strut or the rods may fail first. This technique has been used where heavy loads are on the floor above, or where equipment such as belt drive shafts was suspended from the overhead, or to transfer a load formerly carried on a column.

A tea shop is located on the first floor of a commercial building in Chester, England. To clear the dining area of columns, wooden trusses were erected at the ceiling level supporting the floors above. To the untrained eye they appear to be decorations. A heavy fire in the dining area might precipitate the collapse of the entire structure.

Inspect the old reliable heavy timber buildings for wood or steel trusses, or unprotected steel beams or columns inserted to cope with deterioration or heavy loads.

A huge old brick and timber factory building was converted into shops. It is unsprinklered. A new opening was cut through an interior brick wall. Instead of an expensive arch, an unprotected steel lintel was inserted. The failure of this lintel in a fire could precipitate a massive collapse.

Tactical Considerations

Wherever materials are combined in the structural support system, the fire characteristics of the poorer performer should govern assessment of the collapse potential.

Never get into a position where light-weight interior structures, such as rack storage, can collapse behind you.

Keep advised of alteration permits. Check the structures for alterations which increase the hazard to fire fighters.

Those who tout the merits of this or that construction material are not prone to mention the connections.

Look at the connections and consider the effect of fire on them. For instance, do not be impressed by the slow-burning, three-foot-deep laminated wood beam. It is often supported on an unprotected steel column.

All loads must be delivered to the ground. In all but the very simplest structures, a number of connections, of infinite variety, are used to transfer the load.

Connections

Beam to Girder Connections

The only way to connect wood beams to wood girders, and receive the full benefit of the time it takes the wood to burn through, is to set the beams atop the girder. This is undesirable, however, because it adds height to the walls. All sorts of systems were designed so that the top of the beam could be level with the top of the girder.

Old-time craftsmen are proud of mortise and tenon joints. The beam was cut down from the top and up from the bottom to form a **tenon** or tongue. This was inserted through a **mortise** cut into the girder. A wooden dowel, or **trunnel**, was driven through a hole bored in the protruding end of the tenon. Another method used notched beams on strips nailed to the side of girders. The weakness of any method that reduces the size of the wood is readily apparent. The effective strength of the wood under fire attack is determined by the size of the thinnest portion, not the mass of the member as a whole.

Beam to beam connections must be made whenever an opening is provided in a wooden floor. The joists which are cut to provide the opening, and thus do not reach to the girder or bearing wall, are connected to a header beam which in turn is connected to trimmers. These connections are accomplished in various ways. Mortise and tenon joints are found in older buildings. Notched beams are also common. Joist hangers or stirrups and heavy steel or wrought iron straps shaped to receive the joist and nailed to the top of the header are used. Tests made years ago showed that such stirrups are dangerous, charring the joist and softening rapidly.[15] More recently, very lightweight hangers have been developed for lightweight construction. The rapidity with which such hangers will fail in a fire is an important factor. The small mass of metal and the dependence upon nails do not inspire confidence.

Some joist-to-joist connections are accomplished by simple nailing, either by end nailing or toe nailing. Such connections are very undependable.

Beam to Column Connections

Timber buildings present many options for connecting girders to columns.

Fig. 4-16. Despite the threat of an earthquake, this San Francisco unreinforced masonry building was built with self-releasing floors. Note the "dog iron" that resembles a large staple (arrow) which lightly ties the floor girders together.

Self-releasing floors can be seen in most old timber interior mercantile buildings. The girder sits on a steel shelf which protrudes from the column. A scrap of wood or a **dog iron**, like a big staple, connects the girders and imparts some lateral stability under normal conditions. While self-releasing floors may limit the collapse area, any system which is designed to precipitate collapse must be weaker and more likely to collapse. In any event, many of the buildings fire fighters must cope with are "vernacular" buildings, designed without benefit of Professional Engineers.[16]

The principle that columns should set directly atop one another to carry loads to the ground as axially as possible was widely violated. Where a column is offset, the girder on which it rests becomes a **transfer beam** and is subject to severe shear stresses unless it was designed as a transfer beam. Improperly loaded structural components are more likely to fail than similar components properly loaded.

Cast-Iron Columns

Cast-iron columns create many problems. Good practice dictated that the column be topped by a **pintle**, a solid iron pin of much smaller diameter than the column. The pintle passed through the girder, surrounded by wood, and then enlarged to meet the full width of the col

Fig. 4-17. By contrast, this newly built communication center of the Los Angeles County Fire Department rides on springs to absorb shocks from earthquakes.

umn above. This author has never seen such a pintle on a cast iron column, but has seen such a pintle used to connect wooden columns end to end.[17] In practice, the column passes through the girder at its full width. The ends of the girders were cut in a semicircle to make room for the column. Thus, a relatively small amount of wood rested on the cast flat shelf which extends from the top of the column. The loss of this wood or the softening of the cast iron shelf could cause the initial collapse. Over the years, failures of cast-iron columns have been blamed for many serious collapses. It is true that cast-iron columns can fail, particularly if they are poorly cast so that the material is thin at one point. After looking at a number of buildings with cast-iron columns, the author suggests that the chief cause of their failure is the unsafe connections, rather than the questionable "cold water on red hot cast iron." Very often the cast-iron column is held in place only by gravity so that the slightest lateral movement can cause it to kick out.

In a famous New England department store, the basement ceiling had been cut away at every column to expose the connections, probably for a structural analysis. The tremendous wooden girders, about 24 inches by 24 inches, had been cut in a half moon shape to accommodate the column. The loss of very little wood at this point could cause a collapse. There was an upward draft into the void. Though the

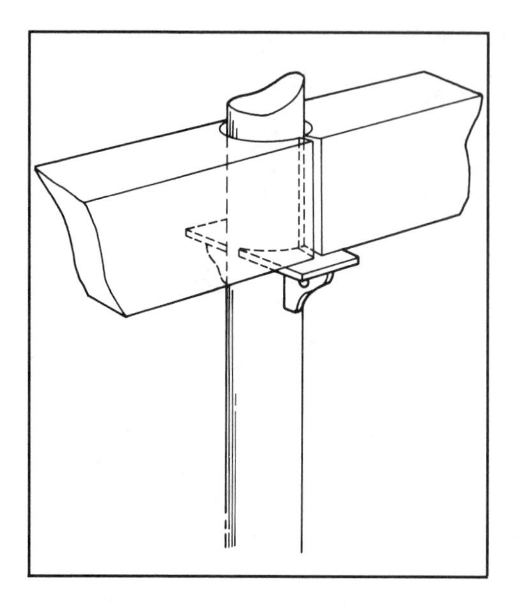

Fig. 4-18. Note the wooden beams that are connected to the cast iron column. Also note the small amount of wood that actually carries the load.

building was sprinklered, it was possible that a fire could extend into the void even before the sprinklers activated, and destroy the building.

Hollow cast-iron columns stacked above one another form a hidden flue for the passage of fire. In fact, early designers proudly used them as hot air heat supply ducts.

Tactical Considerations

Examine heavy timber buildings for the type of connections used. If the building is designed to collapse, either deliberately or unintentionally, plan so that fire fighters will not be in the collapse zone.

Suspended Loads

Balconies, mezzanines, half stories, and other suspended loads, including merchandise racks, are very hazardous. Suspended loads are becoming more common as architects realize the space advantages of suspension over columns. In a fire, the loads may be obscured by smoke, and the suspension system may be subjected to temperatures hundreds of degrees higher than at floor level.

A grocery market was converted to a Japanese-type restaurant in which the guests sit around a grill where the chef performs as the food is prepared and cooked. Such a restaurant requires a heavy steel exhaust hood for each table. The hoods were suspended from the roof. Because of the added dead load, the roof is by no means the same working platform for fire fighters that it was when its only purpose was to keep the rain off the groceries.

Architects are realizing more and more the utility of hanging loads from the overhead. Interior designers are also prone to hanging heavy loads from the overhead. The overhead may be the floor above, or the roof.

The mezzanine of a particular library is supported on a single wooden girder. One end of the girder is hung from a heavy roof beam on a slender rod. The other end rests on a wall. The rod will carry heat into the wood. If the connection fails the entire mezzanine will fall. Some church balconies are similarly suspended from the overhead.

A supermarket designer wanted to have shingled canopies over the checkout stands. If they were supported on posts, the arrangement would lack the required flexibility to accommodate future changes. As an alternate, the canopies were suspended. The ends of steel rods are held by two two-by-fours which extend across the bottom chords of bar joists. A fire overhead, possibly a metal deck roof fire, could either burn the wood away, letting the washer slip through, or cause the restrained bar joist to elongate or twist, dropping the canopy.

Another grocery supermarket was housed in a non-combustible building with a gypsum plank roof. The meat cutting area was in the rear of the structure. The health department required a washable gypsum ceiling rather than lay-in tiles. A substantial wooden structure was hung from the roof bar joists to support the gypsum board. The wood structure also supported the refrigeration units.

An electrical fire occurred in the ceiling void. The only fuel was the wooden structure. Not so much as a bag of potato chips burned. When a truck company was overhauling the completely extinguished structure, they were literally pulling the building down on themselves. Directing their attention to the hazard, I asked the officer to back the fire fighters up and to have one fire fighter gingerly break away the gypsum board so I could get a picture of the poor quality support. As the fire fighter touched the gypsum board the entire refrigeration unit fell.

When the store was rebuilt, it was sprinklered. The wooden structure within the roof was replaced as before. It is unsprinklered. If a similar fire occurs, the results will be similar!

A column may be in an inconvenient location or a girder may need strengthening. The load can be carried upward to a roof truss, to a girder laid across upper floor beams, or to a steel beam supported on exterior bearing walls. The load will be carried on steel rods, which are under tension and very vulnerable to fire.

Floors

When a stairway is relocated, the old opening is usually covered over. The closure is usually much lighter than the floor. In one observed case, two-by-fours were used, whereas the floor was made of three-by-tens. From above, the floor looks uniform. Such an opening has figured in at least one serious collapse.

In some store alterations, the upper floors are abandoned. The stairway is removed to provide ground floor room. Often, a single sheet of metal is placed over the opening. A Washington, D.C. fire fighter died when he fell through a patch of sheet metal over an old opening.

A lightwell is a shaft with windows which provide light and ventilation to enclosed rooms. A lightwell in an old building was floored over. The new flooring was very light and collapsed under the applied load.

In past years, floors were made of sub flooring and a finished floor. In more recent construction, floors are often a single thickness of plywood and carpeting. These floors are much more quickly consumed by fire.

The collapse of a heavy interior suspended load may present an excessive impact on the floor below.

Floor collapse can occur early in the fire. A National Fire Academy executive, Dr. Burton Clark, was a member of a local fire company,

Fig. 4-19. The stairway in this old Savannah, Georgia building was relocated. Note that the patch is of lighter construction than the original floor.

and responded with the first unit to a town house fire. He made a primary search of the second floor on his hands and knees, testing the floor as he went. He went through the floor into a void below. Fortunately, the active fire had moved to another area.

Arsonists have deliberately cut holes in floors where fire fighters would step to delay extinguishment of the fire. Water-soaked contents may collapse the floor. A floor collapse may cause the collapse of a wall.

The old high-ceiling ballroom of a reinforced concrete hotel in Palm Beach, Florida was divided into two floors. The added floor was of steel bar joists. It was impossible to learn whether the floor was a rated floor and ceiling assembly. Even if it is, and assuming it is installed in every detail to conform to the listing by testing laboratories, the potential for failure of such a floor is greater than for a concrete floor, due to the ease of tampering with the membrane ceiling.

A mezzanine was inserted in a high bay floor in Orlando. The original floors are solid sawn joists. The inserted floor is of wood truss construction.

Two floors were added to a sawn joisted floor building in Bethesda, Maryland. The new floors and roof were of wooden I-beam construction.

Be especially cautious of old residences converted to office use. It is likely that the floor loads are excessive.

Fig. 4-20. An eight-alarm fire where two Boston fire fighters died in the collapse of this residence which had been converted to an office. (Bill Noonan, Boston Fire Department)

Roofs

Fire departments are very concerned about equipment such as ropes, aerial ladders, and aerial platforms, which support fire fighters because failure can cost lives. Specifications are tightly written and acceptance and in-service tests are carefully monitored. However, there is one working platform fire fighters use that has cost many lives — it is the ROOF.

One of the basic purposes of a roof is to keep out the rain and enclose the structure. A secondary purpose is to stabilize the walls. In no case is a roof designed or constructed as a fire department working platform.

Generally, fire suppression tactics assume the stability of the roof, until there is warning of collapse. In fact, there are a few warning signs, but they are often too little and too late.

Tactical Considerations

The automatic procedure "get the roof" is no longer valid. The tactical attitude should be reversed.

It is not safe to assume that the roof is satisfactory as a working platform until definite signs of collapse are evident. The basic assumption should be that the roof is questionable, until valid information to the contrary is determined, either from the prefire plan or fire ground observation.

Fire fighters should not be working on any roof structure which is actually burning.

Fire fighters must realize that ventilation increases the air flow to the fire. A roof structure not involved before ventilation may quickly become involved as soon as the roof is opened.

While time-to-failure cannot be accurately predicted, the time the structure has been burning must figure in the fire ground commander's thinking. In one recent case, a mutual aid ladder company which arrived about 45 minutes after the first alarm, was sent to ventilate the roof. One fire fighter died when the roof collapsed. Once the structure itself is involved in the fire, it is deteriorating in an unknown manner and rate.

Fire Characteristics of Conventional Wood Roofs

(The discussion which follows concerned chiefly with roofs supported on solid sawn rafters and beams. Metal deck roofs are discussed in Chapter 7, Steel Construction, and Truss supported roofs in Chapter 12, Trusses. The roofs of tilt slab concrete buildings are discussed in Chapter 8, Concrete Construction. The hazards of single-ply membrane roofs are discussed in Chapter 2, Principles of Construction.)

The sole information offered in one prefire plan lists architectural roof types: flat; butterfly; lantern; gable; hip; gambrel; shed; mansard; dome. Such information, while useful, is incomplete. Structural characteristics as well as the appearance of the roof must be known beforehand!

Solid-sawn wood contains "fat" — wood not necessary to carry the imposed load. When only this part of the wood is burning, roof strength is not greatly affected. As it continues to burn, however, the beam gradually weakens, and the roof becomes spongy. This may or may not warn fire fighters to get off the roof in time to avoid disaster.

When exposed to the NFPA 251, *Standard Fire Exposure Test*, solid-sawn beams lasted about 5 minutes per inch of thickness. This is not to be used as a rule of thumb. Today's fires deliver fire loads more intense than the standard test. The fire ground commander is rarely aware of how long the beams have been burning. The intermittent impact load of a working fire fighter will almost certainly stress the structure more than the distributed static load applied in the typical fire test.

Roof Hazards

A Los Angeles city fire fighter was killed in a roof collapse. In making an addition to an existing building, mortar was picked out from between bricks, and shims were driven into the gaps. A ledger board was nailed into the shims. The roof beams were supported on the ledger board.

The best roof is one in which the roof beams rest on girders. Any sort of hangars or other metal connections make the roof more vulnerable to failure. In some localities steel **U** channels have been used for rafters. The effect of elongation and failure should be evaluated.

Roofs apparently supported on heavy wood beams or laminated arches can be much less reliable than they appear. Long wooden beams are scarce. Apparent long beams are often several beams spliced together, with metal connectors, which can heat up and burn out of the wood.

In a 90-year-old building being renovated in Orlando, the roof joists were exposed. The joists in the center section between the wood girders (any beam supporting other beams is a girder) were short. They did not rest on the girders. They were nailed to joists which overhung the girders. This use of such overhanging and drop-in beams is not uncommon. The nailed connections could fail very early in the fire, and collapse the center section of the roof.

Even if beams resist collapse, the roof boards may be burned away.

Energy conservation and rapid completion are two important considerations in today's buildings. Some roofs are made of large panels in which foamed plastic is sandwiched between sheets of plywood. Even if the plastic is inhibited to discourage combustion, it may melt away in a fire, allowing the roof panel to fall.

In an attempt to make use of natural light, corrugated glass-fiber-reinforced plastic panels are made to the same dimensions as corrugated steel. A fire fighter could easily step through the plastic when visibility is poor. Exposed to fire the plastic will burn away. The glass fiber matting, however, may stay in place and obstruct ventilation.

Remember that a code designated non-combustible building can have a combustible roof. The roof is weaker than the floor because the designed loads are less.

Ventilation tactics can accelerate collapse. Be aware that the impact of an axe may be the straw that breaks the camel's back. Some think that impact is avoided by the use of saws. Not so! There is no such thing as no impact. A fire fighter stepping onto a roof as lightly as possible still imposes a momentary load on the roof *at least* twice his/her at-rest weight.

Excess live loads on roofs can accelerate collapse. Three fire fighters ventilating a roof died in a collapse. There was heavy ice on the roof. When they opened the first hole the roof beams were seen to be burning. This should have been a clear warning that much of the roof's normal resistance to collapse had already been expended. Unfortunately they went on to open another hole.[18]

Water trapped on a flat roof because of blocked drains has caused many collapses. Snow on a roof has caused many collapses. Sometimes wind piles snow up in one area, causing a local concentrated load and resultant collapse.

Tactical Considerations

Be wary of requests to assist in snow or rain overload situations. Your operations may be held in court to have hastened the collapse.

Air conditioners and similar equipment are often mounted on roofs. Sometimes grillage is added to the roof to support the additional load. If all the grillage is affected by the fire as is usually the case, the extra support is ineffective in a fire.

In some cases, equipment is erected on steel beams which are supported on exterior walls. This is a superior construction method, but the supporting beams should be watched for any signs of overheating, and be cooled with hose streams if there is direct flame exposure.

Two fire fighters were injured when a roof collapsed due to steel plates added to the roof to prevent burglary. Three fire fighters died in New York City as they overhauled a ceiling. They literally pulled a roof-mounted air conditioner down on their heads.

Except for fire-resistive buildings, the roofs of buildings today have little or no inherent fire resistance to early failure. This even includes the roofs of non-combustible buildings. The typical non-fire-resistive roof today is usually supported on wooden I-beams, wood trusses, steel trusses, concrete T-beams, or in some cases, cored concrete slabs which are susceptible to failure.

Many roofs are surrounded with high parapets to create the impression that the building is taller than it is. This can deceive fire fighters not thoroughly familiar with the building.

Second or even third roofs may be built over an original roof. At times the reason is simply to improve the appearance of the building. A peaked roof is considered more attractive than a flat roof. Intractable leaky roofs are often overcome by adding an entirely new roof. A so-called "rain roof" was a factor in the deaths of six New York City fire fighters in a supermarket fire.

Tactical Considerations

Too many roofs are not safe to be on, once the roof itself is involved to the slightest degree in fire. Aerial apparatus of the sectional type, which permits a bucket to be placed down on the roof, should be considered.

A roof safety sector might be set up, with the sole responsibility of determining whether the roof is safe to use as a platform.

Horizontal positive pressure ventilation should be studied as an alternative to roof ventilation.

A ventilation opening acts like a chimney. Chimneys exhaust smoke and heat but they also accelerate the fire.

Collapsing roofs often bring down masonry walls. Where a roof collapse is anticipated, fire fighters should be withdrawn beyond the wall collapse area.

Treated Plywood Roofs

One effort in the struggle against fire proposed treated plywood as a fire barrier. Builders have long objected to cost and leakage problems of parapetted fire walls. Many code authorities in the eastern U.S. accepted the use of fire-resistant treated plywood (FRTP) roofing extending four feet on each side of the fire wall which extended to the underside of the roof. Based on this author's observations that delamination of ordinary plywood permits fire extension over fire walls, I expressed doubts over the reliability of this untested method.

At least some chemicals used in FRTP react and deteriorate in ordinary temperatures. The plywood is destroyed. The roof leaks and is unsafe, even for normal use. A fire fighter or a worker walking on such a roof may fall into the structure. Unfortunately, in many cases, the FRTP was used not only along the fire barrier wall, but also randomly throughout the roof. Estimates of the number of homes affected range up to one million. In the Washington, D.C. area alone, one builder is replacing about 10,000 roofs, at a cost of $65 million. Such problems have been reported with roofs of schools, nursing homes, prisons, and small shopping centers.[19]

Tactical Considerations

All fire departments should become aware of the extent to which FRTP was used in their area. Roofs should not be used to support fire fighters unless it is known for certain that the roof does not contain dangerous FRTP.

Void Spaces

As in frame buildings, void or concealed spaces are an inherent part of buildings of ordinary construction. The voids may be totally combustible, as in a wood-joisted building with wood floors and lath, and plaster ceiling. Partially-combustible voids are exemplified by wood stud construction with gypsum interior finish.

Tactical Considerations

Those who train fire fighters must make every effort to present the invisible hazards of void spaces with as much importance and drama as the live visible fire.

The fire fighter who is surprised by fire bursting out of a void space may gain tremendous experience, but may not have the opportunity to apply it or to pass it on.

A modern commercial building, with steel bar-joist roof beams, non-combustible ceiling tile, and a gypsum roof, has a non-combustible void. Non-combustible void spaces often contain some combustible material, such as insulation, construction trash, or wood blocking. Continued smoldering of combustibles can create a smoke problem long after the main fire is suppressed. The various voids may be interconnected in a manner difficult or impossible to detect.

During construction, requirements for firestopping may not have existed or may have been ignored. Even if firestopping has been installed, the execution may have been imperfect, or it may have been removed to permit the installation of ducts or wires. (See Chapter 3 for a further discussion of firestopping.)

As has been noted, void spaces are a major fire fighting problem. The existence of combustible voids has caused many serious losses even in sprinklered buildings because of failure to comply fully with NFPA 13, *Standard for the Installation of Sprinkler Systems*. The fire can be extinguished only if it breaks out of the void or if the fire department breaks into the void.

If circumstances are favorable, expansion of steam (created by hose streams hitting burning material) into the void may accomplish extinguishment.

The fact that a particular void was not as noncombustible as similar adjacent voids was responsible for the fire loss of a new high school.[20] Low-density combustible fiberboard had been applied to the studs behind gypsum wallboard on interior walls to provide acoustical treatment. A fire started by a plumber's torch found a fully combustible void space in a wall and extended quickly to the void space above the ceiling. The roof was an insulated metal deck. The noncombustible acoustical tile ceiling proved to be a hindrance to the fire department's efforts to deliver the water to the underside of the roof, the vital place to have absorbed heat.

Interior Sheathing

There is a concept, never definitively stated but acted upon nevertheless, that by providing a protective interior sheathing or finish for a structure, a substantial contribution has been made to the building's fire safety. This is true as long as the sheathing keeps a contents fire **out** of the structure. However, when the fire originates behind the protective sheathing or extends behind it, the protective sheath becomes a detriment to the fire fighting effort.

Light Smoke Showing

The expression, light smoke showing, has too often been the first report of what turned out to be a disaster. It is important to recognize with fire that "What you see, ain't necessarily what you've got." Ordinary citizens and inexperienced fire fighters are generally unaware of the tremendous fire threat which can be concealed in voids.

Three fire fighters died in a furniture store fire. Heavy smoke was reported showing from more than a mile away by responding units.

Fire fighters operating inside found only light smoke. In fact, they dropped their $2\frac{1}{2}$-inch line in favor of extending a $1\frac{1}{2}$-inch line. Suddenly fire burst out from a void and killed the three fire fighters.

There were a number of other operational problems which contributed to this tragedy: unfamiliarity with the structure; a missing preplan; and a broken microphone. The full story of this fire[21] is well worth reading.

In another fire, this author arrived almost an hour after the first alarm of an **L**-shaped one-story row of stores. The fire fighting operations were at the short leg of the **L.** A sewing machine store with fabric up to the ceiling was at the top of the **L**, furthest from the visible fire. The store was absolutely clear. A brown spot appeared on the metal ceiling. Gradually it turned red and the paint started to burn. Flaming paint fell on the fabrics and the store was soon fully involved. The entire process took less than five minutes.

A large group of people were enjoying an oyster roast in a wooden structure with a combustible fibreboard ceiling. A fire broke out above the ceiling. Some used extinguishers to attack the fire; others stood and watched. The fire broke out so fiercely that 11 of the occupants were trapped and died.

Tactical Considerations

Beware of passing by fire areas in an aggressive attack. Fire can and has broken out of voids furiously. No fire area should be left without attended charged lines. If truck company service is inadequate, engine companies might be assigned the onerous tasks of opening walls and pulling ceilings.

Prepare to be surprised, and perhaps you will not be.

Ceiling Spaces

Older buildings have much higher ceiling heights than those built today. False ceilings commonly are added to conserve heating and cooling costs and to modernize the building. It is not uncommon for a building to be modernized several times in the course of its lifetime with a new and lower ceiling each time.

Lowering the ceiling of old buildings is most easily accomplished by constructing a new ceiling. This provides a convenient void for the new utilities, air ducts, and so on. Firestopping is usually non-existent in these situations. Vertical voids connect the horizontal voids so the building is honeycombed with what amounts to one big interconnected non-firestopped void. It seems as if authorities place almost magical trust in the use of fire-rated gypsum board to absolve all sins of omission and commission. The fallacy of this concept is discussed in Chapter 5, Garden Apartments.

Fig. 4-21. Fire burning in interconnected voids is shielded from the actions of the fire department. Collapse potential dictated a defensive attack. (Gerard Elkins)

Usual construction methods create a huge three-dimensional void across the ceilings. In the rehabilitation process, the buildings are gutted and apartments are created by removing and/or adding walls. Dropped ceiling frames are hung, and the new interiors are lined with plasterboard panels. The walls are not firestopped at the ceiling line, and the vertical stud spaces are unprotected above the ceiling. Old dumbwaiter and pipe shafts are retained for utilities.

Such buildings burn furiously. Fire fighters have to pull down the ceilings to find the fire and occasionally are met by a carbon monoxide explosion, sometimes with enough force to blow them out of the room.

Fig. 4-22. This gypsum board partition between stores is penetrated by the bar joists. Such an arrangement follows the dubious concept that a "rated ceiling" (there is no such thing in this author's opinion) meeting a "rated wall" provides adequate compartmentation.

Fig. 4-23. This commercial building is being renovated into apartments. The new dropped ceiling creates a huge combustible void.

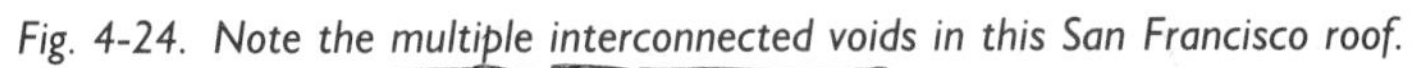

Fig. 4-24. Note the multiple interconnected voids in this San Francisco roof.

Many codes apparently do not adequately address this problem. Authorities are sometimes reluctant to impose requirements which will increase the cost of rehabilitating old housing.

One old factory in New York City was converted into apartments. Sixteen full-strength truck companies (officer and five fire fighters) were required to open up all the voids. One Orlando building showed two tin ceilings, one below the other. There are two voids, and thus a double barrier between the fire department and the fire.

Joist Spaces

Important void spaces in any multi-story, wood-joisted building are the joist spaces. Containing many square feet of exposed fuel, these spaces are protected from hose streams by their construction and the ceiling below. The modern use of wood truss floors has immeasurably increased this problem.

As the fire gains in volume, it spews torches of flaming gas out of every hole, opening, and utility service in the floor above.

Tactical Considerations

Old-timers had a simple technique for stopping this fire extension from floors. They put about two inches of water on the floor above. The fire coming up met the water coming down. Those who never saw a fire when it was out of control, and modern fire chiefs, who believe in "putting only water on the fire," would probably join hands to wail over the unnecessary water damage. There is, strangely enough, no water damage at a total loss.

If a sprinkler system operated on the exposed floor and stopped the fire it would be credited with stopping the extension of the fire. When an "improvised" sprinkler system does the job because none was installed, it is criticized. Strange!

Combustible Gases

The ignition of combustible gases accumulated in voids within the building may provide the fuel for a devastating explosion, even though the building is ventilated in the accepted manner.

The generation of carbon monoxide in concealed places can be as much as 50 times greater than what it would be in the open. The carbon monoxide flammability range is from 12.5 to 74%. Its ignition temperature is 1128°F.

At a stable fire in New York in 1938, the author witnessed an explosion of carbon monoxide that had accumulated in the cockloft (void between the ceiling and roof). The violent explosion took place an hour and a half after the first alarm. It was pre-signaled by dense clouds of boiling black smoke. A longtime observer of fires noted, "It's going to blow." The blast caused the collapse of a side wall and the loss of one fire officer's life.

There was actually a detonation. Apparently, the gas-air mixture was just in the right proportion. The building had been vented according to standard procedures, but removing the skylights did not vent the cockloft. It was incomprehensible to the investigating committee that a backdraft explosion could occur 90 minutes after the first alarm in a vented building. It was reported that the wall simply collapsed.

In a more recent fire, the author observed a fire in a row of stores. The building was vented. The front window of the store was completely out. There was bright fire which was suppressed with hose streams. The wind shifted. Heavy black smoke boiled out. Like the old-timer mentioned above, the author said to a companion, "It's going to blow." It did. Fortunately, the fire fighters were just knocked down.

Tactical Considerations

There is an old fire fighter's saying, "Don't be in such a hurry — get there after it blows." This exemplifies the myth that if there is going to be a backdraft, it will take place early in the fire. A backdraft can take place any time the conditions are right. One sign is heavy, black, boiling smoke. Unfortunately, with the prevalence of plastics today, this is not a sure sign. If there is no known reason for the smoke, prepare for a blast.

Large Voids

Churches, schools, town halls, and other buildings often include vast void spaces. Often there is a basement full of combustibles, non-firestopped walls, and sometimes hollow columns. The voids are so extensive that hand lines cannot reach the fire. Access for heavy-caliber streams is limited so there are large areas which cannot be reached by water. Some valuable fire protection features were never required by code. They resulted from provisions for some other purposes, and the fire protection benefit, though real, was unintended. An excellent example is that windows invariably were provided at the front and rear of a building and skylights in the roof. These wall openings are absolutely necessary to the control of any advanced fire in a building of ordinary construction, but were never legally mandated.

In an effort to modernize the typical downtown store, designers often close up the front windows of the upper floors. Air conditioning and fluorescent lights substitute for windows and skylights. The skylights and rear windows are closed up to foil burglars, and the often heavily fire-loaded upper floors become, in effect, one big inaccessible concealed space, completely impervious to fire department streams. A billboard covering the front of a building can serve also to convert upper floors into what amounts to an inaccessible void. What might have been a slight fire in earlier days becomes a major loss as heat and smoke fill the void and destroy the contents. An advanced fire is an impossible problem for manual fire fighting since "you can't put it out if you can't hit it."

These buildings essentially become volcanos. The attack is simply heavy-caliber streams from above after the roof vents itself.

Sixteen patrons died in a Brooklyn, New York store which had been modernized in this fashion.

In some cases, building code authorities have recognized the fire department's problem and required fire department access panels. These are usually of limited size.

Since many codes require sprinklers for inaccessible cellars or concealed spaces below mercantile buildings, it is logical to press for sprinkler protection when the void is built above the first floor. The problem is the same. The sprinklers may prevent the loss of the building, but a substantial fire loss is almost inevitable.

In some cases, modernization is accomplished at less cost as the windows are merely painted over. In other cases, wood or metal is substituted for glass in blocking off windows. In one Richmond, Virginia warehouse windows have been bricked up behind the glass panes.

On the exterior of some restaurants, several windows are visible. Windows are also to be noted inside. In fact, there is no relationship between the two. The exterior windows are where the architect wanted them. The interior "windows" are where the decorator wanted them. Time wasted in "discovering" this readily observable fact could cost lives in a fire where rescue operations are required.

Huge voids can be found in newer buildings, particularly those with wooden attics. One church, built in 1970, has reasonable construction except for a wooden roof of wood decking on wood joists supported on unprotected steel trusses on steel columns. The steel is protected from a fire in the auditorium by a suspended ceiling. It is not protected from a fire in the wooden roof above. The freestanding walls are composite brick and block, supporting only their own weight, but probably tied to the steel columns for stiffening. A fire in the roof void could cause the distortion of the steel and possibly the collapse of the walls.

Tactical Considerations

Truly, such a building is "designed to burn." This is another case where the pre-fire plan should be shared with the management of the property. Improvements may not be undertaken, but at least the fire department is on record as making the management aware of it.

Fire Extension

In ordinary construction, rarely is any provision made to prevent the extension of fire through the stairways and halls. Even if the stairways are enclosed in response to legislation passed after some terrible disaster, there are often many bypasses. The interconnected voids, pipe closets, utility shafts, and elevator shafts provide fire paths.

Tactical Considerations

Undress the building. In your mind's eye, see all the interconnected voids. Position units to get ahead of the fire. Playing catchup can be fatal.

Ten guests died in a motel fire. Fire raced up a stair tower which was located at the intersection of two non-firestopped-trussed, gable-roofed attics. The fire raced both ways through the combustible attics.

A four-story business block, which housed six businesses, was damaged by fire. The two top floors had been removed. A common undivided cockloft was thereby created over the entire building. Fire in the basement was stopped from coming up the stairs by the fire department. However, fire reached the cockloft via interior stud walls. The loss was about $500,000.

Interior Structural Walls as Fire Barriers

In large buildings of ordinary construction, masonry walls may provide the interior load-carrying structure. Sometimes closures of more or less adequate fire resistance have been provided at openings. These can be useful. However, since the purpose of the wall may not be to stop fire but only to carry the interior floor loads, it would be unusual to find such a wall carried through the attic space. The combustible attic may provide a path by which fire, initially confined to one section, may pass over the top of the wall.

Tactical Considerations

Such walls may be useful in providing a location at which the fire fighting forces can make a stand. If this situation is recognized, prefire planning might call for maximum removal of the roof over the fire section to permit heat to escape. It might also be practical to plan to vent certain windows, while keeping others intact, to relieve the heat through a safe path.

Voids in Mixed Construction

Many buildings are composites of older sections, possibly frame or ordinary construction, and newer sections of protected steel or concrete construction. Unless the architect is aware of the hazard, or an alert fire department can present a convincing analysis of the total fire problem to the building department, the new fire-resistive addition may be completely at the mercy of the old building, particularly when the architect wishes to make the result look like one building.

An industrial building in New England appears to be of fire-resistive concrete construction. The view of the building from an upper story of a hotel nearby shows that it was built around an older building with a huge peaked-roof attic on which air conditioners are mounted. It is almost certain that the attic is of wood construction. This would be quite a surprise at a fire.

Tactical Considerations

Get up high and look down on your problems.

A 1962 fire in the Barbara Worth Hotel in El Centro, California provides a good illustration. The main section of the hotel was four stories, brick and wood-joisted, with undivided voids both vertically and horizontally. A concealed vertical opening between the top story and attic, caused by the arrangement of a wall of the new fire resistive section, admitted fire to the common attic. The fire completely destroyed the original building and badly damaged the annex.

Cornices and Canopies

Earlier, the fatal injury of a Los Angeles County fire fighter caused by a wooden facade collapse was noted. The facade had been added to the building. It was not securely attached to the structure, but reportedly was just nailed to the wall. In addition, it was covered with tiles which weighed nine pounds each.

In the 19th and early 20th century, many buildings were not complete without a cornice. A cornice is a structure of wood, metal or masonry which tops the wall and projects from it.

Collapse of cornices has caused many fire fighter fatalities and injuries. In one case a collapsing cornice severely injured fire fighters in the bucket of a tower ladder. The more functional architecture of the 1920s eliminated the use of cornices. In recent years, however, there has been another change in style and tastes. Cornices, fake mansards, overhangs (sometimes called eyebrows), and other projections are being installed on many new buildings and added to old buildings to improve their appearance. They are usually of wood and sometimes poorly attached to the structure. They present a collapse hazard.

A cornice can interconnect with void spaces and false ceilings. It then can extend the fire and distribute smoke from area to area.

A noncombustible motel in Pennsylvania is decorated with a huge mansard which extends around the building. The fuel load of a motel room is sufficient to pour heavy fire out the window. This fire could ignite the mansard and destroy the motel. It would be interesting to ask the architect, "Did the owner give you a criterion that a single room fire should put the entire motel out of business?" This question might interest the owner.

Closely allied to the cornice is the sidewalk canopy. In some cases, this is of cantilever construction, thus providing a serious collapse potential. Even when not cantilevered, the cornice and canopy represent substantial loads, usually supported by combustible members.

One canopy, several hundred feet long, was protected on the underside with cement plaster on metal lath. Furring strips were laid at right angles to the joists. The lath was nailed to the furring strips. The joist channels were, therefore, all interconnected and fire could spread unchecked throughout the entire length of the structure. Far from being an advantage, the metal lath and cement plaster would handicap efforts to open the void. Since the void was interconnected with the interior, steam generated within the canopy by fire streams could spread into the building proper, possibly doing damage far beyond the fire area.

Tactical Considerations

If it is vital that fire fighters pass under a burning cornice or canopy, use a solid stream to knock off wood about to fall, and drive the fire back. This may buy a short time during which it is somewhat safer to pass under the cornice.

Masonry Bearing Walls as Fire Barriers

A fire wall, by definition, is erected for the sole purpose of stopping a fire, and the truly adequate fire wall can often accomplish this function unaided. Masonry bearing walls may serve to stop the fire, but conditions generally make a bearing wall less than a fire wall. When defects are recognized, tactics may then be planned to take advantage of the assistance offered by the masonry wall.

"Main Street, USA" is made up of buildings of ordinary construction, usually brick and wood-joisted, built side by side. If each building has a 12- or 16-inch unpierced bearing wall, these two walls together form a barrier to the passage of any fire that can be generated. The fire may go over or around such a pair of walls but not through it. Unfortunately, often the walls are pierced.

Fire Doors

When openings have been made in walls to connect buildings, adequate fire doors should be provided. The local code may not require them, however, and even if installed they may be ineffective.

In some cases, the originally planned openings were properly protected by fire doors, but additional openings were made later without proper protection. A Virginia city has many restored buildings. In one case, two older buildings were combined to form a restaurant. The opening was protected with a fire door. Alongside it there is an unprotected window so the manager can see the dining room from the office.

Tactical Considerations

When fire doors are observed during an inspection, the entire wall should be examined for unprotected openings. Steps should be taken to get the openings closed up or properly protected.[22] *The openings should be noted on the preplan for engine company coverage. If fire gets past the opening, the value of the wall is lost.*

Fire fighters should be trained in how to inspect a fire door for proper operation. Particular attention should be paid to keeping the door path clear at all times of any displays or equipment or anything which would block the door.

Prefire plans should indicate the presence of fire doors. It is a truck company's function to see that they are closed.[23] *As much as fifty percent are inoperative at any given time. Water spray systems, installed in lieu of fire doors, have proven ineffective when flaming aerosol cans are involved.*

Tactical Considerations, continued

In buildings more ornate than warehouses, fire doors may be artfully concealed. In Rochester, New York an overhead rolling fire door is concealed in the ceiling. The tripping fusible link is not conspicuous. The door lies across the attack path of the first due engine company.

Fire units should never pass through a sliding or overhead rolling fire door, without blocking it temporarily. Closed sliding fire doors can be very difficult to open; overhead rolling doors may be impossible to open. A sudden burst of fire may trigger the door and the fire fighters could be trapped.

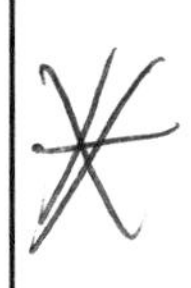

In bygone years, adjacent buildings may have been connected by doorways at one or more floors. The fire department may be totally unaware of openings that were abandoned in the past when the occupancies were separated. The closure at an opening may be an ordinary door, an inoperative fire door with merchandise stacked against it, fiberboard lath and plaster, or properly constructed masonry to restore the full integrity of the wall.

An unsuspected old basement level connection between two buildings at right angles to one another permitted a multi-alarm fire in New York's now fashionable Soho district. The fire broke out in the exposed building as units were leaving the scene of the fire in the first building.

Fire originating in one building may first show up as smoke in the connecting building. This was the case in a fire in New York City in 1966, in which 12 fire fighters were killed. The fire was first reported from an adjacent address. The two buildings were connected by a basement door. Do not assume that the fire first reported is the original fire.

Another method of interconnection is to cut through the wall so that the stairway of one building can serve the upper floor(s) of an adjacent building. This permits the elimination of the stairway in one building and the enlargement of the most profitable, rentable area, the street floor.

Tactical Considerations

Keep in touch with old-timers, downtown real estate brokers, the alteration contractor, and the staff of the building department. All these people may have valuable information about buildings. Look at buildings from across the street. Note the businesses which occupy more than one building. Most likely they are interconnected. Determine the value of the protection, if any, on the opening.

Protection from Exposures

When adjacent buildings are of identical dimensions, and the bearing walls are unpierced and parapeted, the ideal fire barrier exists. Unfortunately, the situation is often less than ideal.

When one building is taller than its neighbor, the owner of the taller building may install windows in the upper stories overlooking the smaller building. Fire coming through the lower roof may extend to the adjacent building via the side windows. Burning material may fall out of the upper windows on to the adjacent roof.

Old, hidden windows may provide a surprise path for fire to travel from one building to another. A key factor in a $1 million fire in Gloucester, Massachusetts was a window in a brick building that had been boarded up. The window was next to the wooden exterior wall of a sprinklered department store. The fire entered the ceiling of the store above the sprinkler piping. The sprinklers collapsed and the store was lost.

If two buildings are not of the same depth, the rear windows of one and the side windows of the other may expose one another. In such a case, the rear exposure may require protection, even before the fire is attacked. Otherwise, a frontal attack may drive the fire out the back windows into the exposure.

Narrow alleys between buildings present difficult defense problems against exposures. Often sheet or corrugated steel, or steel-clad wooden fire shutters protect windows facing the alley. Their worth is suspect. Wired glass is of limited value against radiant heat.

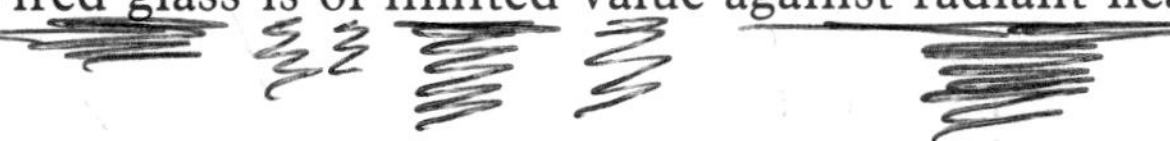

Tactical Considerations

Fire fighters operating lines in a narrow alley are in serious danger. Study the problem in advance. Lines directed downward from the roof of the exposed building or ladder pipes directed into the alley from the street are two methods of attack.

At times, outside sprinklers or spray systems are installed to protect against exposure fires. The fire department should be fully familiar with their operation. In most cases they must be turned on manually. The valve is located inside the building. The system should be tested at least annually. The system will do the work of one or more engine companies in covering the exposure, and without risk to personnel.

Party Walls

In many cities, it is quite common to erect party walls, structural walls that are common to two buildings.

Fig. 4-25. The demolition of one building shows how the party wall supported the floor beams of both buildings in a common socket.

In the area of the original city of Washington, D.C., fire protection regulations issued by President George Washington required that masonry party walls be built, and that the builder of the subsequent building pay the first builder for half the cost of the wall.

Usually, party walls are established by mutual contract between the owners. To the best of the author's knowledge, the Washington situation is unique.

In examining the building, determine the thickness of the wall. A party wall is thinner than two separate walls. Because the joists of both buildings are supported on the same wall, the builder often found it convenient to support the joists in the same socket. Sometimes the joists were overlapped to provide greater strength. Some codes forbade this practice.

In some cases, row townhouse fire walls are actually party walls. A common socket supports doubled beams around the stairways in both houses. Look for this where the floor plans are flipflopped, so that the doors to both units are adjacent.

A typical party wall fire extension occurred at a fire in Ottumwa, Iowa involving two stores separated by a party wall in which joists were lapped in common sockets. The fire in the original store was under control. No examination was made of the ceiling void of the adjacent store. Suddenly, the ceiling void of the adjacent store was found to be involved. A backdraft explosion drove the fire fighters out, and the building was lost. The fire had passed into the ceiling void of the adjacent store by way of the joist openings.

The right of a party wall to exist continues until terminated by agreement of both parties who share the wall. An interesting situation arises when a skeleton frame building is to be erected on the former site of a party wall building, while the other is to remain. The party wall is picked up on the frame of the new building, as a curtain wall. If the older building is thereafter demolished, the party wall remains as a protrusion beyond the line of the framed building.

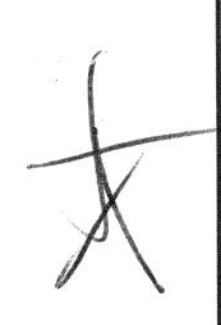

Tactical Considerations

Extension through party walls is sometimes hard to detect. Examination of the adjacent premises early in the fire may show nothing. It may take time for the fire to work its way through. Smoke, which might ordinarily give cause for alarm, goes unnoticed in the general smoky condition of the fire. Without some evidence, officers are often reluctant to order the necessary opening of ceilings. When the need becomes evident, it is often too late.

Partition Walls

Some modern rows of stores approach the definition of noncombustible construction, but they can present problems of fire extension similar to those of party walls.

Partition walls between stores may be of unpierced concrete block. The top chords of steel bar joists rest on the wall. Take a close look at the next such building you see under construction. Note that the design of the bar joists provides a gap about three inches in height between the top of the wall and the roof. Some building departments require that this gap be closed.

Depending on the local building code, there may be no ceiling, a combustible ceiling, or a noncombustible ceiling. The ceiling may be over all or only part of the store. If the ceiling or roof is combustible, like an ordinary metal deck roof, the extension of fire is almost a certainty. If the void is totally noncombustible, smoke and steam may pass to do extensive damage.

The partition wall may be gypsum board on wood or steel studs. In many cases, it extends only to the ceiling. There is a faulty building code concept which regards compartmentation as complete if the "fire-rated wall" abuts a "fire-rated ceiling." The so called fire-rated ceiling may meet only flame spread standards. In any case, any failure of the ceiling admits fire to the void, and there is no further barrier to its extension.

"Fire-Rated Ceiling" is a misnomer. There are rated "Floor and Ceiling Assemblies." UL cautions, "Fire resistance ratings apply only to assemblies in their entirety. Individual components are not assigned a fire resistance rating and are not intended to be interchanged between assemblies but rather are designated for use in a specific design in order that the rating may be achieved."[24]

The building code may require fire separations between occupancies to extend to the underside of the roof. These are usually of gypsum board on steel studs. Often the bar joists pass right through the fire barrier. It is impossible to close the openings. Often the gypsum board above the ceiling line is untaped and the nail heads uncovered in the mistaken belief that this is cosmetic. In fact it is required for fire resistance. Any movement of the bar joists will displace the wall. Such a barrier can be useful if defended but it is unreliable if left alone. The partition wall can also be bypassed by the now popular yet combustible "eyebrows" or other false fronts.

The solid masonry wall parapeted through the roof is the most dependable fire barrier. However, just because you can see a wall extending through the roof, don't assume that it is an effective fire wall. It may be decorative, or it may be pierced.

In some cases, fire walls are made of combined elements. Consider a high bay steel-framed warehouse. A main girder is put in place with an upward camber. The parapet brick wall is built above the girder. Concrete block is built up to the girder. The girder is sprayed with fireproofing. There is no proof that these disparate elements will function together as a fire wall. If steel attached to the girder is heated, it may move the girder.

It is important not to take anything for granted in ordinary construction. Every building should be considered as one interconnected fire area, unless it is otherwise known. If fire officers are trained to regard the entire building as the fire area, rather than just the portion where fire is visible, there will be fewer disastrous surprises.

Risk Analysis

When reports of fire fatalities from any type of building are studied, certain questions keep coming to the fore. "Why were those fire fighters where they were? What was the potential benefit from the risks undertaken?" Too often the answers to these questions have been unsatisfactory. If the answers were completely truthful, it might be learned that the real reason was an antiquated macho attitude on the part of fire fighters. This must be changed.

Fire fighters have been injured and have died unnecessarily in many fires. Taxpayers have been saddled with unnecessary expenses because fire fighters were operating in buildings that could not possibly be saved. It may provide a good feeling to say, "We made a fine stop," when fire fighters extinguish a fire in an already half-destroyed structure. However, fire departments are organized to provide a service, not to make people feel good.

If the fire service is to be regarded as truly professional, analysis of the relationship between risk incurred and benefits obtained must be undertaken.

There are many unavoidable risks in fire fighting, and they are all the more reason for the fire officer, who must consider the lives of the fire fighters, to be well informed about construction features of buildings that have been available for study for many years.

Tactical Considerations

Preplanning the Fire

An officer preplanning fire ground tactics for buildings of ordinary construction should consider the following interrelated factors.

- *The probability of exterior collapse.*
- *The potential for interior collapse.*
- *The existence and interconnection of void spaces in buildings.*
- *The potential for extension to or from adjacent buildings.*

Size-up

Even where prefire planning was not accomplished, much can be learned at the time of the fire and deduced from a good knowledge of similar buildings. Walls which are obviously deteriorated, braced, or tied certainly indicate that a floor collapse will probably bring the walls down.

The volume of fire gives a clue as to how long the floors will last. Ordinary wood-joisted floors are not formally rated by any standard fire-resistance test, but it is dangerous to trust them for more than 10 minutes. It's important to note that the 10 minutes may have expired before the arrival of the fire department.

Heavy volumes of boiling smoke persisting even after the visible fire has been well controlled indicate fire in inaccessible voids. However, light smoke or no smoke at all does not necessarily mean that there is no or little fire in the voids.

A serious error is not to prepare to use heavy volumes of water. The common old-style building made with ordinary construction methods and materials is a combustible structure surrounded by masonry walls. Master streams used in the initial attack may reduce the heat production of a fast-growing contents fire and hold the fire

Tactical Considerations, continued

out of the structure of the building. Proper training of company officers will prevent a ridiculous initial attack with a booster line on a fire already far beyond the heat-absorbing capacity of the small line.

Danger to Exposures

In the case of row buildings, expect extension to the adjacent exposures. Get them opened up early and get charged hose lines into position inside the exposed buildings. If a shortage of personnel does not permit such lines to be kept operative, leave the charged lines in place with a patrol. In too many cases, the fire department "suddenly found that the fire had extended," and by the time operations could be rearranged, the exposed building was fully involved.

Anticipate the possible extension of fire as a result of collapse. The collapse of a grain elevator extended a fire across a wide waterfront slip to a fire-resistive warehouse in Jersey City, New Jersey in 1941. The author watched this tremendous radiant heat source force open wire glass factory windows on the warehouse.

Safety for Fire Fighters

Reduce the number of fire fighters in hazardous positions to the barest minimum. All lines which might require sudden abandonment should be in master stream devices. Be sure that all fire fighters have a path of escape and that they are aware of it.

If the fire is not being substantially reduced in volume, the operation is not a standoff — the building is being destroyed. Positions which might have been relatively safe at an earlier stage may now be unsafe.

As with all other types of buildings, every fire department should have a clear, readily understood signal which means "Everybody Out Now" — without hose lines or other equipment. Drills should be conducted on this tactic, and the use of the signal should not be restricted to the fire ground commander.

Overhauling

When the fire has been controlled, do not be in too much of a hurry to commence overhauling. Make a thorough survey of the building to determine the extent of structural damage. Seek professional assistance from the building department. Use fresh alert units, not personnel whose judgment may be weakened by fatigue. Wait until daylight.

Fireground Safety

Fire fighter safety at the scene is a critical factor that all departments must face—not just those charged with fighting certain occupancies.

One method which could significantly improve fire fighters safety at the fireground is for the fire department to set up a Board of Building Review. Experienced officers, who have studied building construction, assisted by consultants from the building department or private sources, could survey major buildings of ordinary construction and typical smaller buildings. Considering the available fire load, the type of construction, the nature of connections to adjacent buildings, water supplies, protection provided, and other relevant circumstances, an upper limit in minutes could be put on the time that fire fighters were permitted to stay in the building. The time rating of the building would be part of the prefire plan. The fire ground commander, of course, would have the power to order the fire fighters out before the expiration of the rated time, but strong restrictions should be placed on authority to keep them in the building after the expiration of the rated time.

The fire ground commander would be substantially relieved of an unfair responsibility, i.e., determining the boundary between stability and collapse under fire conditions. In a fire the boundary between building stability and instability is a thin line. Ignorance of the boundary is no excuse when a failure occurs.

This author recalls an unfortunate statement made by a fire chief. "Nobody is going to tell me how to fight MY fire." The problem is that it wasn't HIS fire. The day of a chief fire officer not being accountable is over.

The following quote from "Notes on Fire Ground Command" by Chief Alan Brunacini in *Fire Command*, January 1981, is still appropriate.

> "When the FGC decides to change from an offensive to a defensive mode, he announces the change (as emergency traffic), and all personnel must withdraw from the structure and maintain a safe distance. In such cases, the FGC must maintain an effective organization with adequate, well-placed sectors. This system is designed to control both the position and the function of operating companies (when the FGC says "out," he better have an effective organization already in place). In these retreat situations, sector and company officers assemble and account for all personnel on the outside of the building (roll call time). Interior lines are withdrawn or abandoned, if necessary, and repositioned in a defensive mode. Lines should not be operated in door and window openings, but should be backed away so that personnel are in safe positions."

Mill Construction

Mill Construction is a special type of heavy timber and masonry construction. It developed in the textile mills of New England. Mutual insurance companies created a fault-tree analysis of the factors which caused massive losses in factories. A building design was developed which eliminated each of the serious faults.

- Exterior bearing and nonbearing walls are solid masonry, usually either brick or stone.
- Columns and beams are of heavy timber with cast iron connectors used to cover joints where the fire might obtain a hold. The columns are chamfered with beveled corners for the same reason.
- Floors are of thick grooved, splined or laminated planks.
- Roofs of thick splined or laminated planks are supported by beams or timber arches and trusses.
- Openings between floors are enclosed by adequate fire barriers.
- The ends of girders are fire cut to release in the event of a collapse without bringing the wall down. Sometimes a cast iron box is built into the wall to receive the end of the girder. Sometimes the beams are set on a corbelled brick shelf.
- There are scuppers or drains in the wall to drain off water. The combination of waterproof floors and scuppers is provided to reduce water damage on lower floors. If the scuppers do not drain, there may be a serious increase in the floor load due to retained water. The scuppers should be checked for debris. A broom handle will hold the scupper open and increase the flow.

In removing the water which flowed from a ruptured sprinkler riser in a huge navy warehouse at Norfolk, Virginia, fire fighters discovered that the concrete floors had been graded *uphill* to the scuppers.

Note that scuppers were provided in sprinklered buildings to remove water, but not unsprinklered buildings. What did the designers think the fire department was going to use to extinguish the fire? Fire protection is illogical at times.

- Concealed spaces are eliminated. The finish is "open" without voids, such as can be created above dropped ceilings or behind wall sheathing. If a paneled office or display room has been constructed within the open floor area of a mill building, it may provide a destructive hiding place for fire. Such spaces can prevent detection and extinguishment.
- The protection of vertical openings and the division of the building into sections by fire walls are vital. If fire walls are to have any meaning in limiting the loss to the area in which the fire starts, openings in them must be protected by operable, self-closing or automat-

ically closing fire doors suitable to the intensity of fire to be anticipated.

- Most important is an automatic sprinkler system with a water flow alarm and an alarm connection to the fire department. A dependable system of supervisory locks, alarms, and checks is necessary to make sure that the sprinklers are continually in service. If repairs must be made, special fire prevention and fire suppression precautions must be taken.
- Special hazards should be located in detached buildings.

Properly built and maintained, a mill construction building can be a structure in which fires can be brought under control before the building is involved in the fire. If one or more of the vital characteristics is absent, either by original design or because of the way the building was modified, the building can become involved in the fire. Once this happens to any substantial degree, the heavy timber characteristic becomes a liability, not an asset. The so-called slow-burning characteristic is of no value once the fire department must fall back to defensive tactics, and in truth becomes long burning. It is hard to understand how a living tree in the forest which has not been rained on in three months presents "serious fire danger," while a tree trunk used in a building which hasn't been rained on in over a hundred years is "slow-burning."

Conversions, Modifications, Preservation

Many of these sturdy old buildings have been around for a long, long time, and often efforts made to remodel them have detracted from their original fire characteristics. For example, steel trusses or girders may have been inserted, making it possible to remove some columns. The new trusses, by their very nature, do not have the inherent resistance of the massive wooden columns and beams they replaced. If the new trusses were of steel, 1-hour fire protection would be required for them in many codes . . . but there has been many a slip, twixt the book and the building.

Many such buildings are no longer used for their original purpose. They are used for tenant factories, storage, discount stores and, in some cases, are being subdivided into apartments. In many instances, the basic principles set forth earlier are seriously compromised. Fire loads are often beyond the capacity of the installed sprinklers or the sprinklers are turned off or even removed. Unsprinklered areas are created. The result can be and has been disaster.

There can be a question of who is responsible for the sprinkler system, particularly in a multi-tenant building.

In many of these conversions, assistance is provided by local government in order to better local economic conditions. Great pressure may be put on fire authorities to waive red tape and unrealistic regulations in a "slow-burning" building.

The fire safety of a heavy timber building depends on stern maintenance of the features listed earlier. If the authorities persist in scrapping necessary fire precautions, the deficiencies should be clearly noted so that there will be no misunderstanding of the potential for a disastrous conflagration.

Realistic pre-fire scenarios, forecasting severe economic impacts of such a disaster, presented by top fire department offices to plant management and city officials might get attention, particularly if the loss of jobs and tax revenue are emphasized.

Code-Classified Heavy Timber Construction

In many codes, there is a classification called heavy timber construction. Unfortunately, a building conforming to the code definition of heavy timber construction probably lacks at least one or more of the features which the mill construction designers had learned were vital for fire safety. Typical departures from the true mill construction concept are cast iron or unprotected steel columns, steel or part-steel trusses, unsprinklered void spaces, highly flammable contents, unprotected vertical openings, and inadequate or no sprinkler protection.

Tactical Considerations

Fire Departments should place "maintaining the city's revenue and employment base" high on the list of priorities. If an old mill is destroyed, the businesses located there will either go under or find another old mill in another town. Frankly, this economic base approach may have more appeal to city managers and politicians than vague references to saving life and property.

The ONLY defense against a raging fire in an old mill of heavy timber construction is a sprinkler system — in full service. Reliance on manual fire fighting is inadequate. The fire department must know the status of every critical sprinkler system and maximum pressure must be exerted to get inoperative systems back in service. In the interim, extraordinary measures should be taken to provide some special protection. If the pump is out of service, the fire department might make a spare pumper available for hook up to keep up the water supply. The city might consider paying the cost, and assessing a claim against the structure.

Tactical Considerations, continued

Many fire departments back off from checking private fire protection. The attitude is "It's their responsibility, not ours." Fixing responsibility runs a poor second to putting the fire out.

Many fire departments' preplan programs are either primitive or non-existent. The preplan should provide all the available information which the fire ground commander needs. It should include a provision for the huge water supplies that will be crucial to suppression. Red paint and chrome does not put out fires.

This author recalls a conversation with some wreckers who pointed out that once the "goodies" were removed from a building, an "accidental" fire was the cheapest way to clear the site. Fire departments should review their procedures for dealing with buildings being demolished. Special on-site precau˜tions might be justified for the last few days or weeks of the demolition process.

Review NFPA 241, Standard for Safeguarding Construction, Alteration, and Demolition Operations.

By far, more fire fighter lives have been lost in ordinary construction fires than any other type. The information and suggestions presented in this, and other chapters, may serve to reduce this tragic fire death toll.

We must no longer accept this comment by a chief officer after a three-fatality fire. "I didn't learn anything from this fire. If we fought the same fire again, I would have followed the same tactics."

Case Studies for Ordinary Construction and Heavy Timber Structures

These case studies of fires from the fire service press illustrate some of the material presented in this chapter. The case studies are not just abstracts. In many cases, this author's comments go beyond the thrust of the article. Many lessons learned at fires are relearned by other fire departments, sometimes at great cost. A great deal of very useful information is published in the fire service press. This author and Deputy Chief Vincent Dunn FDNY contributed many articles to the fire service magazines. The essence of many of these articles is contained in this text and Chief Dunn's book *Collapse of Burning Buildings*. Thus no articles by either of us are included. The first group of case studies involve fires in ordinary construction, the second group illustrates fires in heavy timber structures.

Ordinary Construction

Anderson, Ross, "Eight Alarm Blaze Threatens Richmond Historic Area," *Firehouse*, July 1985, p. 34. Several century-old brick and timber buildings were destroyed or damaged. The loss was limited because units immediately deployed large diameter hose and heavy caliber appliances. This fire reinforces the point "big fires take big water." This was immediately noted by the acting officer on the first unit to arrive.

Best, Richard, "Two Major Shopping Center Fires," *Fire Engineering*, May 1984, p. 38, 43. Shopping centers in Denver, Colorado and Brunswick, Georgia were heavily damaged by fires which spread through unfirestopped and unsprinklered overhead voids. In the Brunswick Mall fire, the building had been operated for several years without a certificate of occupancy. The fire chief had been unable to get a correction. The article contains photographs of the Brunswick Mall in which a heavy laminated timber beam fell because its supports gave way, even though the beam was only lightly charred. Do not overestimate the value of any specific material. Look at the entire connection system.

Birr, T., "On The Job, Oregon." *Firehouse*, June 1990, p. 90. Note the fire extension due to joists in common party wall opening.

Brennan, T., "Building Collapse Shouldn't Be a Surprise," *Fire Engineering*, January 1984, p. 16. This is an excellent summary of collapse factors. Color illustrations show fire fighters operating on a building apparently unaware of the impending collapse which occurred 20 seconds later. Note the use of the terms "heavy timber" and "mill construction" as synonyms. They are not. See this chapter's discussion of Mill Construction.

Brunacini, Alan, "Fire Command," *Firehouse*, June 1985, p. 82. Note the heavy emphasis on fireground safety. It is an unfortunate fact that some fire chiefs think that safety is incompatible with fire ground operations. This author has been told point blank, "I don't want my guys hearing your stuff about the hazards of buildings. I want them to go where I want them to go." It will probably take a few personal judgments against chief officers to change this philosophy.

Brunacini, Alan, "Fire Ground Command," *Fire Command*, January 1985, p. 19. The command problem when the fire is not clearly offensive or defensive is discussed. Chief Brunacini stresses the need to keep evaluating the situation. This author's study of fire ground fatalities leads to the conclusion that in these situations there is often a sort of paralysis of the thinking process, and a failure to realize that there are no standoffs. If you are not winning, the fire is.

Burns, R., "Flashover," *Firehouse*, July 1988, p. 64. Dramatic story of fire bursting out of hidden spaces left behind by advancing units. Quick thinking saved a fire fighter trapped upside down at a window.

Caggiano, Joseph F., "The Hotel Gregory Fire." *WNYF*, Third Issue, 1984, p. 4. Fire in the voids of this brick- and wood-joisted hotel gave little indication of its intensity to arriving units. Almost a half hour into the alarm a backdraft explosion occurred in the cockloft.

Caulfield, H., "Voluntary Compliance," *Fire Engineering*, June 1986, p. 44. This article quotes a chief officer narrating, apparently with justification, how he ordered a company into a building he knew was full of fire. One of the fire fighters jumped from a window and landed at his feet.

Colella, B., "Get Ready for the Big One." *Fire Command*, March 1990, p. 28. A good discussion of getting the proper size lines into action.

"Cover Photograph," *Fire Command*, August 1984. The cover picture is an excellent shot of fire fighters fleeing a falling wall.

Cox, Bradley, "Disastrous Fire Halts Greensboro Renewal Project," *Firehouse*, September 1985, p. 121. Another fire involving a group of old downtown brick- and wood-joisted structures being rehabilitated. Signs of a backdraft were noted and personnel were removed just before the carbon monoxide explosion blew all four walls down. Personnel were caught, but fortunately were not seriously injured.

Dalton, L., "$8.3 Million Arson Blaze Guts Grantsville High School," *Firehouse*, May 1984, p. 36. Fire started in a dumpster, extended to the overhang and destroyed the school after entering the ceiling void. Dumpsters are common arson targets and should be located far enough from buildings to prevent fire extension. Fire entered the four-foot deep ceiling roof void and could not be stopped.

Ditzel, P., "Five Alarm Fire Destroys Century Old (Buffalo) Church," *Firehouse*, May 1987, p. 30. Another vacant large combustible building. The building was scheduled for demolition. After some time interior operations were discontinued. Shortly thereafter collapse started. The damaged structures were demolished.

Eisner, Harvey, "Church Challenges," *Firehouse*, April 1984, p. 65. Two church fires illustrate the hazards of 19th Century-style churches. Huge voids and collapse potential dictate defensive operations from safe positions.

Eisner, Harvey, "Immense Fire Devastates Brooklyn Church," *Firehouse*, March 1985, p. 34. This is a good example of the tremendous exposure hazard created by a 19th century ordinary construction church. Many such churches are located in residential areas where the water supply is inadequate for the huge fire which can be anticipated. Fire department pre-plans should provide for the delivery of huge quantities of water, and exposures must get full attention early in the fire. This church had stood for 125 years. It was destroyed in less than two hours.

Featherstone, J.J. III., "The Problems With Converted Buildings," *Fire Engineering*, January 1984, p. 14. An old school is converted into apartments for elderly people. A severe potential life hazard is created because the converted apartments are located on the second floor. An open stairway leads from the first floor where the kitchen and recreation room, both likely fire origination spots, are located.

"Granny Flats" are permitted in existing one-family homes. The local code provides that there must be no change in the exterior appearance of the house, so there will be no clue as to the occupancy.

Feldman, Diane, "Dispatches." *Fire Engineering*, August 1989, p. 60. Failure of steel beams pushed a masonry parapet out of line. Two fire fighters were trapped in the rubble.

"Fire Record," *Fire Journal*, November 1985. In California, a ballroom had a huge roof-ceiling void 12 feet deep. The fire could not be located. As units were being pulled out, fire burst out of the void. One fire fighter was killed. An Iowa college had an undivided wooden attic over four three-story buildings. Fire starting on the roof involved the attic. Damage was $10 million. A California auto repair shop attic fire caused a roof collapse which brought down the rear wall and endangered fire fighters.

Gentleman, G., "Use of Firefighting Basics Thwarts Mall Fire," *Fire Engineering*, February 84, p. 20. A serious fire in a partially sprinklered mall was cut off by aggressive fire suppression tactics. The author notes difficulty in determining exactly where to vent the roof due to the absence of "landmarks" common to both inside and outside of the mall. The fire started from a fluorescent light fixture too close to a combustible structure.

Hall, George, "Smoky Five Alarm Blaze Consumes Warehouse," *Firehouse*, August 1985, p. 39. The warehouse was loaded with urethane furniture. Fire spread through partially-open fire doors. Serious fire door deficiencies are usually called minor violations because it often takes only effort, not big money, to correct them.

Harlow, David, "New Roofs Over Old," *Fire Engineering*, November 1987, p. 51. A good discussion of the hazards of adding a new roof over an old roof. This is a common practice to solve the leaky roof problem or to change the architectural aspect of the structure.

Hart, Roger, "Attic Blaze Rips Historic Adrian Church," *Firehouse*, August 1985, p. 25. Fire involved the attic of what appears to be a wood-frame brick veneer church. The fire was confined to the attic. Little smoke was showing on arrival, so no hydrant hookup was made. I keep saying, "What you see ain't necessarily what you've got." Heavy fire can be concealed in a void with little or no smoke showing. Intense heat was found when the door was opened. Fire fighters got out just in time. A backdraft blew out a portion of the brick gable of the attic. Note that the story speaks of water damage to the church organ. Let's get rid of that term. All damage is fire damage. If it were not for the water, the organ would be entirely destroyed. There is no water damage at a total loss.

Herzog, C. and Broehm K., "$2 Million Loss Occurs in Shopping Center," *Fire Command*, August 1988, p. 18. Combustible framing in a continuous void above the sprinklers burned and did extensive smoke and heat damage to the contents. The fire fighters were wary of collapse of the steel trusses under heavy air conditioners. They apparently did not anticipate collapse of the masonry facade, possibly due to the elongation of steel.

Lieper, R. D., "Lessons From School," *Fire Engineering*, November 1987, p. 22. False ceilings provided voids for hidden fire to spread throughout the second floor of a high school. Fire burning down through the floor spread through the first floor-ceiling voids. Fire fighters were endangered by hidden fire above and below. Intense heat and heavy smoke on the second floor convinced fire fighters to withdraw in time to prevent serious injury or death.

Louderback, J., "On The Job, Pennsylvania," *Firehouse*, November 1988, p. 71. A doorway cut through the masonry walls to connect two stores and common basements allowed fire to spread.

Mc Fadden, P., "Residence Fires Starting in the Cellar," *Fire Engineering*, May 1987, p. 8. The author describes the difficulty of cellar fires.

In most codes a cellar has half its height below the grade of the street, a basement has at least half its height above the grade of the street. Usually basements but not cellars can be used for living space. From the suppression point of view there is no distinction. Ask the owner where the gasoline for the lawnmower is kept. If it is in the cellar, react as if you were told "Dynamite." High expansion foam is sometimes very effective on cellar fires if it can be confined. Fumes from some burning plastics will destroy the foam structure. The fire may be pushed up into the joist spaces and continue to burn. Be aware of the potential for fire extending up into the voids in the house above. This is almost a certainty in older homes of multi-story balloon frame construction.

Milnes, Robert, "Dade County Firefighters Battle Shopping Center Blaze," *Firehouse*, May 1985, p. 40. This shopping center had been altered a number of times. Many hidden voids were created. It appears that the alarm was delayed because a security guard called the shopping center manager instead of the fire department.

Mittendorf, J., *American Fire Journal*, July, August, September, November 1986. In a four-part article, a Battalion Chief of the Los Angeles Fire Department discusses the hazards of various types of roofs.

Mittendorf, J., "Victorians In The '90s," *Fire Engineering*, April 1990, p. 40. While the Los Angeles structure described is of wood construction, brick Victorians differ only in the exterior walls. Interior walls are balloon frame. Excellent practical information.

Nelms W., Jr., "Wall Collapse Injures Four Metro (Nashville) Firefighters," *Firehouse*, January 1986, p. 39. A typical wall collapse of a fully involved vacant building which had been damaged by an earlier fire. Bricks flew as far as one hundred fifty feet from a four-story wall collapse.

Phelps, Burton, and Mc Donald, Edward, "Pre-fire Planning," *Fire Engineering*, October 1989, p. 16. This planning guide provides information on the construction of the building up front in the plan.

Rakestraw, R., "Reflections on the Death of a Firefighter," *Fire Command*, April 1986, p. 14. The author, Chief of the Petersburg, Virginia Fire Department, shares his experiences and feelings after the death of a fire fighter in a collapse. It is a fascinating account. (The following remarks are absolutely unrelated to the fire which cost the life of fire fighter Goff.) This article should be must reading for those who denigrate safety, who want aggressive fire fighters who charge in, and who don't want their people hearing anything about building hazards. Put yourself in the shoes of a chief who must go to the fire fighter's relatives or children to tell them that their loved one is dead. Will you be able to say that the death was not only unavoidable but worth the risk?)

Rose, Kerry K., "Building Construction Aids in Fire Control," *Fire Engineering*, November 1984, p. 54. Vertical and horizontal voids in this brick- and wood-joisted commercial building defeated an interior attack. Fire-cut joists and unpierced masonry bearing walls helped prevent extension.

Simmons, A., "Burbank and Glendale, California Fire Fighters Attack Multiple Alarm Blazes." *Firehouse*, June 1988, p. 41. Good description of tactics used in fighting major multiple-alarm fire.

Simmons, A., "'Get Out,' Falling Wall," *Fire Command*, September 1984, p. 26. A roof collapse pushed down a one-story masonry wall. Several Los Angeles fire fighters escaped injury due to a timely warning. Apparently some fire fighters were in the collapse range after roof operations were discontinued due to possible collapse. A roof collapse may well bring down the walls, as it did in this case. Captain Simmons' illustrations are dramatic.

Simmons, A., "Hotel Fire Threatens Downtown Burbank, California." *American Fire Journal*, May 1989, p. 24. The first due captain correctly orders evacuation of old brick building due to potential wall collapse.

Skarbeck, J., "Vacant Structures: The Sleeping Dragon," *Fire Engineering*, December 1989, p. 35. An excellent discussion of the hazards of vacant buildings by a working fire commander. There is a very good list of questions to be answered before commencing an offensive attack.

Smith, J., "Collapse Zone Boundaries," *Firehouse*, August 1988, p. 38. This is a good discussion of tactics to be employed in the face of potential collapse.

Smith, J., "The Complexities of Church Fires," *Firehouse*, August 1989, p. 14 (Also p. 90 and 92). A good discussion of this complicated subject. In this same issue: "E. Provident Fire Fighters Battle Landmark Church Fire" and a photographic story of a church fire that from arrival to destruction took 66 minutes.

Stronach, I., "Multimillion Dollar Convention Center Saved in Fire (Montreal)," *Firehouse*, March 1986, p. 69. This vacant brick and wood structure was so heavily involved on arrival that a fifth alarm was ordered in three minutes. The fire threatened the nearby fire-resistive convention center. In about 50 minutes the entire structure collapsed. The damage to the convention center was about $3 million. Approximately 1000 windows were broken and the openings had been covered with plywood. The use of large areas of plywood or other readily combustible sheeting to cover openings can create the hazard of very rapid flame spread.

Heavy Timber Structures

The following case studies include fires which involved heavy timber structures. In many of these cases the loss was due to lack of or impaired sprinkler protection.

Best, R. *et. al.* "Demolition Fire Ignites Bank." *Fire Service Today*, July 1983, p. 14, and Best, R., "Demolition Exposure Fire Causes over $90 Million Loss to Minneapolis Bank." *Fire Journal*, July 1983, p. 60. These articles refer to the most financially damaging blaze in a building being demolished which occurred in Minneapolis, Minnesota. The building was not even a true heavy timber building. The exterior framing was steel or concrete. The interior had wood and steel columns. The flooring and partitions were wood. The fire exposed the 16-story unsprinklered Northwestern National Bank Building. This building was gutted from the sixth to sixteenth floor. A meeting of concerned city departments was held before the demolition started. There is no indication that the exposure hazard to the bank was ever considered. Fire was spotted in the exposed building seven minutes after arrival. Five minutes later several floors were involved.

Biswanger, W., "$8 Million Blaze Guts Fitchburg (Massachusetts) Warehouse." *Firehouse*, January 1987, p. 38. An old mill was being converted into apartments. A welder was cutting away pipes. A spectacular multi-alarm fire resulted. This scenario is repeated far too often. The article notes that nothing was known about the sprinkler system in the building.

Biswanger, W., "Nine Alarm Fire Ravages Lowell (Massachusetts) Mill Complex." *Firehouse*, September 1987, p. 40. Another huge daytime fire in an old mill. The sprinklers in the original fire building had been turned off after a problem two months earlier. Apparently, this was not known to the fire department. The loss was at least $30 million — a substantial blow to the city's tax income. Some 400 persons lost jobs.

Burns, R., "Nine Alarm Fire Razes One Building, Damages 17 Others." *Firehouse*, July 1984, p. 19. The six story, brick- and wood-joisted Harrison Court building was under reconstruction. A spectacular daytime fire destroyed the building, and extended to 17 others and threatened to destroy downtown Philadelphia.

Corbett, G. P., "Conflagration in Passaic," *Fire Engineering*, November 1985, p. 29. A vandalism fire extended to 17 houses and 60 businesses. The loss was one life, $400 million, 2200 jobs, and 17 homes. Water supply in the fire area was a major problem. The initial attack failed because a 100,000-gallon tank intended to protect the industrial area had been out of service for three years. The article presents a number of excellent conclusions.

Doctor, E., "Built to Burn." *WNYF*, Second Issue, 1984, p. 4. Six 19th century loft (tenant-factory) buildings were altered to what appeared to be a single new apartment building with 116 units. Huge horizontal voids, as much as 4500 square feet, were created by the dropping of ceilings. Utility shafts interconnected all the voids. There was no fire evident from the exterior on arrival but it appears that the fire had extended to eight apartments before arrival. There were 24 engine companies and 16 full-strength truck companies operating. This fire points outs the necessity for extensive truck company work at such fires. Departments weak in truck units should consider training some engine companies to function as truck companies.

Eisner, H., "Fast Moving Fire Destroys Fall River (Massachusetts) Mill Complex." *Firehouse*, April 1987, p. 46. There were 800 jobs lost when fire destroyed a huge old textile mill which had been converted to miscellaneous tenant occupancies. The fire totally destroyed four buildings, including two fire-resistive structures. The entire complex was sprinklered, but there were no fire department connections. Open air ducts ran from the basement to the upper floors. There was no explanation why the fire reached disaster proportions before the first alarm. The fire department planned to initiate a comprehensive fire prevention program for the remaining 92 mills, and more in-service inspections by fire companies.

Eisner, H., "Three Alarmer Strikes Jersey City Warehouse." *Firehouse*, November 1987, p. 26. A typical industrial-area heavy-timber warehouse fire where the Btus greatly outnumbered the H_2O. The fire department was not aware that the sprinkler system was out of service, and that there were no stairways in the building.

Eisner, H., "Trapped." *Firehouse*, June 1980, p. 30. One fire fighter was killed in a fire in a dilapidated multi-story brick and timber building used for baled rag storage. Two others were killed in the collapse of the roof of an exposure. A "light smoke showing" fire suddenly flashed over, apparently from another fire in an area beyond a fire wall. One officer fell from a window to his death, another was injured, others escaped. The fire extended to the roof and into the interior of an adjacent three-story sprinklered, mill-constructed warehouse. Two fire fighters operating on the third floor were killed when the top story of the firewall collapsed on them. Many cities have permitted such "disasters waiting to happen." In this case, the city had appropriated $324,000 for the demolition of the building three days before the fire.

Przybylowicz, H., "Three Alarm Fire Consumes Waterfront Warehouse." *Firehouse*, March 1984, p. 57. This is another fire which demonstrates the high hazard of unoccupied brick and timber buildings, in which the only fire load is the interior structure. Justified fear of collapse of the brick walls dictated an exterior attack. This building was on the waterfront and no exposure problem is mentioned.

Taylor, K. "Burning Down the School," *Fire Journal*, May/June 1990. A fine summary of five years of school fire experience.

Wentworth, P., O'Reilly, J., "Backdraft and Blaze Gut Landmark (Providence, Rhode Island) Building." *Firehouse*, March 1987, p. 49. Another fire in a huge vacant combustible building in a congested area. Because the building was unheated, the sprinklers were shut off. First alarm at 6:31; all fire fighters evacuated at 7:18 due to intense heat and heavy smoke. At 7:43 there was a tremendous backdraft on the fifth floor. There were multiple layers of ceiling. There was no fire department siamese connections on the sprinkler system. Fire fighters succeeded in opening a valve in the basement. The article stated there was no warning of the backdraft. The tremendous heat and heavy smoke which forced withdrawal seem like a certain signal.

Williams, D., O'Farrell, C. "Preplan for Collapse," *Firehouse*, June 1988. Atlanta Fire Department credits preplanning for avoidance of injury in a collapse. Note that the interior sector commander ordered evacuation and clearance of collapse area on his own authority when he could not raise the fire ground commander on the radio.

References

The addresses of all referenced publications are contained in the Appendix.

1. Condit, C. W. "American Building," *The Chicago History of American Civilization*. Boorstein, D. J., *ed*., University of Chicago, Chicago, Illinois, 1968, pp. 64 and 131.

2. Adams, R. "The Importance of Being Thorough." *Firehouse*, June 1990, p. 14.

3. Terrenzi, W. "Official Report of the Trumbull Street Fire." Oct. 1, 1964, Boston Fire Department, Boston, Mass.

4. Ullman, K. "The Fatal Bowman St. Blaze." *Firehouse*. June 1990, p. 50.

5. Feld, J. *Construction Failures*. Wiley, New York, N.Y., 1968, p. 56.

6. McKaig, T. *Building Failures*. McGraw Hill, New York, N.Y., 1962, p. 203.

7. *Engineering News Record*, Oct. 8 1973.

8. McMillan, A. B., and Ross, Steven, *ed*. *Construction Disasters*. McGraw Hill, New York, N.Y., 1989, p. 287.

9. Steckler, K., Quintieri, J. and Klote, J. "John Sevier Home, Johnson City, TN Fire: *Letter Report*," NIST Center for Fire Research.

10. Pitts, W. M. "Long Range Plan For A Research Project On Carbon Monoxide Production And Prediction." NISTIR 89-4105.

11. Isner, M. "Elderly Housing Fire." Johnson City Tennessee, Dec. 24 1989. NFPA Fire Investigation Report, Quincy, Mass., National Fire Protection Association.

12. Stronach, I. "On the Job: Montreal." *Firehouse*, November 1987, p. 42.

13. "Fire Tests of Building Columns." Published jointly by Underwriters Laboratories Inc., Factory Mutual Insurance Companies, and The National Bureau of Standards, U.S. Department of Commerce, 1920.

14. Brannigan, Francis. *Building Construction for the Fire Service*, 2nd ed., Quincy, MA, National Fire Protection Association, 1982, p. 379.

15. Kidder, F., and Parker, H., *eds*. "Weakness of Wrought Iron Stirrups When Exposed to Fire." *Kidder-Parker Architects and Builders Handbook*, 18th ed., Wiley, New York, N.Y., 1954, p. 890.

16. Hool, G. S., and Johnson, N. C., *eds*. *Handbook of Building Construction*, 1st ed., Vol. 1, McGraw-Hill, New York, N.Y., pp. 308 and 378, 393, 394.

17. Eisner, H. "Three Firefighters Killed in Collapse." *Firehouse*, May 1985, p. 58.

18. *Washington Post*, "Razing the Roofs," Sept. 16 1990, p. H1.

19. Parkin, L. "FRT Plywood, Not As Safe As It Sounds." *Fire Engineering*, May 1990, p. 27.

20. Peterson, C. E. "Fire Delays the Opening of Two New Schools." *Fire Journal*, Vol. 62, No. 2, March 1968, pp. 19-22.

21. Peige, J. "On the Job, Baltimore County." *Firehouse*, Dec. 1984, p. 34.

22. Cote, A., *Ed. in Ch. Fire Protection Handbook*, 17th ed., National Fire Protection Association, Quincy, Mass., 1991 Section 6, Chapter 6, p. 6-85.

23. Sachen, J. "Multiple Alarm Blaze Rips Warehouse." *Firehouse*, May 1984, p. 154.

24. Underwriter's Laboratories, "Design Information." Fire Resistance Directory, 1990, p. 7.

5

Garden Apartments and Other Protected Structures

For ten years this author investigated low-rise, combustible multiple dwelling fires for the Center for Fire Research of the National Bureau of Standards (NBS). The NBS is now known as National Institute of Standards and Technology (NIST).

The purpose was not to uncover the fire cause, but to determine why the fire extended beyond the area of origin in a building of so-called protected combustible construction.

Much of the material in this chapter is based on these fire investigations, as well as the study of buildings under construction in many distinct jurisdictions.

Protected Structures

Combustible multiple dwellings include garden apartments, row houses, town houses and similar structures. For convenience, the term garden apartments will be used to describe these structures.

Many motels and small office buildings share the same construction characteristics. Consequently, the comments and Tactical Considerations presented here are valid for these types of structures also.

The term condominium, while commonly used for many structures, is a financial/legal term. It is not usefully descriptive for such structures. It may be important legally, however, in determining ownership or where the building code holds individually owned units to a higher standard than rental units.

The fallacy of considering condominiums as a separate type of building is shown by so called "lollypop" condos. Additional floors are built above an existing non-condominium building. By law the condo must be connected to the ground. The elevator shaft is "commonly owned" and serves as the tie to the ground.[1] In another case of condominium ownership, this author has a friend who has a boat in a condominium marina. He "owns" a piece of the Chesapeake Bay.

Characteristics of Garden Apartments

Exterior structural walls of garden apartments are made of various materials:

- solid masonry
- brick veneer over platform wood frame
- partially solid masonry, partially brick veneer on wood
- wood

The usual height limit for these structures is three stories. By taking advantage of terrain, and the fact that building codes measure the height of a building from the street, such buildings can exceed this height and reach four stories on one elevation.

Many fire departments may not be able to reach victims at top-floor rear windows of such apartments. Ladders may not be long enough; response strength may be inadequate to raise long ground ladders; terrain conditions may make it impossible to move an aerial apparatus into position.

Tactical Considerations

The fact that some victims may be unreachable by the fire department should be made known to political authorities. This may provide added justification or impetus for residential sprinkler ordinances.

Usually individual living units are confined to one floor. However, town houses and row houses often are multi-floor units. Some apartment houses have duplex or triplex units with rooms in a single unit being on two or three floors. In some cases, the same structure may have both one floor and multi-floor units.

Fig. 5-1. This photograph, taken by a fire inspector who lived across the street, shows fire in two apartments as the fire companies are being alerted. (W. LeMay)

Balconies are customary in many such apartments — either extended or within the exterior bounds of the building. Balconies may be made of combustible or non-combustible construction. The balcony may also provide normal access to the apartment. If the balcony is cantilevered, an interior structural fire may destroy one end of the cantilever ("see saw") and collapse the balcony. Combustible and even partially-combustible balconies have been responsible for many vertical extensions of fire.[2]

In many of these structures, gable roof attics extend over the entire structure, with areas separated to the maximum permitted by the local code. Attic fire barriers are frequently as inadequate as the code or inspections will permit.

Fire department operations on peaked roofs are dangerous.[3] While some buildings appear to have flat roofs, they must have a pitch to drain rain water. This pitch creates a void between the tops of horizontal ceiling beams and the sloping roof. Fire can spread laterally through this unfirestopped space.

Regardless of exterior construction of garden apartments, the interior construction is almost totally of wood, using construction techniques similar to those used in single-family homes. When these dwelling units are stacked atop one another, this multiplies the fire extension potential through the voids inherent in combustible construction. When steel is used in these structures it is often not protected by any listed material or approved technique, and so must be considered unprotected.

In garden apartments, plumbing fixtures are vertically aligned, one above the other for economy and speed of construction. The necessary piping is run through vertical voids. Plumbers often weaken structural members during installation by cutting into them. The result is that a fire which starts in or penetrates this void could extend rapidly, both vertically and horizontally, as well as to the attic above. The weight of fixtures, structure cuts, and the air supply due to interconnected voids make early collapse a real hazard in these structures.

The firestopping this author has observed in garden apartments is generally deficient. See chapter 3 for a complete discussion of this subject.

A person who must escape from a single floor ranch home is infinitely better off than the person who must escape from the top floor of a combustible multiple dwelling. In too many cases, stairways and enclosures as well as attics overhead are combustible. The floor truss void may be extended to the stairway platform with only minimum fire barriers.

Such stairways are not a place of refuge for occupants nor a safe operating platform for fire fighters. Roof air conditioners mounted above a stairway collapsed on first-arriving units at an apartment fire. The hazard of combustible stairways was recognized almost a century ago. The New York City Tenement House act of 1903, the purpose of which

Fig. 5-2. Fire reached the attic by way of the hole around the sewer pipe. Why must the hole for a pipe be twice the size of the pipe?

was to provide life safety in combustible apartment buildings, required that stairways be non-combustible, and be enclosed in a masonry tower, with self-closing metal doors on every apartment.

Educating the Management and Tenants

The same right of privacy exists for those in multiple dwellings as it does in detached single-family homes. This is one reason why fire prevention education can have only a partial effect in multiple dwellings. Most fires start in tenant spaces. All tenants in apartment complexes are at the mercy of an individual tenant who, for whatever reason, disregards basic fire prevention measures, or who even starts a fire. In addition, tenants are at risk from fires starting in the public or common use spaces, whether accidental or incendiary.

Tactical Considerations

Public fire education should deal not only with fire prevention but also specifically with fire alarms. It is important for fire prevention or protection staff to recognize that, in many apartments, the manager is the ultimate authority. The manager decides whether to evict or not evict. For this and other reasons, tenants often hesitate to call the fire department when a fire is discovered, but run to find the manager. The manager, in turn, usually goes to the scene to determine that there is a fire. By that time, the fire is often well advanced.

Workers in an apartment complex will usually report a fire to the manager or attempt to use fire extinguishers before they phone in an alarm. Well-meaning citizens often attempt rescues and leave doors open. Discuss the limitations of extinguishers. Point out that TV shows often give a false impression of the toxic smoke a fire can generate, or what an ordinary citizen can do with an extinguisher. Discuss the negative effects of opening doors or breaking windows "to get the smoke out."

Try to talk to the manager and staff together. Point out the short time it takes for a fire to get out of control. Show a film such as NFPA's Fire, Countdown to Disaster, *so the manager and staff can see how fast fire can grow. Have the manager tell the staff that calling in the alarm is to be done first. Respect the manager's authority. If you tell the staff what to do, it is likely they will be told to do the opposite as soon as you leave. Try to help managers overcome the fear of publicity. Point out that small fires do not always make the evening news, but big ones surely do.*

Fig. 5-3. Too many apartment fires look like this on arrival. The fire department should work diligently on preventing delayed alarms.

Tactical Considerations, continued

Don't be vague, get your points over directly and forcefully.

Further, there should be a clear understanding with the local police and any other security or guard services on site that the fire department is to be called immediately on the report of a fire, smoke or gas leak. These forces usually feel impelled to investigate before calling the fire department to prevent "false alarms."

Work with the tenants and their associations. Reduce their fear of calling authorities. Explain that fire departments want to be called immediately for smoke or gas odors. It is important for all fire departments to abolish these terms: unnecessary alarm, smoke scare, or justified false alarm. These terms send the wrong message. It is the fire department's function to distinguish between non-hostile and hostile smoke. Thus, how can such a determination be called "unnecessary"?

There should be three classifications of alarms: fires, malicious false alarms, and emergencies. The latter would include every non-fire occasion for which citizens need assistance.

Try to promote the idea that if the fire department gets to the fire when it still is confined to the room of origin, there is a good chance of successfully extinguishing it and minimizing the loss.

General Advice to Tenants

This author has always advised tenants to help themselves in four basic ways:

1. Be fully insured for the value of all personal property.
2. Keep property of unique value in a bank vault.
3. Call the fire department immediately if a fire or gas leak is suspected.
4. In a fire, evacuate immediately, even if the fire seems inconsequential.

Fire Department Problems

Parking

Parking often presents a serious problem at these structures. Developers usually want to limit space allocated to parking. Plan review should include this parking problem. If company officers on alarm runs or inspections observe illegal parking that would interfere with fire fighting, the police should be summoned immediately. The effort to eliminate this hazard should be vigorous and followed up on until occupants understand and comply.

Building Location

Building location can be very confusing, and present problems not necessarily related to street address. Even though maps are provided by the management, a map drill makes a good indoor planning exercise. The officer draws an outline of the area on the board. The members fill in units. Special difficulties, such as gullies and fences which make it difficult to get from point A to point B, should be discussed and solutions worked out beforehand.

Mutual aid plans prepared by superiors should be reviewed so that those who will carry them out are familiar with all the details. Full-scale exercises may also expose deficiencies or false assumptions which would make the plan unworkable.

Gas Service

Gas service can provide special hazards in these structures. Where gas service is provided the system should be studied. Often the service layout is made for the convenience of plumbers and meter readers, with little or no thought for the safety of fire fighters. Usually all the gas meters in an apartment complex are located in one basement location. Individual lines are run, usually on the basement ceiling, from these meters to each apartment. Grouped together, these pipes represent a substantial weight. The hangars are often flimsy.

In one case, gas lines were located below the ceiling of the passageway between the chicken wire lockers provided to each tenant. A fire engulfed some of the lockers. One of the hangars gave way. The 20 gas pipes overhead looped down to the floor. Fortunately, the Silver Spring, Maryland fire units were on the near side, rather than the far side of the collapse. The pipes failed; the gas was released and ignited. Had the units been on the other side of the collapse, a catastrophe would have occurred. The mass of pipes formed a barrier.

Fig. 5-4. The fire department should be fully informed about the gas-delivery system.

In Brighton, New York, some years ago, a gas distribution system regulator failed in an apartment house. The flames on gas stoves rose to the ceiling and scores of houses were ignited at the same time.

Where there is no utility gas service, the apartment developer may install a large LPG storage tank. With this, the gas is distributed to the tenants much as it is by a utility. In Anne Arundel County, Maryland, an apartment complex was served from a single large gas tank. Its regulator failed, and many fires arose simultaneously throughout the complex.

Tactical Considerations

These fires show the need for dispatchers to be alert that multiple calls are not always necessarily for the same building. The first-in company officer should be aware of this potential scenario. As soon as citizens report several fires, adequate help should be dispatched immediately, without wasting valuable time waiting to check out each report.

Water Supply

The hydrants in many garden apartment developments are often on private mains. All hydrants should be checked periodically. Sometimes this raises objections because it may cause rusty water to flow. Often the fire department is forced to abandon testing.

Tactical Considerations

The right to test hydrants should be written into the contract with the water supplier when permission is given to build. Before testing, flyers should be distributed notifying tenants of what is being done, how necessary it is for their safety, and advising them to avoid washing clothes until the water settles down. This inconvenience is a small price to pay for being sure that the necessary water will be available for fire fighting. Water, not apparatus, extinguishes fire.

Protected Combustible Construction

If you were to ask the designer or building management how a fire in one apartment would be confined and not extend to the neighboring apartments, the answer probably would be: "by the fire-rated gypsum board sheathing or shell of the structure."

Gypsum has excellent fire protection characteristics. It is the only construction material which does not yield heat when burned in pure oxygen. When binders and paper are added to make the familiar gypsum board, it yields some heat, however.

Effect of Fire on Gypsum Board

When gypsum board is heated by fire, it starts to deteriorate as it gives up its moisture. This process, once started, appears to be irreversible. Sometimes insurance adjusters will order the burned gypsum board painted over. In a short time paint will flake off the deteriorating walls. Removing all burned gypsum board in overhauling makes the most sense. If you do so, however, avoid charges of needless destruction by explaining the benefits of this practice.

Fire Rating of Gypsum Board

Wood-stud and floor beams combined with gypsum board can be used to construct floor and ceiling assemblies or walls which pass the NFPA 251/ASTM E119 Fire-Resistance Test and earn a one-hour or greater rating. Unfortunately, the use of fire-rated gypsum board in fire-resistive enclosures has created a myth about an almost magic ability of gyp-

Fig. 5-5. Gypsum board often does a fine job of confining the fire. Penetrations and improper installation too often destroy its effectiveness.

sum board to limit the extension of fire. The rating of the gypsum board cannot be separated from the test structure of which it was a part. By itself, the fact that gypsum board is rated means only that it is better than non-rated board.

Gypsum board is a good material but its applications do not always produce the results many have come to believe that it will. Nor can it overcome a building's sins of omission or commission.

Underwriter's Laboratory warns that its rating is assigned to the entire tested and listed assembly, not to individual components.[4] There are no rated ceilings, only rated floor and ceiling assemblies. The rating also assumes the structural integrity of the supporting structure.

Deficiencies of Gypsum Board

There are deficiencies in gypsum used for the fire protection of wooden structures. This is despite the fact that gypsum undergoes fire-resistance testing.

Fire-Resistance Testing

The "One Hour Fire Resistance" rating of a floor and ceiling assembly is achieved in the following manner.

- A gypsum board manufacturer constructs a floor and ceiling assembly of wood and gypsum board at a testing facility.
- The laboratory staff places the assembly on top of the test furnace.
- The standard NFPA 251/ASTM E119 Test is run.
- If the assembly passes the test, it is rated as one-hour fire floor-ceiling assembly.

A similar test is conducted on wall assemblies.

A finish rating is given in the UL listing. This is the time the wood stud or joist reaches an average temperature of 250°F, or an individual temperature of 325°F.

This author believes that the NFPA 251/ASTM E119 Test does not provide a realistic appreciation of what happens to a protected combustible structure under real fire conditions.

The NFPA 251/ASTM E119 Fire Resistance Test was developed around 1916. At that time there were no wood/gypsum assemblies. The floors tested were brick arches, hollow tile or concrete. Such floors are similar from top to bottom. It appears logical that a test fire attacking the floor form below is only adequate for such floors.

In 1950 this test was applied to wood and gypsum floor and ceiling assemblies at Underwriter Laboratories. At that time, the typical apartment room contents would not generate sufficient heat at the floor line to penetrate the floor. The situation has changed. Today's contents generate hot, fast fires which rapidly attack a wooden floor.

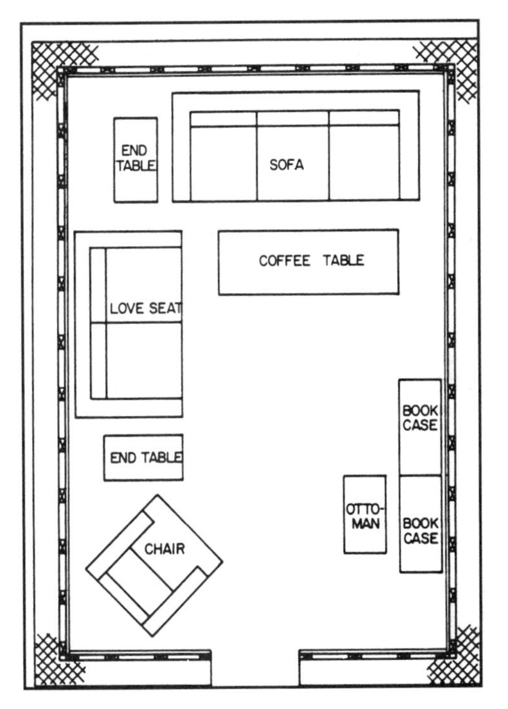

Fig. 5-6. Shown are the furnishings and layout of the test room used in the 1980 tests of fire propagation by the then National Bureau of Standards (now NIST). The room contains typical items that might be found in an apartment. (Courtesy National Bureau of Standards)

Fig. 5-7. In less than four minutes after ignition, the fire in this test room pushed heavy flames out the doorway. (Courtesy National Bureau of Standards)

An April 1975 *Bulletin of the Society of Fire Protection Engineers* warned against the increased hazard of fires in contents. "The rate of fire development can create a condition that may tax or overpower traditional fire defenses." It is reasonable to believe that the rate of fire development in today's contents is even greater.

In 1980 the NIST conducted tests on fires in contents of typical residential rooms. These tests showed very rapid development of total room involvement and temperatures sufficient to penetrate floors very early.[5]

More contemporary resources show additional realistic fire results.

The NFPA film *Fire Power* shows burning plastic dripping from a couch. The NFPA film *Fire, Countdown to Disaster* shows rapid fire growth of a typical bedroom fire resulting in a room fully involved down to the carpeting on the floor in less than three minutes.[6]

The test structure is tightly fire stopped with a limited supply of air. The finish rating established earlier is set to end the test if the temperatures are reached on the top surface. They indicate ignition is imminent. The limited air supply must act to limit combustion of the wood members in the test. Typical finish ratings of 1 hour-rated assemblies range from 11 to 30 minutes.

Rarely are actual buildings so well firestopped that they limit the supply of air. It is reasonable to anticipate flaming combustion of wood members early in the fire.

This author has observed fires which originated in the combustible void from wiring, heat ducts, and defective metal chimneys. The test is not intended to evaluate such ignition sources.

The test also does not account for a fire which burns into the void laterally, as in the case of an exterior fire which burns through the skin of the building. It doesn't account for a fire which rolls out a window and ignites the combustible exterior, such as a plywood ceiling of a porch and this penetrates the structure.

In the test furnace, fire gases are removed by exhaust. There is no buildup of explosive carbon monoxide gas, as can be the case in many fires. (See a more complete discussion of testing and rating in Chapter 6, Principles of Fire Resistance.)

In addition, the test structures are built without penetrations.

Installation

There are four principal deficiencies in the installation of gypsum board.

Gypsum board commonly is nailed up over voids with a large or even infinite air supply behind the board. As heat penetrates the board and ignites the wood, there is ample air to continue combustion. As noted above, this condition is not reflected in tests.

Nail heads often are not properly cemented over. Unfortunately, this heat barrier is often regarded as merely cosmetic. All nail heads in tested assemblies are covered with cement.

In some assemblies, joints are not properly taped. This also is often regarded as cosmetic and not required for unpainted surfaces. Taping is part of the fire-resistance system.

Deviations from the listing are permitted by building departments. Yet, every item in a listing can be crucial. This becomes clear when we understand the procedure. If a one-hour rating is desired and the sample lasts one hour and 15 minutes, the assembly is redesigned to just earn the one-hour rating. This eliminates the added economic penalty of making the assembly "too good." Every item of a listing was put in by the manufacturer to reach a specific rating. There is nothing added for extra margins of safety.

Protective Sheathing

Though never stated explicitly in any code, the concept behind the use of gypsum board is that it provides a protective sheathing or membrane to protect the combustible structure from a fire in the contents. As noted, testing is conducted on a surface without penetrations.

Some membranes can tolerate a small leak, but the gypsum sheath of a combustible structure must be as sound as the plastic bag which carries the goldfish home from the pet shop. A single pinhole can cause disaster. The penetrations of the gypsum sheath range from pinholes to sizable openings. Any penetration allows the fire to spread to the structure, thus converting a contents fire to a structural fire.

Penetrations

The penetrations commonly discovered in structures are varied.

One commonly cited penetration in the sheathing is caused by failure to close the gypsum sheath around utilities such as electrical or gas service. There are listed devices or methods for passing utilities through fire-resistive barriers but these are rarely used in residential construction.

Failure to install the gypsum sheath behind the bathtub presents another penetration problem. Heat is conducted through the steel. Often bathtubs in adjacent apartments are located back to back. This provides a common void and breaches the supposed compartmentation. In some cases, where plastic bathtubs are used, the code often requires that the sheath be complete, but numerous untaped joints and uncovered nails have been observed.

Open doorways are sheathed in gypsum. When a door is installed, the thin wood door casing is the only sheath. This can fail rapidly.

Interior vents, such as from bathrooms and kitchens, not only penetrate the sheath, they deliver fire to the vertical voids of the building.

Fig. 5-8. The protective sheath of the building has many pinholes which permit a contents fire to become a structural fire.

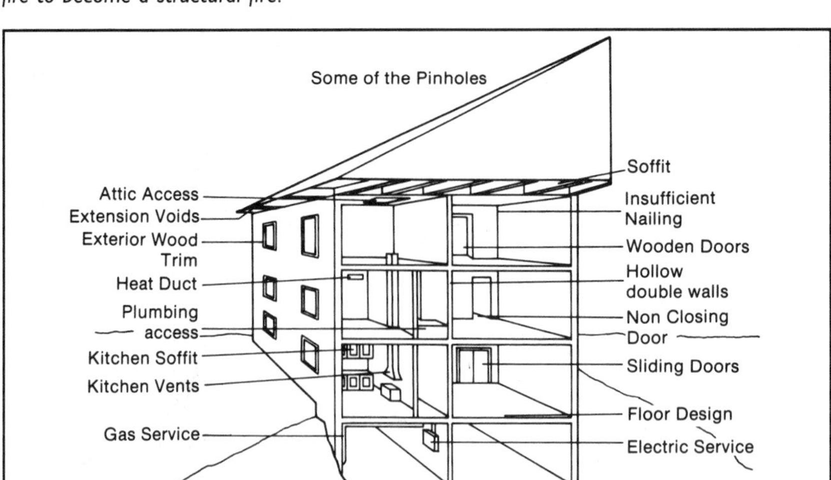

Attics are ventilated to prevent condensation. The lower vents are located in the soffit below the projecting eaves. There is a constant natural air flow upward through the lower vent, out the upper vent, generally at the highest point of the gable ends. Fire reaching out the window will ride this air flow "escalator" into the attic, thus bypassing the protective sheathing or shell. An attempt to amend the BOCA code to require that attic vents be located at least three feet off the line of the windows was voted down. This is required now in Prince Georges County, Maryland, where many serious garden apartment fires have occurred.

Where a fire-rated partition wall meets a fire-rated floor, the "compartmentation" satisfies the building department. However, the floor is easily penetrated downward by a fire in today's fuels. The fire enters the floor void and comes up on the other side of the partition. Further, soffits, such as the box dropped down to provide built-in kitchen cabinets, connect the wall voids to the floor voids.

The fire-rated gypsum-board myth allows for huge interconnected voids. The creation of dangerous, hidden voids is prevalent in the rehabilitation of older buildings, especially when the ceiling heights were greater than today's acceptable eight feet which easily allow for renovations by dropping a ceiling. Carbon monoxide gas can accumulate in these voids. When the ceiling falls, the gas can ignite, either as a backdraft or a detonation.

Fig. 5-9. The Bladensburg Fire Department in Prince Georges County, Maryland expertly cut off extension to the attic with their squirt. The stream was directed into the attic ahead of the fire to cut it off, rather than into the fire to drive it. (Joel Woods, MFRI)

Fig. 5-10. This steel beam is not protected by any rated system. A void fire might well cause it to fall. Its size suggests that it supports a substantial load.

Often the biggest void is the attic. There is no such thing as a rated top floor/attic ceiling. The constant flow of air makes a rated structure impossible here.

Attic access closures are usually made of plywood. Some are located in apartments. In such a case, a room fire can gain total possession of the attic before fire units arrive.

Fig. 5-11. The attic is just a big lumberyard, set up to burn.

Fig. 5-12. The attic was falling on the occupants before they escaped from the building. A stairway is rarely constructed to be a place of refuge.

Protected Combustible Is Not Fire Resistive

From time to time, buildings which incorporate floors and ceiling assemblies and interior walls which are rated according to NFPA 251/ ASTM E119 Test methods are sometimes referred to as being "fire-resistive." Such characterization should be emphatically rejected. The limited application of a complete gypsum sheath or shell, and the many defects cited here, should make it clear that the term "fire resistive" should never be applied to such a structure. In fact, the term "protected combustible" appears to be overly generous in describing this situation.

In a recent article, an author lumped "protected combustible" with "fire resistive" because it "helped the learning process." Such oversimplification might lead to serious consequences on the fireground.[7]

Tactical Considerations

Since the effectiveness of the protective sheathing or shell is so questionable, any fire which has progressed to full room involvement should be regarded as involving the entire structure. Unfortunately, sometimes fire officers concentrate all their attention on the visible

Tactical Considerations, continued

fire target and, as a result, get behind the fire and play catchup—very often unsuccessfully.

Any fire beyond the minor-contents stage should be handled as a total building fire. The fire building ends only at a solid, parapetted fire wall or open space.

"Room Off" or "Apartment Off" really means "The Building is Off."

Even the best code does not provide compartmentation which can be relied on to confine the fire.

Probably the single most important item is whether the building has truss floors, or wooden I-beams. If the fire stopping capability inherent in sawn-joisted floors is absent, the potential for fire spread and early collapse are tremendously increased. (These problems are discussed move fully in Chapter 12, Trusses.)

Fire extension can be subtle. This potential must be present in the thinking of all officers, even those with specific limited tasks.

Fire which extends to vertical voids will reach the roof void unless checked. The fire will spread laterally, even under a flat-joisted roof because of the pitch of the roof. Fire which reaches a gabled attic finds a huge lumber yard of fuel and unlimited air supply. Ventilation of the attic before it becomes unsafe is very important.

Once the attic is fully involved, the use of the ladder pipe for suppression is questionable. If the attic is opened up by the fire, and the ladder pipe can get a good shot into the attic void, the tactic may be successful. If not, the result may just be increased water damage on the lower floors. The roof will burn away in a short time. The sooner it goes, the better, because it will then be safe to operate on the top floor. It might be better to use the ladder pipe to protect the fire wall and extinguish any fire spreading on the roofing. Hand lines could extinguish any fire which drops into the apartments below.

The stairway is not necessarily a safe working platform. It may be of non-combustible construction, but in most cases, the attic void extends over the stairway. Fire in the attic has dropped flaming roof and truss sections down onto a stairway. If the truss floors are extended to act as the stairway platform, a void fire may involve the stairway very early on.

Undress the building! Picture the hidden vertical and horizontal voids through which fire is traveling. If sufficient fire fighting forces did not respond initially, call for help early. Don't wait until the need is all too evident. Prepare for collapse once the fire seriously involves the structure.

Fire Walls/Barriers - Draft Stops

Fire Walls

Fire walls are often used to separate units in multi-family residential structures. Ideally, a masonry fire wall should be able to stop the extension of the fire without assistance from the fire department. Unfortunately, there are many possible defects in such walls. These require prompt action by the fire ground commander to reinforce the fire wall and cover its weaknesses.

One of the primary defects involves not bringing a masonry fire wall through the roof with a masonry parapet. Builders resist this because it is costly, and unless the wall is flashed properly, there can be leaks. In North Carolina, a builder finished off a block fire wall with brick veneer merely rested on the wood roof. It collapsed suddenly into top floor apartments.[8] There is a heavy push for code provisions requiring the fire wall to be limited to the underside of the roof.

It is almost impossible to fit masonry tight to the roof. There will be at least small gaps. In many fires, expanding gases can force fire through the smallest opening. Furthermore, plywood delaminates and can lift up and enable the fire to pass. (See the discussion of the record of FRTP in Chapter 4, Ordinary Construction.)

Permitting overhangs or mansards to project beyond the fire wall is another defect in fire walls. This provides a gap for fire to pass around the end of the wall. Apparently, some codes read "the fire wall must extend to the exterior wall," and, unfortunately, this is interpreted as a limitation.

Permitting the fire wall to end at the interior of a combustible exterior wall simply enables the fire to pass around it. Often there is no exterior evidence of the fire wall's location. The fire wall should be extended straight out or in a "tee" or "ell" shape to cut off the fire.

Utilities are passed through the fire wall. As discussed in the Penetrations section of this chapter, openings around pipes pass fire. The best practice is to run utility mains parallel to the building with branches into each unit.

Unprotected openings frequently are cut into fire walls, particularly in attics for maintenance purposes. These compromise the fire walls.

Openings at the basement level provide access to storage and laundry areas. These are usually designed and built with proper self-closing doors. All too frequently these doors are blocked open. Ironically, in some cases, the door closer has been removed and used as a door stop.

Permitting the fire wall to act as a party wall, or structurally as part of both buildings creates problems. Party walls often have beams or

Fig. 5-13. The "fire wall" is in fact a party wall. Fire passed through an opening like this one and destroyed 16 units. The builder said, "Those fire walls were great. They were the only thing left standing."

girders from both sides in the same opening. These common openings provide a path for fire extension. Look for this in townhouses, particularly where units have adjacent doorways. Stairway headers may be in the same opening.

Older row-frame buildings often had brick laid in the party wall stud voids as a fire wall. It doesn't work because the barrier is incomplete. The brick noggin does not block the floor or attic voids.

Fire Barriers and Draft Stops

Fire barriers and draft stops are intended to limit the combustible void area in the attic to which the fire has access. In some cases, plywood or chipboard is used for the draft stop. By far, most construction makes use of gypsum board on wood studs.

Some barriers are now being made of 2-inch gypsum plank in vertical sections 2 feet wide. The sections are held in a steel channel. They are cut in the field to meet the slope of the roof. However, the cuts are often only approximate, producing openings for fire access.

The fire barrier may range in effectiveness from temporarily reliable to totally useless. In no case, however, can a fire barrier be regarded as even approximately equal to a parapetted masonry fire wall.

There are a number of defects in these fire barriers and draft stops. The first exists even in the best constructed fire barrier. The others are found to a greater or lesser degree in almost all fire barriers.

Consider a fire barrier which extends completely from one side of the attic to the other, including the overhang. Nails are set, joints are backed by two by fours and taped, with no penetrations at all. Yet a fire delaminates and raises up the plywood so that the fire passes over the barrier. A similar problem exists when the fire wall is ended at the underside of the roof.

Another defect exists when the barrier does not extend out to the eaves but stops at the wall line, thus allowing the fire to roar around the end.

Gypsum board improperly backed by wood at joints and intermediate points presents a compromised fire barrier.

Omitted nail covering and joint taping, utilities or structural elements passing through, doorways cut into fire barriers, and the use of scraps of gypsum board all affect the integrity of fire barriers.

When fire barriers have been placed above the mid point of a room, both sides of the barrier are exposed to fire coming out the windows. This comes about because of code provisions limiting the size of the attic void. The builder wants the maximum area in each section, and the fire barrier is so placed even if it doesn't continue a fire separation below.

There are many other defects in fire barriers produced by a wide variety of builders, maintenance personnel, and tenants. It is wise to examine buildings in use and under construction for other fire barrier inadequacies.

Tactical Considerations

The term "compartmentation" is often used to describe the fire protection provided to protect one tenant from another. This author believes compartmentation "goals" are seriously flawed in practice.

These defects must be considered beforehand and covered early in the fire. It may be more important to commit units to defending the fire wall than operating on the fire. If the fire gets past the fire wall, there is a whole new fire.

Party wall extension may be subtle. Smoke in an adjacent unit may be from the original fire. Where common joist sockets exist, open and examine the ceiling of the exposed unit and set up a watch line.

Tactical Considerations, continued

Admittedly, diverting resources may increase the loss in the area of the fire, but the problem would not be the fire department's fault. It would be caused by defective design and inadequate code compliance.

A Word about Sprinklers

Automatic sprinklers which protect occupied spaces will most likely only extinguish contents fires. They will probably not control any fire which originates in, or extends to, the voids. However, as far as tenant life safety is concerned, a NFPA 13R sprinkler system is a tremendous asset. (See Chapter 13, *Automatic Sprinklers.*)

Tactical Considerations

Few fires start out big as with a bucket of gasoline. Most fires start small. It is worth repeating that if you can get there while the fire is still in the room of origin, you have a good chance of holding it. Early reporting of fires by residents or workers can be a significant factor in this effort.

Know Your Buildings!

Fig. 5-14. The fire department should be aware of all the potential for extension. Get ahead of the fire and cut it off. Playing "catch up" leads to disaster.

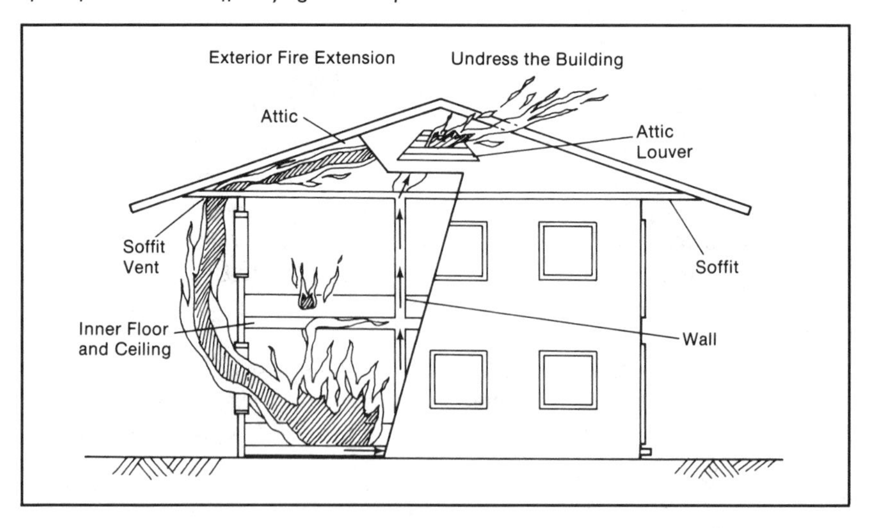

Fig. 5-15. Fire fighters delayed entry for 45 minutes until heavy electrical power was shut off. They then charged in as they would on first arrival. The floor collapsed. Three fire fighters were in the fire at the time this photograph was taken.

Serving the Citizens

There is probably no class of building in which good fire planning and training can pay off more than in garden apartments. Fire departments which study why multiple-dwelling fires spread and develop tactics to cut these fire off, serve the citizenry well.

Those who own their own homes usually have personal property or homeowner's insurance, often purchased to satisfy the mortgage holder's requirements. Renters' insurance is often too inexpensive to be actively sold. The result is that a high proportion of fire victims often suffer crushing financial blows — due to a wide range of causes including inadequate building codes, the builders' negligence, the owners' incompetence, other tenants, or inadequate fire suppression efforts.

The knowledge the fire department gains from the day an apartment development is planned should be disseminated and become the organized property of the fire department. A system should be in place to make the information available to the incident commander at the fire scene. This information and the ability to make effective plans from it provides increased safety for all concerned.

Case Studies

Fires in combustible multiple dwellings under construction are extremely destructive. Access to the property, road conditions, open wooden construction, dry weather and high winds contribute to massive fires. In some cases the building under construction presents a massive hazard to adjacent exposures.

Two spectacular exposure fires occurred in Los Angeles, California.

Boucher, D. "Desert Disaster Strikes L.A. County." *American Fire Journal*, Sept. 1990, p. 12. In Los Angeles County a helicopter made air drops to assist ground units fighting a fire in multiple dwellings under construction. There was no wildland involved. A total of 84 out of 100 units in 24 buildings were destroyed.

American Heat, Vol. 5 program, January, 1991. Ventura County, California suffered a massive loss of buildings under construction in November 1990. Twenty-two buildings containing about 350 units were destroyed. The units destroyed were all wood frame, without stucco or gypsum board. As is often the case, a plumber's torch started the fire. Very dry conditions and high winds accelerated the blaze. As is customary, the water supply was sized for the risk *after* the buildings were finished—not for a number of fully involved buildings. In addition, water was lost when the 2-inch plastic service pipes burned.

Helicopter drops were used. In discussing the fire on *American Heat* videotape, a chief stressed that there must be close coordination so that the water is not dropped where fire fighters are working.

Ventura requires two means of access to all projects—even during the construction. This is very useful because a single access could have aided fire spread.

Future projects will limit the number of structures in a bare wood condition.

Ditzel, Paul. "Tested Under Fire," *Firehouse*, March, 1985. A fire-resistive Los Angeles high-rise retirement apartment building was severely exposed by fully-involved combustible multiple dwellings under construction. Nine floors of the retirement apartment building were ignited. Fortunately, trained residents reacted properly and without panic. A heavy fire department commitment was made early in the fire.

Meadows, M. "*On The Job in California*," *Firehouse*, June, 1990. The largest commitment of units to a structural fire in Los Angeles was required for a fire which started in a five-story combustible multiple dwelling under construction. The fire extended to a 14-story fire-resistive highrise, a three-story apartment building, and about twenty additional structures. The total loss was estimated to be between $30 and $40 million.

White, D. "Fire Ravages Apartment Complexes in College Station, Houston," *Firehouse*, October, 1984. Two fires in wooden apartment complexes under construction are described. The "lessons learned" from this account are very pertinent.

Mines, R. "$5 Million Fire Strikes Private Key Largo Club," *Firehouse*, February, 1984. A wooden top floor was being added to a three-story apartment building. A plumber's torch started the fire. Massive, long-distance mutual aid helped limit the loss.

Morris, Gary P. "Strategic Approaches to Fires in Apartments under Construction," *Fire Engineering*, November, 1984. A thorough discussion of the problem of wooden multiple dwellings under construction. The author notes the necessity for immediate use of heavy caliber streams.

Rigamer, G. "Six Alarm Blaze Destroys Buildings Under Construction," *Firehouse*, January, 1985. Eight of 29 wood frame, three-story apartment houses under construction were fully involved on arrival. Despite difficulties, fire fighters confined the fire to the units initially involved.

Simmons, T. "On the Job in Phoenix," *Firehouse*, October, 1986. An incendiary fire at the construction site of a wood-frame apartment complex shows nothing significant on arrival, then erupts furiously. There were serious exposure problems and great difficulty with a chain link fence. The familiarity of the fire fighters with the construction site was an asset.

Steenbergen, H. "In A Fire, Seconds Count," *Fire Command*, July, 1989. Wyoming, Michigan fire fighters successfully protected an occupied apartment unit from a raging fire in 30 units under construction. The pictures are excellent.

Tactical Considerations

Units should keep on top of the changing conditions at construction sites. Water supply must be in service before any units are built. Work with plumbing and roofing contractors to be sure all fires will be reported promptly, especially those "extinguished" by workers.

Ulley, L. "Four Alarms in Four Minutes," *Fire Fighting in Canada*, Sept. 1990, p. 6. Montreal, PQ fire fighters transmitted four alarms within four minutes for a four-story apartment house under construction which had extended to several exposures.

Halstead, D. "Wiring Ignites Fatal Fire," *Fire Command*, September, 1989. An excellent detailed report on a fire which started in a wall void. When it broke out, heavy fire involved the second floor and the attic. There were two fatalities. This article provides excellent insight, particularly for fire departments without much experience, into how difficult such fires can be.

Kyte, G. "Tragedy Strikes Biloxi," *Fire Command*, May 1987. Two fire fighters died in an old house converted into seven small apartments. Fire was in concealed interconnected voids. "No smoke showing" was the first report. The fire was reported at 2:55 AM. It spread into the habitable spaces at about 3:38 AM. The victims were not immediately missed.

Fire lurking in void spaces is like a tiger perched in a tree. It is hard for many fire fighters to realize the amount of fire which may be concealed, or the speed with which the fire can grow when a void is opened up, by the fire or by the fire department.

Tactical Considerations

Many fire departments need improved discipline on the fire ground. The British Fire College at Moreton-on-Marsh teaches procedures for keeping track of every fire fighter who enters the fire structure. SCUBA diving teams do so also. Is entering the water really any more dangerous than entering a burning building, particularly when fire is hidden?

Not enough personnel you say? The first consideration should be the safety of the fire fighters. If this means one person less for fire fighting, so be it. Is a structure ever worth lives?

Fetete, A. "We have a Roof Cave In," *Fire Command*, January, 1984. Air conditioning units were located directly above the stairway in garden apartments. Five minutes after the first alarm, the air conditioning units collapsed onto the first-due engine company. Later, the truss roof collapsed on fire fighters operating in an apartment where there was only light smoke showing.

Cover Photograph, *Fire Engineering*, April, 1985. The cover photograph shows an aerial view of an apartment house. Note the carefully concealed air conditioning unit supported by the roof.

Cole, B. "Flashover Fire Scorches Apartment," *American Fire Journal*, January, 1989. Sacramento County, California fire fighters were operating in the attic of an apartment. A flashover drove them out. Fire destroyed the roof area. The fire stops were made of chipboard.

Tactical Considerations

Roofs are hazardous. Today's truss roofs are especially so. Tactics which include automatically "getting the roof" and waiting for signs of weakness before evacuating the building must give way to preplanning in order to evaluate the roof beforehand as a fire department working platform.

Isman, W., and Ward, M. "Natural Gas Explosion Threatens Apartment Complex," *Fire Command*, April, 1985. Children playing on gas pipes pulled a fitting apart. Leaking gas exploded and caused heavy fire and collapse. Shoring was necessary but materials were not readily available.

Tactical Considerations

Fire departments should arrange a means to get necessary materials from commercial sources at short notice.

References

The addresses of all referenced publications are contained in the Appendix.

1. *New York Times*, June 1, 1986, Section 8, P. 1.

2. (See Poage, R. "On The Job, Kentucky" *Firehouse*, Sept. 1989, p. 28). A typical account of vertical fire extension. Note also that there was a delayed alarm. The roof collapsed early in the fire.

3. Dunn, V., "The Peaked Roof," *Fire Engineering*, March 1990.

4. Fire Resistance Directory, 1990, Underwriters Laboratories, p. 6. "Fire resistance ratings apply only to assemblies in their entirety. Individual components are not assigned a fire resistance rating and are not intended to be interchanged between assemblies but rather are designated for use in a specific design in order that its rating may be achieved."

5. In 1980, The National Bureau of Standards (now NIST) published "Fire Development in Residential Basement Rooms" (NBVSIR 80-2120). Don't be misled by the specificity of the title. The rooms were typical living rooms, or office reception rooms. A sketch of a typical room layout and a picture of a typical fire pouring out the full doorway in less than four minutes are shown in the text. The fires were ignited in newspaper spread over the back of the couch.

   Figure 21 of the report illustrates heat fluxes at the floor level in the range of 100 to 160 kW/M2, in the time period from 6 minutes until 12 minutes. I have been told that this energy level is sufficient to ignite a wooden floor.

   Figure 25 shows a comparison of the fire exposure curve derived from test 9 with the ASTM standard curve. The test curve peaks at over 1000° Celsius (over 1800°F) in ten minutes.

   Conclusion 1. "The rate of development and intensity of real fires involving the burning of typical furniture and interior linings in a room during the first 20 minutes may be significantly greater than those defined by the ASTM E119 standard time-temperature curve.

   A more realistic time-temperature curve for residential occupancies is presented in this report. This curve is considered suitable for testing exposed floor construction, floor-ceiling assemblies, wall assemblies, columns or doors."

Not long after this report appeared, the Reagan administration began an unrelenting effort to abolish the fire research funding of NBS, alleging that private industry should do the job. Only the united opposition of the fire protection community (including private industry) saved the function, year after year.

Readers are urged to write NIST for the full report.

6. "Fire, Countdown to Disaster," 1985, NFPA. Pictures and some data were published in the August 1986 *Fire Command*, p. 21-23.

7. Smith, A., "A Quick Guide To UBC Building Types." *American Fire Journal*, May 1990, p. 22.

8. Letter (1990) from W. Le May, Assistant Fire Marshal, Mecklenburg Company, North Carolina.

6

Principles of Fire Resistance

Toward the latter part of the 19th century, it became apparent that building structures with "fireproof" or non-combustible materials provides no guarantee against tremendous fire losses. As steel frame buildings emerged, the necessity for protecting the steel from the heat of fires became understood. A 1921 book on *Fire Prevention and Fire Protection as Applied to Building Construction*[1] gives fascinating information on the early designs of fire protection for such structures, some of which, like the Government Printing Office in Washington D.C., are still standing. At that time, construction engineers had freedom to design "on the job." Destructive fires in these and other unprotected buildings built with innovative but untested techniques helped to bring about the realization that adequate standards and test procedures for fire-resistive construction were necessary. This chapter discusses the basic principles of fire resistance from the early history of fire tests to applicable standards and estimating fire severity, as well as combustibility and preplanning.

Early Fire Tests

Floors

Floors received the earliest attention as columns were apparently considered to present no problems. Chapter 4, Ordinary Construction,

The author expresses appreciation to Mr. Jack Bono, then President of Underwriters Laboratories Inc. for having Chapter 6 reviewed for errors of fact. Any opinions expressed are, of course, the responsibility of the author.

pointed out the two schools of thought on the behavior of wood floors in a fire. One school maintained that if floors were designed to resist collapse, they presented the risk of a general collapse of the building.

The other school believed that if a floor were designed to collapse early in a fire, the walls would be preserved. The dilemma could be resolved by designing fireproof floors.

In 1890, the first fire test of a fireproof floor assembly in the United States was conducted for the Denver Equitable Building Company.[2] Hollow tile floors were tested. The test determined that porous hollow tiles set in end construction (tile cells at right angles to beams) were superior to dense tiles set in side construction (tile cells parallel to beams). The floors were subjected to load, shock, fire and water, and continuous fire tests (24 hours at 1300°F).

Six years later, the New York City Building Department conducted a series of tests on fireproof floors using brick kilns as test furnaces. The central panel of the floor was loaded to 150 pounds per square foot (psf). A wood fire was built on the grate and maintained at 2000°F during the last four hours. After the fire, a hose stream was applied and the floor was reloaded to 600 pounds per square foot for 48 hours, with the final load resting on the arch rather than a beam.

Other fire tests were conducted at Columbia University. Later, gas or oil fires were used in tests for better control. Further, in addition to structural integrity, requirements that the floor must not permit passage of fire were developed.

Columns

Between 1896 and 1916, not much fire testing was done on columns. However, testing that was conducted showed cast iron to be superior to unprotected steel.

Standards for Fire Resistance

New York's Parker Building fire in 1908, the Equitable Building fire in 1912, and the Baltimore conflagration in 1904, which involved many "fireproof" buildings, convinced many that there was a dire need to develop standards and test procedures to ensure truly fireproof construction.

The effort to develop standards for fire resistance brought together the National Bureau of Standards, (now known as National Institute of Standards and Technology (NIST)), Underwriters Laboratories Inc., the National Fire Protection Association, and both capital stock and mutual insurance interests. Fire resistance tests were conducted during 1917 and 1918, and early standards were developed which have existed substantially the same ever since.

Materials and assemblies are classified by their fire resistance, or, more acurately, fire endurance, capabilities according to standards which are designated by different organizations. Committees from each organization keep the various texts conformed. The standards are:

*NFPA 251, *Standard Methods of Fire Tests of Building Construction and Materials* - National Fire Protection Association.

*UL 263 - *Fire Tests of Building Construction and Materials*, Underwriters Laboratories Inc.

*ASTM E119 - *Methods of Fire Tests of Building Construction and Materials*, The American Society for Testing and Materials, E-5 Committee on Fire Tests. The standards are essentially equivalent. They also are ANSI standards.

What Does It All Mean?

It is important to understand both what fire resistance intends to provide and what it does not.

- Fire resistance of columns is concerned with resisting collapse.
- Fire resistance of floors is concerned with resistance to passage of fire and collapse.
- Fire resistance of walls is concerned with passage of fire and collapse.
- Fire resistance of fire doors is concerned with passage of fire.
- Fire resistance is not specifically directed at life safety. Many lives have been lost in fire-resistive buildings.
- Fire resistance is not specifically directed at smoke control. For example, metal-clad wood fire doors generate toxic carbon monoxide as they burn. Furthermore, while some earlier fire-resistance systems provided smoke containment, it is not part of the standard.
- Fire resistance is not concerned with the dollar loss due to fire. There have been huge losses in fire-resistive buildings. The First Interstate Bank in Los Angeles resisted a spectacular fire, but the loss was in excess of $100 million.

Fire-Resistance Testing

Essentially, the provisions of the standard require a reproducible test fire, consistent method for conducting tests and classifying results, and specific instruction on the selection and preparation of test specimens. The object of these provisions is uniformity in testing.

The reproducible fire used in the testing procedure follows a standard time-temperature curve (Figure 6-1). The temperatures listed are those of the exposing fire.

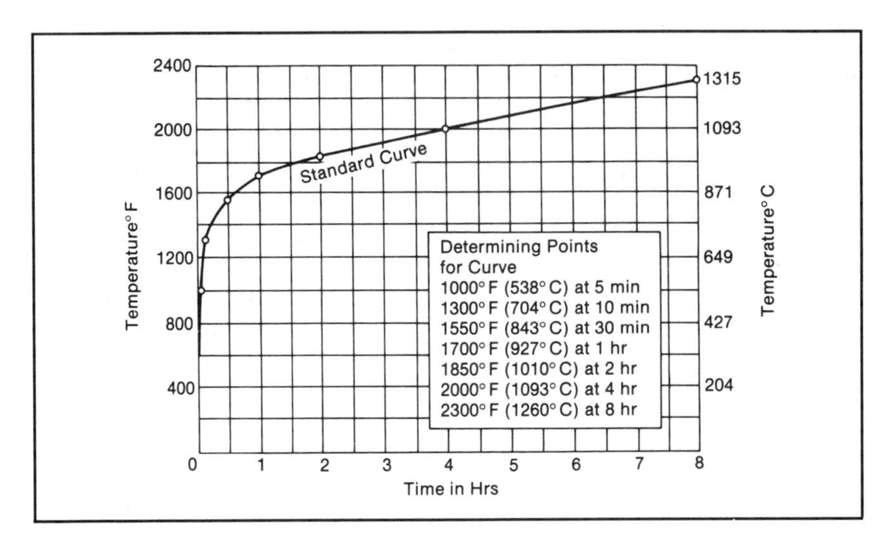

Fig. 6-1. The standard time-temperature curve.

In the test, samples of typical structural elements are exposed to the standard fire. The minimum sizes for test specimens are: columns, 9 feet; beams and girders, 12 feet; partitions and walls, 100 square feet; and floors and roofs, 180 square feet.

The load-bearing structural elements are loaded with the prescribed weight for the assembly being tested. The unit passes the test as long as it successfully resists its superimposed load and, in the case of partitions, walls, floors, and roofs, resists the passage of fire. Full details of test requirements are found in NFPA 251 (ASTM E119 or UL 263).

The principal requirements of the test are summarized below.

Columns have several thermocouples peened into the column. When the average temperature exceeds 1000°F or any one thermocouple exceeds 1200°F, the test is ended.

A floor structure must not develop conditions which would ignite cotton waste on the unexposed surface or permit an average temperature rise on the unexposed surface of 250°F. Under some circumstances, a hose stream is applied to the assembly after the test.

Assemblies which pass the test are rated in units of hours, for example, one-hour, two-hour, etc., depending upon the time the test unit survived the test fire. Rated units are listed in references such as the UL Fire Resistance Directory which is published annually. Based on such listings (not approvals) local authorities may permit use of specific materials and assemblies.

Caution:

There is no point in a material supplier achieving a rating for a non-standard time such as two hours and twenty minutes, since the assembly will receive only a two-hour rating.

Fig. 6-2. Steel columns being prepared for testing of fireproofing material. (Courtesy Underwriters Laboratories Inc.)

Fig. 6-3. A test of fire doors and hardware. (Courtesy Underwriters Laboratories Inc.)

Minimum Requirements

In short, the listed assembly meets the minimum requirements the manufacturer could develop to pass the test. Therefore, all components assembled as specified in the listing are important. Many believe that the standard is developed by engineers who include redundant precautions. Nothing could be further from the truth. By the nature of the testing process, the listed assembly is the minimum quality structure capable of passing the test. Not requiring assemblies to achieve this minimum standard could give a supplier a market advantage.

At times, building departments give approval to variations from the listing. Any departure from the listing means, however, that most probably the assembly installed is not the equivalent of the listed assembly.

The listings are not like restaurant menus from which one can select one item from column A and another from column B. Components should not be considered apart from the tested and listed assembly. A very common fallacy is the belief that fire resistance can be achieved just by nailing up some ⅝-inch "fire rated" gypsum board.

The unit being tested may be seriously damaged or effectively destroyed in passing the test. Fire-resistance ratings are often misinterpreted. A two-hour fire resistance rating does not mean that the building will be relatively undamaged for two hours. Fire damage can be quite severe for even short exposure.

However, fire resistance does mean that if the building reacts to the fire in the same manner as the specimens did for that particular type of construction in the test, and if the actual fire load does not exceed the test fire load, and if the duration of the fire does not exceed the time specified for the components involved, then the structure should survive the fire.

However, the time ratings given to fire-resistive assemblies do not relate to real time in an actual fire. Rather, the ratings provide a relative *measure* when compared to other assemblies' degree of endurance.

The Legal Path

The path by which fire-resistance standards are legally enacted and used in buildings is worth tracing.

- A municipality decides that it should have a building code to help protect the general welfare of its citizens.
- The municipality may examine several nationally recognized codes and consider other alternatives.
- By the appropriate legal process, the municipality may decide to:
  - Adopt one of the model codes verbatim,

- Adopt one of the model codes, with its own modifications, or
- Write its own code.

- The code the municipalty has selected becomes law when it is adopted by the appropriate legal procedure.

In this municipality, a builder wishes to construct a building with steel columns. To continue our example, we will follow the code requirements that certain columns have three-hour fire resistance.

- Manufacturers of fire-resistance systems have anticipated this need and have had their system tested at a recognized testing laboratory.
- The manufacturer's staff working in the laboratory puts the fire-resistive material on the test column.
- When the column is ready (wet materials such as concrete must cure), the test is conducted in accordance with the NFPA 251 test standard (ASTM E119 or UL 263). When the limiting temperature, cited in the NFPA 251 or UL-263 standard, is reached in the column, the test is ended.
- Since the time from start to finish of the test was three hours, the tested system receives an appropriate time rating of three hours.
- A listing of the fire-resistance system is placed in a directory along with other systems which have passed the test. Every significant detail of the assembly which passed the test is stated.
- The builder indicates to the building department which fire-resistance system (listed by the laboratory), is to be used in the building.
- The building department accepts or rejects the fire-resistance system. However, the building department is not obliged to accept any particular system. If the department objects to a system, it need not accept it for use in its community. (This decision is subject to the customary appeal system.)

Tests for floors and walls follow a similar path except that the entire assembly is built in the laboratory by the manufacturer.

Another Caution:

Fire-resistance ratings for building materials must be carefully distinguished from flame spread ratings. It is common practice to speak of "fire-rated tile." This term is meaningless.

Fire-resistance ratings are concerned with the length of time a particular building assembly will continue to perform its structural and/or barrier function, in the face of an assault by a standard fire test.

Flame spread ratings are concerned with the rate at which fire spreads over the surface of a material, the smoke the material develops, and the fuel it contributes to the fire (see Chapter 9, Fire Growth).

The two terms are not interchangeable, but in some cases, a single material may be evaluated for either its fire-resistance rating or its flame spread rating, depending on the way it is used. As an example, when acoustical ceiling tile is part of a fire-resistant floor and ceiling assembly, its ability to act as a barrier to passage of heat is of prime importance. However, when acoustical tile is used solely for its sound-deadening and decorative qualities, such as when tiles are directly adhered to the underside of a concrete floor, then its fire-hazard rating is of prime importance. As part of the interior finish of a building, its ability to spread flame is a life safety concern. In many cases both fire resistance and flame spread may be a concern.

Tactical Considerations

It is dangerous to relate a building's rated fire resistance to a projected collapse time. The following concept is not correct: "The building has a two-hour rating. If the fire is not controlled in one hour and 45 minutes, prepare to withdraw and operate from outside."

Depending upon the fire load, the rate of fire development, and many other factors, such as possible failure of a key structural element or fire barrier, the building may be in distress, or the fire may communicate beyond a fire barrier, in much less than two hours. The converse may also be true.

Since there are so many variables, the limit of confidence that one can have is more nearly the concept that a four-hour fire-resistive building is more fire resistive than a two-hour fire-resistive building, and a one-hour fire-resistive building is less fire resistive than a two-hour fire-resistive building. It is not simple to relate the building ratings to an anticipation of what will occur during a fire, either preplanned or actually in progress.

Estimating Fire Severity

There is a rule of thumb which is useful in estimating the severity of a building fire. It involves an estimate of how much of a fire exposure certain amounts of fire loading (the amount of fuel per square foot of floor space) will produce. For example, a fire loading of five pounds of ordinary fuel per square foot of floor space in an office building has a potential of releasing 40,000 Btu or the equivalent of a 30-minute fire exposure under the time temperature curve used in fire tests. Higher amounts of combustibles per square foot of floor space produce higher corresponding equivalent fire exposures.

Figure 6-4 relates combustible contents to heat potential and to equivalent fire severity. Note that the "pounds" are of ordinary combustibles estimated at 8000 Btu per pound. For plastics, rubber and flammable liquids, use 16,000 Btu per pound.

Combustible Content Total, Including Finish Floor and Trim Lb per sq ft	Heat Potential Assumed* Btu per sq ft	Equivalent Fire Severity Approximately Equivalent To That of Test Under Standard Curve For The Following Periods:
5	40,000	30 min
10	80,000	1 hr
15	120,000	1.5 hr
20	160,000	2 hr
30	240,000	3 hr
40	320,000	4.5 hr
50	380,000	7 hr
60	432,000	8 hr
70	500,000	9 hr

*Heat of combustion of contents taken at 8000 Btu per lb up to 40 lb per sq ft; 7600 Btu per lb for 50 lb and 7200 Btu for 60 lb and more to allow for relatively greater-proportion of paper. The weights in the table are those of ordinary combustible materials, such as wood, paper, or textiles. Data applying to fire-resistive buildings with combustible furniture and shelving.

Fig. 6-4. Estimated fire severity for offices and light commercial occupancies.

It is possible, therefore, to estimate the fire severity of a given occupancy by roughly calculating the fire exposure for that amount of loading, and then comparing it with the fire-resistance rating of the building. The result will be at least a rough determination of the possible effect on the building of a severe fire in the contents. A study of file rooms and supply rooms in a one-hour rated office building might provide a real shock.

Combustibility and Fire Resistance

Fire resistance does not necessarily mean non-combustibility. There are listings of assemblies which include combustible elements, typically wood beams or studs. In early editions of the UL *Fire Resistance Directory*, the word "combustible" appeared in the UL listing immediately after the rating. This has been dropped in the listings of many combustible assemblies. The dropping of the word, for whatever reason, does not change the combustible nature of the assembly.

The fallacy of arguing that fire-resistant assemblies represent a safe working platform for fire fighters lies in the fact that the ASTM E119 standard was developed for noncombustible floors assaulted only from below. Combustible structures are attacked through the floor above and by fire starting in or pushing into the combustible void, factors unevaluated by ASTM E119.

This deficiency is of vital importance to fire fighters. It is fully covered in Chapter 12, Trusses.

Containment

An important element of fire control is the physical containment of the fire. There is a common assumption in building code interpreta tion that if a rated fire-resistive assembly abuts another such assembly, containment is assured. This has not been true in a number of instances.

* A rated wall assembly that is combustible is erected on a rated floor-ceiling assembly that is also combustible. Some believe that this provides fire containment. This is not true. Fire can and has burned down through floors and passed under walls (See Chapter 5, Garden Apartments).
* A steel bar-joist rated floor and ceiling assembly joined to a steel column for protection is held to be adequate. The column is thought by some to need no protection in the plenum or void space. In fact, the column is vulnerable to failure if there is any failure of the ceiling (See Chapter 7, Steel Construction).
* Fire walls built of several elements, such as a concrete block wall built up to a spray-fireproofed, steel girder topped by a brick parapet wall, are questionable (See Chapter 7, Steel Construction).
* Perimeter firestopping of fire-resistive structures, with prefabricated wall panels, which are not fire-tight to the floor is sometimes attempted by the use of gypsum board interior surface panels (See Chapter 11, High-Rise Construction).

Fire Intensity and Duration

In reality there are no "fireproof" buildings. The measure of the resistance of a structure to fire involves determinations of both the intensity of a fire and its duration. The difficulty lies in anticipating how big a fire can be expected and how long it will last.

L. G. Seigal points out in his book[3] that a high-intensity short-duration fire may be more severe than a low-intensity fire that lasts a longer time. Small increases in temperature cause large increases in intensity, because radiant energy is proportional to the fourth power of the absolute temperature. Thus, a 2000°F absolute fire lasting 20 minutes could have eight times the radiant energy potential of a 1000°F absolute fire lasting for 40 minutes.

This contradicts the long held belief that if the "degree hours" in different cases are equal, the severity of the fire is equal.

Preplanning

Before we can preplan fire department operations intelligently, we must accurately estimate the size and severity of possible fires. The First Interstate Bank fire in Los Angeles is a spectacular example of the sustained heavy fire load which can be encountered in a modern office building. Fortunately, the cementatious fireproofing was excellent. There are other buildings which would have probably suffered structural failure with the same fire load.

Some essential elements of preplanning, relative to fire resistance, are:

★ Be aware that, generally, only buildings required by law to be fire resistive are so built. In some cases, the designer may choose to use some fire-resistive elements, or elements which resemble fire-resistive elements, but which are not.

★ There are many substantial public buildings, institutions, and churches that are classified as fire resistive but which were built to earlier inadequate levels of fire resistance.

★ Recognize that not all fire-resistance systems which receive the same rating are equal. There are more potential deficiencies in a steel floor-ceiling assembly than in a cast-in-place concrete floor, even though both achieve the same rating in a carefully controlled test environment.

★ Examine the building for deficiencies such as columns which are unprotected in the void space, poor perimeter fire stopping, the potential for widespread smoke pollution through ducts, and so on.

★ Compare the fire load in high fire-load areas to the fire-resistance rating in pre-planning for a fire.

The Ultimate Test

The fire-resistance rating of a building is an accumulation of ratings of scores or hundreds of components. These are built and installed by persons of all levels of competence. The building as a unit is never rated for its total fire resistance. The failure of only one of many elements may be disastrous.

The fire is the ultimate test of the fire-resistive structure.

References

The addresses of all referenced publications are contained in the Appendix.

1. Freitag, J.K. *Fire Pevention and Fire Protection as Applied to Building Construction*, 2nd ed. New York, New York: John Wiley, 1921, pp. 369-370.

2. Shoub, H. "Early History of Fire Endurance Testing in The United States," National Bureau of Standards, *STP 301 Fire Test Methods*. American Society for Testing and Materials, Philadelphia, Pennsylvania, 1961.

3. Seigal, L.G. "The Severity of Fires in Steel Frame Buildings." *American Institute of Steel Construction Journal*, October 1967, p. 139.

Additional Readings

Cote, Arthur E., *Fire Protection Handbook*, 17th Ed. Section 6, Chapts. 2, 5, 6.

7

Steel Construction

Steel is the most important metal used in building construction. It is generally available and relatively inexpensive in the U.S. Without it construction would be limited to massive all-masonry buildings with arched floors, or masonry wall-bearing buildings with wooden floors.

Steel is tremendously strong. Its modulus of elasticity (a term which measures its ability to distort and restore) is about 29 million psi — far more than any other material. Steel's compressive strength is equal to its tensile strength. Its shear strength is about three-quarters of its tensile strength. (See Chapter 2, Principles of Construction, for definitions of compression, tension, and shear.) This great strength enables steel members of relatively small mass to carry heavy loads, particularly when used in trusses.

Fire resistance, however, is a function of mass. Such strong but lightweight members have little inherent fire resistance.

Fire Characteristics

Steel has several important characteristics to consider regarding its behavior in fire. This chapter will discuss these characteristics and how they affect the fire performance of buildings. It will also look at the deficiencies, if any, of methods used to overcome them.

An excellent treatise on the benefits and advantages of steel construction is *Fire Protection Through Modern Building Codes*[1]. However, this chapter deals with the problems construction steel presents to fire fighters.

The characteristics which concern fire fighters are:

- The coefficient of expansion of steel is such that substantial elongation can take place in a steel member at ordinary fire temperatures (about 1000°F). This elongation may cause the disruption of other structural components, such as masonry abutting the ends of the steel.
- If the steel cannot elongate because of restraint, it will buckle or overturn. This can be particularly significant when other components rest on a steel member. Often steel members, usually beams, are inserted into an otherwise combustible building to provide a greater span than a wooden beam could provide or to support an opening in a brick wall. Such unprotected steel beams can be the origination of a collapse.
- When heated to higher temperatures (above 1300°F), which are common at major fires, the yield point of steel is drastically reduced. At this temperature, steel members may fail, bringing about a collapse of the structure.
- The temperature above refers to structural steel. Cold-drawn steel such as steel tendons used for tensioned concrete and for excavation tiebacks and elevator cables will fail at about 800°F. These characteristics reveal that steel is truly a thermoplastic material.

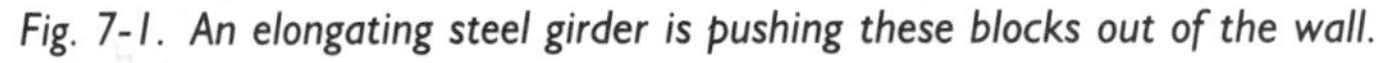

Fig. 7-1. An elongating steel girder is pushing these blocks out of the wall.

Fig. 7-2. These steel I-beams supported heavy air conditioners. A fire started while a car was being refueled. The steel failed before fire fighters could cool it.

- As steel is a good thermal conductor it transmits heat readily. This is a characteristic that is beneficial under some circumstances. A large massive steel member is a heat sink of considerable capacity and thus is capable of redistributing heat. This capacity is not unlimited, however. Many destructive fires have demonstrated that failure temperatures can be reached at one point in a steel structure when the heat input rate exceeds the rate at which the heat can be redistributed.

The thermal conductivity characteristic is also important in that heat can be transmitted by conduction through steel to combustible material that otherwise is unexposed to the fire. This characteristic is important both in the development and extinguishment of metal deck roof fires (discussed in detail later in this chapter). It is also significant in ship fires.

Unwarranted Assumptions

The fact that steel is non-combustible leads to unwarranted confidence in its "fireproofness" and suitability for all applications where fire is a problem. "I never would have believed it," is a common reaction by otherwise well-informed people when they see the ruins of an unprotected, steel building.

Fig. 7-3. The finance officer of this Arizona county was probably proud of this 19th century steel safe. However, because the door is just steel without insulation, it would transmit heat readily.

Fig. 7-4. This chart shows the effect of temperature on the strength of structural steel.

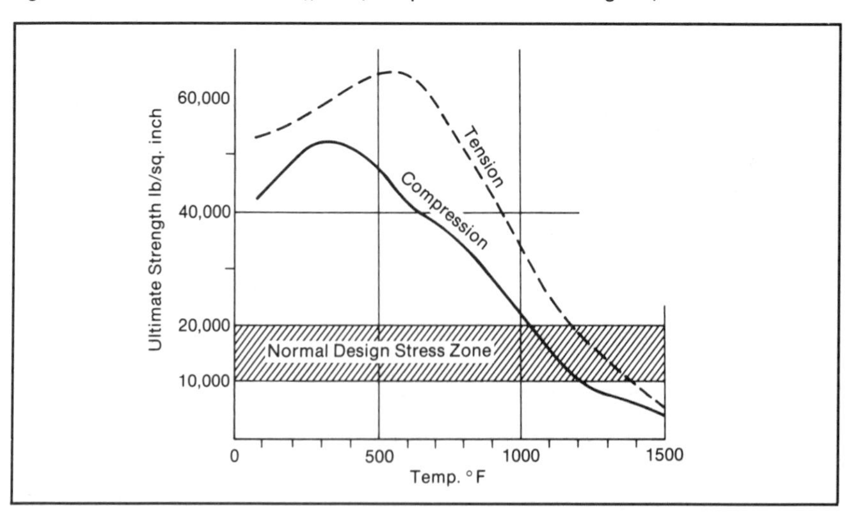

Water on Hot Steel

The heat evolved by a fire can be **triaged** or considered and treated according to priority needs. In reverse priority this includes:

- Heat leaving the structure. Let it go. Every Btu which leaves is one less to keep the fire going. This is sometimes contradicted by the "Chicken Little" school of fire fighting which apparently believes that the sky can catch fire and the world will come to an end.
- Heat being evolved from contents which are burning. This is of secondary importance. This concept often comes as a surprise to those whose sole objective is to "put the wet stuff on the red stuff." Contents partially damaged are, in effect, totally destroyed. In fact, much of overhauling consists of throwing out partially damaged contents.
- Heat being absorbed by contents or structural elements which will be ignited, or caused to fail. This heat is the most important heat to be removed. In the case of unprotected steel, failure can occur early in a fire and the consequences can be catastrophic.

The fire department's heat removal medium is water. It seems apparent that water should be used to remove the destructive heat from the steel. Unfortunately there is a myth which is fairly widespread in the fire service, and which the author has seen in print, that water should not be thrown on heated steel because of possible dire results — *This is simply not true.*

If the steel is elongating, the cooling effect of water will draw it back to its original dimensions. Many draw bridges, elongated by the summer sun, refuse to go back into position until the fire department cools the structure.

If the steel is failing, the water will simply freeze it in whatever shape it is in. When I published these statements in the first edition of this book in 1971, they were questioned by several people. I have since checked with the American Iron and Steel Institute, and was assured that I am correct.

IFSTA supports this position and has published the following statement.

> "Contrary to widespread belief, heated steel beams do not present a hazard when cooled with streams of water. They will not buckle, but will return to their original shape, or if past the point of strain, will keep their deformed shape.
>
> The effects of expansion, however, may present some hazards. Steel expands more readily and to a greater extent than most other building materials. There is the possibility that the expansion of a steel beam could crack masonry walls, making them unsafe. Another point to remember is that all the beams used in constructing one building might not be of the same size. Smaller beams will expand or contract quicker

> than large beams. If two sizes of beams are bolted or riveted together, this difference could result in the bolts or rivets shearing."

This author observed a Texas fire department fighting a fire in a steel hay shed. All attention was given to throwing water on the hay. No attention was paid to cooling the steel. The shed was severely damaged. The wet, half-burned hay was useless.

Tactical Considerations

It is very likely that the most important use of water in some fires is to remove the heat from the steel so that the structural integrity is maintained. It can be argued that this will create water damage. Note, though, that if sprinklers were installed they would be doing the same thing the hose stream does, although more efficiently. Think of the hose stream as an improvised sprinkler system, necessary because the owner failed to install one. Also consider the argument that there is no water damage at a total loss.

Plans for the cooling of steel should provide for a safe location for personnel. If a safe location is not available, this should be pointed out to the owners of the building so that they may be forewarned to the probable consequences of the fire. They might choose to provide protection for the steel.

It is often difficult, even for technically trained people, to understand how fast steel can be destroyed. Pictures of steel damage might be convincing. The distortion of the steel due to fire may alter the structure of the building so that it will not be able to carry the superimposed loads or unexpected stresses, particularly torsional, that may be created due to buckling. However, structural steel that has been exposed to fire and water will not have significantly different metallurgical properties than steel unaffected by exposure to fire.

The possibility that elongating bar joists may push a block wall out of alignment and when shrunk back may cause the roof to fall off the wall is sometimes advanced as a reason not to cool any steel. The possibility is remote and is guarded against by cooling steel from a safe location.

Definitions

Some necessary definitions of steel construction members are given here.

Angles: These steel members have two legs at right angles to one another. They are L-shaped in cross section.

Bars: Plates less than six inches in width are called bars. Bars are also made square or round.

Box columns: Large hollow columns built from steel plates are known as box columns.

Box girders: These are large girders, hollow like box columns, and often used for highway bridges.

Channels: Steel structural components which have a square U-shape cross section are called channels.

I-beams: These are beams shaped like the letter "I." The top and bottom of the "I" are called flanges and the stem is called the web. The top flange resists compression; the bottom flange resists tension. The web is necessary to keep the top and bottom apart. The principle has been copied in light-weight wooden I-beams (discussed in Chapter 3). Composite beams also exist with wooden flanges and corrugated steel web.

Plates: Flat pieces of steel are called plates.

Rolled or built-up members: Steel structural members can be either rolled or built up. Rolled members are one solid piece of metal. Built-up members are made up of different sections riveted, bolted, or welded together. A girder built from steel plate with angles riveted on each side of the top and bottom to form flanges is often called a **plate girder**. A column made of vertical units connected with diagonal pieces is called a **lattice column**.

Tees: A standard I-beam cut lengthwise through the web forms two such beams with T-shaped cross sections. If the end of the cut-off web is thickened, it is called a Bulb Tee.

Tubes: Steel structural members that are rolled in cylindrical, square or rectangular shapes are called tubes. They are most often used as columns. Some circular steel columns can be made fire resistive by filling them with concrete. There are hazards with these Lally columns. (See Chapter 8, Concrete Construction.)

Weight: The weight of steel sections is usually given per running foot. Thus, a 730-pound column weighs 730 pounds per foot. (This is a very heavy column.)

Wide-flange shapes: These are I-beams which have flanges wider than standard I-beams. Some are H-shaped and, because they are square, they are more suitable for columns. The letters WF are part of the designation of wide-flange shapes.

Zees: As the name implies, these are members with a Z-shaped cross section. They are not often used in structures. Short lengths of Z sections, fitted over the edge of a subway platform, provide a base to shift a subway car away from the platform.

As steel technology developed, mills were built which could roll larger and larger sections. Therefore, built-up structural members, of a size for which rolled members are used today, typically will be found in older buildings.

There are some additional relevant structural terms which are usually associated with steel.

Purlins: Beams, usually channels, set at right angles to trusses or roof rafters to provide support for light-weight roofing.

Spandrel girders: Girders that tie wall columns together in a framed building. They also carry the weight of the panel wall, which can be considerable when masonry is used. In addition, spandrel girders sometimes are connected to the columns and stiffened to help resist wind shear.

Steel as a Construction Material

Often the first structure which comes to mind when steel is mentioned as a construction material is the high-rise building. The development of steel framing as an engineering technique, and of steel fabrication, made it possible to erect tall buildings. The example of a steel-framed structure which most clearly shows its nature is the universally recognized symbol of Paris, Gustave Eiffel's remarkable tower.

Not so well known is the fact that New York City's Statue of Liberty is the first covered steel-framed structure. Eiffel designed the interior steel frame which supports the statue's copper skin.

The strength of steel, the consistency of its structural characteristics, and its ability to be connected to other structural elements so that loads can be adequately transferred are all important to the use of steel as a building material. Only a few specialized buildings, such as fiberglass buildings which are "invisible" to electromagnetic radiation, or less affected by corrosive atmospheres, are built without steel.

Aerial trams and ski lifts are supported on cold drawn-steel cables. Often the stations are of wood.

The effect of fire on steel can be crucial to the stability of almost any building during a fire.

Steel in Non-Steel Buildings

Steel has many uses in non-steel buildings. Steel is used for connectors of wooden structural members. Chapter 4, Ordinary Construction, notes

Fig. 7-5. A fire in the wooden station could cause the cold-drawn steel cable to fail in this chair lift high in the Canadian Rockies.

the small minority of buildings standing in which wood connections, such as mortise and tenon joints, were used.

In simpler days buildings were completely made of wood, and ordinary construction consisted of masonry walls and wood interiors. Both types of buildings now can have substantial unprotected steel members, particularly girders. These are hidden in combustible voids. The gypsum board interior finish of the building is not adequate to protect the steel.

Connectors, such as nails, screws, gusset plates, lag screws, threaded bolts, joist hangers, dog irons, rods, wire and cable, struts, and ties, are all made of steel. There are steel lintels over windows and doorways. Steel channels, spreaders and tie rods and cold-drawn steel cables brace insecure walls. Steel struts support high masonry parapets. Steel supports roof loads such as signs and water tanks.

Interior floor, beam, and column systems may be completely or partially constructed of steel, either by initial design or through alterations to strengthen a building weakened by age or overloading. Steel framing to provide structural stability has been erected within wooden or ordinary buildings undergoing rehabilitation.

Steel fire escapes are provided on many buildings. If in good condition they can be very useful, not only in evacuating occupants but also for fire department access. Fire escapes can be hazardous too.[2]

Fig. 7-6. Fire escapes are important not only for the occupants but as working platforms for fire fighter attack. All fire escapes should be inspected regularly. (Bill Noonan, Boston Fire Department)

Steel reinforcing bars and cold-drawn steel tendons are vital to concrete construction in providing the tensile strength that concrete lacks. Steel is also used in concrete flooring systems. Corrugated steel provides "left in place" forms, which are often designed to react together with the concrete under load, thus forming a composite.

Fig. 7-7. An alert student fire fighter noted the unprotected steel columns being installed in a rated fire-resistive mall. The columns were brought to the site prepainted the same color as the paint on the walls. Fire fighters should be eternally curious.

Steel is used to repair failures in concrete buildings. Even when a building is required to be fire resistive, the added steel is often unprotected.

In one case, an alert fire fighter noticed unprotected steel columns being installed and painted to match the walls in a fire-resistive rated concrete mall. The building department was notified but did nothing for years, until a building permit was requested for another job.

In any fire, the fire characteristics of the steel may be the key factor in collapse.

Steel Buildings

Steel framing is used for many commercial and industrial buildings, usually of one-story construction. To achieve a peaked roof, the framing may consist of columns and beams with triangular trusses . The columns may support deep parallel-chord trusses spanning wide areas, with smaller trusses, often called bar joists, spanning the main trusses to support a flat roof. Space frames (three-dimensional trusses) provide huge clear spans in some modern buildings.

The steel in such buildings is almost universally unprotected. At best, such buildings can only be classified as non-combustible. In fact,

Fig. 7-8. British fire fighters are having a tussle with this asphalt asbestos-protected metal building.

steel is rarely protected from the effects of fire until the building is of such height or size as to be required by law to be fire resistive.

Protected non-combustible sprinklered construction is found occasionally. In such buildings, major structural elements are provided with some fire resistance, but not enough to qualify the building as fire resistive. The concept is that the protected members will resist collapse until the sprinklers can control the fire.

Rigid frames are a cousin to the arch and are used to achieve wide clear spans. In the rigid frame, the column is narrow at the base and tapers to its widest point at the top where it meets the roof girder. This girder is also tapered so that it is narrow at the ridge and wide where it joins the column. This wide haunch resists the outward thrust of the roof. Rigid frames often are tied together under the floor to keep the legs from moving outward. If there is a basement, and the ties are exposed to a basement fire, failure of the ties may cause failure of the building.

Rigid frames can provide clear spans of about 100 feet. The area of the structure can be increased indefinitely by using Y-shaped columns, common to both frames. If such a column is distorted by fire, the area of damage may be doubled. Rigid frames can also be built of reinforced concrete or laminated wood.

Many steel-framed buildings are prefabricated. The Butler Company is a prominent manufacturer of steel "prefabs." Such buildings are often referred to as "Butler buildings" regardless of the manufacturer. Originally, the light-weight steel-framed building was exclusively for industrial use, but in recent years, schools, churches, and many other types of buildings have been built with similar designs.

Huge Spans

When huge spans are achieved by using rigid frames, trusses, or space frames, collapse can be sudden, general, and tragic. In buildings with huge spans, adjacent bents are tied together to resist wind load. Tying the steel units together, however, means that if one part of the building is distorted by fire, torsional or eccentric loads beyond the designed capacity may be placed on the balance of the building. This can start the progressive collapse of the building, often far beyond the area involved in fire. As a matter of fact, the better the building is tied together to resist wind load, the more likely it is to suffer progressive collapse due to fire distortion.

Designers are working wonders in developing wide-span trusses made up of very light sections. Recently a hangar was built in Spain with a span clearing 787 feet, 272 feet deep. The push to design large open areas with a minimum of mass is fraught with the potential for disaster. Designs may be pushed beyond the limits of steel. Hasty field changes or errors in construction can have catastrophic consequences.

The collapses of the C.W. Post Auditorium in 1978, a 171-foot span steel and aluminum space dome located on Long Island, New York; the Hartford Coliseum in 1978, a 300-foot span steel space frame; and the Kemper Auditorium in Kansas City, Missouri in 1979, a 324-foot steel truss roof suspended from a space frame, caused no fatalities. Considering the thousands of people who had occupied these buildings, it is easy to see that there was a disaster potential which could dwarf the problems faced by rescue personnel in the Kansas City Hyatt Regency Hotel skywalk collapse in 1981, in which 113 persons died and 186 were injured.

Walls of Steel-Framed Buildings

A steel-framed building can be provided with a variety of walls. Some common wall materials and their fire characteristics are:

- Cement-asbestos board: This material is non-combustible and is often used for friable construction. **Friable construction** is used where an explosion is a possibility. It will break away readily, relieve the pressure, and not itself provide missiles during an explosion. While such a scene appears to have great devastation, the loss is much less than for a building which would resist the blast.

Concern for asbestos hazards may require monitoring before overhauling, or investigating a fire where this material is involved. (Transite, a trade name of such material, is often used as a generic name.)

Some sandwich boards of cement-asbestos "bread" and low-density fiberboard "meat" are used. Fire can burrow into the fiberboard and smolder.

- Glass-fiber reinforced plastics: While glass fiber is non-combustible, the resinous binder most often used with it is flammable. This is not generally known, and the material is often thought of as noncombustible. Some reinforced plastic panels with good fire characteristics are available.
- Aluminum: This material (aluminium in Canada) is non-combustible, but has a low melting point and little mass per unit of area, so it disintegrates rapidly in a fire. This is not always a disadvantage. In some instances, it may provide needed venting and access for hose streams to the interior of the building.
- Precast prestressed concrete panels: Panels of this type are erected in large sections, and their collapse can be particularly hazardous to fire fighters. Study the relationship of the steel frame to the wall. If the steel expands under fire conditions, will it deflect the wall section from the vertical? If deflected, will the wall section fall freely or will it be restrained from falling? Precast concrete panels formed the side walls of McCormick Place, the large Chicago exhibition hall destroyed by fire in 1967. Distortion of the steel roof in the fire pushed the wall sections out of alignment and several of them collapsed.
- Wood siding: A combustible material, wood siding is not common, but partial wood siding may be used for aesthetic effects. Wood siding and any other combustible siding must be considered when estimating the effect of heat on the steel.
- Masonry walls: The most common walls for unprotected steel-framed buildings are made of concrete block or a composite of concrete block and brick. Natural or artificial stone walls are also seen. Usually, the walls are only curtain walls. The exterior surface may show unbroken masonry; the interior wall may show panels of masonry between steel wall columns whose interior surface is approximately flush with the masonry. In other designs the wall is independent of the steel frame and is tied to it only for wind resistance. Occasionally, in framed construction, one wall, or part of one wall, is a bearing wall. In either case, it is important to analyze the effect of the expansion of the steel frame on the wall. For instance, if a girder rests on a bearing plate, it may be free to slide, and thus disturb only masonry above. On the other hand, a beam tied to a girder that is incorporated longitudinally into a wall may, by pushing on the beam, bring down all or a substantial part of the wall.

This author has seen several sheet steel-walled buildings "improved" by covering the steel with a brick veneer. Any movement of the steel would likely bring down the brick since its stability depends entirely on the steel.

The signs of masonry deterioration, which were mentioned in Chapter 4, are equally important in the steel-framed building. However, the absence of deterioration does not decrease the possibility of collapse of the masonry. A very probable cause of collapse is a steady lateral push by expanding steel.

- Galvanized steel walls: These walls are an industrial standby where heat conservation is not important. Sometimes the metal is weather-protected by an asphalt coating. This is **asphalt asbestos protected metal** (AAPM), often called Robertson Protected Metal (RPM), one proprietary name. The tar coating is combustible. (See Fig. 7-8.)
- Metal panels: Prefabricated metal panels are often made up in a sandwich construction to provide one unit combining thermal insulation and interior finish. In earlier construction, low-density fiberboard will be found. Later, mineral and glass fibers were used. Today, plastics are most often found. Even when the outer layers are noncombustible, the insulation, vapor seal, or adhesive in the panels may be combustible. It may be necessary to open such panels far from the main fire area if the interior of the panel becomes ignited. This author recommends that those overhauling the structure work slowly and patiently, and cut with the circular saw, rather than with the axe, to avoid excessive destruction.

In one type of construction, polyurethane insulating panels are protected by gypsum board and stainless-steel sheathing. It is quite possible that, many years hence, alterations will be undertaken by persons not aware of the insulation. A cutting torch may be used, and a smoky, destructive fire may result.

At a research establishment, a steel tunnel-like structure was lined with polyurethane insulation. A crew installing a door cut a hole with a torch and a serious fire resulted.

Aluminum sandwich panels with foamed polyurethane are also available. Some are listed by Underwriters Laboratories Inc. for low flame-spread ratings but their "smoke-developed" ratings may be quite high.

Metal wall panels can be used on any framed building and, in fact, are used on many concrete buildings.

The design of panel walls, the method of installation, and the degradation of insulation or expansion of metal under fire conditions are some of the reasons the closure of the wall panel to the floor slab can fail and permit the extension of fire from floor to floor. This can be particularly significant in high-rise buildings beyond the range of effective ground attack.

Prefabricated metal panels forming the front wall of some motel rooms have neoprene gaskets between the panel and the wall separating the rooms. A fire in one room would quickly destroy the gasket and extend toxic gases and fire to the adjacent rooms.

High-Rise Framing

For many years, steel framing stood unchallenged as a method for high-rise buildings. In recent years, concrete is finding more use for framing high-rise buildings, and the two materials compete. The reinforced concrete South Wacker Tower in Chicago is 946 feet high. The principles of construction are the same, however, and much of what is discussed here applies equally to concrete-framed high-rise buildings.

When steel-framed buildings were first built, builders were hesitant about featuring the fact that it was the frame which supported the building. Traditionally, most people associated strength and permanence with solid masonry. Consequently, tons of brick, stone, and terra cotta, all structurally unnecessary, were loaded onto framed buildings. The heights of buildings were also suspect. Rather than arrange the exterior so that vertical lines led the eye upward, as is common today, buildings were **belted** with masonry corbels or trim. Different materials were used on successive floors, and all possible architectural devices were used to reduce the apparent or perceived height of the building.

The masonry served another purpose in early construction. The **spandrel space** (the distance between the top of the one window, and the bottom of the one above) was made of masonry. It was sufficient to prevent the extension of fire from floor to floor on the outside.

Today, the pendulum has swung the other way. Glass and metal exterior panels are common, and the firestopping is dubious. High-rise buildings should be evaluated for the dependability of perimeter firestopping. Tactical plans should be developed to deal with the problem. (This is discussed further in Chapter 11.)

Steel-Framed Buildings Under Construction

Except for the problem of breaching shear walls around elevator shafts (as noted in Chapter 2), the method of wind bracing a building is of no concern to fire fighters after the building is completed. While the building is under construction, however, wind forces must be resisted, as the building is not fully connected. Often, cables are strung diagonally across the steel work, and at times, these braces are not properly installed. If there are no high winds, nothing happens and the perpetrator of the dangerous bracing has another "good experience."

Experience is the best teacher — of bad habits too. In one case, a bracing cable ended in an eye splice to be connected to the end of a turnbuckle attached to a column. The cable eye was passed through the turnbuckle and "secured" with a scrap piece of wood through the cable eye. The "artisan" who accomplished this engineering triumph then fixed three others the same way.

Tactical Considerations

Carefully examine the temporary bracing before going up to extinguish a fire on a steel building under construction, particularly if the weather is windy. Be aware that it is obviously impossible to make all the connections as soon as steel is placed. Temporary bolts are placed in some rivet holes; thereafter, rivets are driven in the remaining holes, then the bolts are replaced with rivets, or permanent bolts. The practice is known as "field bolting," and it is good to give a wide berth to field-bolted structures during high winds.

"Plastic" Design in Steel Construction

In an effort to seek economical steel design, and to compete with concrete, plastic design has been developed. The typical steel-framed building is pinned. In a pinned building, loads are delivered to the nearest columns. In plastic design the connections are built to transfer loads beyond the column. This is analogous to monolithic construction in concrete. Because of the rigid connections, a load which might be locally excessive is redistributed over the structure. As a result, the weight of steel in the building can be decreased. Beams are lighter and columns are smaller than they would be otherwise. This is of more than academic interest. The lighter the steel, the less fire resistance. Thus, the provision and maintenance of adequate fireproofing for the steel is even more important in a plastic designed building than in one of conventional design.

More on the Fire Characteristics of Steel

At this point, the comments made in the beginning of this chapter concerning the fire-significant characteristics of steel will be expanded upon.

Steel Conducts Heat

The fact that steel transmits heat is well known. A suit of armor is non-combustible but no one would attempt to fight a fire in it. Despite

Fig. 7-9. The columns of this Bethesda, Maryland building are unprotected in the plenum space. Additionally, the building is of plastic design. All steel members are smaller than they would be in pinned construction. Smaller members heat up faster.

this, many building codes required (and still may) so-called **tin ceilings**, which are actually embossed steel, in certain occupancies. These can transmit fire in either direction.

In earlier years, sheet steel doors were used commonly on vaults — some of which still guard valuable governmental and private records. The chief disastrous result of using steel doors could be the loss of records. Across the country, businesses and homeowners alike seem to rest secure in the knowledge that their vital records are "protected" in uninsulated steel files. Tin boxes are often sold in stores as fire-resistant chests.

While it may be acceptable for an individual to hazard his or her own records, it is a different case when vital records of others are involved. The next time you go to the dentist or the doctor, ask how your records are safeguarded. Most often you will find them on open shelves or in non-fire-rated files. There are rated safes and safe files available, designed to protect records against a standard fire for a rated time.

Magnetic tapes and computer disks fail at temperatures much lower than the 325°F permitted in the interior of ordinary safe files. Much heavier special safe files are required for their protection.

The conductivity of steel can be a factor in spreading fires. For example, heat was conducted through **steel expansion joints** in a concrete floor at a fire in a multi-story post office. Mail bags, resting on the joints in the concrete floor, were ignited by the heat.

Steel sheets directly attached to combustible surfaces to provide fireproofing in many cases have acted to transmit or retain heat. In one case, a steel plate was attached to the wood-joisted ceiling above the end of a rotary kiln. When it was ordered removed, plant personnel were surprised to find that all the wood behind it had carbonized to charcoal.

It is generally understood that metal smoke pipes which lead from a furnace or stove to a chimney or metal stock must have clearance from wooden members. However, this concept is often ignored in the case of other metal ducts which may have fire in them, despite requirements to the contrary. Grease ducts are a good example. A grease duct fire often extends to adjacent combustible construction by conduction.

Ships are not buildings, but the practice of using ships as buildings, like the Queen Mary in Long Beach, California, is growing. A number of ferry boats have been converted into restaurants.

Ships have steel walls known as **bulkheads**. Many people associated with ships confuse non-combustibility with non-conductivity. Typically, welding operations will be performed on one side of a bulkhead without any concern for heat transmission through the steel to ignite combustible material on the other side. This can cause a problem when the other side is covered with wood paneling which could be ignited from within the concealed space behind the paneling.

Self-storage facilities built of steel have many of the characteristics of ships. The Tualatin Valley, Oregon Fire and Rescue Department fought a stubborn fire in a self-storage building which involved 97 units. Some of the units contained flammables and explosives. The fire spread from unit to unit by conduction and radiation. Scalding steam was generated when water hit the hot steel sheets.[3]

Steel Elongates

Steel will expand from 0.06 percent to 0.07 percent in length for each 100°F rise in temperature. The expansion rate increases as the temperature rises. Heated to 1000°F, a steel member will expand 9½ inches over 100 feet of length. You will see later, at temperatures above 1000°F, that steel starts to soften and fail, depending upon load.

Elongating steel exerts a lateral force against the structure which restrains it. If the restraining structure is capable of resisting the lateral thrust, the expanding steel structure may be even stronger, at least for a period of time. Some floor assemblies have a different fire resistance rating depending on whether they are restrained or unrestrained.

If steel beams are restrained, as by a masonry structure, and the temperature of the fire is sustained around 1000°F, the expansion of the steel may cause the displacement of the masonry, resulting in a partial or total collapse. A one-story commercial building in Dallas was constructed of concrete block walls and a steel joist roof. The steel joists were restrained by the walls. A metal deck roof fire sent heat rolling through the roof void. Elongating bar joists pushed down the walls, dropping six fire fighters into the structure. Fortunately, all six were rescued.

In hot, fast fires, failure temperatures are reached rapidly and the lateral thrust against the wall is minimized. In a one-story and basement non-combustible building, a heavy girder was supporting a one-wythe thick decorative brick wall in the reception lobby. There was a hot fire in the basement. The steel beam attempted to elongate but could not because of restraint. It took the only other course to absorb the increased length — it twisted, dropping the wall completely across the lobby. Fortunately, no one was in its path. This overturning can be anticipated when unprotected steel girders are used to support wooden floors in combustible buildings.

Steel Fails

When steel is raised to temperatures above 1000°F, it starts to lose strength rapidly.

Temperatures above 1000°F are quickly reached in fires. Recognizing this, the standard fire test, NFPA 251 (ASTM E119), used to test building materials and assemblies, reaches 1000°F in five minutes. Temperatures developed in National Institute of Standards and Technology (NIST) tests of typical basement room fires reached 1500°F in five minutes, reflecting today's heavy plastic fire loads, rather than any special significance of basements.

In standard tests of the fireproofing of steel columns, the test is ended when a temperature of 1200°F is exceeded at one point or 1000°F is exceeded on the average in the column.

How fast the fire temperature will be achieved in the steel varies.

A principle variable is the weight or mass of the steel unit. For instance, a bar joist or light-weight steel truss will absorb heat very rapidly. In a fire equivalent to the standard test fire, the bar joists generally will fail in about seven minutes. In one observed fire in a one-story and basement bar-joist building, the bar-joist supported floor was collapsing before the people could get out of the first floor. On the other hand, very heavy steel sections might have survived longer.

The amount of ventilation may also be a significant factor. It appears that if a fire is well ventilated, the temperature in the steel may be less than if the fire is confined. This may help to explain observed differences in the effect on steel in different fires.

The heavier the steel is loaded, the faster it will fail.

Fig. 7-10. Some people don't like to say that steel fails. In this case, the bar joists developed "undesigned hinges."

Overcoming The Negative Fire Characteristics of Steel

There are a number of options open to the designer to deal with these characteristics of steel.

- Ignore the problem.
- Rely on an inadequate code.
- Take a calculated risk.
- Fireproof (insulate) the steel.
- Protect the steel with sprinklers.
- Fireproof the steel with an internal water cooling system.
- Locate the steel out of range of the fire.

Ignoring the Problem

The potential for fire damage to steel buildings is not clearly understood. Often, unwarranted confidence is placed on the fact that the steel is non-combustible. A major insurance carrier found it necessary to conduct demonstrations of actual fires for plant managers in which the only fuel was the hydraulic oil in presses.

During World War II the War Production Board, which controlled all automatic sprinkler installations, refused to sprinkler a 400,000 square foot Navy warehouse with combustible contents because it was of steel construction.

An article on "Paper Mill Fires" in *Fire Engineering*, September 1990, p. 48, by J. R. Hall, makes an excellent presentation of the hazards and problems of fighting paper fires. I would supplement the article by pointing out that the machinery and structures are worth far more than the paper and, where practicable, water should be used to cool any yet undamaged steel and thus save the structure and the machinery.

A metal building intended as a site for thawing frozen drums of uranium residue was under construction at an atomic energy facility. The shed and all the roller conveyors were destroyed by a fire. The only fuel was the 1-inch fiberboard lining of the building. The builder was astonished. He had been so convinced that there was no fire problem that he had not insured the job.

Unprotected light-weight steel trusses called bar joists are susceptible to early failure (See Chapter 12).

Tactical Considerations

Stay alert. Be aware of the hazards to fire fighters of a fire in an unprotected steel building.

Steel Highway Structures and Bridges: Bridges and overpasses on highways provide a good example of unprotected steel which is vulnerable to an occasional gasoline truck fire. Such fires can result in enormous damage, possibly accompanied by major traffic disruptions. For some interchanges, an analysis would prove that protection is justified, but any effective action is most unlikely.

A bridge on the Trans Canada Highway in Nova Scotia was melted when struck by a fuel truck that exploded. The bridge was out of service for 69 days and the repairs cost $350,000.

A bridge in Pittsburgh passed over a plumbing warehouse. A 10-alarm fire in the warehouse severely damaged the bridge. It was out of service for a year. The total loss was estimated to be $10 million, taking into account the losses suffered by the 18,000 people who had used the bridge daily.

A Texas fire department fought a fire in a lumberyard. The lumber extended under the structure of a highway which spanned a ravine. The highway was out of service for several months while expensive repairs were made to the bridge. A single engine company committed to cooling the steel might have prevented the damage.

In New Jersey, an illegal construction debris business existed under the elevated portion of Interstate 78 — ten lanes wide. A fire in the combustible debris did major damage to the highway. Heavy steel girders elongated, knocking over rocker bearings, and then sagged. It was closed for months and caused massive traffic disruption.[4]

A gasoline tank truck was on fire below a steel highway overpass. The fire department used the available water to make foam. The fire was extinguished but the bridge was ruined. The removal of the unburned gasoline was a serious problem. It might have been better to use the available water to cool the steel, as the gasoline burned out. (The use of water to cool steel was fully discussed earlier on page 259.)

The Chicago Fire Department fought a truckload of magnesium on fire under a concrete overpass. The structure was protected by hose streams as the fire burned out. The bridge was saved.

Seventy years ago, when prefire planning was generally unknown, the New York City Fire Department developed very specific preplans to get water as soon as possible onto fires on the East River bridges, recognizing the vulnerability of the unprotected steel. If assigned units are unavailable, the dispatcher must cover them and notify the replacement units of their specific function.

Tactical Considerations

Fire departments should survey all highway structures for vulnerability to building fires.

Prefire plans for buildings which expose highway structures should take into account protection of the structure. The highway structure may be far more valuable than the building, and cooling the steel might be more important than extinguishing the fire. Tactics not normally acceptable, such as driving the fire down into the building, may be necessary.

Accidents under highway structures may result in massive fires. Fire departments should preplan to protect the steel (or concrete), and anticipate a flammable liquid fire. Adequate water supply must be preplanned. Generally, it must be delivered by units located off the highway. Water used as foam to suppress the fire may not be as effective as water used to cool the steel or concrete of the bridge structure.

If water is used to cool steel and mixes with gasoline, it may provide a flowing fire. Will the fire flow harmlessly, or will it flow into nearby buildings? In the latter case, the use of water may not be recommended due to the potential extension.

Tactical Considerations, continued

In any case, preplanning is essential. The initial dispatch should include all necessary units because of traffic tie-ups. The police should be included in preplanning to assure the diversion of traffic as soon as possible.

Hazards of Concentrated Fire Loads: Even when average fire loads are low, an unprotected steel building may be endangered by a highly concentrated fire load in one area. Occupants of the buildings are not always as enamored of the big open space as the designer. While unprotected steel industrial buildings are usually only one-high story, it is not uncommon to find a mezzanine built into the building to provide office or storage space.

The term **one-high story** is used to designate buildings of greater than usual height from floor to ceiling. A description such as "one equals five" describes a one-story building of a height equal to five ordinary stories.

Often a mezzanine is built entirely of wood. Where problems such as noise, temperature, pilfering or privacy exist, structures may be built within the building to provide office space, controlled environment, privacy for secret processes, or security for valuable merchandise. Such structures are often combustible. The structure represents a hazard not only to the building but to that which it is designed to protect.

The experimental hall at a scientific research facility is often built of light-weight steel construction to provide the greatest free floor area at the least cost. The steel is usually unprotected; the fire load is expected to be minimal. In recent years, however, the practice of purchasing ordinary house trailers to use as offices or house telemetry equipment inside such halls has grown. These provide privacy and comfort. From the fire protection point of view, however, they provide a serious potential source of heat. Grouped closely together, the trailers expose one another, increasing the possibility of a serious loss.

Prefabricated metal buildings are sometimes used in place of the trailers. Often these are of sandwich construction with an insulator placed between the inner and outer walls. The nature of the insulation should be determined. Many kinds of insulation present severe fire problems.

A major athletic arena is tied together by unprotected steel cables. This author has been told that an agreement was made with the county that the management would never permit a serious fire load, such as might be provided by an exposition, into the building.

Excavation Bracing: Because of the requirement for underground parking, building excavations are being made much deeper than in previous years.

Fig. 7-11. Unprotected cold-drawn steel cables hold this arena together. A fire in an area with a high fire load might cause massive collapse.

When a hole is opened in the ground, the earth under the adjacent structures may tend to shear sideways into the hole, causing a structure to collapse. Often it is necessary to shore up the excavation. The shoring literally holds up the neighborhood.

Rows of vertical steel beams called **soldier beams** are driven into the ground. They are tied together by a horizontal beam called a **waler**. Diagonal columns called **rakers** brace the entire structure. These columns may be loaded to twice the load that would be permitted if they were permanent structures. They are, therefore, subject to very early failure in a fire. The consequences of such a failure could be catastrophic.

An excavation is often loaded with combustibles, such as plywood, plastic, and fuel.

This array of rakers is a nuisance in the erection of the building. If ground conditions permit, **tiebacks** are used. Cold-drawn steel cables are inserted in holes driven into the rock, and anchored with epoxy. They are then tensioned, similar to post-tensioned concrete (See Chapter 8, Concrete Construction). The end of the cable sticks out and thus can be easily heated. If the tensioned portion of the cable reaches 800°F, it will fail. This author is not aware of any precautions taken to protect this vital steel from fire.

The unprotected steel tower crane is similarly vulnerable. The hazard does not appear to be covered in any code. It should be explained to the building department and to the builder. "Nothing like that has

Fig. 7-12. In this San Francisco excavation, the bracing is keeping the adjacent building from falling into the hole. Such bracing is usually permitted to have a heavier load than the permanent steel structure will have, thus it will fail faster. In a fire, immediate cooling is necessary.

Fig. 7-13. Soil conditions at this location permit the use of tiebacks, thus freeing the excavation of inconvenient bracing. The tiebacks are made of cold-drawn steel. If the steel beyond the anchor reaches 800°F, the tiebacks will have lost their tensile strength. They should be cooled continuously from a safe distance if exposed to fire.

ever happened. You guys are always thinking up busy work." are likely responses to this warning. However, the fire department will be in a much better position if the disaster occurs.

The Redmond, Washington Fire Department, with the help of mutual aid units, fought a fire in a concrete building under construction. The fire heated the tower crane and put it as much as four inches out of plumb. If uncontrolled this could have caused collapse. Heavy hose streams reduced the out of plumb gap to two inches. Water lines were directed on the crane. Surveying instruments monitored sagging.[5]

Tactical Considerations

Preplans should provide for the rapid cooling of all exposed steel in an excavation fire. Apparatus placement and operating positions should take into consideration the fact that adjacent buildings may collapse into the excavation. The occupants of such buildings should be evacuated.

Buildings Under Construction: During construction, steel, which will later be "fireproofed," may be unprotected for extended periods.

Fig. 7-14. This steel was many stories up in a high-rise building under construction. It was heavily damaged by a fire in the framework that had been built to apply the concrete fireproofing. (H. Spaeth)

However, because it is not yet carrying its design load it will resist fire failure for a longer time.

When the World Trade Center was under construction, original plans to use fire retardant treated plywood for form work were scrapped in favor of ordinary plywood and $1 million in savings. When fire protection of the project was discussed at an NFPA meeting, the Chief of the New York City Fire Department, John O'Hagan, warned the Port Authority of New York and New Jersey about the hazard of such a "seven block-long lumber yard" to the structural steel of the World Trade Center. Fortunately, no fire occurred.

Tactical Considerations

When construction operations present unacceptable hazards to unprotected steel, fire departments are well advised to put building owners on notice — even if there is no applicable code.

All structural steel exposed to fire should be cooled with hose streams.

Relying on Inadequate Codes

Building codes generally classify steel buildings as "unprotected non-combustible," or "protected non-combustible." In this latter case, major columns and beams have some protection. This is often in conjunction with sprinkler protection. The concept is to hold the building and the sprinkler system together until the sprinklers can reduce the heat production of the fire. Nevertheless, non-combustible buildings have been destroyed by fires in their contents.

There is another problem with a building that meets the code definition of non-combustible — the roof. Some codes permit a wooden roof on a non-combustible building. Is a combustible metal deck roof permitted? If it is, the building is subject to destruction independently of the contents due to the roof materials.

In studying a potential fire in an unprotected steel building, the first consideration should be the basic type of construction. Column, girder and beam construction is used for both single-story and multi-story construction. If a masonry bearing wall is substituted for some of the exterior columns, the building is wall-bearing. Column and girder buildings are characterized by relatively short spans, so the failure of one column may affect only one portion of the building. As with all such generalizations, however, watch for the exceptions.

A Maryland church is built of steel columns which support steel roof trusses. The roofing is wood. The 40-foot-high curtain walls, which do not carry the load of the roof, are composite brick and block. Should there be a severe roof fire, the heat absorbed by the roof steel would cause it to elongate, pushing the columns out of alignment. The moving columns might bring down the walls.

The author examined a church under construction in Louisiana. It is code-classified non-combustible. The framing is steel. There is a huge fire load of combustible fiberboard sheathing and a wooden balcony. The structure is brick-veneered. The roof is a conventional (therefore combustible) built-up metal deck. The interior of the structure and the underside of the wood-floored balcony are "protected" by ⅝-inch fire-rated gypsum board.

In Chapter 5, Garden Apartments, the widely-held fallacy of extending fire ratings achieved in the laboratory on specific test structures to gypsum board itself, was noted. The *1990 Fire Resistance Directory* of Underwriters Laboratories Inc. states, "Fire-resistance ratings apply only to assemblies in their entirety. Individual components are not assigned a fire-resistance rating and are not intended to be interchanged between assemblies but, rather, are designated for use in a specific design in order that the rating may be achieved."

Another unfortunate assumption about fires is that they will only burn upwards. Fuel in a church balcony might include hymn books and cushions of foamed plastic. A fire in these fuels would readily burn down into the void. A fire involving the wooden balcony or the metal deck roof could well cause the steel framing to move and thus cause one of the brick-veneered walls to fall. A well-advanced fire in such a structure should be treated with caution.

Steel High Above the Floor: In almost all building codes, steel used to support roofs at a certain distance above the floor (usually about 20 to 30 feet) does not require protection. It is generally believed that flame impingement is not likely at this height. The concept is probably valid if there is no significant fire load in the area below the steel. Such a situation is unlikely, however. For example, the situation which existed at Chicago's McCormick Place is far more common.

The McCormick Place Fire[6]: McCormick Place was a huge Chicago exhibition hall built in 1960. It burned on January 16, 1967, with a loss of $154 million.

The main exhibit area provided a clear area of 320,000 square feet. The roof was supported on 18 steel trusses, each 60 feet on center. Each truss had a 210-foot central span and a 67-foot cantilever on each side from the column to the exterior wall. The trusses were 16 feet deep at the column, tapering to 10 feet at the center of the building. They were made of heavy steel members.

The columns were trusses themselves. In accordance with the previously mentioned provisions common to many building codes, the columns were fire protected up to a height of 20 feet above the floor. The roof trusses were unprotected.

The exterior walls consisted of free-standing steel columns supporting precast concrete sandwich wall panels. The sandwich consisted of concrete interior surfacing, foamed plastic insulation, and exterior sculptured concrete.

At the time of the fire, preparations were being completed for the exhibit of the National Housewares Manufacturers Association. Exhibit booths crowded the floor. Some exhibits were two stories high. The fire loading has been estimated at 80,000 to 120,000 Btu per square foot. In addition, the combustible material, both by nature and arrangement, was fast-burning, and provided a high rate of heat release. Fortunately, the fire started in the early morning hours. Loss of life might have been staggering had the fire occurred during the day.

The fire was discovered early, but soon spread beyond the capabilities of the maintenance employees. There was about a six-minute delay in the alarm. A second alarm was transmitted on arrival. The first-due engine company penetrated the hall a few feet when they heard a great roar as the flames approached. At this time their water supply failed. It became known later that the hydrants in the area which were out of service were the responsibility of the exhibit hall management. In retrospect, at least one close observer thought this was a blessing. Had the water been available, it would have made little or no difference to the loss, but it is possible that a number of fire fighters would have been in serious danger advancing lines into a collapsing building.

Within a half hour after the start of the fire, portions of the massive roof trusses began to fail. The failing trusses pushed out a number of the wall panels, making close approach hazardous. Several of the panels collapsed.

McCormick Place Fire Report: After the McCormick Place fire, there were a number of studies and reports.

The following excerpt from the report by the American Iron and Steel Institute[7] is pertinent.

> "For the size and occupancy classification of the building, the construction of McCormick Place apparently met the fire protection requirements of the Chicago Building Code and most other widely recognized building codes. Most contemporary codes would classify the building as an assembly occupancy. However, at the time of the fire, it more closely resembled a mercantile occupancy such as a department store, discount house or shopping center. In this type of occupancy different fire protection measures would have been required by most modern building codes.
>
> If viewed as a mercantile occupancy, modern building codes would have required a building of this size to be equipped with an automatic sprinkler system throughout. Emergency heat and smoke venting and

> draft curtains at the roof level would also have been required but additional or supplementary fireproofing of the structural members would not have been required.
>
> Automatic sprinklers installed throughout the exhibit areas, draft curtains and emergency roof vents probably could have limited the spread of this fire and enabled the fire department, with a properly operating water supply, to bring it under control with a minimum of damage."

A building which can accommodate 50,000 people and is the equivalent of a tinder-dry forest cannot be protected by mere compliance with codes which did not contemplate such a building. Neglecting installed equipment or assuming that employees would take proper emergency action also critically affected fire protection at McCormick Place. At least this became apparent after the fire.

How apparent is it to your local armory, hotel ballroom, or exhibit hall management? Is your fire department satisfied that the extent of their responsibility is to see to it that all "rubbish" is removed?

The New McCormick Place: In the new McCormick Place, structural steel is protected with directly applied fireproofing delivering one-hour fire resistance. The entire building, except electrical enclosures and enclosed stair towers, is sprinklered. Sprinkler systems protecting the high fire load exhibit area are hydraulically designed to deliver 0.30 gpm per square foot over areas as large as 6000 square feet. Ninety-eight percent of the floor area can be reached with 100 feet of hose from standpipes. Elaborate provisions have been made for smoke venting. Supervisory control has been provided for water supply. Training programs for employees and prefire planning with the Chicago Fire Department are provided.

Important Test Experience: Testing undertaken at Underwriters Laboratories Inc. in the aftermath of the fire examined whether sprinklers would have controlled the fire.[8] The data developed in these tests is worth examining, especially as it relates to the temperatures developed at quite ordinary fires.

The fire tests were conducted in a building 30 feet high, the height above a floor at which it is presumed that steel in the roof needs no protection. The fuels used were light-weight, readily-combustible materials typical of those used for exhibition hall displays. Plywood, tempered hardboard, cardboard cartons, and packing materials made up the fire load average of 20 pounds per square foot. The materials had a high surface-to-mass ratio and thus would produce a fast, hot fire. The display material was set up over a 20- by 30-square-foot area. Thus, the total weight of combustibles was about 12,000 pounds, which would yield almost 100 million Btu if completely consumed. This was a rugged test, but not unrealistic. (For information and comparison purposes make some calculations for yourself at the next trade show or exhibit you attend.)

The first of the test fires is of great interest. The plan was to keep the sprinklers turned off for six minutes after ignition, corresponding to the reported gap between ignition and time of discovery of the McCormick Place fire. The report gives no indication that even the experienced engineers running the test had any qualms about the plan of holding back the water for six minutes. After five minutes and 45 seconds, temperatures of 1500°F were recorded at the ceiling. The sprinklers were turned on at this time to avoid damage to the test facility.

Steel beam temperatures were monitored by thermocouples peened into the surface. A bar joist at the ceiling reached 1540°F and an I-beam reached 1355°F at a little over five minutes. Temperatures at the ceiling level exceeded temperatures in the booth area by a slight margin.

The fire load used in the test cited is not excessive either in quantity or rate of heat release. There are many unprotected steel buildings with such a great fire load that collapse will have started prior to the arrival of the fire department, even with prompt alarm and normal response.

This author can recall a situation with a similar potential for disaster. A newspaper article described a recreational vehicle show at an armory. The location was unclear, so I called the local fire department's communication office. They had no knowledge of any such show. The show was in a National Guard Armory — jammed with people. The fire load included about 40 closely-spaced trailers and RVs. The roof was supported on unprotected steel. There was only one narrow unlocked door. The armory officer in charge had only one concern, "No tape on the walls, it pulls off the paint." A fire in one of the vehicles would have spread quickly. Units responding to a "fire at the armory" might have been horrified to find hundreds of casualties buried under a collapsed roof.

Despite the McCormick Place fire, and the tests cited above, buildings are still being designed in accordance with the concept that steel only below a certain height needs to be fireproofed.

The following paragraph is from an Engineering News Record description of a new air terminal at Tampa, Florida.[9]

> "Maintaining the true arch configuration, and simply lowering the trusses would have required fireproofing the structural members, defeating the architectural conception of the exposed structure."

Tactical Considerations

Fire fighters must be ever alert to what is going on in the community. They should particularly note activities which assemble large crowds in unusual places.

Tactical Considerations, continued

Plans should include the necessity of using hose streams to cool affected steel.

Taking Calculated Risks

The lack of fire protection for unprotected steel may be the result of a calculated risk.

For purposes here there are three classes of calculated risks:

- Financial or economic
- Engineering
- Forget it

The basis of the calculated risks must always be closely examined.

Financial Calculation: One calculation for a government-owned building went like this:

This non-combustible building is of low value and is slated for demolition in the five-year plan. In the meantime, it is used for the reserve storage of foamed-plastic shipping containers. The value is low and the loss of the containers would not hamper production. The cost of any special protection could not be justified.

"In case of fire — let it go."

Tactical Considerations

Since that is what the owner thinks, the fire officer should be guided accordingly. A high fire load, in an unprotected steel building, particularly of truss roof construction, can bring the building down unbelievably fast. In such a building, there is often no time for routine aggressive interior attack tactics. Any serious fire in such a building should be fought with the safety of fire fighters uppermost.

Engineering Calculation: The fire load in this non-combustible building is minimal. Unpackaged non-combustible heavy metal sections are stored on pallets. The combustible load averages about two pounds per square foot, and is dispersed. There is no apparent source of ignition. No special protection is recommended.

This calculation is adequate.

The steel industry is in constant competition with the concrete industry. When steel is required to be fireproofed the cost shows up as a line item in the budget. The cost of fire-resistive concrete, on the other hand, is buried in the cost of the concrete. This gives concrete a competitive advantage.

Some steel-framed parking garages were large enough to require fireproofing under some codes. The steel industry conducted fire tests to demonstrate that steel-framed parking garages did not require protection of the steel.[10]

As a result, some codes permit unprotected steel if a substantial portion of the garage is open to the air. This permits the escape of heat. This engineering calculation could not be applied to the parking garage in a Texas city in which the floors are supported on bar joists, rather than on heavy steel.

In another example, a steel garage was built over a railroad track which carried flammable liquid tank cars. It was decided rightly that the steel columns of the garage should be protected. This was done literally—the columns and steel directly at the track were protected. If the anticipated fire occurs, the fire will roll out beyond the protected elements and destroy the building. The same sort of questionable engineering was seen on a Canadian car-passenger ferry. Originally a deck was fitted out with passenger cabins. The overhead steel in the cabin area was spray fireproofed. The beams extended out to form the overhead of the outside deck. This steel was not protected. When the cabins were removed to make more car carrying space, the entire area was enclosed. The newly enclosed steel was not protected. Thus only half the length of each beam is protected; the other half is bare.

Forgetting It: In making "calculated risks," keep in mind the well-known computer programmers' acronym "GIGO" — garbage in, garbage out. No calculation is any better than the information used in it.

Fire protection interests are often hampered by a management obsessed with the short term results. Industrial managers may defer fire protection expenditures to show a better quarterly profit picture. Politicians may vote to defer the expense of sprinkler installation to erect a new structure that will carry a bronze plaque.

Some managers are fond of using a calculated risk as an excuse to do nothing, particularly when the cost of providing protection interferes with other objectives. They get upset when it is pointed out, however, that they do the calculating while others take the risk.

A southern state was ordered to improve conditions, including fire protection, in its mental health facilities. The state, pleading poverty, did nothing until the judge threatened to seize and sell state property to provide the funds.

This author recalls another such calculated risk. An Atomic Energy Commission official described the proposed shipping of a radioactive material disolved in water in a container without a relief valve, because a leak or valve operation might necessitate an expensive decontamination and clean up. When it was pointed out that a fire in the container would explode (BLEVE), he replied "We are taking that calculated risk." My reply (modified) included "Who are you to decide to endanger some fire fighter, just so there won't be a contamination problem?" The container was redesigned with a one-hour fire-resistive jacket.

Fireproofing (Insulating) the Steel

Steel is fireproofed by one of several systems. Steel qualifies for a fire-resistance rating if the protection system previously passed a standard fire test (NFPA 251, *Standard Methods of Fire Tests of Building Construction and Materials*, also ASTM E119).

While there is no such thing as a truly fireproof building, the term has survived as the designation of the system by which steel is insulated. A subcontractor contracts to fireproof the steel. The material used is known as fireproofing. Steel which has been fireproofed is known as protected steel. Confusingly, in the building industry, sheet steel which has a weather-protective coating is also known as protected.

Types of Fireproofing

Fireproofing of steel is classified as individual or membrane.

Individual fireproofing provides protection for each piece of steel.

- In one method, steel is encased in a structure. Concrete, terra cotta, metal lath and plaster, brick, and gypsum board are materials commonly used. This method is called encasement.
- In another method, fireproofing is directly applied to the steel, usually by spraying. Materials are asbestos fiber (no longer used) an intumescent coating containing non-combustible fibers which swell and char when exposed to flame, and cementatious coatings.

Membrane fireproofing does not protect individual members.

- In one method, wire lath and cement plaster are used.
- In another method, the fireproofing of the floor is accomplished by a rated floor-ceiling assembly.

Fireproofing is applied to meet the standards required by the local building code. If, for instance, the code requires three-hour fire resistance on columns, the designer selects a system that has been approved by the local building department. As a supplement to the building code, the building department will indicate which systems tested at which laboratories are acceptable.

The building department may require modifications. Recall from Chapter 6, (Principles of Fire Resistance), the procedure in which a manufacturer develops a fire-resistance system which is tested at a laboratory to determine whether or not it meets the standard.

The efficiency of fireproofing depends first on the competence of the subcontractor and the willingness of the builder to demand quality work. It further depends on the building department staff who inspect the original installation, and on the fire department inspectors to determine that fire proofing and fire protection are not compromised. This is particularly important in the case of spray-on fireproofing and membrane-floor and ceiling assemblies. There are advantages and disadvantages of each method.

Encasement Method: Terra cotta tile was used early for encasement. The cast-iron columns of the Parker Building in New York City (destroyed by a fire even though it was regarded as the best type of fireproof building) were protected with three-inch terra cotta tiles. Electrical conduit was laid against the columns inside the tile. When the heat elongated the conduit, it tore the fireproofing off the column. A major column failed catastrophically. (See Chapter 11 for a discussion of this fire.) There are buildings still standing which may have this same defective design.

Another error in the encasement was to leave the bottom web of beams unprotected. After a number of failures, this error was corrected by the development of tiles which were shaped to fit around the steel. The tiles (called **skewbacks**) below beams are easily removed to access the beams during alterations, such as to hang equipment from the overhead. As a result the fireproofing is often seriously defective.

Fireproofing that is easily removed is at the mercy of the person making alterations who often has no knowledge of why this material is present. In altering an operating department store for the construction of a subway entrance, a contractor removed the protection from a major column. The sprinklers were shut off in this area. About a hundred cylinders of propane gas were stored adjacent to the column. A fire in the propane could have collapsed the building. Fortunately there was no fire.

It is important to look in on rehabilitation work on buildings. You may pick up vital information. This author noted tiles had been removed from the bottom of steel beams to hang sliding doors.

Concrete became quite popular as a protective covering for steel, particularly where concrete floors were being used. The wood falsework required for forming concrete provides a high fuel load and has been involved in a number of serious construction fires.

Concrete has the advantage of being the most permanent fireproofing because it is so difficult to remove. However, concrete can be knocked off beams and columns by vehicle impacts.

The disadvantage of concrete is its weight. In the effort to reduce the dead weight of a building, and thus its cost, fireproofing is often a tempting target.

Encasement systems of gypsum board or wire lath and plaster were developed to save weight. Wire lath covered with cement plaster has been used for fireproofing for many years.

Some circular steel columns can be made fire resistive by filling them with concrete. See the section on Lally columns in Chapter 8.

Sprayed-on Fireproofing: The search for lighter fireproofing led to various sprayed-on coatings. Sprayed concrete spalled badly when exposed to fire.

Many combinations of asbestos fibers and other materials were developed. While some of these materials can pass laboratory tests, some serious questions are raised about their reliability in the field.

Writing in the May 1965 *Fire Journal*, Robert Levy, Superintendent of the Bureau of Building Inspection for San Francisco, pointed out that when sprayed-on assemblies are tested, the steel is cleansed with carbon tetrachloride or some other solvent. In the field this is impracticable. Ensuring that the material is applied at the specified thickness and density is another problem with sprayed-on coatings.

Sprayed-on material is often easily knocked off by other trades doing their own work. The author was once invited by an insurance executive to the 30th floor of the nearly complete Bank of America Building in San Francisco. The fireproofing had been stripped from all the columns by the plasterers.

There was a serious fire in the Time and Life Building, a modern high-rise office structure in New York. The steel members of this building were protected by sprayed-on fibers, but they were washed off in several places by hose streams. The loss of fireproofing either prior to or during the fire can have serious consequences.

A state office building in California was being sprayed with fireproofing material by a contractor who had no appreciation of the importance of the work. Overspray was swept up from the floor, laid on the lower web of the beams, and covered with a fresh coating. The building was not insured, but an insurance engineer took pictures of the work. When a fire occurred in the building it cost almost as much to repair the damage to the steel as it had to build the building.

On the other hand, the First Interstate Bank Building of Los Angeles,[11] which suffered a multi-story fire in 1988, was found to have excellent spray-on cementatious fireproofing. Had the fireproofing been as

Fig. 7-15. The building isn't finished yet, but this sprayed-on fireproofing is already coming off.

poorly applied as observed elsewhere, a catastrophic collapse might have occurred.

On more than one occasion sprayed fireproofing has been found generously applied to all exposed steel surfaces in a building, including the sprinklers.

Asbestos fiber fireproofing is under attack on another front — there is a serious health hazard in its use. It is difficult if not impossible to sell a building with asbestos fireproofing. Health regulations have eliminated the use of asbestos fireproofing, and asbestos is being removed from existing buildings. Published accounts are invariably silent as to what is being done to replace the asbestos.

Tactical Considerations

The fire department should be aware of asbestos fireproofing removal, and its replacement. Pre-fire plans should be amended accordingly.

Asbestos in fire debris may impede a fire investigation. While the Dupont Plaza Hotel fire in Puerto Rico was being investigated, it was discovered that the hotel had been contaminated by asbestos.

Membrane Fireproofing: Membrane fireproofing is accomplished by the use of a suspended ceiling that provides a membrane

Fig. 7-16. Tiles had been removed from a storeroom to replace damaged tiles in public areas of the building.

under the entire area to be protected. The membrane may be of cement plaster on wire lath, but in recent years the use of a suspended acoustical tile ceiling as the membrane has become popular. The steel structure requiring protection is left bare.

Testing and Rating: In rating such a fire protection system, Underwriters Laboratories Inc. tests and lists a floor and ceiling assembly or a roof and ceiling assembly.

In such floor or roof assemblies, the ceiling, hangers, electrical fixtures, bar joists, left-in-place formwork for the concrete floor (usually corrugated steel), air ducts and diffusers, and the concrete floor are tested as a whole unit as it would be built. Therefore, it is incorrect to speak of a fire-rated ceiling. The ceiling is one part of the total assembly.

The assemblies are tested under the NFPA 251 (ASTM E119) Standard Fire Test.

The original designers of the ASTM E119 test did not envision an assembly which provides a void trussloft (like a cockloft) in every floor. Fire can and has extended through this void. The designers of the test also did not anticipate the problem of otherwise unprotected columns passing through the trussloft. However, the UL Building Materials List states:

> "All ratings are based on the assumption that the stability of members supporting the floor or roof is not impaired by the effect of fire on the supports."

Hazards of Floor-Ceiling Assemblies: Buildings with floor-ceiling assemblies, particularly where the columns are unprotected passing through the plenum space above the ceiling, can present a serious menace to the safety of fire fighters who are, generally, unaware of the dangers. The observations and considerations which lead to this opinion are many.

- The building trades sometimes disregard detailed instructions, particularly when they do not understand them or find them inconvenient. A fire inspector found aluminum wire being used to support the ceiling channels. "What's the matter with aluminum?" asked the mechanic.
- The value of listing is contingent upon the field assembly being accomplished exactly as performed in the laboratory. Few building inspectors have sufficient expertise and time to follow and examine every item.
- The ceiling system is at the mercy of all owners, operators, tenants, tradespeople, and mechanics who have reason to remove ceiling tiles. Access to utilities and additional storage space for tall items such as rugs or ladders are only two common reasons to remove tiles. Where the plenum space is part of the air-conditioning supply, employees soon learn that displacement of a tile improves airflow at their location.
- There are no legal provisions requiring that the membrane protection be maintained. Even if there were, enforcement would be difficult. Replacement acoustical tile, which appears to be identical to the original tile, may be combustible, rated for fire hazard only, rated as part of a fire-resistance system but not the system installed in the

Fig. 7-17. Because of a water leak, spaces exist where ceiling tiles have been removed.

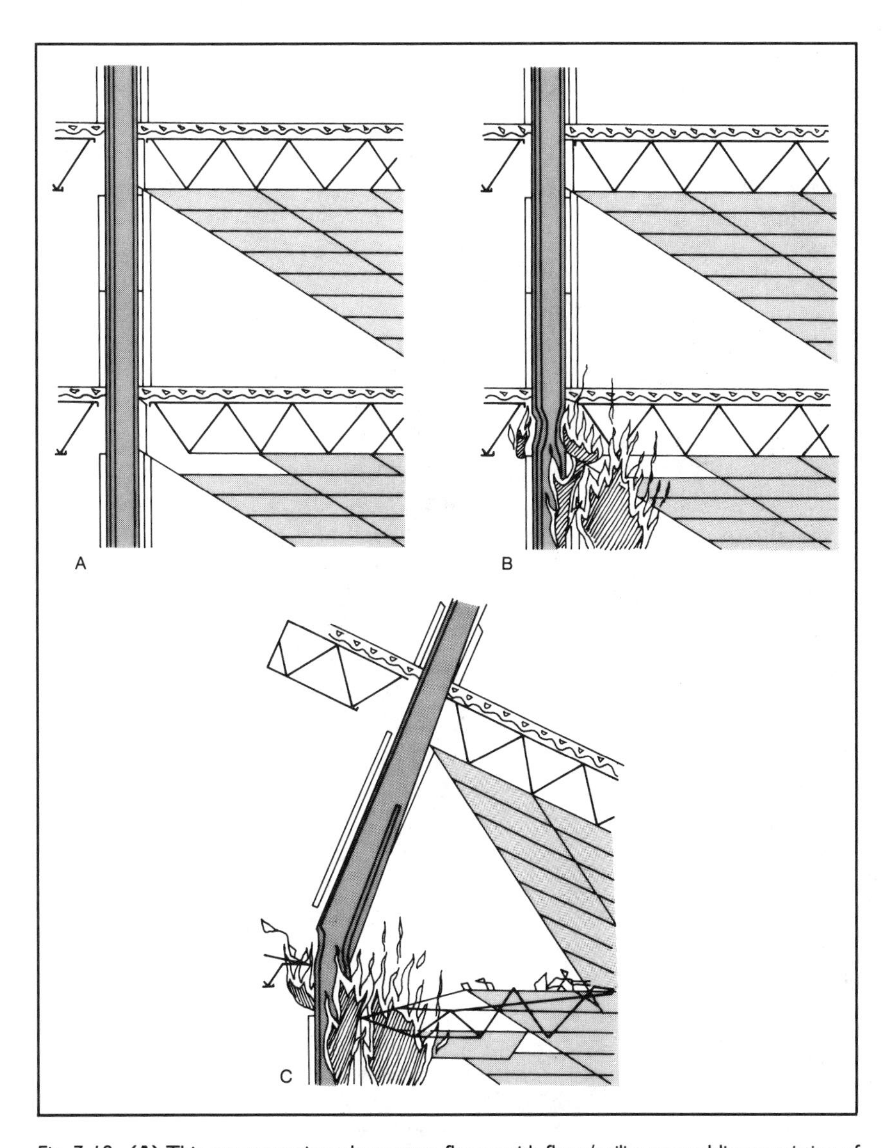

Fig. 7-18. **(A)** This cut-away view shows two floors with floor/ceiling assemblies consisting of *a concrete floor atop corrugated steel resting on bar joists, and a grid ceiling below the bar joists. The steel H-shaped column is fireproofed from floor to ceiling but not within the floor ceiling void. One tile is missing.* **(B)** Fire from below passes through the defective ceiling and heats the column. **(C)** The column fails and the structure collapses.

building in question, or may be the proper tile, which meets the specification of the listing. All electrical fixtures, air-duct openings, and other penetrations of the ceiling must be rated as part of the ceiling system. Proper fire protection and building codes require that fire ratings be maintained. Unfortunately there are many lapses between code directives and actual practice.

- The term "fire-rated" is used quite often in the fire protection and building construction fields. By itself it is non-specific and, therefore, meaningless. With respect to ceiling tiles it may refer to tiles which have merely met the requirements for flame spread, or tiles which are part of a listed fire-resistance system. As has been noted elsewhere, no part of a listed fire-resistance system stands by itself.
- Owners, managers, and occupants — even of buildings where fire protection equipment is well maintained — are usually totally unaware of the significance of absolute integrity of the ceiling system. A common practice is to remove tiles from the storeroom to replace tiles in public locations. (Who needs a fancy ceiling in the storeroom?) Tiles are also damaged by water leaks. Tiles are removed by tradespeople. Holes are cut through tiles. Displays are hung from the metal grid. Floor-ceiling systems are not tested with superimposed loads.
- The list of ways in which the integrity of a floor-ceiling assembly can be compromised is endless. There are many pictures to prove it.
- Laboratory fire tests are conducted under a slight negative pressure to remove smoke and fumes. Fires generate positive pressure, and lay-in ceiling tiles may be easily displaced by fire pressures. When a tile is displaced by a worker it is often dropped back down into place without any restraint to upward motion.
- The addition of insulation that is not part of the specifications of the listed ceiling assembly, particularly in a roof and ceiling assembly, can significantly affect the fire performance of the assembly. The insulation will cause heat to be retained in the channels supporting the tiles, causing them to fail earlier than expected.
- A membrane protection system must be perfect. The failure of even a single tile is the equivalent of a pin hole in a water-filled plastic bag.

It was noted earlier that in effect there is a cockloft between the ceiling and floor. In one incident, fire starting in one room, traveled across a hallway above the ceiling and came down through joints in the tile ceiling of another room to ignite books on a top bookshelf. Only alert fire fighters prevented the full involvement of the floor.

Some code provisions provide for firestopping, particularly along demising walls or tenant boundaries and around columns, but the use of the plenum space for various services makes it probable that the firestopping will only continue to conform to the definition of legal firestopping (see Chapter 3, Wood Construction).

- In some buildings, the use of deep, long-span trusses to provide clear floor areas creates plenum spaces several feet in height. Wasted space is often intolerable. What is more logical than to provide access to this space and use it for storage? Engineers closely associated with hospital construction call this void interstitial space. Whether the membrane fireproofing is provided by suspended tile or plaster on metal lath is immaterial, the fire load has been placed adjacent to vulnerable unprotected steel, where fire fighting will be almost impossible.

- Not all floor-ceiling arrangements are intended to be fire resistive. A steel bar-joist floor with concrete topping with flame-spread-rated tiles below may appear to be fire resistive. In fact, it may only be non-combustible construction.
- The fact that tiles are obviously missing does not necessarily mean that a fire-resistance system has been violated. The building may not be required or intended to be fire resistive. It may be simply of non-combustible construction. In such a case, ceiling tiles are at the option of the owner and need merely to meet flame-spread requirements, if any.
- The building may be of concrete construction. There are some listed concrete assemblies in which suspended tiles are incorporated, but, in general, the suspended ceiling is installed to provide a hidden void for utilities, as well as an acoustical and aesthetic treatment.

Up to this point possible failures which would involve sections of a floor, or an entire floor, have been discussed.

There is a much more serious condition permitted by some building codes which might result in a catastrophic partial or total structural collapse of a supposed fire-resistive building — column collapse.

Fig. 7-19. Note the effect of fire on a steel column after a wooden platform burned in a subway station. The street above dropped a foot. This would be much worse in a structure. The adjacent straight column is a timber to shore up the ceiling.

Potential Column Collapse: Some jurisdictions now permit the omission of fire protection for columns as they pass through the plenum or void space. The concept is that the bare area of the column is protected by being within the rated fire-resistive floor assembly. The deficiencies cited above conclusively demonstrate that this is dangerous.

A report of laboratory tests justifying the omission of column fireproofing in the plenum space carries this vital caution qualifying the protection of a column by a floor-ceiling assembly: "the column supports only a single floor or roof." This clearly excludes multi-story buildings.[12]

There is a clear warning contained in *Fire Protection Through Modern Building Codes*[13] that protective materials should extend to the full-column height. There should be no discontinuities, either by cut-offs at the ceiling lines, or at other constructions that may abut the column.

Factory Mutual requires three-hour fire protection on all columns. Unprotected sections are not permitted.

It is very probable that the removal of one tile in a high fire-load area, near an unprotected column, is the most serious hazard, since this would concentrate the heat output of the fire at a single vulnerable point. Imagine the huge volume of fire pouring out the windows of the First Interstate Bank Building in Los Angeles being concentrated on an unprotected steel column.

A single 50-year-old temporary wooden platform burned in the 42nd Street subway station in New York City. Serious failure of the unprotected steel columns resulted in the collapse of the street above. Similar damage to a high-rise building might well cause catastrophic collapse.

Deficient Column Protection: In some observed cases, the column protection in the plenum consists of gypsum board enclosures. True to the widespread erroneous trade belief that the taping of joints and nail-setting are merely cosmetic, the nail heads and joints are sometimes left untaped. Exposed to fire, such incomplete assemblies will fail rapidly.

Smoke and Gas Movement

The open connection between the ceiling void and the column will allow smoke and toxic gas to move from floor to floor through the void space (re-entrant space) in the columns. There have been some feeble attempts at firestopping with a handful of rock wool. It may misdirect fire fighters by causing a smoke detector on a distant floor to be the first to operate on a fire.

Tactical Considerations

The fire department should be aware of all buildings in which the columns are bare or inadequately protected in the hidden area of the plenum. This information must be available to the fire ground commander for use at a serious fire. The information cannot be left in the files of the building department or fire prevention office, where it will surface only after a disaster has occurred.

Experience: It is not sufficient to say that there has been little bad experience with rated floor/ceiling assemblies. There was no very bad experience shipping oil out of Valdez, Alaska until the Exxon Valdez ran aground in 1989, causing an environmental disaster.

The potential for serious problems exists. There are present dangers, and the fire department must be prepared to present its point of view to code authorities, train its own personnel, and properly educate the owners and occupants of membrane-fireproofed buildings.

Equivalent Fire Resistance? It is not true that all assemblies which get the same fire-resistance rating are identical in response to actual fires.

For instance, a heavy concrete floor would probably be a greater heat sink than a bar joist assembly, simply because of the greater mass. Every Btu absorbed by the structure is one less Btu available to extend the fire. In addition, it is more difficult to interfere with the fire-resistive effectiveness of the concrete construction.

Protecting Steel with Automatic Sprinklers

As an alternate to fireproofing of steel, automatic sprinklers are installed to keep the steel temperature down below a hazardous level. Automatic sprinklers can provide protection to steel unprotected by fireproofing, provided that the water supply is adequate in quantity and distribution. Otherwise, it is quite possible that the early failure of the steel structure will cause sprinkler piping to break, resulting in loss of the sprinkler system. (Further discussion of automatic sprinklers is contained in Chapter 13, Automatic Sprinklers, and Chapter 14, Rack Storage.)

There are several variations which relate to steel construction that are noted here.

Automatic sprinkler protection is provided with the dual function of suppressing the fire and cooling the steel. At times, in addition to the normal sprinkler distribution, separate sprinklers are provided to protect major structural members. In one case, automatic sprinkler protection is provided to cool the steel of the roof structure, while high-expansion foam is provided to suppress the fire.

At the huge Sawgrass Mall in Sunrise, Florida, sprinklers are provided specifically to protect all columns.

Foam/water sprinklers or deluge-type sprinklers are installed where there are flammable liquid hazards.

Due to excellent experience over the years the fire service has become complacent about automatic sprinkler protection and has a tendency to regard fires in sprinklered buildings as representing only a minor problem. The experience of the past is not directly applicable to the present because of changes in the buildings.

Chapter 14, Rack Storage, discusses seven fires (between 1978 and 1988) in sprinklered warehouses which resulted in losses of more than $750 million. In some cases, the early collapse of a steel roof broke the sprinkler piping.

Fireproofing with Water

Internal combustion engines have long been fireproofed with water. The heat of combustion is removed from the cylinder wall by water, thus preventing distortion. The water releases its heat to the atmosphere and recycles to the engine block.

The United States Steel Company wanted to make its new headquarters in Pittsburgh a showcase for steel, particularly its Corten® alloy. Corten weathers to a dark brown color, from an iron oxide coating, which, unlike ordinary rust, does not penetrate beyond the surface. The material does not need to be painted. The advantages of such an alloy are lost if it must be buried under fireproofing.

The columns in the building are hollow boxes. Water with antifreeze added has been placed inside the columns. Columns are cross connected with piping so that heated water can circulate by gravity. Though not formally rated, these columns probably have an almost infinite fire resistance. Several other buildings have been designed similarly.

The structural supports of tanks containing flammable liquids present serious problems of possible collapse during a fire. It would be simple to design such supports as piping systems and provide for water circulation inside the pipes.

Locating Steel Out of Range of Fire

The architectural interest in exposed steel elements, and the constant search for methods to reduce weight and cost, led to a design in which important steel elements are located out of reach of the fire.

A 54-story building was built for United States Steel Corporation. Major parts of the exterior are formed by 70-inch deep exposed steel spandrel beams. The beams are fireproofed on the interior surface, and 14-gage steel covers are provided for the top and bottom flanges for flame shields and architectural cladding. The exterior web is left exposed. Tests showed that the steel temperature would not exceed 640°F.

Suspension Cables: The floors of the Central Bank of Ireland building in Dublin are suspended on steel cables from beams cantilevered out from the top of the central core. This author has been told that an analysis made by a prominent British fire researcher showed that the cables would not be overheated by a fire. If the bank floors are loaded with computer work stations, however, as was the case in the First Interstate Bank fire the fire load might well be reevaluated.

The exterior cables suspending a dormitory at a west coast university is another example. They are placed right against the building. Dormitory rooms often have very high fire loads. A room fire might locally overheat a cable with possibly serious consequences. Cold-drawn steel cables can resist temperatures up to only 800°F. It would be necessary to heat only a small section of cable to induce failure.

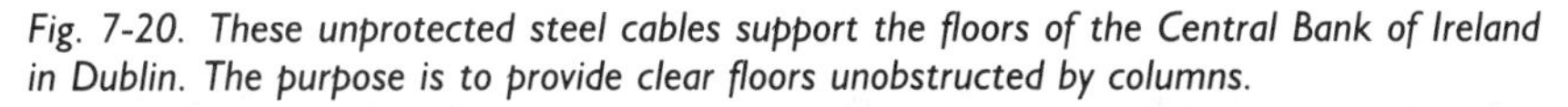

Fig. 7-20. These unprotected steel cables support the floors of the Central Bank of Ireland in Dublin. The purpose is to provide clear floors unobstructed by columns.

Insulated Metal Deck Roof Fire Problem

In August 1953, the General Motors transmission plant at Livonia, Michigan burned. Its loss of $32 million was the largest industrial fire loss to that date. The metal deck roof was the principal contributing factor to the destruction of the plant.

Metal deck roofs consist of metal sheets laid over steel bar joists. The sheets are crimped together making a joint which is not gas-tight. Usually insulation is added, often a low-density fiberboard, though plastics currently are making inroads.

Insulation which has absorbed moisture is useless, so it must be protected from moisture driven through the roof by capillary attraction. An adhesive is necessary also to secure the insulation to the roof deck and to prevent loss of the roof covering in a windstorm. A bituminous coating serves as the adhesive and sometimes as a moisture-stopping vapor barrier. On top of the insulation are successive layers of bituminous material and roofing felt. From the successive layers grows the term, built-up roof.

Fig. 7-21. A hose stream directed into the overhead cooled the steel and stopped this metal deck roof fire. Note the frozen icicles of dripping tar.

Some roofs are built with a layer of concrete laid on the corrugated metal and the roof built up above the concrete. Such roofs are not of concern here.

An "approved roof" is the accepted term for roofing which meets Underwriters Laboratories standards for roofing resistant to the propagation of fire from building to building by flying brands. A roof can be UL listed but still be a combustible metal deck roof. The two problems are entirely different. The UL listing points out that fire under the roof is not a consideration.

When a fire occurs below a combustible metal deck roof the metal deck heats up. Tests show that 800°F for five minutes is enough to start the process. The heat is conducted through the deck to the bituminous adhesive. The adhesive liquefies and then vaporizes. The gas cannot escape through the roofing material, so it forces itself down through the joints in the deck. When the gas mixes with the air below, it ignites from the fire below.

Such a gas fire, which the author has observed to be as much as six feet deep, rolls along under the roof, heating additional roof areas which generate more gaseous fuel. Almost unbelievable quantities of thick black smoke are generated. At the Tinker Air Force Base fire in Oklahoma, the building was 35 feet high. On arrival, the smoke was down to the 6-foot level. This type of fire becomes self-sustaining, independent of the original fire, and spreads rapidly in all directions.

The roofing insulation and some of the roofing felt may burn, showing fire on top of the roof. This is of little consequence; the problem is at the underside of the roof.

There are two solutions open to the builder to prevent such occurrences.

- Use Factory Mutual Class I roofing or a UL Classified Roof. These have been tested in accordance with NFPA 256 and will not provide self-sustained combustion—the problem with the ordinary combustible metal deck roof.[14]
- Provide adequate automatic sprinkler protection for the roof, even though the contents may be non-combustible. If a ceiling is installed the sprinklers must be located above and below the ceiling.

Several shopping malls in Florida have sprinklers above and below the ceiling.

Fires of Interest

In 1985, U.S. taxpayers suffered a $195 million loss in a fire at a production building at the aforementioned Tinker Air Force Base.[15] The facility was sprinklered! The fire was started by roofers. The sprinklers were below a wire lath and plaster ceiling. The water did not hit the underside of the roof deck, and the fire burned unimpeded. An attempt was made to cut the fire off by removing the built up roofing, but the fire passed the cuts.

In 1956, the Atomic Energy Commission (AEC) plant at Paducah, Kentucky, suffered a $2 million fire loss in a metal deck roof fire. It serves as a classic example. The fire started as an explosion in process equipment, and in less than 10 minutes the fire had spread to the roof. The insulation was glass fiber so essentially only the vapor seal was involved. There was practically no combustible material in the plant. The burning tar which fell from the roof found no fuel to feed on. The 70,000-square-foot roof was completely destroyed and, as it collapsed, it pulled down the walls.

The smoke from this type of fire is extremely dense. A fire fighter jumped from the second level through a process hole, knowing he was falling about 30 feet to escape the smoke.

Even Class I roof construction, if combustible, may not always be acceptable. For example, a fire in a large concentrator building at the Wabush Mines complex in Labrador badly damaged a metal deck roof 90 feet above the floor. Investigators theorized that, in this instance, if the roof had been of Class I construction, damage still would have been severe. Failure of unprotected steel supports undoubtedly would have let the roof sag, exposing to the fire asphaltic materials in the built-up covering over the roof insulation.

The term "Class I" may be taken to mean "completely satisfactory under all circumstances" by persons not fully familiar with this subject. If a Class I roof has any combustibles at all it may not be satisfactory for use over a high-value installation, such as a computer area. Total non-combustibility should be sought.

After the fire in Paducah reported above, the AEC studied the problem of fire protection for thousands of acres of metal deck roofs covering some of the most vital production plants in the country. Many proposals were examined. The final decision was to add sprinklers to the plants. Over $20 million was spent on sprinklers and water supply. When a fire similar to the first Paducah fire occurred in another sprinklered building at Paducah, it was confined to the area of origin. The fire was a very hot chemical fire, and the sprinkler operation caused a steam cloud which rolled through the plant. Over 2000 sprinklers opened, causing a water flow of millions of gallons. Thereafter all sprinkler heads in these plants were replaced with higher temperature heads.

A bowling alley was destroyed by fire in a Maryland community. The fire started in the concession stand and extended upward to the metal deck roof. The roof burned very rapidly and collapsed. There was essentially no fire on the bowling alley floor level except for the concession stand.

In another example, this author examined the damage at a U.S. General Services Administration warehouse in Fort Worth, Texas which had a metal deck roof. The sprinkler system was a dry pipe system in apparently good operating condition, with an excellent water supply. The contents were paper goods. A contractor was cutting holes in the roof for skylights. Hot metal ignited the contents. The fire ran away from the sprinklers. The holes in the roof might have exhausted enough heat to slow the sprinkler operation. Moving steel collapsed two fire walls. The Fort Worth Fire Department made a determined stand at the third fire wall and stopped the fire.

An Unrecognized Problem

Careful reading of fire reports indicates that the metal deck roof is a factor in many losses where its significance is not understood by the fire suppression group.

In May 1977, the Beverly Hills Supper Club in Kentucky burned with the loss of 164 lives. This author made an extensive study of the circumstances of that fire and is convinced that a very significant factor in the huge loss of life in the Cabaret Room was the tremendous amount of black, choking smoke developed by the fire in the metal deck roof. The initial investigators discounted the possibility of fire moving through the voids, because they were cut up by walls. It is true that the roof was pierced by parapetted walls. The building had been added to many times, but undoubtedly there were penetrations through the walls. A metal deck roof fire does not need a big opening. A tongue of flame passing through a small hole, impinging on the metal deck of the next section would be sufficient to extend the fire. (Factory Mutual tests on metal deck roof assemblies have shown that it takes only 800°F for five minutes for heat impinging on the surface of steel decking to start a self-sustaining roof fire. This fire is independent of the original fire.)

It is not at all clear that if the supper club building had been sprinklered, according to the usual practice, that the loss of life would have been substantially reduced. Under typical practice, the void would be regarded as non-combustible and left unsprinklered. Since it was above the ceiling insulation, a separate dry-pipe system would have been required to prevent freezing. Of course, had code requirements for exits and fire resistance (a fire-resistive structure would not have had a metal deck roof) been met, there would probably have been few lives lost.

It appears that in most jurisdictions, an ordinary combustible metal deck roof would be permitted on a non-combustible building. Sprinklers below the ceiling in such a building will not protect against a metal deck roof fire moving through the void. However, they might

Fig. 7-22. These two identical steel beams were subjected to the same fire. One beam was protected by hose streams; the other (arrow) could not be reached. It overturned and dropped the roof.

prevent a contents fire from reaching the metal deck roof. They would also operate on contents fires started by dripping, flaming tar.

Two fires illustrate these hazards.

A fire penetrated the ceiling of a sprinklered department store and involved the roof. The store was sprinklered below the ceiling line. The void space above was 11 feet deep. The store was totally destroyed. Fire fighters familiar with the fire stated it was a typical combustible metal deck roof.[16]

In another fire, four elderly persons lost their lives in a one-story nursing home in Dardenelle, Arkansas in 1990. The building had been described as a protected non-combustible structure by the Arkansas Department of Health. The building had a combustible metal deck roof. Extremely thick black smoke, typical of a metal deck roof fire, drove the staff from the west wing. The NFPA Fire Investigation Report declared that "Flammable vapors from the heated roof assembly contributed to the fire extension in the concealed space, and the smoke and toxic gases generated in the concealed space filled the occupied patient rooms."[17]

Fighting a Metal Deck Roof Fire

This author first became aware of the metal deck roof problem in 1946. A fire occurred in a 250-foot section of a 1500-foot by 100-foot warehouse at the Marine Corps Supply Depot in Norfolk, Virginia. The warehouse was loaded with field supplies. The fire was in the center of the section, involving stock on both sides of the center aisle. I was supervising handlines inside the building, hitting fire in the burning stock. Blue flames extended down six feet or more from the underside of the roof, accompanied by dripping, flaming tar. When the heat and tar became hazardous, the streams were directed upward to wipe out the fire. We made a "good stop," cutting the fire off right in the piles of stock 20 feet high.

The pictures of the building, however, told a different story. The roof was gone from fire wall to fire wall and it appeared as if the fire department had played no effective part in stopping the fire, but that rather the fire walls had done the job. After much discussion and study, the conclusion was that the roof had burned overhead, independently of the main body of fire in the contents.

It appeared that a fire involving a metal deck roof could be controlled only by the continuous application of water to the underside of the roof. By cooling the steel, the tar would be prevented from generating gas. Literally, the fire would be extinguished by removing the fuel.

The chance to try the plan came in about 15 months. In a similar warehouse section, the Marines had built a plywood office, sick bay, and theater, while the balance of the area was loaded with combustible stock. Like the first fire, this fire was well advanced upon arrival, and an immediate second alarm was sounded.

Looking up along the underside of the roof, from the door at the end opposite the fire, I observed the characteristic blue waves of fire several feet deep advancing rapidly. A hose wagon with deck pipe was ordered into the building, taking a position in the center aisle opposite the door. The orders were to direct a master stream at the underside of the roof, keep it moving, and ignore the burning contents. The operation was successful. The warehouse looked as if a knife had cut through the roof and walls; the building was lost only in the area where the huge body of initial fire had done its work before our arrival.

Tactical Considerations

A prefire plan for a metal deck-roofed building should be based on taking a safe position where all parts of the underside of the roof can be reached with heavy stream equipment of sufficient reach. Don't hesitate to throw water into smoke—even though the target will probably be invisible.

A ladder pipe or ladder tower stream directed downward onto the roof is useless. The roof is designed to keep out rain, and the fire is always many feet beyond the point at which the roof is opened up. The fire that can be seen on top of the roof is in the fiberboard or plastic insulation or the upper layers of the built-up roofing and is unimportant. Stop the progressive heating of the "frying pan" from

Tactical Considerations, continued

underneath and the roof fire will die out or decrease to easily manageable proportions.

The ladder pipe or ladder tower can be very useful if it can be maneuvered into a position where it can deliver the heavy stream onto the underside of the roof.

The metal-deck roof fire cannot be cut off by ventilation. The fire can move very fast. Tests show that only huge vents far beyond the capability of fire ground tactics would be effective. Working on a roof over a metal-deck roof fire would be extremely hazardous. An article discussing collapse of wood trusses makes the point that wood trusses collapse while steel trusses sag. This is not necessarily a blessing. The sag may cause fire fighters on the roof to slide downhill on a slick, liquid tar surface.

A combustible metal deck roof doomed a large manufacturing plant on Nun's Island, Verdun, Quebec in August, 1990.[18] *Ventilation was out of the question once the roof was involved and sagging.*

Heat moving through the truss void elongated the steel bar joists in a Dallas, Texas fire. The joists pushed down the block wall, dropping six fire fighters who were on the roof. The collapse area can extend far beyond the fire area.

A Different View

Battalion Chief John Mittendorf of the Los Angeles City Fire Department offers another tactic in a Fire Engineering article cowritten with this author.[19] *He has designed and tested a procedure which involves cutting the layers of built up roof. The procedure is summarized here.*

* Use a chain saw or a rotary saw with a wood blade, make two parallel cuts, 3 to 6 feet apart and ahead of the fire.
* Just cut the roofing material, not the corrugated steel.
* Cross cut every four feet.
* Remove the cut sections. This will permit the heated gases to vent.

Chief Mittendorf points out that the procedure takes time and requires several blades. It is noteworthy to point out that the cutting should be done beyond the area where heat is elongating the bar joists, as it did in the Dallas fire cited earlier.

Obviously, the safety of fire fighters is the paramount consideration.

Metal Decks on Non-Metal Buildings

It is important to note that the metal-deck roof is not found exclusively on steel-framed buildings. Many are built on masonry buildings but the problem is the same. In December 1970, in Kensington,

Maryland, fire fighters responded to a serious fire in the contents of a one-story building housing a janitor supply business. There was heavy smoke, and the ceiling was not visible through the smoke. The chief felt hot tar dripping on his neck. Later he said: "I immediately thought 'metal-deck roof,' and though we couldn't see the overhead, we directed a big line up to the ceiling area to cool everything."

When the smoke cleared away we were able to see and photograph icicles of chilled tar hanging from the joints in the steel.

The metal-deck roof fire is not always readily recognizable, particularly where a ceiling is installed. An unsprinklered shopping mall in Winter Park, Florida, was constructed of one-story high store sections with metal deck roofs on either side of a high-bay central mall of precast concrete. A fire in a store extended to the metal-deck roof. The steel of the lower section elongated and moved the supports of the high-bay section. A section of the concrete T-beam roof fell, fortunately without causing injury. An alert fire officer pulled the ceiling of a walkway in the one-story section and saw the fire moving under the roof. The officer placed a monitor nozzle in position and cut off the fire.

Fire Walls

The material in this section is principally drawn from *Fire Walls in Modern Industrial Buildings*, published by Factory Mutual.

The structural steel frame provides the stability for an unprotected steel industrial building. If a fire wall encases some of the steel frame, or is fastened to it, the fire wall can be destroyed by a fire which distorts the steel frame.

Steel members which sag due to fire will try to carry their loads as suspension members. This causes large horizontal forces. If these are transmitted to the fire wall, it can be destroyed. Few walls are designed to resist such a horizontal force. The wall anchor may not be strong enough to transmit it, and the anchor may fail.

A basic example of the fire wall is the freestanding masonry wall as used in large wooden buildings. It is completely independent of the building, unsupported by the framing, and carries none of the load. Such walls are rarely found in steel-framed buildings.

It is important to be aware of the various types of fire walls in order to assess the reliability of a wall in the event of fire.

Tied Fire Walls

These are walls which are fastened to, and usually encase, members of the steel building frame. The pull of collapsing steel on one side must be resisted by the strength of the unheated steel on the other side. Should there be heavy fire on both sides of the wall, and not necessarily in the immediate vicinity of the wall, it will almost certainly be in danger of collapse. If there are only a few bays of steel tied to a wall on one side and none on the other, it will have very little strength to resist a horizontal pull from the other side, and is a one-way wall.

Freestanding Walls

These walls are not connected to the building at all. The steel frame on each side must be quite close, however, so that the unexposed steel can resist the lateral push on the wall by the heated exposed steel.

Partition Walls

These walls may not be designed as fire walls but may serve very effectively as defensible barriers in conjunction with hose streams, if their deficiencies are recognized.

Combination Walls

This author recalls observing a one-high-story, steel-framed building under construction in which the fire wall parapet of brick was completed, even though there was no wall inside the building below. A steel girder had been placed with an upward camber, and the weight of the parapet, after it was built, flattened out the girder. A concrete block fire wall was then built up to meet the girder. The girder was then protected with directly applied fireproofing. Heavy unprotected steel girders were connected to the fireproofed girder. Any elongation of the unprotected steel would displace the girder. Such a wall should be watched for signs of failure of the protection on the steel which might precipitate the collapse of the parapet. The reliability of such a wall under fire conditions is uncertain. Be very watchful of it.

Tactical Considerations

Preplanners should attempt to determine the type of fire wall.

In any case, a fire wall in a major building might well be made a sector, with a commander assigned to monitor its integrity and stability.

Fire Doors

Practically all fire walls are equipped with fire doors. Recent studies have confirmed that only about half of them would operate in a fire. Inspectors must make management aware of the vital importance of proper maintenance of fire doors. It is critical that fire doors operate properly at a fire. If the fire gets past the door, the fire wall is useless, and may be dangerous.

An article in the Factory Mutual *Record* offers excellent guidelines on fire doors.[20]

Water-spray systems have been substituted for fire doors in some situations. They have proven ineffective in stopping the extension of fire by "Molotov cocktails"— spray cans with flammable propellants which become self-propelled flaming rockets.

Tactical Considerations

In some fire departments checking that fire doors operate properly is a routine ladder company responsibility. In any case this responsibility should be pre-assigned and not left to chance.

Fire doors, both horizontal and overhead-rolling, are triggered by fusible links, sometimes located high up in the overhead. A burst of heat might trip the fusible link. It may be impossible to raise an overhead fire door and it can be very difficult to open a sliding fire door.

Fire fighters should never advance through a fire door without blocking the door to prevent it closing behind them. The blockage should be removed when all fire fighters have returned through the doorway.

Types of Protection of Steel Structures

Prefire planning for steel structures must begin with a determination of the type of protection provided for the steel.

Steel structures can be divided into the following types:

- Unprotected
- Dynamically protected
- Passively protected
- Passive/dynamic combination protection

Unprotected Steel

Unprotected steel structures can be extremely hazardous because of the potential for early collapse.

Trusses are covered in Chapter 12, but the story of a fatal collapse should serve as a grim warning of the dangers of buildings in which clear spans are achieved by steel trusses.

Four fire fighters lost their lives in the collapse of a Wichita, Kansas automobile showroom which occurred about 10 minutes after they arrived. The showroom had repair bays which required large open areas and was built of light-weight steel truss construction.

Fire was visible to units turning out, and a second alarm was transmitted immediately. This fact alone should have signaled caution. A large amount of heat was being produced. The steel would absorb the heat. Early collapse should be anticipated.

Unfortunately, the fire fighters followed the standard aggressive offensive procedure of advancing a handline to the seat of the fire. From what has been shown, collapses are often predictable and, therefore, the hazard may be avoidable.

By constant training, personnel will become accustomed to the critical need to cool all heated steel. The quantity of water needed at any point is not excessive. Water supplies for spray systems designed to protect tanks and steel supports from flammable liquid spill fires are calculated on a requirement of 0.25 gallon per minute per square foot. It is most important to cool all the steel that is within reach of hose streams, and to give special attention to columns. The stream should be kept moving; overcooling in one spot obviously means undercooling, perhaps disastrously, someplace else. It is important that the stream reach the steel. When a long reach is required, a solid stream tip might be better than a fog tip, (even on the fog tip's most solid setting). A fog stream may absorb a greater amount of heat in penetrating the flame zone. Converting all the water to steam as it passes through the fire, leaving none to cool the steel, is not the best use of water, particularly when the fire is uncontrollable.

A Washington, D.C. junkyard consisted of several identical unprotected steel buildings joined together without walls. A fire in waste paper exposed two identical steel girders above the fire. One was successfully cooled with a master stream. The other, shielded by the first one, heated, overturned, and dropped the roof. Fire fighters conditioned to hit the fire sometimes neglect to protect the steel.

Safe locations for cooling the steel should be preselected. Personnel should not be endangered by being placed in the collapse zone which may cover a much greater area than the fire zone.

Cautionary Note

It seems that there is a passion for cookbook answers to serious problems. "How long is it safe to stay in the collapse zone?" is a universal question. The answer cannot be calculated. There are too many factors. The only rule of thumb comes from tests in which bar joists failed in seven minutes in the standard fire test. Failure temperatures were reached in the UL tests after the McCormick Place fire in less than six minutes. These times begin with the start of the fire, not when the fire department arrives.

Water Damage

The use of water in the manner urged here might bring the criticism of unnecessary damage. However, the damage potential was created by the person who placed the combustible loading in a thermoplastic building. If the building can be saved, it will probably be possible to save some of the contents. If the building collapses, the contents are lost too. In this day of close attention to municipal revenue sources, note that it is the building, rarely the contents, on which the tax base rests.

To the argument that "we never throw water except when we see fire," the reply can be offered that "If the owners had sprinklered the building, as they should have, water would be discharged on the steel, whether or not the sprinklers could see fire."

Where cooling-stream tactics are deemed necessary to prevent collapse, it would be beneficial to explain them to management in advance. Then, if the tactics are successful, the water damage to contents may be severe. The necessity of using the tactics might not be apparent after the fire, and the fire department may find itself ridiculed, or even sued.

The possibility of severe water damage from hose streams might cause management to take a second and closer look at providing automatic sprinklers. The secondary objective of the preplanning for any major hazard should be to reduce the severity of the hazard.

Curiously this also works in a reverse way. Highly Protected Risk (HPR) insurance companies usually require sprinklers for combustible metal-deck roofs, even over non-combustible contents. This surprises management who often determines that it would be better to install a more costly roof that would not sustain combustion, to avoid the initial cost and maintenance of sprinklers.

Tactical Considerations

The chief consideration in unprotected steel buildings must be the safety of personnel. The possibility of sudden collapse of large areas

Tactical Considerations, continued

must govern all actions. The pell-mell rush to get a handline into the building must be replaced with studied action based on the awareness that the building can collapse in as little as five minutes. The minutes start to tick off when the fire starts, not when the fire department arrives.

Unprotected Steel in Wood and Ordinary Construction

Deputy Chief Smith of the Philadelphia Fire Department relates the story of a fire in a one-story brick building, 75 feet wide. Elongating steel roof beams caused visible cracks in the walls. All personnel were kept beyond the collapse zone. The wall collapse brought down the roof.[21]

In 1987, two Montreal fire fighters lost their lives, and several others were injured when the wall and roof of an early 20th century church fell on their aerial ladder. The frame of the church had unprotected steel columns concealed in the wood lath and plaster walls. They supported an unprotected steel truss roof. The thick stone walls were supported on a grillage of three I-beams that rested on short steel columns. These columns were fully exposed in the crawl space. Above the crawl space was a heavy timber floor. The fire buckled the short columns. The steel was not apparent. The structure appeared to be of heavy timber and masonry. The fire is well worth studying.[22] This article is an excellent guide for anyone writing about a fire where the construction is a factor.

It is not easy to always "Know Your Buildings" but it is vital if deadly ambushes are to be avoided.

More and more steel is being used in the construction of what appears to be wood and ordinary buildings. Since these buildings are classified as non-fire resistive, the steel is not required to be protected. Gypsum board is a component of a number of tested and listed rated enclosures for steel. However, the fact that gypsum board encloses a combustible void in which steel and wood are combined does not provide rated protection for the steel. No such structure has ever been tested.

Steel cables, steel framework and trusses are sometimes artfully concealed in rehabilitated buildings. A historic New England church was leaning badly after 200 years of exposure to the wind. A steel structure was artfully inserted into the church to take the wind load and prevent further distortion.[23]

Fig. 7-23. **(A)** *Montreal fire fighters ascend an aerial to combat a church fire. They did not know that fire in the crawl space below the floor was heating the steel columns which supported the masonry walls.* **(B)** *The steel failed. The wall collapsed. Two fire fighters died. (Both photos: Ian Stronach)*

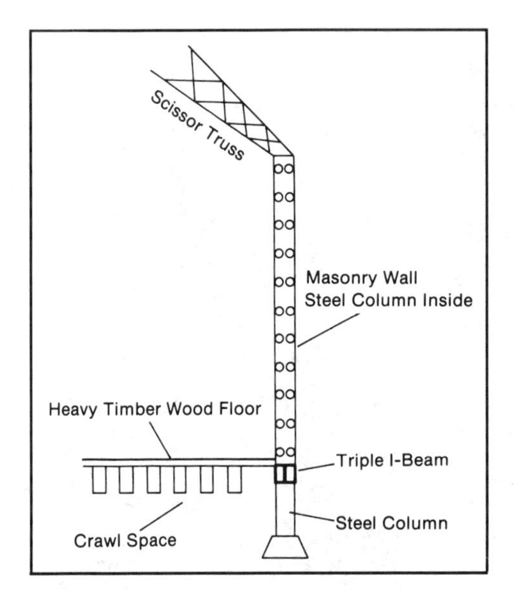

Fig. 7-23 **(C)** *The drawing shows how the wall was built.*

Dynamic Protection

Dynamic protection is generally accomplished with various types of automatic sprinkler systems. In some cases a building is simply conventionally sprinklered. This may not be adequate to protect the steel.

It is important for the fire department to know whether the sprinkler system is adequate for the fire load. Rolled paper has been the fuel for a number of fires which destroyed sprinklered steel buildings. In one case the building collapsed on the piles of paper, but the sprinkler pipes didn't break and the sprinklers operated continuously.

Hydraulically calculated sprinkler design can produce water flows adequate for the fire load, but this is no guarantee of success. Sometimes the procedure is used to design the system down to the bare minimum to keep costs low. Increases in fire loads overwhelm the system.

Heavy structural steel is sometimes protected by special lines of sprinklers. Flammable liquid hazards have long been covered by deluge sprinkler systems. More recently fog/foam systems have been installed to provide a foam blanket over the flaming liquid on the floor.

Passive Fire Protection

Passive fire protection is provided by protecting steel in one of the ways discussed earlier in this chapter.

If a building is rated as fire resistive, two considerations are vital:

1. Is the legally required level of fire resistance adequate for the fire load as it exists in the building?
2. Has the protection of steel been completely provided and is it maintained?

If the answers to these two questions are negative, there are three other critical questions:

1. Is there any legal relief?
2. Is it up to the fire department simply to do the best it can in the event of a fire?
3. If the fire department estimate of the situation indicates potential or inevitable disaster, who, if anyone, is notified?

All personnel making inspections should be made aware of the requirements for fire resistance as applied to specific buildings. The manner in which fire protection of steel can be degraded by the actions of persons completely unfamiliar with its vital purpose is also crucial.

As noted, sprayed-on fireproofing and membrane protection are the two methods most likely to be degraded. Sprayed-on protection is often removed by mechanics accomplishing other work or disposing of an asbestos hazard. The fire department should insist on its being replaced with an adequate safe material. It is also vulnerable to being washed away by hose streams. Caution should be exercised when returning to areas from which the fire department was driven by the fire.

There are so many ways a membrane ceiling can be tampered with that a prefire plan for any building in which one is installed should draw special attention to the possibility that collapse might occur. This is particularly a possibility where steel columns are unprotected in the plenum space.

One scenario for catastrophic damage to a column or columns could go like this: An interior stockroom has a characteristically high fire load. A ceiling tile has been removed for replacement, used elsewhere, or the ceiling fails. The column is heated to failure along a limited length and the column drops a foot or two. The effect would be staggering.

Tactical Considerations

Fire fighting plans might well subordinate the small line, low water-use tactics which fire departments customarily employ in fire-resistive buildings, in favor of heavy-caliber stream attacks on the fire to absorb the maximum Btu output before the heat is absorbed by the steel.

Tactical Considerations, continued

Where heavy stream tactics are deemed necessary to prevent collapse, it would be extremely beneficial to have a record of your pre-fire discussion with the building management and owner.

The fire department should be aware of the extent to which membrane fireproofing is permitted under the local code. Discussions should be held with the building department and local installers to reinforce the need for strict adherence to installation details as specified by the manufacturer's listing. In one case, poke-throughs for wiring permitted a fire to spread to several floors of a computer installation. There was extensive collapse of bar joists.

All must be aware of the difference between fire hazard or flame-spread listings of tile and the fire-rated tile as part of the fire-resistance system of the floor and ceiling assembly. Similar-appearing tile is often used in these two quite different categories. Not all suspended ceilings are a part of the fire-resistance system. The ceiling may be installed for any reason, such as to reduce the room volume for better air conditioning, to hide ducts and pipes, to conceal an old ceiling, for acoustical treatment, and so on. At one fire it was found that the fire-resistive membrane was provided by suitable lath and plaster, that the visible tile ceiling was only acoustical, but that to install the tile ceiling, many holes had been punched in the fire-resistive ceiling to accommodate the support wires.

If the ceiling, including light fixtures and air diffusers, is an integral part of the fire-resistance rating of the building, the owner and occupants should be put on notice that the fire department or building department must be consulted before any changes are made. Inspecting and preplanning personnel must be aware of the hazard of tampering with the ceiling, and an appropriate item should be provided on checklists.

Tactical Considerations

Particular note should be taken of high fire-load areas, where fast, high-heat output is possible, or where inaccessibility may provide a lower but long-sustained Btu output. Plans should provide for fast, high-heat absorption capacity lines. The fast, hot fire or long, slow fire may exceed the fire rating of the structure. (See "Estimating Fire Severity" in Chapter 6, Principles of Fire Resistance.)

Passive/Dynamic Protection

Protected combustible construction combines partial static protection of the steel with automatic sprinklers. One major insurer approves a spray-on coating which does not meet the standard for rated fire resistance, but would insulate the steel for a period of time which might be adequate to permit the sprinklers to control the fire.

Code Problems

In a discussion of prefire planning, it is difficult to separate fire-fighting procedures from fire protection requirements for steel structures, or proper maintenance of the passive or dynamic protection.

The fire department must develop competence in the many facets of the overall problem. It is impossible for all officers to be equally competent and informed. One person or a small group should be designated. Becoming familiar with current and past local building codes, and most importantly, the manner and degree of this enforcement, are prime areas of concern to effective fire departments.

Some building officials are less than enthusiastic about fire department interference in "their business." Incredibly, in one metropolitan-type county, fire inspectors were not permitted to inspect a building until the certificate of occupancy was issued. Not only does this ignore the hazard of fires in buildings under construction, but many defects can be hidden from view in a completed building.

Some fire prevention managers are equally unenthusiastic about any interest by fire suppression managers in "their business."

The ultimate fiasco is to have information which is vital to fire suppression locked up in Fire Prevention Office files at 3 a.m. when a fire occurs.

Code Variances

Building and fire prevention officials sometimes take the authority given in a code as carte blanche to grant exceptions to the code, based solely on their own undocumented action. Political influence or worse may be involved. All variances granted should be fully documented. The fire suppression forces should be involved in the process. It is possible that the beneficiary of the unrecorded variance may at some future date drag the municipality into a lawsuit. Without proper documentation, the municipality is very likely to be found liable.

Preplanning Your "McCormick Place"

With the advantage of hindsight, and a study of the McCormick Place fire, consider a possible preplan of a fire in an unprotected steel building being used as an exhibit hall. For discussion purposes, assume the building is a typical armory, i.e., brick-bearing walls with unprotected steel trusses supporting the roof. Also assume there is a balcony running down both sides which is used for sports events but not for exhibits. Recommendations for automatic sprinklers have not been carried out. No money is available. The state will not permit the city to dictate to it, and finally, most of the time the fire load is low.

The first move might be to set up a bruising battle for either sprinkler protection or denial of the facility to high fire-load exhibits. The battle is set up not to be won, but with the full expectation that it will be lost. In losing, however, the fire department may get some concessions. Without being too specific, get in writing an agreement that the fire department can take whatever steps necessary to protect life and property during exhibits. Having beaten back the greater financial threat, the managers of the armory can only be magnanimous and yield a point to the fire department. When they learn the proposed precautions, they may be quite upset, but the fire department has the written agreement. In such a public battle, it is not the fire department who is willing to risk the lives of our citizens.

Consider a worst case scenario in this armory. An exhibition, such as a housewares, home, or furniture show is scheduled. Calculations of the combustibility of typical exhibits together with the floor area develop a fire load in the range of 20 pounds per square foot. The surface-to-mass ratio of the material and its arrangement will provide a fast fire with early high-heat release at a rapidly increasing rate. The same fuel load in a more restricted area might give off heat more slowly due to restricted airflow, but an unlimited supply of oxygen can be assumed in the exhibit hall.

The fact that a hot, fast fire is anticipated indicates a more severe test of the structure than a slower fire. The radiant energy of a fire is proportional to the fourth power of the absolute temperature (actual thermometer reading in degrees Fahrenheit plus 459); thus a 2000°F absolute fire of 20 minutes' duration could have a radiant energy potential 16 times that of a 1000°F absolute fire of the same duration. Because of the large void above the bottom chord of the trusses, a huge smoke storage area is provided, and for quite some time there might be little smoke to protect the bottom chords of the trusses from radiant heat. If a bottom chord fails, the truss fails.

Tactical Considerations

If the facts are accurate, it becomes painfully apparent that even instant notification of a fire in an exhibition hall by a fire fighter equipped with a walkie-talkie radio is not fast enough to assure that the fire department can cope with the fire.

Only one plan remains. Get set up ahead of time. Provide a temporary "sprinkler system" by pre-positioning hose streams. In a building of ordinary ceiling height, fire fighters are accustomed to the operation of one to three sprinklers on the average fire. On a high-bay fire, expect the operation of a greater number of sprinklers even under the best of conditions.

In Underwriters Laboratories Inc. tests simulating an exhibition hall occupancy, the first sprinkler opened two minutes after the start of the fire, and a total of 12 sprinklers opened over an exhibit booth in a section with a 30-foot-high ceiling. Since the sprinklers were on 100-square-foot spacing, an area of 1200 square feet was wet down to control the booth fire. If the building were sprinklered, the minimum area which would receive water damage in the event of fire would be, based on the tests, approximately 1200 square feet.

Hose lines would be prepositioned in the balcony, feeding fixed nozzle holders. Portable deluge guns could be used, or simple nozzle holders could be fabricated in the fire department shops. The point is that there is a nozzle covering every part of the hall. If there is a standpipe in the hall, it would be used to supply the lines. If not, large-size supply lines would be laid as temporary fire mains and "water thiefs" used to take off the supply to the nozzles. A fire department pumper would be hooked up to a hydrant and to the supply lines. Lines would be charged and then depressurized. Of course, good hose, tight fittings, and good gaskets are a must.

One or more fire fighters would be assigned to the fire watch along with a pump operator. If fire should break out, the fire watch, equipped with a radio, would sound the alarm and order the operator to provide pressure. The fire watch then would move to the nozzle which covers the fire area. Depending on local planning, 60 to 90 seconds might be allowed for exhibitors or the maintenance staff to bring the fire under control. If then the fire is not obviously under control, water should be started into the overhead at a point directly over the fire.

On first glance such a plan appears ridiculous. "Fire fighters and equipment cannot be tied up to fight a fire before it happens. The precautions would upset the patrons of the show. We never did anything like that before. Who will pay for the time? We have a fire, we call you, you come. That's the way it's always been."

An Idea Is Adopted

The concept described above may not be as farfetched as it seems.

"Since many tests have shown that a number of sprinklers will undoubtedly open on any fire in a high-bay area, perhaps the engineering of protection should be rethought. Instead of overhead piping, perhaps the concept of an automatic system similar to the manual system described in Tactical Considerations above should be investigated."[24]

The Hoosier Dome in Indianapolis has installed just such a system. Photoelectric smoke sensors direct beams across the upper part of the arena. When the safeguards built into the system are satisfied that there is a serious fire, water is started from one of the fixed nozzles located around the perimeter of the area.[25]

The San Diego, California Convention Center has a fabric roof which could not support a conventional sprinkler system. A system similar to the one in the Hoosier Dome was installed in 1989.

The suggestions made above thus are no longer in the realm of fantasy. If a fixed system can be installed, it is certainly reasonable to improvise a similar system.

In San Jose, California, a department store has a fabric roof protected by sprinklers suspended from the roof cables.

If the building is sprinklered, study the potential water supply requirements, particularly against the possibility of the opening of ten to twenty sprinklers. It might well be advisable to make a standby hookup to the sprinkler system siamese connection so that the supply could be augmented in both quantity and pressure immediately upon the start of a fire.

Be Proactive

This author can recall lecturing to Navy personnel. "There is no law saying you must have a fire before you open a can of foam. Foam can be

Fig. 7-24. One of the "water cannons" installed in the Hoosier Dome in Indianapolis. Since such fixed systems have been found to be practical, it is not unreasonable for the fire department to set up a temporary system to accomplish the same level of protection. (Nick Chergotis, Pyrotronics Corp.)

used to suppress a flammable liquid hazard before the fire starts." While it was a revolutionary idea then, now the use of foam to control a flammable liquid spill is commonplace.

There is nothing the human intellect resists more than a new idea. Step by step, logic leads inexorably to the practical solution — prepare for fire before it happens. Learn from those fires that have occurred.

Fire fighters should be eternally curious. Just about everything that happens has some impact on the fire protection situation. Evaluate it and take action to contain or perhaps eliminate any problem.

References

The addresses of all referenced publications are contained in the Appendix.

1. Boring, D., Spence, J. and Wells, W. *Fire Protection Through Modern Building Codes*, 5th ed., American Iron and Steel Institute, Washington, D.C., 1981.

2. Dunn, V. "The Dangers of Fire Escapes." *Fire Engineering*, May 1990, p. 38.

3. Washburn, J. "Four Alarm Fire Sweeps Through Self Storage Complex." *American Fire Journal*, Oct. 1989, p. 24.

4. "Fire At Illegal Building Debris Dump Heavily Damages Elevated Highway." *Engineering News Record*, August 17, 1989, p. 8.

5. Coggen, R. and McCullough, A. "Profile Redmond, WA F.D." *American Fire Journal*, July 1991, p. 22.

6. Juillerat, E. and Gaudet, R. "Chicago's McCormick Place Fire." *Fire Journal*, May 1967, p. 15.

7. "The McCormick Place Fire, Chicago, Illinois." American Iron and Steel Institute Report, February 15, 1967, Washington, D.C.

8. Webb, W. E. "Effectiveness of Automatic Sprinkler Systems in Exhibition Halls." *Fire Technology*, Vol. 4, No. 2, May 1968, p. 115.

9. Brown, S. "Arched Pipe Trusses Span New Air Terminal at Tampa." *Engineering News Record*, October 22, 1987, p. 13.

10. "Automobile Burn Out Test in an Open Air Parking Structure." Gage-Babcock Associates, Westchester, Illinois, 1972.

11. Klem, T. "Los Angeles High Rise Bank Fire." *Fire Journal*, May/June 1989, p. 72.

12. Sauer, G. "Suspended Ceilings as a Fire Protection for Columns." *Building Standards*, Part I, July-August 1970, p. 6.

13. *Fire Protection Through Modern Building Codes*. p. 126, (See Ref. 1).

14. Sotis, L. "The FMRC-Approved Roof: A Proven Roof." *FM Record*, Sept/Oct 1990, p. 16.

15. Goodbread, J. "Fire in Building 3001." *Fire Command*, July/August 1985, p. 34.

16. "On The Line, Monroe, N.Y." *Fire fighters News*, June/July 1990, p. 29.

17. Isner, M. "Fire Investigation Report: Nursing Home Fire, Dardanelle, Arkansas." National Fire Protection Association, Quincy, Mass., March 13, 1990.

18. Ulley, L. "First General Alarm Fire for Montreal Island Community." *Fire fighting in Canada*, October 1990, p. 12.

19. Brannigan, F. and Mittendorf, J. "The Metal Deck Roof Debate." *Fire Engineering*, March 1988, p. 58.

20. Viera, D. "Can You Count on Your Fire Doors." *Record*, July/August 1991, p. 12.

21. Smith, J. "Fire Studies." *Firehouse*, August 1988, p. 38.

22. Stronach, Ian. "Montreal, Death Knell for Fire Fighters at Church Blaze." *Firehouse*, November 1987, p. 42.

23. De Boer, C. "Trinity Church." *Modern Steel Construction*, No. 1, 1989, p. 33.

24. Brannigan, F. *Building Construction for the Fire Service*, 2nd edition, National Fire Protection Association, Quincy, Mass., 1982.

25. "Stadium Shoots Down Fires." *Engineering News Record*, April 19, 1984, p. 47.

8

Concrete Construction

Concrete is a cementatious material produced by a chemical reaction of portland cement and water to which inert materials called aggregates are added.

Shortly after it is mixed, concrete sets into a solid mass, but it continues to cure indefinitely. Construction specifications set a date by which concrete must reach its required compressive strength. For instance, concrete required to reach design strength in 28 days is sometimes referred to as "28-day concrete." "High early strength concrete" achieves full compressive strength in a shorter time.

While curing, concrete generates heat of hydration. During its initial curing, concrete must be protected from freezing. Low temperatures retard the curing of concrete, and freezing is harmful to the material.

Good concrete results from proper handling of carefully controlled materials. Because there are so many possibilities for a poor concrete product, high factors of safety are used in concrete design.

Concrete is very weak in tensile strength and has poor shear resistance. Its compressive strength is good, particularly when compared to the cost of steel to resist the same load.

Fire Service personnel should avoid the common error of using the term "cement," when "concrete" is the proper term. Cement is a component of concrete.

Concrete Structures

Until the post-World War II era, structural concrete was regarded as suitable only for massive, low-rise utilitarian structures in which aesthetics played little part. High-rise buildings almost universally built of steel frames were often fireproofed with concrete and used cast-in-place concrete floors.

Precast concrete principally was in the form of cinder blocks. Cinder blocks use cinders as the aggregate to produce a light, relatively inexpensive block. Concrete blocks use other materials for aggregate. Heavier concrete blocks usually have smoother surfaces than cinder blocks.

Underwriters blocks are concrete blocks produced under Underwriters Laboratories' classification. The manufacturer is authorized to issue a certificate giving the type and number of units delivered to a specific job. These blocks are used in assemblies that must meet fire-resistance standards.

Concrete warehouses and factories built before World War II are easily recognizable. Their panel walls, usually of brick or brick-and-block composite, were built on each floor between the exterior columns.

Today, there is a bewildering variety of mix-and-match building construction elements. Steel-framed buildings now often have cast-in-place concrete floors made with forms that are removable or which can be left in place. The floors may or may not contribute to the stability of the building. Precast concrete and prefabricated metal wall panels are common.

In some cases, an architect may wish to emphasize concrete construction. Instead of considering form-marks unattractive, architects have accentuated them in some building designs. In other cases, decorative brick veneer is used to conceal the concrete construction.

There have been several cases where substantial brick veneer was lost from a reinforced concrete building. High early-strength concrete tends to shrink. The brick veneer does not. If proper provision is not made for the contraction, the brick veneer may crack and fall off.

Steel vs. Concrete Framing

Many designers prefer to design in steel, while others prefer concrete for framed buildings. There are some buildings which present a unified exterior appearance, yet have one section concrete-framed, and another section steel-framed. Building owners are sometimes less interested in the material of construction than in speed and cost of construction. When a building is put out for bids, the price often depends on which contractor has readily available personnel and other resources.

Fire Department Problems

The problems which fire departments face with concrete construction can be divided into three distinct areas.

- Collapse during construction with no fire.
- Fire during construction.
- Fire in completed, occupied buildings.

These problems will be discussed after you study concrete construction itself.

Types of Concrete Construction

There are two basic types of in-concrete construction: cast-in-place, and pre-cast.

Cast-in-place concrete includes plain concrete, reinforced concrete, and post-tensioned concrete.

Precast concrete includes plain concrete, reinforced concrete, and pre-tensioned concrete.

Definitions

Aggregate: This is material mixed with cement to make concrete. Common aggregates are both fine and coarse. Fine aggregate is usually sand. Coarse aggregate may be one of a great variety of materials, depending upon availability and the characteristics desired in the finished product. Concrete often is described by the name of the aggregate used.

Crushed stone, gravel, cinders, breeze (a coke by-product no longer used), expanded slag, shale, slate, clay, vermiculite and perlite (natural light-weight materials), light-weight stones such as lava and pumice, and fly ash reclaimed from boiler plants have all been used as aggregates.

Cast-in-place concrete: This concrete is molded in the location in which it is expected to remain.

Casting: This is a process of placing the fluid concrete into molds, generally called forms, in which the concrete is permitted to harden to a certain shape.

Chairs: These are small devices designed to keep the rods up off the surface of the form, so concrete will flow underneath the rods.

Composite and combination columns: These columns use steel and concrete combined into one unit. As in all composite construction, the different elements must react together, and any failure of the bond between them creates a distressed column. Composite columns were formerly used on the lower floors of a heavy-duty building to avoid surrendering floor space to large-diameter concrete columns made solely of reinforced concrete.

Composite construction: This is a different term than a composite structural element. Reinforced concrete is one of several composite structural elements. See Chapter 2 for further information. The term composite construction is often used to describe buildings in which different load-bearing materials are used in different areas, such as exterior columns being made of concrete and interior columns constructed of steel.

Sometimes the top floor of a concrete building is constructed of light-weight steel. The extra story is "free." In one case, the top floor of a concrete high-rise in Florida was made of wood. This lighter structure may be called a penthouse, which is any structure above the roof line. The term penthouse apartment properly refers only to an apartment built on the roof. Apartments on lower floors which have terraces are properly called terrace apartments.

The search for less costly buildings has brought the use of corrugated steel forms for floor slabs. This steel becomes a left-in-place form and the steel and concrete, reacting together under the load, provide a composite floor.

Many substances are not suitable for close contact with concrete. Aluminum, for instance, as used in electrical conduit, reacts with concrete and can cause decomposition. Some natural stones cannot be used for aggregate because of undesirable reactions.

Continuous casting: This is a process for casting or pouring concrete without interruption from start to finish. This avoids the problem of joining new concrete to concrete already poured. If such a joint is not properly made, cold joints, which are planes of weakness, result.

Continuous slipforming: This entails pouring concrete continuously as forms move upward so that continuous casting may be accomplished.

Drop panels: These are thicker sections of floor near columns to assist in resisting the natural tendency of the floor to shear off at the column.

Footings: These are thick concrete pads, usually heavily reinforced, that transfer the loads of piers or columns to the ground.

Lally® columns: These are steel pipes filled with concrete to increase their load-carrying capacity. In the 19th century, cast-iron Lally columns were usually drilled to provide a vent for steam generated in the concrete. Today, steel pipe concrete columns are only sometimes drilled. Exposed to heat, they may explode as their undrilled predecessors did. This author has seen "deck houses" where such columns were used without holes being drilled. Building officials I spoke to seemed unaware of any hazard.

Such columns also have been tested for fire resistance. This would be advantageous structurally because the steel would act as a form, and surface fireproofing might not be required.[1]

Tactical Considerations

If such columns are found without drilled vent holes, refer building department officials to Lie's article referenced above.

In a serious fire, column failure and subsequent collapse should be anticipated.

In one example in a fire in a concrete plant, expanding steam from concrete blew out one end of the Lally columns, creating rockets. Rocketing columns which flew several hundred feet forced the closing of a highway.

Lift slab: This is a type of construction in which initially the columns are erected to their full height. Then the first floor slab is cast on the ground. Each subsequent floor slab and then the roof slab is cast, one directly on top of the other. A bond breaker is used between each slab so there will be no adhesion. Starting with the roof, the slabs are successively lifted into position. Steel deck roofs have been assembled on the ground and lifted into position in a similar manner.

L'Ambience Plaza, a lift-slab building under construction in Bridgeport, Connecticut, collapsed in 1987 with a loss of 28 lives. A single component in the lifting system failed and caused the collapse.

Monolithic construction: A method where all the concrete in a building is properly bonded together. The resultant structure can be likened to one piece of stone (Greek, monos-one; lithos-stone).

Mushroom caps: These are tapered extensions at the tops of columns which assist in transfer of loads from floor to column. Columns in heavy-duty concrete construction may show both mushroom caps and drop panels.

Plain concrete: This term refers to concrete that has no reinforcement, except possibly light reinforcement to resist temperature changes. A sidewalk slab or a concrete base for a child's swing set are examples of plain concrete. Some early concrete bridges, such as the Taft Bridge in Washington, D.C., were built of plain concrete. Necessarily, the entire bridge is in compression, as were early stone arch bridges developed by the Romans.

Pre-tensioning and post-tensioning: These are processes by which steel tendons or rods are placed under tension, drawing the anchors together. The tensioned steel places the concrete in compression. This is explained more fully later in this chapter.

Precast concrete: This concrete has been cast at a location other than the place where it is to remain. The precasting may be done at a plant miles from the construction site, at the construction site, but remote from the ultimate installation location, or immediately adjacent to the point of use.

The common cinder block or concrete block is precast, plain concrete. Other precast concrete units are reinforced or tensioned.

Reinforced concrete: This is a composite material made of steel and concrete. Steel provides the tensile strength that concrete lacks. Steel may also be used to provide compressive strength. The term reinforced concrete is a misnomer. It creates a belief that concrete has a certain strength and that steel reinforces that strength. This assumption sometimes leads to a failure to recognize the hazard of a failure of the bond between the concrete and the steel. Reinforced concrete is a composite, and both elements are equally important.

In order to accomplish the designed function of reinforced concrete, the concrete and steel must remain in contact. Most importantly, exposed steel reinforcement lacks fire resistance, as does any unprotected steel.

An article[2] describes concrete with strengths as high as 20,000 psi. This results in use of less concrete, and greater usable floor area. The article does not discuss fire resistance, though the basic element of fire resistance is mass.

Tactical Considerations

When a bond between two equally important elements of a composite material is broken, the structural strength ceases to exist, and all that remains is deadweight.

Fig. 8-1. The reinforcing steel in these bridge piers was found to be inadequate. The bridge was closed and external tension rods were installed.

Reinforcing bars or rods: These are steel rods or bars that are usually intentionally deformed. A raised pattern in the surface of the steel aids in the transmission of stress from the concrete to the steel. It would be more accurate to call them "tension rods."

There is also interest in glass fiber reinforcing rods. The effect of fire on fiberglass reinforced concrete is uncertain. A concrete swimming pool with bamboo reinforcing rods was built by Thomas Edison in Ft. Myers, Florida early in this century. It has been filled with water ever since.

Slipforming: This is a technique by which forms are moved, usually upward, as the concrete is poured.

Spalling: This term describes the loss of surface material when concrete (or stone) is subjected to heat. It is due to the expansion of moisture in the concrete. Some concrete and certain aggregates are more subject to spalling than others. Concrete made with ground-up fire brick is very resistant to spalling and is often used at fire testing facilities. Explosive spalling occurs violently, throwing out bits of concrete projectiles. Note the correct pronunciation is spall-ing, not spall-ding.

Temperature rods: These are thin rods installed near the surface of concrete, usually at right angles to the main reinforcing rods to help the concrete resist cracking due to temperature changes.

However, failure of the bond between the temperature rods and the concrete does not materially affect the strength of the structure.

Tilt up or tilt slab: These terms describe casting wall panels on the ground along the outside perimeter of a building. When the panels are cured, they are tilted up into place and attached. The more common practice is for the precast slabs to be delivered from a fabrication plant.

Concrete Structural Elements

Columns

Since a chief virtue of concrete is its high compressive strength and low cost, concrete is used for columns, even in structures otherwise built of steel such as highway bridges. A short column with a large cross section in proportion to its length is called a **pier**. It may be of plain concrete if it is loaded only compressively. Otherwise, columns are of reinforced concrete to provide the tensile strength necessary for loads created by eccentric or torsional loading.

The steel reinforcing rods not only provide necessary tensile strength but also carry some of the compressive load. The compressive strength of steel is many times that of concrete. In early concrete construction, it was the practice to design larger-diameter columns successively on lower floors to cope with the increasing loads. At the lowest level, the size of the columns might be so great as to interfere with the use of the building. In some cases, standard steel columns were used within the concrete to take advantage of the far greater compressive strength of steel, and reduce the width of the column.

In modern buildings, increasing column sizes are unsatisfactory. The useable area would vary from floor to floor. This is overcome by increasing the size of the reinforcing steel as the loads increase so the outside dimensions of columns are identical on all floors. In some cases, the columns have so much steel in them that a special additive is mixed into the concrete to permit it to flow into the relatively small spaces between the steel.

Reinforcing rods are long with a relatively thin diameter. Sleeves, reinforced overlaps or welding is used to connect the ends of rods and transfer the load. The rods in a column are joined by lateral reinforcements called **ties** or **hoops**. (In round columns, the ties connecting the outer rods are often formed in a helix. Such reinforcement is wrongly termed spiral reinforcement.)

Fig. 8-2. The rods in columns carry some of the comprehensive load. They are columns in their own right. Note how the ties, in effect, divide the column into many shorter columns, thus permitting it to carry a much greater load.

The ties or hoops serve to cut the long slender column up into a number of relatively short columns one on top of the other. Recall from Chapter 2, Principles of Construction, the effect of lengthening a column section by the loss of bracing. Together with the concrete to which they are bonded, the rods form a composite structural element. If the unity of the composite fails, such as by loss of concrete either under normal circumstances or as a result of fire, the steel rods may buckle under load. Buckling will cause the rods to protrude from the column.

Tactical Considerations

Protruding rods are a serious sign of possible collapse and should bring about an immediate evacuation. A column failure may have catastrophic results. Fortunately, column failures are rare.

Beams and Girders

Plain concrete is strong in compression, weak in shear, and does not have any tensile strength. As noted, when a beam is loaded, it deflects.

This deflection brings about compression in the top of the beam and tension in the bottom of the beam. Reinforcing bars are placed in the bottom of the concrete beam to provide the necessary tensile strength. Precast beams often have the word "top" cast into the top of the beam to ensure that the beam is installed right side up.

In a cantilever beam, however, the tension is in the top of the beam and thus the reinforcing (tension) rods are in the top of the beam. The far end of the beam bends downward like a diving board. In one case, the then new cantilevered balconies of a concrete high-rise building all required replacement because an old-time supervisor "knew" that the plans were wrong, and installed the rods in the conventional manner.

In a continuous beam supported at more than two points, there is tension in the top of the beam in the area over the tops of the columns, and tension in the bottom of the beams between columns.

In all cases, reinforcing must be properly designed. Vertical reinforcing bars in concrete beams designed to prevent cracking under shear stresses are called stirrups. Some tensile reinforcing bars may be bent up to accomplish the same purpose.

T-Beams

Concrete below the neutral plane (where there is neither tension nor compression in the beam) serves very little purpose. Sufficient concrete is needed to bond with the reinforcing rods for the relatively slight compressive stresses, and to protect the steel from fire.

This leads to the characteristic T-beam design in which the neutral plane coincides with the bottom of the wide, thin floor slab. Double T's are floor slab-and-beam combinations with two beams. Floor-beam combinations with four beams are also manufactured.

As in columns, where space requirements are paramount, steel may be substituted for some of the concrete to handle compressive loads. For any unit area, steel has 15 times the compressive strength of concrete. Steel is, however, more costly.

Concrete Floors

There are many varieties of concrete floors used in concrete and/or steel-framed buildings and in ordinary construction. Concrete floors may be cast-in-place or precast. Precast floor units are prestressed. Prestressing is discussed in the next section.

Concrete was first used for leveling brick and tile arch floors. This led to segmental and flat concrete arches sprung between unprotected wrought iron or steel beams.

Fig. 8-3. The formwork has been removed from this waffle concrete but the concrete has been reshored. Reshoring indicates the concrete is not yet up to full strength, or there may be an extra heavy load above.

Other early floors were built of individual beams supporting a floor slab, all poured at the same time. The appearance of the underside is reminiscent of wood joists. This method is still used.

In so-called **waffle concrete**, closely-spaced beams are set at right angles to one another, and unnecessary concrete is formed out. The lower side resembles a waffle.

Hollow tiles: Were used in early efforts to lighten concrete floors by eliminating concrete below the neutral plane. The tiles are visible from the ceiling side of the floor. This technique often puzzles modern engineers.

Many concrete floors are designed as **continuous beams**. The entire floor is one big beam, just as a wooden floor might be made of one huge sheet of thick plywood. In heavy construction, mushroom caps and drop panels are used at columns to cope with high shear stresses.

In lighter construction, the floor may be just a **flat plate**, with no projections below the floor line. This gives a smooth surface, easily finished, and it is popular for offices and apartment buildings.

When concrete floors are cast onto corrugated steel, the steel is a **left-in-place form**. The steel provides necessary tensile strength. If the

Fig. 8-4. The corrugated metal provided the tensile strength. When it was removed, only the temperature reinforcement served to barely hold the concrete together.

bond fails, the floor section may fail. In one parking garage, the steel was eaten away by deicing salt, and a collapse resulted.

When precast T-beam units are used, additional concrete is often cast-in-place on top of the units, which are left rough on top to ensure a bond. The entire unit becomes an integral beam-and-floor element.

Precast **cored concrete** floor units have cylindrical openings cast lengthwise through the units to remove unnecessary weight and to provide channels through which building utility services can be run.

Concrete floors may be simply load-bearing, or may be an integral part of the structural stability of the building.

Building codes in some cities require that the street floor of buildings of ordinary construction be of "fireproof" construction. Consequently, a concrete floor or, more likely, a concrete topping over a wood floor, may be found in a building of ordinary construction. In one case, a concrete topping over wood beams was required by health department regulations for a washable floor in a restaurant. This concealed the destruction of the beams by fire. Four fire fighters died when the floor collapsed without warning.

Some 1-hour rated designs of wood floors include a light-weight concrete topping as much as 1½ inches thick. This retards the passage of heat through the floor and delays the temperature rise on the unexposed surface — a condition of the NFPA 251 *(ASTM E119) Fire Resistance Standard.*

Fig. 8-5. Concrete in the top of this cored slab provides compressive strength. Concrete along the neutral plane is not vital and can be left out. Concrete below the neutral plane bonds the tension strands to the slab and protects them from fire. Note what appear to be black dots. They are the ends of prestressed tendons.

Tactical Considerations

Concrete-on-wood floors require special attention in a fire. The insulating concrete may deceive fire fighters above the fire as to the fire below their feet. The added dead weight also may be a factor in the collapse of the floor-ceiling assembly.

Building a cast-in-place concrete floor into ordinary wall-bearing masonry buildings sometimes provides a hazard during construction, and perhaps for the life of the building. In construction, the wall is carried up a story or more beyond the point where the floor is to be located. A slot is left in the wall at the point where the floor is to be cast. The concept is that concrete will fill the slot, thus uniting the floor and wall.

If a windstorm occurs during the time that the slot is open (usually a few unmortared bricks set on end are relied on for temporary support), a collapse may result. A ladder or hose line over the top of the wall may also cause collapse. For all practical purposes, the wall is undercut, as is done in felling a tree.

Unless the concrete is carefully pushed up into the gap when the floor is poured, the wall will always be weak at this point. The finished floor may give no evidence that the continuity of the wall is interrupted by a void which creates a permanent plane of weakness.

Concrete floors in steel buildings may be precast, or cast-in-place. They may be only load-bearing, or they may provide structural stability by being designed to resist lateral loads such as wind. In the latter case, studs are welded to the steel beams and the reinforcing bars are attached to the studs.

Concrete floors in cast-in-place, concrete-framed buildings are cast integrally with columns, providing a monolithic rigid-framed building.

Precast units may be pinned, or can be connected as a monolithic unit. In this case, reinforcing bars are designed to protrude from the finished precast slabs or other units. When the slabs are laid down, a space is left between them. The protruding bars of one slab extend past the ends of the protruding bars of the other slab. They overlap without touching. An 18-inch space, for instance, may be left between slabs with bars that are usually hooked, protruding 12 inches. A form is made, and concrete is poured to level. The sections, thus joined by a wet joint, become monolithic.

Concrete floors in precast, pinned concrete buildings may be only load-bearing and may not contribute to the structural stability.

Precast columns are often built with haunches or shelves cast into their sides on which beams are set. Steel plates imbedded in the concrete may be welded together. This, of course, is pinned, not monolithic construction. In some cases, the beams are held in place only by gravity. The cast girder connection was missing from a column in a Seminole, Florida mall. A bracket was installed with lag bolts. There was a massive collapse shortly before the mall was to open.

Prestressed Concrete

Conventional reinforced concrete has been used since the invention of the reinforcing bar in 1885. In recent years prestressed concrete has been developed. Prestressing places engineered stresses in architectural and structural concrete to offset the stresses which occur in the concrete when it is placed under load.

Consider a row of books side by side. Used as a beam across two chairs, such a row will fail of its own weight, without any superimposed load. There is no shear resistance between the books. However, drill a hole through the row of books laterally, pass a wire through the books, and tighten the wire against the end books. The row of books would be compressed by tensile stress in the wire and compression in the books. The "beam" could then be placed across two chairs and loaded. The "beam" has been prestressed sufficiently to counteract the stresses placed in it by the load.

Special high-strength, cold-drawn steel cables, similar to those used for suspension bridges, or alloy steel bars are commonly used. In building construction they are technically known as **tendons**, but are called **strands** or **cables** by those working with them.

High-tensile-strength wire, ordinarily used for prestressing, is more sensitive to high temperatures than structural steel. There is virtually complete loss of prestress at 800°F. This is less than the temperature of a self-cleaning oven.

Protection of the tendons from fire is of paramount importance.

Prestressing Methods

There are two methods of prestressing: **pretensioning** and **post-tensioning**. The prefix *pre* and the descriptor *post* refer to whether the concrete is poured before or after the tension is applied.

Pretensioning is done in a plant. High-tensile-strength steel strands are stretched tightly between the ends of a form. Concrete is then poured in the form. As the concrete sets, it bonds to the tensioned steel.

When the concrete reaches a specified strength, the tensioned strands are released from the ends of the form. The stretched strands attempt to draw back, thus compressing the concrete. This creates a built-in resistance to loads which produce tensile stresses. The completed units, such as the common T-beams, are shipped to the job site.

Post-tensioning is done on the job site. High-tensile-steel strand wires or bars, encased in plastic or paper tubing, or wrapped to prevent any adhesion between steel and concrete, are positioned in the forms. The concrete is poured. After the concrete is set and reaches a specified strength, the steel tendons are stretched and anchored at the ends of the unit. Technically this is called stressing the tendons. On the job site it may be called jacking the cables. A major advantage of post-tensioning is that floors can be thinner. Thinner floors will allow several more rentable floors in a tall building.

Sometimes post-tensioning is performed in increments, as the load of the building increases with additional stories.

Fig. 8-6. This huge tendon that will be in a transfer beam is a bundle of rods. It is sheathed to prevent contact with the concrete. It was pulled "live" in several increments as the building grew in height. When tendons are pulled from both ends simultaneously they are "pulled live." When pulled against an anchor they are "pulled dead."

Fig. 8-7. Note how thin the floors are in this Vancouver, British Columbia parking garage. The floors are cantilevered over a distance longer than the length of a car. Such is the strength of prestressed concrete.

Some bridge girders are tensioned enough to make shipment possible, then post-tensioned after being placed.

After tensioning is completed, some designs call for pressure grouting in which a cement paste is forced into the space between the tendon and the concrete to provide a bond.

Some tensioned floors consist of small precast units through which tension cables are run, then tensioned.

Tactical Considerations

It may be a matter of life and death for fire fighters to understand critical fireground information such as, "They haven't jacked the cables yet."

Reinforced Masonry

In the earlier discussion of ordinary construction, it was noted that there is a built-in limit to the economic height of an ordinary brick bearing-wall building. This exists because of the requirement that the walls increase in thickness as the building's height increases. The limit is generally about six stories, but there are brick bearing-wall buildings as tall as 16 stories. In recent years, it has become possible to build brick bearing-wall buildings 20 or more stories in height, with no wall thicker than 12 inches. Concrete block can be used in a similar fashion.

In one method of construction, two wythes of brick are built. The width of one brick is left between them. Reinforcing rods are placed vertically in the cavity. After the brick masonry sets, concrete is poured into the void. Floors in one such building are of precast-tensioned, concrete double T slabs, but other systems can be used.

Reinforced masonry construction is widely used to resist earthquakes, particularly on the West Coast. Unreinforced masonry buildings collapse readily in earthquakes. Reinforced masonry is unsuitable for multistory buildings in which large clear spans are required, but can be used for buildings where multiple interior walls are acceptable. The interior and exterior walls are equally load-bearing. Apartment houses and

Fig. 8-8. This reinforced masonry wall is composed of concrete block for the lower courses, which will be below grade, and brick for the upper courses. Concrete is poured into the voids within the block and the space remaining between the bricks. The result is a 12-inch reinforced masonry wall.

motels with their characteristic inflexible, repetitive box construction are well adapted to this method. Office buildings require more flexibility. The high level of soundproofing is another benefit of this type of construction.

The 23-story reinforced-masonry Excalibur Hotel in Las Vegas is the tallest wall-bearing concrete block building.

In low-rise buildings up to about 70 feet, some recent designs have eliminated reinforced concrete in the wall. High compressive-strength bricks and special mortar are used along with masonry reinforcement trusses and placed between courses and to tie cross walls to the exterior walls. Together with precast concrete floors integrated into the building by the use of wet joints, these materials can produce a masonry wall-bearing building several stories high with no wall thicker than eight inches.

Similar construction, usually of concrete block, has become popular for some resorts. Some recently built seaside motels consist of individual concrete boxes. When these are serviced by outside open-air stairways and balconies configured in accordance with the Life Safety Code (NFPA 101), or the egress-related provisions of the building code in effect, the epitome of life safety in a multiple-unit occupancy is achieved. The occupants are almost completely safe from all folly but their own.

Tactical Considerations

Conventional wall-breaching techniques would be ineffective, possibly dangerous, if used on a reinforced masonry wall. It is important to know whether the interior walls of a concrete residential structure are of reinforced masonry or drywall on steel studs, which can be readily breached. In some cases walls between rooms are alternately reinforced masonry and drywall.

Know Your Buildings!

Collapse During Construction

From time to time, concrete structures under construction collapse. The fire department is almost invariably called to rescue construction workers. The following suggestions are offered as aids in planning a department's role in rescue operations when a building collapses and there is no fire.

While the authority of the senior fire officer present at a fire is often clearly spelled out, the authority to act or to direct others at emergencies other than fires is rarely defined. The fire officer should be well informed on the legal position of the fire department.

Many building codes assign the power to order the removal of a dangerous structure to the building commissioner or similar official, thus making it quite clear that the fire department has no right to demolish a structure just because it represents a hazard in the view of the senior fire official.

Politics, grandstanding, glory grabbing, turf protection, and potential movie rights have all played a part in disaster operations which often are improvised.

Every community should have clear plans setting forth areas of responsibilities at emergencies other than fire.

After many disputes, the fire department and police department of New York City established a protocol in 1990 for emergency situations. The fire department is in command of rescue operations at building collapses, hazmat situations, and utility emergencies.

The police department is in command at diving operations, vehicle extrication, and bomb threats.[3]

Lawsuits are very common today. Often after the dust settles in a building collapse, there may be a series of legal actions as the owner, architect, general contractor, subcontractors, and the victims attempt to determine financial responsibility. Collapses do not just happen — they are caused — and some of the people at the scene may well know, or at least fear, that the blame may be put on them. They may actively advise or assist the fire commander.

Be careful. Somebody may be trying to cover up. It may not be your duty to prevent this, but it certainly is not part of your duty to assist. Be wary of such suggestions as: "Let's knock down the rest of this before it falls." A study of the surviving portion of a structure may show why the collapse occurred.

An Industry Warning

Some experts have warned of the collapse hazard of concrete structures. They argue that design engineers should use construction loads as governing loads in structures. At times, the loads on buildings under construction exceed the design load and can approach failure load.[4]

Problems of Falsework

Falsework is the temporary structure erected to support concrete work in the course of construction. It is composed of shores, formwork, and lateral bracing. Formwork is the mold that shapes the concrete. Shores are members that support the formwork. Lateral bracing is usually made up of diagonal members that resist lateral loads on the falsework. Falsework can represent 60 percent of the cost of a concrete structure.

Concrete falsework is designed without the extra strength calculated into a building to compensate for deterioration. It is usually built at the lowest possible cost, out of low quality material, by people rarely versed in the basic engineering and mechanical principles. With falsework, the builder hopes to contain a fluid load that can provide a head pressure of up to 150 pounds per square foot for each foot of height. Large fluid loads are placed at the tops of slender columns, and the loads are then vibrated. The structure must often absorb the impact load of the sudden stopping of motor-powered buggies carrying heavy loads of concrete. It is not surprising that falsework failures occur; what is surprising is that they are relatively rare.[5]

In 1975, the concrete roof of a Cupertino, California shopping mall collapsed under construction. A trade publication article noted that the failure of a single four by four in the falsework precipitated the collapse.

Falsework for walls or columns must have adequate strength to resist the tremendous pressure of the heavy fluid concrete. As concrete sets, the pressure is reduced due to internal friction. If a wall is designed for several successive pours and an overpour occurs, collapse may result.

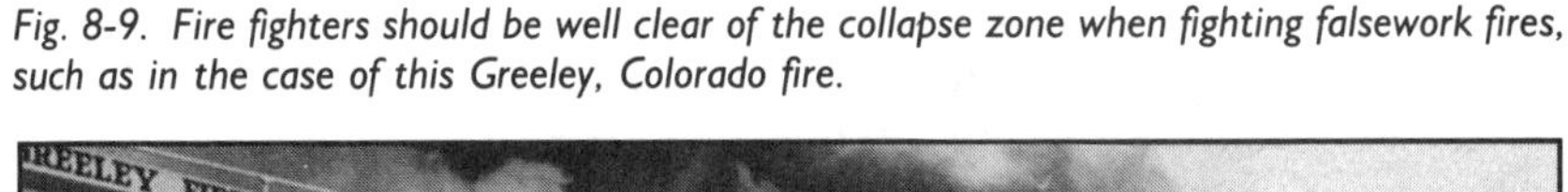
Fig. 8-9. Fire fighters should be well clear of the collapse zone when fighting falsework fires, such as in the case of this Greeley, Colorado fire.

The setting of concrete is temperature dependent; cold weather slows it down. Concrete poured at 50°F will develop a third more pressure than at 70°F.[6]

Tactical Considerations

When a wall formwork bursts, it usually goes out at the bottom. Workers may be trapped, particularly between falsework and the sides of the excavation. There will probably be no reinforcing steel to interfere with rescue. Rescue involves uncovering faces, supplying oxygen, digging the victims out, and avoiding further injury.

If it is necessary to retard the setting of concrete, as much sugar as can be obtained should be worked into it. Concrete containing sugar will set slowly, if at all. Bear in mind that the concrete can transmit pressure. Avoid standing in concrete which might transmit the weight of your body to the victim.

Reshoring

Concrete requires time to cure; the length of time depends upon the type of cement used and the temperature during the curing period. After a proportion of the design strength is reached, falsework is removed for use elsewhere. Some shores are set back in place in order to help carry the load of the still-curing concrete. This is called **reshoring**. It is an indication that the concrete is not yet up to full strength. It may be unable to handle large superimposed loads, such as debris removed from a collapse, excess numbers of people, or additional apparatus or construction equipment. Avoid adding equipment and construction material to existing loads.

Collapse of Floors

Many collapses involve falsework supporting newly cast floors. Sometimes the collapse is of framework supporting a high bay floor, where two sets of normal floor-high shores are used one on top of the other. Such structures have all the instability characteristics of long, narrow columns. Proper cross bracing is vital as is proper footing.

Often, falsework rests on the ground, sometimes on unconsolidated backfill. The planks on which the shores rest are called **mudsills**. If water liquefies the soil and turns it to mud, the bearing may be inadequate. Repeated vibrating of the concrete to provide a dense smooth surface may be the straw that breaks the camel's back, and cause a collapse.

Fig. 8-10. If a disaster like this one at Skyline Towers in Fairfax County, Virginia happens, your community should be prepared with a plan which has determined in advance who is in charge of what. In this incident workers knocked out the shoring. The effect of fire burning the shoring would be the same.

Tactical Considerations

In a collapse, most of the victims will be on top of the falsework. Use bolt cutters to cut reinforcing rods to manageable lengths. Thicker rods may require torch cutting. Rods are tied together in a matting. Further harm may be done by attempting to remove too-large sections.

Note well that the suggestions for cutting reinforcing rods do not apply to post-tensioned cables or rods which are under tension, and will whip dangerously if cut.

When concrete is being poured, almost invariably there are carpenters and helpers under the structure, bracing and wedging any shores that appear to be distressed. Determine from the supervisor whether all of the workers are accounted for. Missing workers may be under the falsework. In one case, the fire department, assured that all were accounted for, was preparing to leave the scene, when a worker wriggled out from under the plywood.

A Widely Believed Fallacy

In the first edition of this book, it was stated, "Reinforced concrete which has set hard to the touch usually has developed enough strength to be self-supporting, though it may not be capable of handling superimposed loads." This was based on information from experienced concrete designers. It is still widely believed.

This was disproven in the Skyline Towers collapse in Arlington, Virginia in 1973. In this collapse, shoring was removed from the topmost floor. The floor collapsed, and the collapse was progressive, ending at the ground level.

The removal of the shoring by laborers is no different than the removal by fire. The result would be identical. In short, any concrete falsework failure presents the likelihood for catastrophic collapse, though the potential is even more pronounced in the case of post-tensioned concrete. Few concrete buildings are designed to withstand the collapse of one floor onto another. When this occurs, progressive collapse is almost a certainty.

Hazards of Post-Tensioning

The tendons for post-tensioned concrete are laid out on the formwork, before the concrete is poured. When the concrete is being poured, samples are taken in standard sample cylinders and are sent to a testing laboratory. When the concrete reaches a designated fraction of its ultimate strength, hydraulic jacks are used to tension the tendons or jack the cables. Until this is accomplished, the entire weight of the concrete is on the falsework. There is no bond between the tendons and the concrete—quite the contrary, the tendon is contained in a sheath to prevent bonding.

The weight of the concrete is transferred to the columns only when the tensioning is completed, several days after the concrete has been poured.

A cured concrete bridge awaited tensioning in California. Apparently, failure of the falsework caused the total collapse of the bridge, killing a man driving underneath it. An adjacent bridge in the same stage of construction was immediately and successfully tensioned. An article describing fire department operations mentioned cutting cables with a torch. If the cables are under tension, the use of a torch to cut them would cause the cut ends to whip violently.[7] Construction workers have been killed by such whipping tendons.[8]

The 22-story Skyline Towers mentioned earlier was made of conventional reinforced concrete. The garage, located behind the Towers, was of post-tensioned construction. A crane hit the corner of the garage. Both floor slabs sheared completely off the columns. Shear resistance of post-tensioned slabs is poor.

In California, the Las Lomas bridge suffered a post-tensioning failure as tendons were being tensioned to 850,000 psi. The concrete had set for more than 28 days and exceeded its design strength of 3500 psi.[9]

Collapse of Reinforced Masonry

Since large numbers of masonry units are used in construction, it is easy for workers to overload a portion of a floor with brick or block. This happened in Pittsburgh, Pennsylvania. The excess load caused the partial collapse of several stories of precast floors.

Collapse of Precast Concrete

Precast concrete buildings under construction are very unstable until all connections are completed. Temporary bracing, such as tormentors (adjustable poles set diagonally) or cables, holds units in place. In some cases, wooden temporary shoring is used. The completed structure may be of pinned construction, with bolt-and-nut or welded connections, or may be made monolithic by the pouring of wet joints to join precast units together.

In Montgomery County, Maryland, a three-story parking garage nearing completion totally collapsed because an oversized washer was used on a bolt.

Lift Slab Collapses

The recent fatal collapse of a concrete building under construction in Bridgeport, Connecticut with the loss of 28 lives has focused atten-

Fig. 8-11. This precast concrete parking garage was almost completed. It collapsed because a single connection failed.

tion on this method of construction. Lift-slab construction, also known as the Youtz-Slick method, was first used for the construction of Trinity University in San Antonio, Texas in the 1950s.

In lift-slab construction, the ground floor slab is cast first. The other floor slabs and the roof are cast on top of the ground floor slab. Bond breakers are used between the slabs so they will not adhere to one another. When the slabs reach design strength, they are raised up the columns by hydraulic jacks placed on top of the columns. All the jacks are controlled from a central console. As each floor reaches its final position, it is temporarily connected to the columns.

The Bridgeport collapse was due to the failure of a single connection. Commenting on the collapse, Engineering News Record questioned why critical connection details do not get enough attention from engineers. It further noted the need for greater attention to temporary loads during construction, and designs that provide enough structural integrity so that local failures do not trigger progressive collapse.

Many of the accidents which have occurred while using this method have happened while the slabs were being lifted. In some cases, however, failure occurred when no lifting was being done.

A Cleveland, Ohio garage was almost completed. Steel collars were cast into the floors to be welded to the columns. They were temporarily wedged when high winds struck. The structure was seven feet out of plumb but was successfully jacked and pulled back to plumb. The floors were to be welded to the columns.

In San Mateo, California a roof slab was being lifted 16 feet. The columns were not braced — only bolted to the foundations. As the slab reached the top, the columns were 3 inches out of plumb. As an attempt was made to pull the slab back into place, it collapsed. The designer blamed the lack of bracing and the lateral shifting of the slab for the failure.

As a result of the Bridgeport collapse, OSHA has ruled that only essential workers be on site when lifting is being done.

Tactical Considerations

It is imperative that the fire suppression forces be up-to-date on the state of all buildings under construction, but particularly those where progressive collapse is almost inevitable.

It is reasonable for command to refuse to place fire fighters on or under such a structure until all the floor-column connections are permanent.

Fire Problems of Concrete Buildings Under Construction

Concrete buildings under construction can present serious fire problems. Fire in falsework can easily result in major collapse with potential for massive life loss. There is little reserve strength in falsework. There seems to be little understanding by designers, constructors or even fire officers of the potential catastrophe which could be caused by a fire in falsework. A few knowledgeable people are astonished that fire fighters would place themselves below burning falsework which is supporting the weight of a concrete floor.

Tactical Considerations

How many fire chiefs could make these statements?

"At all times our fire suppression forces are completely aware of any concrete which is supported on fire-vulnerable falsework."

Or, "Our operating procedures are clear and explicit. No personnel are to be in any location where they are endangered by the potential collapse of the structure during a fire in a concrete building under construction."

Few if any departments have these stated policies. While such procedures may be widely adopted as policy after a terrible disaster occurs, they are not universally used. There are always those who do not get the word. Or, as one chief said after a midwest fire with multiple fire fighter fatalities, "I learned nothing. I would use the same tactics again."

The matter of concrete failure cannot wait until the command structure is set up at a fire, or the building department is called. The disaster may involve the first arriving units making a conventional attack.

An Unwelcome Recommendation

It is a tough thing to convince an experienced or lucky fire officer that there are situations which demand defensive operations immediately on arrival — and this is one of them.

Fire chiefs should determine exactly what their suppression forces would do if they responded to a falsework fire in a concrete building tonight.

Could they suffer a terrible disaster by following their usual aggressive tactics?

If the answer is YES, they should change their tactics before they suffer a tragedy.

Tactical Considerations, continued

There are those who do not believe that such a terrible disaster as the loss of scores of fire fighter lives in a collapse of falsework could happen. Just ask an architect or structural engineer.

Fire Potentials

There are numerous fire causes present at a construction site: welding, cutting and plumbers' torches, temporary electrical lines, and arson. Fuels, such as wooden falsework, flammable liquids, and plastics for insulation, are readily available. Glass-fiber formwork is also combustible. This is often a surprise to those handling it.

The most dangerous hazard may be from heating. Sometimes workers burn scrap wood in steel barrels or use kerosene heaters. By far the most dangerous and, perhaps most common heating method, is

Fig. 8-12. The construction crew on this site is in great danger. Note the unsecured LPG bottle in the center of the photograph. To the lower left of it is a lighted open-flame heater (salamander). If the bottle falls, the valve could break off and spill the LPG. A ball of fire would ensue, destroying everything and everybody around it.

the use of liquefied petroleum gas (LPG). In order to retain heat, the building is sheathed in plastic. Inside the heated area, unsecured gas bottles are present with the nearby open-flame salamanders in which the gas is burned. When this is the case, the stage is set for a tremendous disaster.

Most codes for the use of LPG require storage of the gas away from any open flames, for good reason. In 1963, 74 people died in an LPG explosion at the Indianapolis Coliseum. A gas-fired cooker was being used under the concrete stands. A leaking cylinder exploded when the gas reached the flame of the heater. According to a rescuer, one victim was found in the center of the arena with a huge piece of concrete on him. Do not be deceived by the fact that the enclosure is of plastic. It may vent and preserve the building, but all in the area may die.

Two construction workers died in an explosion of propane gas leaking from a fixed system in apartments under construction in Florida. The fire chief commented that before this tragedy the fire department had responded to gas leaks without a second thought about the danger.

In one west coast city, the practice is to store the gas bottles at ground level and pipe the gas with plastic tubing to the salamanders (heaters). An **excess flow valve** should be installed. Such a valve senses a sudden increased flow, as from a broken line, and shuts off the gas.

The situation is not necessarily improved when non-combustible falsework is used. Aluminum falsework in combination with glass fiber forms is a popular combination. The aluminum is subject to failure by melting, as it did in a recent fire in Virginia.

Steel-tube shores are better but can fail from high fire loads such as from gas cylinders or lumber piles.

The fire hazard represented by the massive wooden formwork used in the dome of St. Mary's Cathedral in San Francisco was recognized. Temporary automatic sprinkler protection was provided for the lower portion of the formwork. Some questions were raised as to whether, in such an open structure, a fire would move upward ahead of the opening sprinklers.

Nothing is on record as to the effect of fire or hose streams, alone or in combination, on green or curing concrete.

In August, 1984 The Miami, Florida fire department fought a spectacular fire in a concrete high-rise under construction. Eleven 2½-inch lines, and three and a half hours were required to suppress the fire.

A number of important points were made in an article discussing the fire by R. Milnes in *Firehouse*.[10] They are given below as part of the lesson to be learned.

Fig. 8-13. This spectacular falsework fire occurred in Miami, Florida. (Chief Teems, Coral Gables, Florida Fire Department)

- The standpipe was not in operating condition. Hose was transported to the 10th floor level of the parking garage by car, then by hand to the 19th floor. Lines were also stretched from extended aerials.
- The 19th floor was piled with 280 tons of reinforcing steel and oil-soaked wood forms.
- In an almost unique case, the building department is a division of the fire department. This made fast cooperation possible.
- Burning debris falling down shafts set fires on almost all floors down to the tenth floor. Burning debris also fell on fire units operating in the street.
- The killing of power cut off all elevators.
- Smoke required the evacuation of a nearby hotel.
- The wind was light, lessening the exposure to other buildings from flying brands.
- Fortunately, construction had stopped so there were no workers to be evacuated. Because it was not a workday, traffic and onlooker problems were lessened.
- Chief Teems felt that it was fortunate it was a daytime fire. Darkness would have concealed open shafts and other hazards and possibly caused injury or death to a fire fighter.

The lesson to be learned is that such fires require specialized training and comprehensive pre-planning.

Hazards of Post-Tensioned Concrete

Post-tensioned, prestressed concrete construction presents catastrophic fire collapse potential. It is vital for a fire department to be aware of any post-tensioned structures being built in its area and recognize the clues which indicate post-tensioning. Structures include bridges, parking garages and other structures.

Post-tensioned concrete presents a greater catastrophic collapse hazard during a fire than conventional reinforced concrete. As noted, the entire weight of a post-tensioned concrete floor or beam rests on the falsework until the tensioning transfers the load onto the columns. A falsework fire, therefore, could cause the sudden collapse of an entire concrete floor slab, or the entire structure, onto fire fighters. It is possible that several floors may be awaiting tensioning, as was the case in a South American building. If a fire in such a building was being fought conventionally, and a collapse occurred, the casualties would approach those of the bombed barracks which collapsed in Lebanon in 1983, killing 225 U.S. Marines.

After tensioning, the ends of tendons are left hanging and exposed for varying periods of time.

In the case of floor slabs, several feet of excess tendon may protrude from the socket until cut off and the socket is dry packed.

In the case of beams, the anchors may be left exposed for further tensioning, or due to normal construction delays. In either case this represents a serious threat to the stability of the building during a fire. Hanging tendons can act as heat collectors and will fail at about 800°F.

When a floor is tensioned in increments, the excess tendons are rolled up and attached to a wooden rack. In a fire the rolled-up tendons would act as a heat collector, transmitting heat to the previously tensioned portion of the tendons.

The substantial fuel loads present at construction sites or from exposed buildings may provide the heat to cause tendons to fail. Failure of tendons will cause the collapse of that part of the structure.

In one case, a high-rise motel consists of 13 floors of cast-in-place concrete framing rising above a transfer floor where the column loads are rearranged to provide specified column spacing. There were several huge post-tensioned transfer beams. When heavy loads are carried on tensioned beams, the entire prestress cannot be applied at once. The prestressing takes place in increments as the building load increases. The tension anchors were left exposed for months on the third-floor level of the motel and in pits within the building to facilitate successive tensioning. The ends within the building were particularly vulnerable to an ordinary construction fire. A loss of tension

Fig. 8-14. In a falsework fire, destruction is total. Fire fighters must be extremely cautious. (Chief Teems, Coral Gables, Florida Fire Department)

in the major beams would probably have caused total collapse of the motel. Unfortunately, this is a common practice. (See Fig. 8-6.)

A Texas fire department found several post-tensioned buildings under construction. In one case, wooden falsework potentially exposed partially-tensioned tendons to severe heating. The prefire plan called for apparatus to proceed to a fire in the falsework through a passage under the partially tensioned building. The plan was changed.

Protection of Tendons

There is one opportunity to control a disaster before it occurs. Insist on fireproofing tendon anchors immediately after tensioning is completed, and on the provision of temporary protection for incrementally tensioned tendons. Instruct all personnel in the hazards of to-be-tensioned concrete.

Tactical Considerations

In a fire, it is vital that any exposed steel connected to stressed tendons be cooled immediately for the duration of the fire — if this can be done from a safe position.

A Total Collapse

A post-tensioned building under construction in Cleveland, Ohio suffered a falsework fire. The fire department operated from the adjacent completed section, extinguished visible fire and notified the contractor that he had a problem. A second fire occurred. As the fire department was setting up, the entire 18-story building collapsed. No fire fighters were injured. They had been kept out of the collapse zone.

Tactical Considerations

If a concrete building is to be built in your area, learn whether it is to be post-tensioned, and plan accordingly. You may find that the construction superintendent will be incredulous when you speak of total collapse. Press the superintendent for facts. Much post-tensioning is done. Without planning it may only be a question of time before a disaster occurs of the type discussed here.

Hazard Awareness

Many professionals are unaware of this tremendous potential for disaster. Recently *ISFSI Instructor of the Year,*[11] Chris Naum, an architect, cautioned fire fighters about the hazard of operating under fresh concrete. His article drew a reply from a representative of the Portland Cement Association,[12] who argued that concrete, in place for two to seven days and properly curing, should be able to support its own weight. Extreme caution should be exercised, however, when concrete is less than 48 hours old.

The response made no reference to post-tensioned concrete, which may set for several days before it is tensioned, and the load is transferred to the columns.

Almost twenty years ago, in 1972, this author presented this concrete hazard at an NFPA meeting in Philadelphia. The paper discussing the hazard was summarized in *Engineering News Record.* Nobody from industry contradicted it, but the editor of a fire publication rejected the paper saying, "I can't believe they are building buildings which will completely collapse."

Tactical Considerations

Learn to recognize the signs of post-tensioned concrete construction: protruding cables, coils of tendons, characteristic anchors. If the fire department alarm office functions as a critical information center, "post-tensioned construction underway" should be flagged as a hazard item.

Tactical Considerations, continued

Do not let the emphasis on post-tensioned concrete hazards lead to complacency with the hazards of conventional concrete.

Preplanning

Fire departments should keep updated on what is happening on construction jobs. It is a good idea to meet a designated and qualified representative of the builder. Consult the constructors in developing preplans.

However, bear in mind that many construction accidents have occurred under the supervision of supposedly qualified people.

In my experience, some field construction supervisors resent intimations of possible failure. The construction engineer of the Westgate Bridge in Melbourne, Australia was killed along with 31 workers when the bridge collapsed in 1970. Reports at the time indicated that the engineer had gone out onto the bridge to calm the fears of workers who were concerned about the stability of the structure.

Precast Buildings

When precast concrete units are being erected, temporary bracing or support is provided until the full space-frame relationships of the building are completed. Columns are usually braced by telescoping tubular steel braces (tormentors), but wood braces are also used. Wooden falsework is often used to support horizontal elements until connections are made. The loss or upset of any such temporary support may precipitate a general collapse. Cold drawn steel cables (which fail at 800°F) often provide diagonal bracing in precast buildings.

In Orange County, Florida, a tilt slab building with a steel bar-joist roof was almost finished. The only fuel available was a pile of foamed plastic insulation. A fire occurred and the elongating steel roof pushed the walls out of alignment. The first action taken by the builder was to replace the tormentors since the roof was no longer reliable. Tilt slab buildings require a roof for stability.

Cantilevered Platforms

Cranes are often used to deliver construction materials to buildings under construction. This necessitates a temporary cantilevered platform extending out from the floor. A common method of securing such a platform is to brace it into place by the use of wooden shores to deliver a compressive load to the floor above. A fire involving such a platform would destroy the shores and drop the platform to the street. Keep all personnel and apparatus well clear of the collapse zone.

Fig. 8-15. The elongating steel roof trusses displaced the tilt slab walls, making them unstable. The builder immediately replaced the tormentors to brace the walls.

Tower Cranes

The high tower crane, a common sight in modern construction, seems to climb up the building. As it climbs, it is supported on the building's structural frame. The weight of the crane may be distributed over several floors by falsework. A fire involving this falsework can bring down the crane.

Falsework on an otherwise apparently completed floor should be investigated. It may be supporting the patch over a hole that was left in the building to handle or transport materials, or it may be supporting a heavy load such as the crane.

In one example, formwork for concrete placement burned on the 23rd to 25th floors of a high-rise under construction. Flaming debris fell to the street, and the tower crane operator was trapped in his cab above the 28th floor. Fire fighters protected him with a heavy-caliber stream from a nearby roof until an engine company fought its way through the fire to save him.

Tactical Considerations

Safety First

There is no point in risking personnel in a concrete building under construction. All concrete involved in the fire will be torn out.

Tactical Considerations, continued

It is almost impossible to predict the effect of the loss of even a small portion of the falsework in a fire. Command officers should err on the side of caution. Heavy-caliber streams from safe locations deserve serious consideration. The falsework structure cannot survive the loss of material caused by a fire.

In all buildings under construction, move carefully and deliberately. Floor openings may be inadequately guarded or covered over with a flimsy piece of plywood. Side rails may be nonexistent or of poor quality.

Fire Problems in Finished Buildings

Fire Resistance

When concrete construction was first developed, it was regarded by some as the answer to all fire protection problems — it was believed to be truly fireproof. After a series of disastrous fires it became evident that concrete, like any other non-combustible material, can be destroyed by fire given sufficient fire load and fire duration.

Concrete is inherently non-combustible. It may have been fabricated to meet a fire-resistance standard. Unfortunately, many people, including some who should know better, confuse non-combustibility with fire resistance. Neither is synonymous with fire safety.

Concrete construction carries no guarantee of life safety. The reinforced concrete Joelma building in Sao Paulo, Brazil burned in 1974 with a loss of 179 lives. The structure escaped with relatively minor damage. Factors other than the structure itself, such as interior finish, and the inability of the occupants to quickly reach a safe nontoxic environment, may be far more important to the safety of the occupants.

Using computer techniques, the concrete industry can sometimes induce (negative) bending stresses away from the steel by varying restraint conditions. Channelling the larger stresses to the top of fire-exposed members reduces the hazard of overheating steel.[13]

Concrete construction can be made to deliver different levels of fire resistance. Steel fireproofing (an inaccurate but accepted trade term for the insulator which protects steel) is accomplished by use of a separate material. The fireproofing is a separate line item in the building budget. The fireproofing of concrete is integral, that means it is accomplished by a specified mix of concrete in a specified thickness. Some concrete, particularly that below the reinforcing rods, is not necessary for structural strength but is necessary for fireproofing. Because there is no budgetary distinction regarding the material, this may give concrete an economic advantage.

The ratings given for concrete columns in the National Building Code of Canada were based on 1920 tests. In 1979, the Portland Cement Association and the National Research Council of Canada started a series of 41 full-scale tests on concrete columns. The tests provided basic data that developed and validated mathematical models to calculate fire resistance for a number of variables.[14]

Concrete is made in the field. Its only test of fire resistance is the fire. An experienced concrete supervisor built a concrete fire-demonstration tank for the Navy Shipboard Fire Fighting School in Panama. It survived hundreds of fires — beautifully. His associate, using the same specifications, built a concrete pit. It spalled explosively during every fire.

Fire fighters are indebted to Robert B. Gore, PE, consulting structural engineer, for the following:

> The concrete in a concrete beam carries the compressive stresses at the top of a beam. Reinforcing steel carries the tension loads in the bottom of a beam. The design of a concrete beam is said to be balanced when the strength of the concrete is equal to the strength of the beam. If more steel is put into the beam than is required for balanced design, the steel fails first. Concrete fails without substantial deformation. Steel stretches considerably before failing. The stretch in steel warns of imminent failure.
>
> ACI (American Concrete Institute) 318-77, paragraph 10.3.3, limits tension steel in bending members to 75% of the amount that would produce a balanced design. This provision assures that failure in bending will be preceded by large deflection before collapse occurs.

Tactical Considerations

In evaluating the hazard potential for fire in a concrete building, the first determination should be whether the building was intended to be fire-resistive. If not, then the concrete is merely non-combustible and there is no basis for assuming its integrity in a fire. In addition, the building may not be made totally of concrete. To cite one instance, a non-combustible concrete building can have a lightweight, wood-gusset plate, or unprotected steel bar joist, truss roof.

Signs of Trouble

There are clear signs of possible trouble which can be determined in a prefire plan. Deteriorated concrete, spalling which exposes reinforcing rods, and cracks in concrete which can admit corrosive moisture to the reinforcing rods are indications that the building is in distress.

Many chemical processes give off fumes which can damage or destroy concrete. Chlorine corrosion caused the collapse of a concrete roof of a swimming pool in West Roxbury, Massachusetts. Aluminum also reacts with concrete. Aluminum electrical conduit buried in concrete did severe damage to a sports stadium.

The concrete roof of a paper manufacturing plant in Brunswick, Georgia was corroded by fumes. The roof was replaced with over a thousand foam core panels.[15] They are supported on painted steel purlins. Reinforced plastic structural members were available for this environment but in this case they were rejected due to cost.

Parking garages in areas where salt is used to melt snow and ice are particularly vulnerable to corrosion. The extent of damage is often difficult to determine. In some non-freezing areas calcium chloride added to the concrete has caused problems. Preventive measures include sealing the concrete, providing adequate drainage, and flushing surfaces with fresh water in the spring.

Rehabilitation of concrete has included removal and replacement, installation of cathodic protection and additional steel beams to shorten the span of concrete slabs. Some garages have been demolished.[16]

Concrete deterioration came to attention tragically in a collapse of a Miami, Florida office building. A slab had been added to the roof to surface a parking area. The building collapsed with the loss of seven lives. This prompted a unique program in which the building department inspects the structural integrity of existing buildings. It was found that some of the buildings were beyond economical repair and required demolition. One common cause of the concrete failure was the use of sea sand with a high salt content. The salt corroded the reinforcing rods.

Tactical Considerations

Reinforced concrete is a composite material. Any failure of the bond between the two materials means that the composite has failed. Then you do not have what was intended, you have only dangerous dead weight.

Unprotected Steel

When a concrete structure is in trouble, it is often repaired with steel. In the case of fire-resistive buildings, very often no consideration is given to providing equivalent protection for the steel. If steel cables are used, failure will occur below 800°F.

Fire fighters should be ever watchful about what is being done to buildings. Almost none of what is done to a building after it is completed, except for installation of automatic sprinklers, benefits the fire suppression effort.

A fire fighter student saw steel columns being brought into a three-hour rated fire-resistive shopping mall. The columns were prepainted to match the wall color in the mall. The roof was failing and the columns were provided to shore it up. The owner did not want the building department to know of the problem.

If the columns were structurally necessary they should have the same protection as the structure. After the case was reported to the building department, it took several years to get the columns protected.

It appears that if steel is initially designed into the structure, the proper degree of fireproofing is usually specified. If it is an afterthought, or part of a repair job, it will usually be unprotected. Steel connectors for precast units are often unprotected.

An alteration of an old Florida hotel shows a practice which may be common. In altering the building, it was decided to convert the high-ceiling ballroom into two floors. A steel bar-joist floor and ceiling assembly was used because of its relatively light weight. Altered buildings in particular may not be consistent throughout in their construction.

Heat Absorption and Fire Load

Concrete structures represent a massive heat sink compared to other structures. This is beneficial because every Btu that heats the concrete is one less which is available to support the fire. However, heat fed back from the concrete can seriously affect the fire suppression forces. This problem should be recognized. It was a serious problem in the Los Angeles Central Library fire and the Smithsonian Museum fire.

Tactical Considerations

Delay overhauling the structure until it has cooled down enough to minimize the debilitating effects of heat on personnel.

Fire Load vs. Fire Resistance

In any fire-resistive building, the existing fire loads should be compared with the fire-resistance rating of the structure. Where high fire loads exist, and where fire suppression efforts will be inadequate, severe structural damage is a distinct possibility. Furthermore, because of their great structural strength, concrete buildings can be heavily loaded with combustibles and thus the fire load may be excessive.

Ceiling Finish and Voids

One of the fire protection advantages of concrete construction is that there are no inherent voids. While voids may be created in the finishing of the building, they are not a necessary part of concrete buildings, although they are useful to the occupants.

A fire in the void between the hanging plaster ceiling and the concrete floor of a former department store was difficult to find. The fuel was an eighty-year accumulation of dust.[17]

Waffle slab concrete was originally considered ugly and was "improved" aesthetically with a suspended ceiling. Later architects appreciated the coffered effect and emphasized it. In one mall, imitation plastic waffle concrete is suspended below the structural slab.

Unfortunately, when ceilings were installed, combustible tile with a high fire-hazard rating was often used. The combustibility is not intended; it just happens that combustible tile is cheaper than low fire-hazard rating tile. The interconnected voids above the tile make it possible for the tile to burn on both sides simultaneously.

Tactical Considerations

When inspecting a concrete building under construction or renovation, note if air supply diffusers are hung below the floor above, with the finished outlet in place. This indicates a hung ceiling will be installed in the future. Warn owners against the dangers of combustible tiles and concealed spaces.

When suspended ceilings are installed as part of initial construction, it is more likely that they will have satisfactory fire hazard characteristics. But it is necessary for fire officers to be aware of the local code requirements. The tile usually will be only as safe as the law requires.

Combustible tile need not be suspended to create a serious hazard. Tile applied with a flammable adhesive to a non-combustible surface was a key factor in a fatal hospital fire.

Installing new ceilings which meet flame spread requirements below old combustible tile ceilings present a serious hazard (see Chapter 9, Fire Growth). Of particular note is the 1989 fire in the John Sevier Retirement Center in Johnson City, Tennessee, where 16 occupants died in a fire-resistive concrete building. A key factor was the concealed combustible acoustical tile.

Combustible voids can be created in a variety of ways. A heavy wooden decorative suspended ceiling is located in a restaurant of otherwise concrete construction. Sprinklers are below the ceiling. Fire could burn unchecked in the void. The fire resistance of the ceiling hangers, which are attached to anchors (possibly lead), is uncertain. A fire might cause the collapse of the heavy false ceiling. Decorative or functional wall paneling and screens may also create combustible voids.

While voids are not inherent in concrete construction, they can be very useful. A church was built of concrete without voids. When the church was to be air conditioned, the only solution was to run the ducts on the roof. This created costly maintenance problems. In a modern office building, with its huge communications and other utility requirements, as much as one-third of the height from floor to ceiling may be in ceiling or under floor voids.

Even when the void is non-combustible, it may still present problems. Combustible thermal or electrical insulation, and combustible plastic service piping may be concealed in the ceiling void.

Since the fire resistance of a building is achieved by the concrete construction, the hung ceiling, though it may be identical to the ceiling used for membrane fireproofing of steel, is generally not required for the structural integrity of the building. The owner is free in this case to remove or modify the building within code requirements for flame spread. However, there are some rated fire-resistive assemblies which combine precast concrete elements with suspended ceilings. In these cases the maintenance of the ceiling is necessary for the fire resistance of the floor-ceiling assembly.

Tactical Considerations

Do not fail to examine the ceiling void early on when seeking an unknown source of smoke.

Combustible Trim

The current popularity of wooden interiors is illustrated by the widespread use of nonstructural wooden beams in many buildings. These beams may be solid wood, wood boxes, or easily-ignited and combustible polyurethane made to resemble wood. They are held in place by different methods, many of which may not have any rated degree of fire resistance.

Ironically, the auditorium of the headquarters of the American Institute of Architects, an unsprinklered fire-resistive building, has large wooden beams hung on the ceiling for decoration.

Tactical Considerations

Expect such "architectural enhancements" to come down. Do not be under them when they do.

The Integrity of Floors

If a building is fire resistive, it is presumed that the floor will be a barrier to the extension of fire. More and more, however, codes are requiring sprinkler protection, in addition to the passive protection provided by compartmentation. Many older buildings were constructed under the less costly but inadequate concept that compartmentation was all that was required to limit fires.

The concept of isolating the fire to a limited area has great appeal, and on first glance appears to have great merit.

However, the compartmentation is rarely achieved. Even in a limited area, a high fire load may yield a difficult, costly fire. One high-rise fire was confined to the small area permitted by the local code. Nevertheless, it took 500 fire fighters to confine the fire, and the loss was over $40 million.

Building use often requires that the floor be penetrated for utilities, plumbing, heating, air conditioning, electrical service, and communications wiring. There are proper ways to do this so that the integrity of the floor is not compromised. Unfortunately, they are not always used. Further, some of the methods used are open to question.

This author has seen enclosures which had gypsum board on one side only. The rated enclosures have gypsum board on both sides. Sometimes taping and nail covering are omitted except where the surface

Fig. 8-16. In this telephone/electrical room, the conduit is aluminum, and a handful of rock wool is stuffed into the unused holes in the ceiling. The enclosing partition is broken out to pass wiring into the building void, and the building isn't sprinklered — a typical recipe for disaster.

is to be painted. Taping and covering nails are integral parts of the fire-resistive enclosure.

For whatever reason, enclosures around ducts may be inadequate, and have failed in fires. This can permit transmission of fire and/or smoke to other floors. In addition, failure of the enclosure may expose fire fighters to the danger of falling down the shaft.

"Poke-throughs" are holes provided to draw utility services up to a floor from the void below. The penetration of the floor for poke-throughs may negate the fire-resistance rating of the floor. An article in *Architectural Record*[18] quotes Chief John Degenkolb explaining that even if utility services openings are properly protected in the initial construction, tenants may ask for additional services, and that the common method of providing them is to drill through the floor. Where this is recognized as a hazard, the opening is often stuffed with mineral wool. Tests showed that this is ineffective. In addition, the article cites the case of a fire high up in a high-rise building in California. The water that was used spread downward to all floors via poke-throughs.

In addition, any holes that violate the integrity of concrete floors may provide passage for deadly fire gases. It is not safe to assume that there cannot be an extension of fire from floor to floor in a building because the floors are of concrete. Be aware of the local practices in providing utility services.

As penetrations of floors for services increase, the floor may be unable to resist the passage of fire adequately, and the building department may require a suspended ceiling, which, together with the floor, will hopefully develop the necessary fire resistance. In this case, the owner is not free to modify the ceiling. Many tested and rated assemblies are floor/ceiling assemblies and, therefore, cannot be modified.

Tactical Considerations

If it is known that utilities were installed in this manner, plans should provide for immediate deployment of units to the floor above the fire floor to control extension, and units should be assigned for salvage work on the floor below.

Concrete floors require expansion joints. In one case, steel expansion joints transmitted fire from floor to floor in a huge postal building. Aluminum expansion joints dropped molten aluminum into the lower level of McCormack Place, and extended that great fire. Combustible fiber expansion joints can provide a difficult fire problem.

Concrete shrinks, and this shrinkage may create large cracks. Such cracks may permit the passage of fire from floor to floor, even though the concrete is structurally sound. Shrinkage, distortion, or settlement may put door frames out of alignment, thus preventing the closing of fire barrier doors.

A Management Gap

This discussion points up a common gap in the public management of the fire problem. The building official, who approves the building as built, rarely ever sees it again. The fire official who inspects the building is often uninformed of building code requirements or has no authority to enforce them. Without coordination, an entire building and its unsuspecting occupants may fall through the gap.

Wall Panels

In many modern steel and/or concrete-framed buildings, prefabricated wall panels of various design are hung on the exterior of the building. The resistance of the joint where the wall panel abuts the concrete floor slab is questionable. Expansion of the panel may open a gap along the wall and permit the passage of fire.

Precast and imitation concrete panels have been sealed with foamed plastic at the point where they meet the floor. The plastic, even if inhibited from flame, will melt in a fire and permit extension along the perimeter of the building.

Exterior Extension

Even if the interior compartmentation were to be perfect, the hazard of exterior extension still exists. Those who saw huge volumes of fire pouring out the windows of the First Interstate Bank fire in Los Angeles in 1988, will have no difficulty understanding the potential of exterior extension in an unsprinklered building.

Tactical Considerations

If fire extension to the floor above is a possibility, and if the building is unsprinklered, the prefire plan should consider putting a few inches of water on the floor above if the available forces cannot control the fire. Again, bear in mind that the fire fighting forces are doing only what a sprinkler system would do if it had been installed. Do not allow fear of unnecessary water damage to cause the loss of a building.

Before any fire, it is appropriate to discuss with the owner/occupants the implications of the tactics made necessary by

Tactical Considerations, continued

defective construction. This might result in corrective action. At the very least, it cuts the ground out from the statement, "If you had only told me that, I would have . . . ."

Imitation Materials

Imitation concrete panels are commonly used, particularly on the exterior of buildings. In one case, they are made of two sheets of asbestos board, with a combustible foamed plastic interior. Epoxy glue is painted on the surface which is then coated with small stones.

The top panels of some buildings are genuine precast concrete. The wall panels are usually imitation concrete. At the base, however, concrete block panel walls are painted with epoxy adhesive and stones are added. The entire building appears to be of identical finish. The fasteners that hold the panels on the building are held in the plastic. If the plastic burns or melts, the panels will drop off the building.

Most imitation concrete panels are made of a steel framework with asbestos millboard. Real concrete is applied in a thin coat, or stone is epoxyed on.

Energy conservation has brought about the use of Exterior Insulation and Finish Systems (EIFS) that give a building the appearance of sand-finished concrete. One type starts with exterior gypsum board screwed to steel studs. Glass-fiber insulating batts are placed between the studs. A foam panel is then glued or mechanically attached to the gypsum board and a woven glass-fiber mat is laid over the plastic panels. A layer of adhesive secures the glass fiber. Usually, a coat of organic polymers is added to the finish coat.

Buildings can be finished in this manner when constructed, or EIFS can be added to the building for energy conservation or to change its appearance.

A Ft. Worth, Texas courthouse was finished in this manner to make its appearance more consistent with the 19th-century courthouse next door.

A Victoria, British Columbia hotel was finished this way during construction.

In March 1985, a major fire occurred in the top floor of an old brick- and wood-joisted building in Manchester, New Hampshire. The fire extended to the surface of an EIFS-finished building 20 feet away. The chief said the fire extended to the entire upper wall area immediately. Fortunately, a ladder pipe was in position and the fire was readily extinguished.

In January 1990, a fire involving 24 vehicles in the parking area of a Winnipeg, Manitoba apartment building extended several stories up the EIFS face of the apartment house.[19]

Tactical Considerations

Fire departments should be aware of combustible surface finishes and the possibility of vertical extension. Prepare to protect them with hose streams.

If it is proposed to use such a system out of reach of exterior hose streams, the owner, insurer and building department should confer to determine a course of action in the event of fire.

Partitions

It is unsafe to assume that partitions, whether masonry or gypsum on steel stud, provide a fire- or smoke-tight seal against the floor above, particularly where a suspended ceiling is installed.

Even if a tight seal were designed, installation of utilities may have created penetrations through the barrier. The same problem occurs in the case of firestops.

Concrete's Behavior in Fire

The concrete in fire-resistive construction serves two purposes; it resists compressive stresses and protects the tensile strength of steel from fire. The latter function is accomplished in proportion to the thickness of the concrete cover. By sacrificing itself, the concrete provides time to extinguish the fire; however, it may also lose its strength.

Tactical Considerations

Any concrete subjected to severe fire attack should be avoided. If holes must be drilled or cut through a concrete floor to reach fire below, personnel should be supported independently of the damaged floor panels. Light-weight scaffold flooring might be a useful addition to the fire department's equipment.

Impact loads to damaged concrete, such as from sledges or jackhammers, should be carefully evaluated. All areas where spalling has reached the reinforcing steel should be shored before any work is done in the area. Concrete floors which are seriously damaged on one side may give no clue to the distress on the other side.

Cutting Tensioned Concrete

Fire tactics sometimes include cutting through a concrete floor to operate distributors on an otherwise inaccessible fire. One department's high-rise unit carries a hole cutter.

Fig. 8-17. A typical example of the disintegration of massive concrete columns in the Chicago warehouse fire. (Tom McCarthy)

It is possible to cut a hole in conventional reinforced concrete and, in the process, to cut through reinforcing rods without doing serious structural damage.

This is not the case in tensioned concrete structures. The tensile strength is provided by steel cables under tremendous tension. These cables are not in contact with the concrete. Cutting such a cable would be like cutting any steel cable under tension. The ends might whip about with tremendous energy. The cutting of a cable also might cause a massive structural collapse.

In 1989, a worker was killed in the demolition of a post-tensioned garage in Washington, D.C. He and others were cutting cables when the upper floor collapsed.[20]

According to David Berg, vice president of a contracting company in Baltimore, Maryland, there are safe methods to use to demolish a post-tensioned concrete building. While taking down a garage in downtown Baltimore recently, Berg installed an engineered shoring system to support sagging concrete and steel as the structure was detensioned. Despite the precautions, the structure started shearing at some braces during detensioning, so more shoring was added.

Tactical Considerations

When a concrete building is finished, there are no clues to help the fire officer determine whether the floors are of conventional reinforced or tensioned construction. If the standed cables which signal tensioned concrete are found, they must not be cut. The cutting of even a single cable can be disastrous.

Joe Batchler, well known in fire training circles, worked for some years in construction. He related an instance where the X rays taken to locate tendons were read upside down, and a tendon was cut during alterations. It was a serious mistake.

Fortunately, tendons cannot be cut with bolt cutters as ordinary rods can. Officers in command of units equipped with cutting torches or power saws should be wary of any request for equipment to cut "reinforcing rods." To quote one expert, "It's just like sawing off the tree limb you are sitting on."

The core cutter or concrete blade must never be used to open a hole in a known or suspected post-tensioned floor. Beyond the danger of cutting the cable, tendons are cold drawn steel and fail at 800°F. Even if the tendon isn't cut, opening a hole over the fire would probably subject one or more tendons to temperatures which would cause failure. In short, all cutting into post-tensioned concrete floors in a fire is unsafe.

The National Association of Demolition Contractors (NADC) has been lobbying since 1982, without success, for some type of warning to be placed on tensioned buildings.

The NADC contacted both the American Institute of Architects and the Post Tensioning Institute (PTI) regarding the placarding of post-tensioned structures and retaining a permanent set of plans. They report that the PTI does not feel there is a danger and were not receptive to the request.[21]

As is customary, the fire service must take care of itself. Check your local building department. At least one major building department destroyed all plans after 5 years because it lacked storage space. This might be a convenient way to get rid of what might be embarrassing or even incriminating evidence.

Fire departments should make a permanent record of every post-tensioned building and incorporate this information into the prefire plan system because there is no way to tell them by appearance. It would be too late to discover this hazard after the disaster.

Precast Concrete

A cast-in-place, monolithic concrete building is very resistant to collapse. Because this type of building has a rigid frame, the loss of a column does not necessarily cause collapse. The load will be redistributed and other columns, if they are not overloaded, will be sufficiently strong to carry the load.

Precast concrete buildings are another story. The individual columns, floors, girders, wall panels (load-bearing or non-load-bearing) are usually pinned together by connectors. The fire resistance of the completed structure is very dependent upon the protection afforded to the connectors. Connectors are set into the precast element, and mating connectors are provided on the structures to which it is to be attached. Bolts and nuts may be used, or the connections may be welded. The recess provided for the joints is then dry packed with a stiff mortar. Often no protective covering is provided for the connectors; it may not even be required.

If an ordinary or non-combustible building is permitted by the code, a builder is free to raise a structure which may appear to be fire resistive, but which is not. Typically, in such a case, the connectors are unprotected. This problem should not be overestimated. Connectors must resist wind and other forces, and the fire load usually must be quite severe to cause their failure.

The possible effect of an explosion in a precast, pinned building should be evaluated where such possibilities exist. Such buildings have none of the redundancy which characterizes the rigid-framed monolithic concrete building.

Probably the most spectacular disaster involving precast concrete construction took place in England. It was the so-called Ronan Point collapse. A 24-story apartment building was constructed of precast floor- and bearing-wall panels — literally, a giant concrete house of cards. A gas explosion high up in one corner blew out one wall-bearing panel. This caused the collapse of the floor above which fell to the floor below. The impact load was too much for the next lower exterior bearing wall panel and it buckled. Thus, the corner of the building progressively collapsed down to ground level. There were five persons who died in the collapse.[22]

The gas explosion provided an impact load for which the building was not designed. The immediate safety measure taken was to eliminate gas service from buildings of this type. After the interim investigating report, local authorities were required to notify all residents of tall buildings whether their building was built by this system. Presumably, they could decide for themselves whether they wished to assume the risk. There is much to be said for this novel approach.

In 1984, tenants moved out of eight such buildings when an engineering inspection disclosed that a fire could affect the joints, already in distress. A fire test in an emptied building was stopped prematurely due to deflection of the floor above the fire. The building was demolished in 1986.

Tilt Slab Hazards

There is special hazard in tilt-slab construction industrial buildings which are very popular around the country. When these are being constructed, the builder carefully braces the walls with tormentors or braces. This is required for stability until the roof is in place and tied in, thus stabilizing the building.

The roof may be of any type of construction. If the roof is being lost in the fire, beware of wall collapse as has happened in warehouses from California to New Jersey.

Wood roofs may simply burn away, leaving the concrete panels free standing as a vertical cantilever. Concrete may spall from the bottom of T-beams exposing the tendons to their failure temperature. Steel may elongate significantly before failing. This elongation may push the walls down.

Examine openings in the wall carefully. Ordinary-size doors are usually placed in an opening cast into a panel. Loading doors may take the full width of a panel. The concrete panel above may be suspended from the roof, and subject to collapse.

If the building is sprinklered, careful observations should be made of the density and volume of smoke. If heavy smoke is being generated, then the sprinklers are probably not controlling the fire, and the roof is vulnerable. Massive and, in some cases, total-loss fires in a number of sprinklered warehouses have been especially disturbing to those familiar with the exemplary record of successful sprinkler operation. See Chapter 14, Rack Storage, for an extended discussion of this subject.

Bridges

Bridges are not designed to be fire resistive, and the concrete structure may be no more fire resistive than an equivalent unprotected steel bridge.

In March 1969, the Chicago Fire Department fought a fire in a truckload of scrap magnesium on the Dan Ryan Expressway under the 91st Street bridge. The magnesium burned violently for some time but because of several heavy fog streams, damage to the bridge overhead was averted.

Bridges which are entirely constructed of one huge hollow box girder present an interesting potential fire problem. The sections of bridge are joined with gaskets soluble in gasoline. Gasoline leaking from a tank truck could dissolve the gaskets and convert the bridge into an impromptu gasoline tank. This author was informed that this has already happened in a west coast city.

Fires in Concrete Buildings

The Los Angeles Central Library fire in 1986 was the largest loss of any library fire in the United States.[23] An estimated 200,000 books and numerous periodical collections were destroyed, along with the largest collection of patents in the western United States. Built in 1926, the building was of massive concrete construction. Its bearing walls are as much as three feet thick, with massive columns and thick roof and floor slabs. The book stacks were on seven tiers of unseparated, unprotected steel loaded to an estimated 93 pounds of fuel per square foot. This would be the equivalent of about 11 hours of fire exposure to the standard fire curve.

The fire required a tremendous commitment of apparatus, over 350 fire fighters and seven and a half hours to extinguish.

In a Glenwood Springs, Colorado building, a propane release exploded. The blast lifted the precast concrete roof, the walls fell out, and 12 people died.[24]

In Pittsburg, Pennsylvania, a concrete cold-storage building had been converted to store palletized paper products. Only after jackhammers

Fig. 8-18. The Los Angeles Library was a massive concrete building. Unfortunately, it lacked automatic sprinklers. (Alan Simmons)

provided by the local public works department were available, could the heavy concrete roof be opened. Even then, great heat from the fire forced defensive operations.[25]

In Chicago, Illinois, thousands of old tires had been dumped into an abandoned reinforced concrete building. In a fire, there was massive failure of walls and columns.[26]

Fire-weakened concrete can make secondary missiles when struck by heavy hose streams. The Mid-Hudson Warehouse in Jersey City, New Jersey was a large, heavily reinforced concrete building. In 1941, it was ignited by a massive exposure fire from a grain elevator. The sprinklers were not backed up so the fire involved several floors of the building.

On the second day of the fire, this author watched a New York City fireboat attack the upper floor fires with a stream from its five-inch tipped bowgun. Very shortly thereafter, sizable pieces of the building fell to the street on the other side.

In Sanford, Florida, fire completely destroyed a concrete self-storage building. The concrete was not required to be fire resistive.

A Florida shopping mall has one-story stores on either side of a one-high-story wide central corridor. The stores are constructed with unpro-

Fig. 8-19. These burning tires in a Chicago warehouse of heavy concrete construction caused massive destruction. (Tom McCarthy)

tected steel. The central corridor roof is of prestressed concrete T-beams supported on concrete columns. Fire involved the built-up metal deck roofs and displaced the columns. This dropped sections of the roof, fortunately without injury. The metal deck roof fire was controlled by a heavy-caliber stream set up inside the building.

A Christmas tree fire in a 16-story reinforced concrete high-rise apartment building in Dallas, Texas resulted in $340,000 in damage and the potential for a large loss of life except for competent preplanning by the fire department. Utility and vent pipes were punched through the ceilings. The gypsum and steel stud enclosures extended only to the suspended ceiling, not to the underside of the slab. Lighting fixtures and air-conditioning grills pierced the ceilings. Thus, smoke had an unobstructed path from the eighth floor of origin to the floors above.

The Military Records Center near St. Louis, Missouri was severely damaged in a fire in 1973. The building was of concrete construction, six stories in height. The top floor, where the fire occurred, had over 200,000 square feet of undivided floor space. The incredible fire load included over 21 million military personnel files in cardboard boxes on metal shelves.

The fire destroyed the top floor, which was subsequently removed from the building. The fire caused the roof to elongate several feet almost all around the perimeter. The elongation caused top floor columns to shear off at either the top or bottom. It should be noted that the portion of the roof which collapsed held together for about 22 hours after the start of the fire. The building did all that could be expected of it and more, considering the huge fire load.

In the 1960s, computer installations in the Pentagon were destroyed in a fire which burned through a concrete floor. The computer area had been provided with a suspended acoustical ceiling of low-density fiberboard. A 100-watt bulb set the ceiling on fire. The guard on duty had been provided with a CO_2 extinguisher—ineffective on ordinary combustible material. He probably could have extinguished the fire with a hose line.

This is typical of what to expect when management worries more about damage from the extinguishing agent than from the fire.

In the 1940s, a multi-story Kansas City, Kansas concrete warehouse was filled with rubber tires. A fire destroyed the building — to the surprise of investigators.

Tactical Considerations

This author was a very early enthusiast for high-expansion foam. At a fire school, foam was tried on a pit filled with rubber tires. The fire was controlled as soon as foam filled the pit. If the foam can be confined, it is worth trying on a rubber fire.

A Prince George's County, Maryland mid-rise apartment complex was built with precast, pretensioned, cored concrete floors. Two very ordinary contents fires in the 1970s caused extensive failure of the cored floor units. Concrete sections fell on fire fighters. Tendons were exposed to failure temperatures. Current orders for these buildings advise going to defensive operations as soon as all tenants are evacuated.

In Sao Paulo, Brazil a 1987 fire caused the complete collapse of a 20-story concrete building.[27] Fortunately there was enough warning for all fire fighters to escape. In South America, concrete often is not as well made as in the United States, and greater failure potentials exists.

A mid-rise concrete apartment house was built in a Texas Gulf Coast resort. The stairways are of fire-retardant treated wood. The local officials objected to the wood, but the owner prevailed in court. Fire suppression resources were limited in the area, so an apartment fire actually ate holes in the concrete. The wooden stairways survived. They were not subjected to the fire.

A reinforced concrete and steel grain elevator provided fire fighters with a difficult battle. Cracks in the concrete walls caused the Fire Ground Commander to withdraw all fire fighters. Thereafter there were several interior collapses.[28]

Heavy fire engulfed the eleventh floor of a concrete high-rise Postal Service Headquarters Building. Heat and smoke made an inside attack impossible after 20 minutes. Standpipe risers were located 140 feet from stairwells (a common but dangerous practice in this author's opinion— "following the line back" does not always lead to safety). The risers could not be found because of smoke. Exterior master streams were used to knock down the fire and the interior attack resumed. Two sides were inaccessible. Falling glass endangered fire fighters. Imagine this fire above the reach of master streams. There is no answer except full sprinkler protection, which has since been provided.[29]

Know Your Buildings

For several years many fire departments complained that "They are building so many buildings so fast here, we can't keep up with them." The 1990-1991 recession showed that in many areas of the country there was a glut of buildings of various types, causing a slowdown or even a cessation of construction. This always presents an opportunity to get current on the hazards of specific buildings and to educate officers in the hazards of construction before the building boom picks up again.

My own nightmare is a fire department charging headlong into a formwork fire in a post-tensioned building under construction, which collapses completely with scores of fire fighter casualties. We were given our warning in the Cleveland post-tension collapse but it went unheeded.

Fire fighters must learn not to wait for "experience." Wise old Ben Franklin told us, "Experience keeps a dear school, but fools will learn in no other." In the fire service the price of experience is blood and grief. The post-tensioned collapse hazard is entirely credible. Make sure your department is not the one to provide the convincing.

References

The addresses of all referenced publications are contained in the Appendix.

1. Lie T., Chabot, M. "A Method To Predict The Fire Resistance of Circular Concrete Filled Hollow Steel Columns," *Journal of Fire Protection Engineering*, Oct., Nov., Dec., 1990, p. 111.

2. "Strong Concrete." *Engineering News Record*, March 19, 1987, p. 21.

3. Downey, R. "The Rescue Company, Concrete" *Fire Engineering*, May, 1990, p. 14. See also letter by Brannigan, F., in *Fire Engineering*, July, 1990 supplementing the Downey Article with respect to Post-Tensioned Concrete.

4. Ross S. *Construction Disasters*. McGraw Hill, New York, New York, 1984. "Progressive Collapse Causes Examined," p. 271 and "Sudden Problem," p. 272. This book is a collection of articles from *Engineering News Record*. All who are really interested in construction and structural hazards should read *Engineering News Record*, published weekly. Find it in a technical library or your city building department.

5. Lew H. S. "Safety During Construction of Concrete Buildings, A Status Report." NBS (now NIST) Building Science Series 80, January 1976, Center for Building Technology (NIST).

6. "Design of Wood Formwork for Concrete Structures," *Wood Construction Data No. 3*, National Lumber Manufacturers Association, Washington, D.C., p. 1.

7. "Rescue Operations," *Fire Command!*, October, 1970, p. 27. See also "Bridge Collapse," letter by Brannigan, F., *Fire Command!*, January, 1971, p. 8.

8. "Girder's High Prestress Likely Cause of Failure," *Engineering News Record*, Oct 12, 1989, p. 11.

9. "Beefed Up Webs Antidote for Prestressing Failure," *Engineering News Record*, Feb 15, 1979, p. 16.

10. Milnes, R. "Miami Weekend. Fire Fighters Battle Two Major Blazes," *Firehouse*, December, 1984, p. 68.

11. ISFSI is the International Society of Fire Service Instructors. Annually the society presents the "Instructor of the Year Award" at the Fire Department Instructor's Conference in Cincinnati, Ohio.

12. Naum, Chris. "Safety First," *Firehouse*, October 1987. See also Messersmith, J., Portland Cement Assn. Skokie, Il. "Concrete Facts," Forum, *Firehouse*, January 1988, p. 7. See also Brannigan F., *Firehouse*, Forum, May, 1988, p. 8. This author points out that Mr. Messersmith makes no mention of post-tensioned hazards.
   We in the fire service should be wary of industry representatives whose objectives are sometimes to strengthen their own. While they can provide much valuable information, bear in mind that they are not always working in our best interests.

13. "No High Ground," *Progressive Architecture*, October 1980, p. 84.

14. Lie T. T. "Fire Resistance of Reinforced Concrete Columns," *Journal of Fire Protection Engineering*, Oct., Nov., Dec., 1989, Vol. 1, No. 4, p. 121.

15. "Fiberglass Roof Experiment," *Engineering News Record*, Feb. 22, 1990.

16. "Don't Rub Salt In The Wound." *Engineering News Record*, May 12, 1988, p. 32.

17. Keyser J. "Where There's Smoke," *Fire Command*, February 1990, p. 18.

18. "Makeshift Holes for Utilities Negate Floor Slab's Fire Resistance," *Architectural Record*, Dec, 1968, p. 147. See also Crowder J., and Bletzaker, R. "A Fire In New England Exposes An Interior Insulation and Finish System," *Building Standards*, July/August 1988, p. 26, and Sept/Oct 1988, p. 12. Published by ICBO (International Conference of Building Officials).

19. Timlick J., "Arson Suspected in Apartment Fire," Winnipeg Free Press, January 11, 1990, p. 3.

20. "Perilous Work Produces Unresolved Issues," *Engineering News Record*, March 1, 1990, p. 42.

21. Letter July 3, 1990 from W. L. Baker, Executive Director, NADC.

22. "Systems Built Apartment Collapse," *Engineering News Record*, May 23, 1968, p. 54.

23. Isner M. "Investigative Report Central Library Fire, Los Angeles, California, April 29, 1986," NFPA Fire Investigations Division. This report should be in the hand of any fire department attempting to get sprinklers into a major library and faced with the all too common phrase, "the books will get wet." Amazingly, there are librarians who have never heard of this fire. A summary of this report was provided in *Fire Journal*, Mar/April 87, p. 56. See also Simmons, A. "On The Job Los Angeles," *Firehouse*, August 1986, p. 33.

24. Roessler, A. "On The Job Colorado," *Firehouse*, April 1986, p. 62.

25. Dickinson, C. "Cold Storage Warehouse Turns Up The Heat," *Fire Engineering*, July 1985, p. 32.

26. Prendergast, E. PE, SFPE, "Fire Breaks Down Concrete," *Fire Engineering*, September 1989, p. 32.

27. *Factory Mutual Record*, July/August, 1981, P. 3.

28. Herzog, C. and Broehm, K. "Fire in a Grain Processing Facility," *Fire Command*, March, 1985, p. 42.

29. Irvin, Paul. "Blaze Causes $100 Million Loss to Postal Headquarters," *Firehouse*, February, 1985, p. 35.

9
Fire Growth

Flame spread, or the more accurate description, fire growth, is a particularly hazardous fire phenomenon which has come to be recognized during the author's lifetime. It was not recognized as such two generations ago.

On November 28, 1942, rapid fire growth was responsible for 492 deaths in the Cocoanut Grove nightclub fire in Boston. At the time, flammable decorations were blamed for the rapid spread of the fire. Inadequate exits and overcrowding were blamed for the monstrous death toll. Pictures from the fire clearly show tell-tale globs of burned adhesives used to glue highly combustible acoustical tile to the ceiling.[1]

After other deadly fires with rapid flame spread in a soldier's hostel in Newfoundland and Mercy Hospital in Iowa, fire officials in the late 1940s began to understand that there was a problem with some building materials. Early publications on hazards of certain building materials were criticized.[2] Today, rapid fire growth is still imperfectly understood, even by some fire suppression forces.

In 1982, two Boulder, Colorado fire fighters died in a training fire in a building lined with combustible fiberboard. An article in the fire's aftermath noted "The building was toured by many people prior to the drills and no one noted any problem with the interior finish of low-density fiberboard."[3]

Fire growth is often regarded as a problem for fire prevention staff, particularly where there is a dangerous "Iron Curtain" between fire prevention and fire suppression personnel.

Fire suppression personnel are advised not to skip over this chapter. It is not written just for fire prevention staff. A knowledge of potential fire growth problems is vital to adequate fire preplanning, and may save fire fighter's lives.

Examples of Fire Growth

Serious losses due to fire growth have occurred across the country.

- In Los Angeles, California in 1982, fire claimed the lives of 24 occupants of the Dorothy Mae Apartments. A key factor was flame spread over plywood in the exit hall.[4]
- In a 1984 Beverly, Massachusetts fire, 15 died in the Elliott Chambers Boarding House. Plywood paneling was used for interior finish. Even an activated fire detection system was inadequate in the face of the fast-spreading fire.[5]

Combustible tile ceilings are often suspended below the floor above, creating a void in which explosive carbon monoxide gas can be generated and stored. When this gas ignites, there is often an extremely violent explosion. Three of the following five fires illustrate this phenomenon.

- A magazine article described a fire in which a fire fighter and three civilians lost their lives in a flashover in a fire-resistive building. There was no mention in the article of what flashed over. A call to the author revealed that the fire-resistive building had been "improved" with a combustible-tile suspended ceiling, concealing a void from which the fire could burst out.[6]
- In Silver Spring, Maryland in 1977, this author observed a fire in a one-story row of stores. The ceiling was made of suspended combustible tile, which created a combustible void above the tile. The store was well ventilated. The front windows were out. The surface fire on the tile was extinguished with a big line, but the tile continued to burn within the material and on the back side. Suddenly a huge amount of black smoke under pressure appeared, a signal that a backdraft was imminent. Fortunately the fire fighters were only knocked down without serious injury.
- Some years ago, several Orlando, Florida fire fighters escaped from a store with their bunker coats afire. Fire had burst out of the void above a combustible tile ceiling. Combustible interior finish has been a factor in fatality fires where disabled or elderly people have resided.[7, 8]
- Carpeting served to rapidly spread the fire in the Pioneer Hotel where 28 people died in Tucson, Arizona in 1979.
- Carpeting also spread the fire in the Harmer House Nursing Home where 32 died in 1970 in Marietta, Ohio.

Fig. 9-1. A fire in combustible acoustical tile appeared well knocked down, although it continued to smolder. Carbon monoxide accumulated in the void above. In a short time there was heavy black smoke followed by a backdraft.

Building or Contents Hazard?

Flame spread, or rapid fire growth, can be a problem caused by both the building itself and its contents.

The fire growth building problem can be differentiated by a location characteristic:

* Hidden
* Exposed

The fire growth contents problem can result from:

* Furnishings
* Decorations
* Mercantile stock

These categorizations are not meant to be absolute. As is the case with much of this text, it is impossible to make absolute distinctions regarding many factors in building construction. The same material may be, at different times, a hidden building element, an exposed building element, furniture, or merchandise contents in a commercial building. This can be very frustrating to those who want everything as black or white. Unfortunately, these distinctions are often fuzzy.

In addition to fire growth, high flame-spread materials may contribute heavily to the fire load and to the generation of smoke and toxic products. These are additional threats to the life safety of both occupants and fire fighters.

Hidden Building Elements

There are a wide variety of materials and hidden building elements which contribute greatly to rapid fire growth.

The paper vapor seal on batt-type glass-fiber insulation gives a phenomenal flame spread. (Flame spread ratings are explained later in this chapter.) The California Building Code recently recognized this hazard and requires exposed vapor seal to be "fire-rated to existing standards." A similar policy has been adopted in Washington state.[9]

Batt insulation laid in ceilings must be kept free of light fixtures. The heat from the fixture can ignite the vapor seal. In one case, insulation laid on a ceiling was allowed to be in contact with the top of

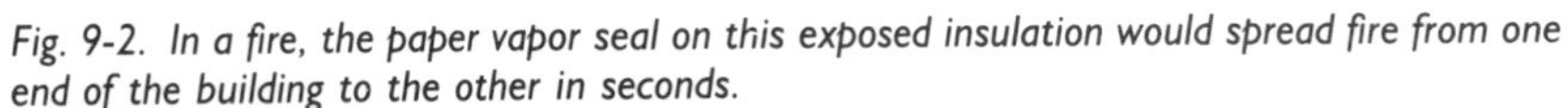

Fig. 9-2. In a fire, the paper vapor seal on this exposed insulation would spread fire from one end of the building to the other in seconds.

a light fixture. It ignited. Fire spread on the paper. Over a hundred sprinklers operated below the ceiling without effect.

A supermarket with a combustible ceiling was fully sprinklered, including the cockloft. A flammable liquid fire extended to the cockloft and spread rapidly onto the paper vapor seal on the insulated underside of the roof. Twenty heads operated in the cockloft and 90 heads in the store.

In another case, an electrical fire ignited the vapor seal on insulation. The fire spread through the huge undivided roof-ceiling void, which extended over the entire building.[10]

Glass-fiber insulation with paper vapor seal was installed around an upstairs bathroom for soundproofing in one of a series of connected expensive dwellings in Lake Benbrook, Texas. The piping was so well insulated from heat that it froze. A torch used to thaw the pipes set fire to the paper vapor seal on the insulation, and ten houses lost their roofs or upper floors.

Motels are often built with back-to-back rooms, opening to the outside. A plumbing corridor is often provided between the rooms. For privacy, batt-type fiberglass is sometimes installed between the studs. The paper vapor seal is exposed. A plumber's torch could ignite the paper and flash into a fire extending the length of the corridor and through pipe channels to the corridors above.

Combustible fiberboard is commonly used as insulating sheathing on wood frame buildings. It is also used as soundproofing. This material can support a fire hidden in the walls. (Combustible fiberboard is discussed in detail under "Exposed Building Elements.")

Foamed-plastic insulation is also used as sheathing, concealed in cavity walls, or glued to the interior surface of masonry wall panels. Foamed plastic applied to walls and ceilings for insulation has been involved in many disastrous fires.[11] When such insulation is installed, it should be protected from exposure to flame by a half-inch gypsum board covering. This protects only against ignition from a small source. In a well-developed fire, as the gypsum board fails, the plastic will be involved, possibly suddenly and explosively as the gypsum fails or falls away.

In what may be one of the costliest fires ever, workers were foaming plastic insulation around wires to prevent cross contamination between two sections of the Tennessee Valley Authority's Browns Ferry Nuclear Power Reactor. Incredibly, workers were using a candle to check for air leaks. The fire started in the foamed plastic and spread over the wiring.[12, 13]

The actual cost of wire burned was about $50,000. The overall financial loss to the taxpayers was estimated at $300,000 per day for the year and a half the reactor was out of service.

Foamed plastic may be manufactured so that its flame spread is reduced, but it still can melt. It lacks dimensional stability. When the plastic is used structurally, this also may be disastrous.

Air-duct insulation commonly installed years ago was usually made of a hair felt with a high flame spread. The presently-used aluminum-faced foil (not aluminum-faced paper), glass-fiber insulation presents little flame spread problem.

Electrical insulation may be self-extinguishing. However, the tests of this material are conducted on wire not under load. When electrical wiring is operated at or above its rated capacity, the heat can break down the insulation and flammable gases can be emitted. The well-discussed McCormick Place fire in Chicago is thought to have started in the flaming insulation of an overloaded extension cord. Large groups of electrical wires can support self-sustaining ignition.

A spectacular, tremendously smoky, multiple-alarm fire occurred in the cable vault of a telephone company building in New York City. The fuel for the fire was the insulation in the cables. A number of fire fighters suffered seriously from the smoke.[14, 15]

Factory Mutual (FM) has developed classifications of cables as to their flammability.[16, 17]

- Group I — These cables resist fire propagation. They may ignite, but the fire spread will be limited to the area of fire origin. Group I cables can be used without sprinkler protection in areas of non-combustible construction.
- Group II — A fire in Group II cables, although slow-burning, can be expected to spread beyond the area of fire origin. Some level of fire protection is required.
- Group III — These cables exhibit very rapid, intense fire propagation that could potentially involve an entire facility. Fire protection is necessary.

Interior Finish

At one time lime plaster was an almost universal finish for ceilings, although some ceilings were also made of embossed steel (tin ceilings) and wooden boards called matchboarding. Plaster does not contribute to a fire. In fact, lime plaster absorbs heat and thus slows the progress of the fire. Today, interior finishes are many and varied.

There are three ways in which interior finishes increase a fire hazard.

- They increase fire extension by surface flame spread.
- They generate smoke and toxic gases.
- They may add fuel to the fire, contributing to flashover.

Common interior finishes include such materials as wallboard, wallpaper, lay-in ceiling tile, vinyl wall covering, and interior floor finish items, such as carpeting. Sometimes used are "bad actors" such as low-density fiberboard and combustible tiles.

Low-density fiberboard: In the 1930s, low-density fiberboard made of wood fibers or sugar cane residue called bagasse came into use. Produced in 4- by 8-foot sheets with a painted surface, fiberboard provided a cheap interior finish that could be installed quickly. Fiberboard sales grew astronomically. The material was used both for initial construction and for rehabilitation. It was substituted for wood sheathing in frame construction. It was also widely used for interior finish in low-cost construction.[18]

The Celotex Company manufactured many different building materials including a low-density fiberboard. The name "Celotex" is widely and inaccurately used to denote fiberboard, regardless of its manufacturer.

When low-density fiberboard is used for sheathing and for soundproofing, it is concealed in walls. A common method of igniting this material is by a plumber's torch. The plumber sees only the gypsum board wall. The fire enters the wall along the copper tubing and ignites the combustible fiber sheathing. The fire often goes undetected until it erupts violently. In December 1990 in Ventura County, California, 350 apartment units under construction were lost in a fire reported to be caused by a plumber's propane torch.

A school under construction in New England had fiberboard on wood studs covered with gypsum board. The fire extended to the metal deck roof. The damage was in the millions in this "noncombustible" building.

In a Des Moines, Iowa shopping center, sprinklers were omitted in one part of the concealed ceiling space. Combustible fiberboard formwork for the concrete roof topping had been left in place. The fire burned the combustible fiberboard. The estimated loss was $700,000.

This author once examined an apartment fire in Gainesville, Florida. The walls between units were doubled with two lines of studs with fiberboard in between. A fire in the laundry room entered the wall void and extended to the roof, then advanced laterally through the attic.

Combustible acoustical tile: When punched with holes, fiberboard acquires desirable acoustical properties, and becomes combustible acoustical tile.

Consider the situation in a building with badly deteriorated plaster ceilings. Replacing the plaster would require removing the old plaster, placing on new plaster, then painting — three messy jobs. An alternative is for workers to come in over the weekend, nail furring strips to the ceiling and add fiberboard tile, producing a new, attractive ceiling with acoustical benefits by Monday morning.

Industry Opposition

Years ago industry vigorously fought attempts at regulating fiberboard. In one case, regulations were characterized as "taking the bread out of the mouths of Louisiana cane farmers." A U.S. Department of Commerce-approved "industry standard" fire test was developed which amounted to a fire in a thimble full of alcohol. A light flame retardant coating enabled the fiberboard to pass this "test."

In April 1949 a tragic fire occurred in St. Anthony's Hospital in Effingham, Illinois. There were 74 deaths in the fire, mostly infants in the nursery, but also nuns and nurses who would not leave them. The *NFPA Quarterly*, in two separate issues, reported on the fire and addressed the hazards of combustible tile. Subsequently manufacturers threatened legal action against the NFPA, but the association stood its ground.[19]

A major change in industry attitude regarding fiberboard standards came about, not from reasoned explanations by fire protection experts,

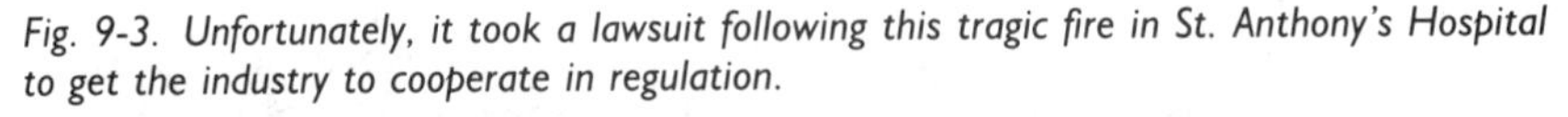

Fig. 9-3. Unfortunately, it took a lawsuit following this tragic fire in St. Anthony's Hospital to get the industry to cooperate in regulation.

but as a result of a successful lawsuit against a manufacturer in the wake of the fire.

Void Spaces

Suspended ceilings of combustible tile form void spaces in which fire can burn undetected until it bursts out furiously. In a 1956 fire, 11 people lost their lives in Anne Arundel County, Maryland when fire involved the attic above a combustible ceiling in an amusement hall. The raging fire overhead did not appear to be serious.

Three fire fighters lost their lives in a furniture store fire with combustible voids. Heavy smoke was reported from one mile away. Light smoke was reported from close in. Close-in positions are not always the best place to see the situation, particularly a multi-story structure.

Fourteen people died in a fire-resistive hotel fire. The third floor where the fire started was the only floor paneled in wood. It had a dropped ceiling, probably made with combustible tile.[20, 21]

Tactical Considerations

It cannot be repeated often enough that hidden fire launches a deadly ambush which traps the aggressive fire fighter. Preplanning should include the potential for hidden fire in voids.

There may be little or no smoke visible below the void. All fire fighters should look for, and act on, signs of fire or heavy smoke coming from the roof, or other distant locations. Ceilings should be opened with caution and with a determined escape path.

Be aware that the tiles are probably burning freely on the unexposed surface. Determine if there are any ducts or other connections from the void above the tiles.

Fire burrows into combustible fiberboard and hangs on with amazing tenacity. Apparently extinguished combustible tile should be torn out and removed to a safe outside location.

Remodeled Ceiling Hazards

A most serious hazard is often created when a building is remodeled. Code requires the installation of a new ceiling which meets flame-spread requirements. No code that this author is aware of requires the removal of the old ceiling.

Fig. 9-4. In California, this Los Angeles County building had a combustible acoustical tile ceiling glued to the concrete overhead ceiling. Suspended below was a newer grid ceiling which met the flame spread requirements of the local code. This is a common but dangerous condition.

In the Roosevelt Hotel fire in Jacksonville, Florida in 1963, a number of people died far above the fire floor. The fire was in an old combustible ceiling that had been left in when the new ceiling was installed. The smoke moved through old air ducts.

When the new fire-rated ceiling is installed below the old ceiling, the dangerous combustible ceiling is left above. Fire can burst down out of that void. Two fire fighters died in Wyoming, Michigan when fire burst out of the ceiling. Even then, the city did not amend the code.

In December 1989 in Johnson City, Tennessee, sixteen persons died in the John Sevier Center, a fire resistive unsprinklered high rise housing elderly citizens.

A prime factor in the loss of life was an old combustible tile ceiling glued up to the concrete overhead and left in place when the new suspended grid ceiling was installed. NIST researchers noted that carbon monoxide generation in such a void could be as much as 50 times that normally generated.[22]

The International Conference of Building Officials, (ICBO) which produces the Uniform Building Code, developed an excellent video tape/text program "Regulation of Interior Wall and Ceiling Finish Materials" in 1991. The tape shows a vivid picture of one of the occupants

Fig. 9-5. In this experimental controlled burn, the only fuel burning at the moment the photograph was taken is the combustible tile ceiling.

of the John Sevier home at a window begging for help. The narrator commented on the hidden tile situation. I am informed, however, that there are no provisions in the Uniform Building Code to deal with this situation, and no code change proposals have been submitted.

Tactical Considerations

When a building with a combustible-tile ceiling is being remodeled, determine whether the old ceiling is to be removed. If the building department does not require it to be removed, make every effort to convince the owner that the tile should come out. Your recommendation should be on permanent record. Code compliance is not a shield in a negligence action. Provide the owner with a graphic description of why it will be necessary to tear down the entire new ceiling, and use heavy-caliber streams.

Be sure all personnel are alerted to such problems. Specific ceiling tile information should be provided in the preplan so there will be no deadly surprise if a fire occurs.

Eliminating Existing Hazards

It takes years to get rid of hazards once they are in place.

Some years ago, this author drove a granddaughter, then a high school student, to a summer program at a well-known southern university. Her room was on an upper floor of a dormitory. All the ceilings were made of combustible acoustical tile.

Plastic rubbish containers were piled high with combustible debris. All the stairways were open. A careless match could have caused a disaster.

Combustible surfaces have been a factor in many multiple-life loss fires.

I immediately purchased a smoke detector for our granddaughter's room and mounted it on the transom. The door to the outside fire escape was opposite her room. I instructed her and her roommate to be prepared to get out promptly via the fire escape.

I wrote to the university. They expressed serious concern but pointed out that they had hired an architect to eliminate fire hazards. The building was reported to comply with an unspecified code. New codes are not always applied to existing buildings.

Sometimes this is the case because of an inaccurate legal perception that new codes cannot, or should not, be applied to existing buildings. In a search of appellate court records for the U.S. Fire Administration, no case was found which barred enforcement of a new code on an existing building.[23] Note also that NFPA 101, the *Life Safety Code*, requires existing buildings to conform to minimum standards.

The fire service must realize that "up to code" is by no means synonymous with fire safety, and that some architects do not understand fire problems.

Efforts to Mitigate the Hazard Are Sometimes Inadequate

These hazards may exist as the result of using a wide range of materials including tiles and fiberboard, paper, fabrics, cork rattan, many types of wood, plastics, and carpeting.

Adhesive

The Hartford Hospital corridor ceilings were made of combustible acoustical tile glued to a gypsum board ceiling. The hazard had been recognized and the tiles painted with a flame-retardant coating. In 1961, a fire, in which 16 patients died, developed in a soiled linen chute, rolled out, and attacked the ceiling. The fire roared down the corridor. Samples of the tile were tested and showed a reduced flame spread not sufficient to support as intense a fire as was experienced. An entire 25-foot section of the ceiling was removed and sent for testing. The flame spread was very high. The difference was due to the adhesive used to attach the tile.[24]

The Clark County, Nevada Fire Department report on the MGM Grand Hotel fire in 1980 noted that it required 12 tons of adhesive to attach tiles to the ceiling of the casino. The flammable adhesive added a large fuel load to the fire.[25]

High-Density Fiberboard

Fiberboard can be manufactured to very high density. Very dense fiberboard was selected for radiation shielding in the Rocky Flats Nuclear Weapons Plant. It was regarded as fire-safe because it was tested with a blow torch. This is a commonly used but inadequate test. Long continued heat from spontaneously ignited plutonium ignited the fiberboard. The resultant fire and contamination problem was extremely costly.

When punched with holes, high-density board is called pegboard. At least one manufacturer (Masonite Corporation) produces such hardboard with a very low flame spread rating. This author has seen advertisements for decorative plaster ceilings made of hardboard. There was no information as to their flame spread characteristics.

Paper

Some years ago, the U.S. Atomic Energy Commission was building an extension on Strong Memorial Hospital in Rochester, New York. Craftspeople were working in the excavation. Winter was approaching. A temporary snow roof was erected to permit work to go on during the winter. The least expensive available material was selected for waterproofing. It proved to be a hemp-reinforced bituminous impregnated paper of the type used to protect finished lumber. It has a very high flame spread rating. A stove for the burning of wood scraps was in one corner of the snow roof. There was a serious potential for sending a sheet of fire up the face of a functioning hospital. The situation was corrected.

Fabrics

The terrible 1944 circus fire in Hartford, Connecticut in which 168 lives were lost, was fueled by flames spreading rapidly on the gasoline-paraffin impregnated canvas. It might be noted that the caged runway

Fig. 9-6. This model, set up at the Ringling Museum in Sarasota, Florida, shows the circus tent as it was set up on the day of the terrible 1944 fire at the Ringling Brothers and Barnum and Bailey Circus. Note the wild animal cage and its chutes which blocked the exit.

for the wild animal show was erected across the main exit "for only 10 minutes." Remember this disaster when you are asked to overlook the "temporary" blocking of exits.

In another example of a potential fabric hazard, the author was giving a slide-illustrated seminar on combustible decorations in the northeast. After the seminar, attendees went to a popular local restaurant. The ceiling was draped in burlap. The fire marshal commented that there should be no problem since the place was sprinklered. He had no idea that fire could race over the burlap much faster than the sprinklers could operate. It is hard to imagine a greater panic-producing situation than fire falling from the ceiling on people, literally igniting hair and clothing.

Cork

A California seafood restaurant has a ceiling decorated with cork floats. It also is sprinklered. The possibility of flame spreading faster than sprinklers can operate is often unevaluated. Cork paneling in an office was a key factor in a five-alarm fire in the Empire State Building.

Rattan

In a south Florida city, a two-story, brick and wood-joisted building had been "rehabilitated" into a botique. The ceiling was of rattan. On a visit a few years later the building was found in ruins. The fire had started on the second floor. In the words of one witness, "It spread so fast it must have been arson!" An equal possibility was the rattan ceiling.

Wood

Wood surfaces are very popular. Flame spread over the surface of plank and beam construction is discussed in Chapter 3.

Plywood

The unexposed side of plywood or any wood paneling can burn unobserved and protected from fire department streams. For this reason it is not sufficient to correct such a situation by flame-retardant surface treatment of the exposed surface.

When Boston's John Hancock Building suffered massive losses of glass window panels, plywood was a temporary substitute. At some point the vertical flame spread hazard was recognized and flame-resistant plywood was used for any further closures.

In Montreal a wood-joisted building was so heavily involved with fire on arrival that a fifth alarm was ordered in just three minutes. The fire broke over a thousand windows in the convention center. The openings were then closed with plywood.[26]

Tactical Considerations

An estimation should be made of the potential for rapid flame spread over interior finishes. For instance, the many churches with exposed wood plank surfaces present the potential for extreme flame spread and, thus, a fully developed fire.

Fire can develop with unimaginable speed. In a Florida church fire, photographs showed only light smoke as units arrived. Within a minute, heavy fire was rolling out every window and door. A fire department that has not preplanned properly for this type of fire may well waste time using handlines that will be completely useless. This fire was suppressed in 14 minutes by three heavy-caliber streams positioned so as to sweep all surfaces.

The streams should be solid, not fog, so as to wet the surface of the wood. Locations for the streams should be preplanned.

Plastics

Some materials are easily recognizable. Others are disguised, such as plastic-imitation wood sheathing and beams, and imitation glass mirrors.

Rigid-foamed polyurethane has been used for interior finish in many houses. In at least one fatal house fire, it was a contributing factor. When the extreme hazard was recognized, an attempt was made to reduce it by plastering the interior.

A restaurant in a Texas city was made into a cave by the use of foamed polyurethane. Reportedly, pieces broken off by students were easily ignited. A 1971 fire in a French nightclub, in which 144 persons died, was fueled by foamed plastic used as interior finish.[27]

Forty-nine teenagers died in a fire in a Dublin, Ireland disco. The Irish government convened a Tribunal of Inquiry to determine the circumstances. The ceiling was of mineral tile, and thus not a factor. The built-in banquettes and the wall covering were of plastic. An expensive, unprecedented re-creation of the fire, commissioned by the Irish government, at the British Fire Research Station showed how a small fire in a foamed plastic seat could spread almost explosively, due to the nearby wall covering.[28] The movie "Anatomy of a Fire" shows the almost unbelievable rapid spread of fire in such an environment.

This Tribunal of Inquiry unfortunately has no U.S. counterpart. It is a judicial body with technical expertise. It has subpoena powers and wide authority. Too often, in the U.S., all the facts of disasters are never revealed, due to sealed settlements of civil law suits, and the narrow scope of criminal investigations.

The first structural use of plastics of which the author became aware is a tunnel where visitors pass through a shark pool. The plastic tunnel supports tons of water. The entire building is sprinklered except for the tunnel. If this clear plastic supports flame spread as other clear plastics do, the lack of sprinkler protection might be serious. Plastic is a poor conductor of heat so the water in the pool would have no cooling effect on the fire.

An interesting feature of Expo '67 in Montreal, Canada was the United States Pavilion. It was a geodesic dome covered in clear plastic. An architectural writer marveled at this beautiful light-weight futuristic building. One article about the Expo said "... the dome may just possibly steal the show."[29]

After the building was turned over to the City of Montreal, a worker accidentally set fire to the plastic with a cutting torch. The clear plastic skin—the equivalent of solidified gasoline—was destroyed in 10 minutes. Had the fire occurred during Expo, the disaster would have horrified the world.

Clear plastic has many inviting uses, but not all designers understand its terrible potential hazard.

This author was told by the plastics supplier that they were reluctant to supply the material and had impressed on U.S. officials the necessity for strong precautions. The supplier said that when the pavilion was turned over to the City of Montreal, the company wrote to the city explaining the hazard.

Some materials are probably too hazardous to use. Administrative controls often are unreliable.

As recently as 1991 an architectural writer, criticizing the U.S. exhibit at the upcoming Seville World's Fair, commended the Montreal Expo's geodesic dome as "exhilarating." The stories of major disasters should be taught to architects. A letter to the Washington Post went unacknowledged.[30]

Fig. 9-7. The shell of the geodesic dome which housed the United States Pavillion at Expo '67 stands as a silent witness to a disaster which by good chance didn't happen. (Chief Inspector Douglas Lion, Cote St. Luc, Quebec Service de Incendies)

The Isle of Man in the United Kingdom is located in the inhospitable North Atlantic. In an attempt to create a mediterranean-type resort, a huge entertainment complex called Summerland was built. A major portion of the walls and roof was acrylic plastic and another large portion was asphalt-coated steel, often called Robertson protected metal. The interior paneling was of pressed fiberboard. This created a combustible void wall. The stage was set for disaster. A glass-fiber reinforced plastic stand caught fire outside the building. The fire extended to the asphalt-coated steel, then to the plastic. Fifty persons died in the fire.[31]

A recent newspaper article described the rehabilitation of an old Washington, D.C. house by its architect-owner. The floor of a child's bedroom was a sheet of clear plastic to allow light to penetrate below. Even when the problem of combustibility is understood, other fire problems that are possible with structures incorporating large amounts of light-weight plastics, are not apparent, even to architects.

The retired dean of architecture at Columbia University once described a proposed pressurized membrane structure which would cover a complex of four to six government office buildings. The structure would encompass 400,000 square feet. The purpose was to provide an ideal year-round environment independent of outside atmospheric conditions. There is no mention of the deadly atmosphere which would be created within the structure by the release of the toxic combustion products locked within the contents inside the environmental envelope. These toxic products could be released with a single match.

Aircraft Interiors

There have been a number of aircraft tragedies where victims survived the crash but died in the resultant fire. In some cases the fuel was the plastic seats and other interior fittings. Effective August 1990, new aircraft must meet new stricter FAA standards for fire and heat resistance.

Acoustical Treatment

Acoustical or soundproofing treatment is another potential source of trouble. Many acoustical installations have been made without consideration of flame spread consequences of materials used.

Open-Plan Offices

Flame spread restrictions in codes are typically strictest for corridors, and less strict for rooms off these corridors. This has created a potential problem in the modern open-plan office. Where does the corri-

Fig. 9-8. Note the accoustical panels hanging below the ceiling of the auditorium. They were installed many years ago and nothing is evident or was known to management about their combustible characteristics.

dor begin and end? One fire department solved it simply. If there is no wall, there is no room; it is all a corridor.

This author has been informed by competent authority that this was not a proper interpretation of the code. Considering the devastating demonstration of the open office flame spread potential shown in the First Interstate fire in Los Angeles, the code interpretation might well be reexamined.

Carpeting

Noted earlier were two carpeting fires which claimed 60 lives.

The use of carpeting has changed in recent years. It is now also used on walls and ceilings. It is not always evident to designers or even to fire officials that the location where a material is installed may increase the rate at which a fire can grow.

A spectacular 1980 fire in the Las Vegas Hilton Hotel spread from floor to floor outside the building because of flammable carpeting on the walls and ceilings of the elevator lobbies. The fire grew into one terrifying fire front extending many stories in height and claimed eight lives.[32]

Tactical Considerations

A carpeting fire will fill a corridor with flame. It should be fought with a solid stream, directed through the fire to wet the carpet beyond the fire, thus stopping extension. A fog stream may push the fire.

Alterations

Fire departments should be aware of any building alterations planned in their jurisdictions and ensure that all pertinent codes are obeyed. The author has observed a number of instances in which interior finish materials that would not have been permitted under the code during original construction, were installed after occupancy. Needless to say, this may be the case in numerous buildings.

A chain of steak houses decided to change the interior decor to have a rough, plastered look. Combustible paneling resembling plaster was used. It was done without consulting local code officials. When the question was raised, the management provided certification from the manufacturer that the material met local (not-too-rigorous) flame spread standards.

Decorations and Contents

Present-day decorations and building contents present a major fire hazard. In 1975, the Society of Fire Protection Engineers warned in a Hazard Alert in its April Bulletin:

> "The rate of fire development can create a condition that may tax or overpower traditional fire defenses. Defenses of the past, both passive and active (evacuation, alarm, ventilation, and manual fire control), have not been designed by engineer or code to anticipate this hazard.
>
> This is primarily a furnishings problem. Many occupancies (residences, office buildings, theatres, hospitals) are affected."

At one time, a popular decoration that provided a real flame spread or rapid fire growth problem was the Christmas tree. While there are still tragedies during Christmas season, such as the multiple loss of life at a San Francisco yacht club a few years ago, the incidence of such fires has been reduced.

It is possible that this has led to complacency. Regardless of precautions, the amount of decorations in places of assembly should be strictly limited to reduce the potential volume of fuel for a fire.

Decorations, usually hanging from the overhead and/or from the walls, and contents such as furniture or stock in trade, are very difficult to control from a fire prevention perspective. There are few applicable regulations. These regulations, where they do exist, are difficult to write in the manner required by law, and are difficult to enforce.

Decorations often represent a serious problem because the hazard goes unrecognized even by people who consider themselves fire-conscious.

One of history's most serious fires occurred in 1859 in a cathedral in South America. The overhead was decorated with fabric to resemble clouds. A fire broke out and about 2000 persons were reported to have died.

Closer to home are the fatal fires that have occurred when well-meaning adults create a Halloween haunted house. Often cotton sheeting is used, which has a high flame spread, in proximity to light bulbs (a 100-watt bulb can ignite cotton) or candles. One man lost his life in assembling a haunted house in a school in Virginia. The disaster would probably have been much worse had it occurred when the children were in the house. The presence of the old combustible tile ceiling above the later-installed non-combustible ceiling was also significant in the spread of the fire.[33]

Students in a fire-resistive girls dormitory at Providence College, in Rhode Island, had a contest with cash prizes for attractive Christmas decorations. Evergreen trees, made of paper and cotton applied with masking tape, provided continuous fuel along the length of the dormitory corridor. Even fire alarm boxes were covered over. Individual rooms were similarly decorated.

Ten students died when a room fire extended through the hall. Dead-end corridors, permitted under the 1938 code, were a serious contributing factor. Additional stairways have since been provided.[34]

Newspaper photos in the February 22, 1990 Orlando Sentinel showed a teacher stapling paper showing the periodic table of the elements to the ceiling. Well-meaning, innovative people can unwittingly set up dangerous conditions.

In the 1942 Cocoanut Grove Night Club disaster, the initial spread of the fire was along combustible decorations at the ceiling line. Later study showed the additional involvement of acoustical tiles.

A small counter-type restaurant in New York had plastic imitation wooden beams and plastic imitation fruit on the ceiling. The author photographed it for a lecture series. Sometime later, the building was involved in a multiple-alarm fire that started in the restaurant. Fire fighters who responded on the first alarm reported that flames were shooting across the wide avenue.

A fire in the Six Flags Haunted House in New Jersey in which eight persons died was reported to have been caused by a visitor touching a lighter to a polyurethane surface. The NFPA report on this fire[35] is must reading. The extreme hazard of foamed polyurethane often goes unrecognized. Samples of this foam ignited after a one-second application of flame from a butane lighter.

Also worth reading is "Testimony on Trial"[36] which presents a detailed description of the testimony and accompanying furor and controversy surrounding the criminal trial associated with this tragic fire. Much other relevant material related to this fire has been published.[37, 38, 39]

A scientist in a research laboratory was annoyed at the reverberations from the metal building. He made pyramidal fabric bags which were filled with urethane scraps and hung them from the ceiling to suppress sound. He was quite surprised when they were ordered removed.

A water treatment plant under construction had a great deal of glass fiber-reinforced plastic installed for various purposes. There appeared to be little fire hazard; the insurance was cut to the bone. After the fire the contractor was paid only a fraction of the loss because he was underinsured. Because of the "glass" in the name, many mistakenly consider the plastic to be non-combustible.

The Consumers Product Safety Commission created a huge acrylic plastic lobby display to show what they were doing for public safety. This pile of "solidified gasoline" was located between the employees' offices and the exits. A young attorney, schooled in fire protection, protested, and it was removed.

A Texas museum was very proud of a walk-through exhibit — a room full of rodeo figures. The larger-than-life-size figures were made of papier-mache. Visitors in the museum walked on pieces of foamed polyurethane covered in burlap. A potential exit was blocked by bales of hay. Neither the creators nor the visitors had any concept of the hazard. A dropped match or a cigarette lighter could have created a tragic inferno.

This author observed a serious fire in the Smithsonian Institution, National Museum of Science and Technology. The heat produced in the fire was unparalleled in the experience of many fire fighters. The source of the heat was not understood until it was learned that huge sheets of clear plastic had hung from the overhead as part of the exhibit. They had been completely consumed by the fire.

Hacienda Heights, California is the home of the largest Buddhist Monastery and temple complex in the western hemisphere. In the complex, there are about 10,000 niches containing polyurethane statues of Buddha. The Fire Marshal ordered them removed or the building sprinklered.

Today's Fire Loads

Fire loads and rates of heat release in current homes, stores and offices are much greater than they were when most fire codes were written. In addition, a further hazard has arisen. The nature of furniture and contents has changed. Solid and foamed plastics have replaced much wood, cotton, wool, and other materials. The hazard presented by these plastics is often unrecognized.

First Interstate Bank Fire

This fire is discussed in Chapter 11, High-Rise Construction, but it is pertinent to note here that the floor of origin had wide open spaces crammed with computers and related equipment. Rapid fire spread over the plastic cabinets and wiring insulation produced a quickly raging fire which was reported from blocks away, while building security staff were busy resetting smoke detectors. The fire growth was thoroughly analyzed by the Fire Research Division of NIST (formerly NBS).

The NIST report should be copied and discussed with the management of similar occupancies. It appears that many are still unaware of the hazard of fire loads. Others are still fearful of sprinklers. Some still believe there is plenty of time to deal with a fire.[40, 41]

Hotel Remodeling

A pile of plastic furniture was stored temporarily in a public area of the Dupont Plaza Hotel in San Juan, Puerto Rico. In December, 1986, disgruntled employees set fire to it. The fire grew to enormous size almost instantly, and resulted in 97 lives being lost. Recent reports have indicated that claims settlements from deaths and injuries from this fire have reached $200 million.[42, 43]

Hotels are remodeled regularly. Mattresses and furniture often are removed from rooms and stored in hallways, a tempting target for arsonists. Carpeting piled in the hallway was ignited in a Fort Worth, Texas Ramada Inn in 1983, where five persons died. Two Maryland hotels recently suffered incendiary fires when arsonists used this very scenario.

Tactical Considerations

Sometimes a hotel floor will be put out of service and used for storage of incoming furniture. In effect, there is a furniture warehouse within the hotel. A fire department prepared only to use the typical 1¾-inch high-rise pack will find that their attack water is hopelessly inadequate.

Any indication that a hotel is to be refurbished, remodeled or spruced up should be a red flag to the suppression forces to determine what changes in fire department tactics are necessary, and what additional precautions should be required of the management.

Regulations

On a national basis, little has been done to control the hazard of furnishings in buildings. However, the State of California has done much, as has the City of Boston. For its own considerable holdings, the Port Authority of New York and New Jersey has laid down fire load requirements. This action was sparked by a devastating fire in a terminal at Kennedy Airport, which was fueled by plastic seats. The New York State requirements recently developed, however, are concerned with toxicity, not flammability.

Certification of Interior Designers

New York State's recently enacted legislation providing for the certification of interior designers elevates that vocation to a new level of "legitimacy."[44]

The reference goes on to point out that selections of decorator items have been made on the basis of aesthetic considerations alone, without attention to fire hazard characteristics. The New York Law (seven other states require certification of interior designers) fortunately does not provide for "grandfathering." Candidates will take two examinations, one on interior design itself, the other on building codes and fire safety. Much will depend on the quality of the state board which will monitor certification.

It would be useful if architecture and interior design writers could be persuaded to take the same instruction that will be necessary for the designers to pass the building and fire safety examination.

Residential Fire Tests

In 1980, the National Bureau of Standards conducted fire tests in the contents of typical residential basement recreation rooms. The word "basement" easily could be dropped from this description as immaterial. The room could be a living room or motel room, a fire station recreation room, a doctor's office, a reception area or elevator lobby in a hotel.

The full report of these tests is a gold mine of information. Do not be deterred by the technical nature of the writing — study it.[45, 46]

Information concerning Test No. 14 is very revealing. The test room was 3.3 by 4.9 meters in size, (10.7 feet by 15.9 feet). The furniture consisted of a sofa, an upholstered love seat, ottoman, two end tables, coffee table, and two bookcases. The carpet was olefin pile with foam rubber backing. The walls were plywood. The ceiling tiles were wood fiber, with a 200 flame spread rating. The fire was started by matches in about a pound of newspaper on the couch.

In less than four minutes heavy flame was pouring out the full height of the doorway. Six minutes after ignition, the average gas temperature was reported as 700°C or about 1300°F.

Films Available

Unfortunately, there are no films or tapes generally available of the NBS tests. However, the NFPA has produced two movies and videotapes, "Fire: Countdown to Disaster" and "Fire Power," which give graphic demonstrations of how fast fire grows in today's interior environments.[47]

In "Fire: Countdown to Disaster" a sparsely furnished bedroom develops a violent backdraft explosion in less than three minutes.

Fig. 9-9. The fire pouring out the window at the end of the test corridor is from the carpeting. (National Bureau of Standards)

"Fire Power" also graphically demonstrates the value of residential sprinkler protection.

Tactical Considerations

There is a commonly accepted belief that it takes fifteen minutes for a fire to flashover. This is no longer the case with today's fuels which provide accelerated fire growth. In "Fire, Countdown to Disaster" a bedroom fire develops to flashover in two minutes and a violent backdraft occurs shortly thereafter.

Some argue that this is artificial because there is no vent in the room. Note that in chapter 2 an article was referenced about dangerous fires in rooms with modern multiple-layer, energy efficient windows, which resist breaking in a fire.

Entirely aside from this film's intended message, about how quickly a small fire can become deadly, the film demonstrates the awesome potential of a backdraft to fire fighters who may never have seen one. This is a hazard too dangerous to demonstrate in

Tactical Considerations, continued

training. This author believes it is impossible to beat down such a fire with a hose stream. Fire fighters flattened to the floor with their faces buried in their arms might be in the best survival position. There are no guarantees.

All fire fighters should see this tape, with the hope that the first backdraft may not be a fire fighter's last.

Difficulties

Materials used in building interiors can be confusing. It is not wise to make any assumptions about them. Specific knowledge is needed. For instance, imitation brick observed in one building materials display was made of vermiculite, a noncombustible mineral material. Imitation stone in the adjacent display was made of glass-fiber reinforced polyester resin plastic. The glass fibers are non-combustible but the plastic is combustible.

Tactical Considerations

Begin with prefire planning. Prefire planning properly starts with a scenario of the fire potential in a building with all its ramifications. The rate at which a fire will grow is one of them. It makes a tremendous difference to know that the fire-resistive building is lined with fast-burning materials. It is likely that the usual, initial small line tactics will be ineffective and a waste of the time that is needed to bring heavy-caliber streams into play.

Company officers must be well informed as to flame-spread potential. Preplanning will forewarn personnel of possible flashover or heavy fire situations. Personnel should be well trained in the possibility of hidden fire. Learning by experience may be fatal.

Tactics and equipment should be developed to deliver the water necessary to control fast-spreading high-heat output fires.

Control of Rapid Fire Growth

There are several possible approaches to control the problem of fast fire growth.

Eliminate High Flame-Spread Surfaces

This is the path taken for U.S. merchant ships. Foreign ships may contain combustible trim and veneer. Reliance is placed on fire suppression from automatic sprinklers whose water supply might fail, particularly during an engine room fire. The rules for U.S. merchant ships severely limit the combustibility of construction materials and surface finishes. The rules for furnishings, however, apparently might not rule out combustible chairs or other contents that can provide the fuel for a huge fire.[48]

Separate the Material from the Combustion Source

A second approach is to separate the material from the source of combustion. In 1946, the author was investigating a major fire at a naval facility. An engineer who had been working there related that when building an office into this naval warehouse, fluorescent light fixtures were recessed into the combustible tile ceiling. He pointed out to workers at the time that a defective ballast in fluorescent fixtures can reach temperatures of 1500°F. His advice to avoid contact with the ceiling had been ignored. He was happy to help the author locate the fixture which caused the fire. Thereafter, the National Electrical Code®[49] was amended to include the following:

> 410-76 Fixture Mounting.
>
> (a) Exposed Ballasts. Fixtures having exposed ballasts or transformers shall be so installed that such ballasts or transformers will not be in contact with combustible material.
>
> (b) Combustible Low-Density Cellulose Fiberboard. Where a surface-mounted fixture containing a ballast is to be installed on combustible low-density cellulose fiberboard, it shall be listed for this condition or shall be spaced not less than 1½ inches (38mm) from the surface of the fiberboard. Where such fixtures are partially or wholly recessed, the provisions of Sections 410-64 through 410-72 [for flush and recessed fixtures] shall apply.
>
> (FPN) Combustible low-density cellulose fiberboard includes sheets, panels, and tiles that have a density of 20 pounds per cubic foot (320.36 kg/cu m) or less, and that are formed of bonded plant fiber material but does not include solid or laminated wood, nor fiberboard that has a density in excess of 20 pounds per cubic foot (320.36 kg/cu m) or is a material that has been integrally treated with fire-retarding chemicals to the degree that the flame spread in any plane of the material will not exceed 25, determined in accordance with tests for surface burning characteristics of building materials. (See Test Method for Surface Burning Characteristics of Building Materials, ANSI/ASTM E84-1984.)

This is a widely violated provision of the NEC®. Unfortunately, while the code provides a technically correct definition of the combustible material, the problem is that most electricians do not recognize it, except by its common name, "Celotex," which also happens to be that of a supplier of combustible fiberboard and many other building materials.

A few years ago there were two, multi-hundred-thousand-dollar fires in Maryland due to the improper installation of fluorescent fixtures on low-density fiberboard.

Cut Off the Extension

A metal-working shop in Richland, Washington had no fire problems, except one. There was a dip tank in one corner. Fumes from the dip tank rose overhead and coated the wooden roof with oil, creating a high flame-spread hazard. A minor fire in the tank spread rapidly under the roof, destroyed the plant and closed the business. The tank should have been isolated from the main part of the plant.

Some codes require a metal door sill separating more flammable carpeting, as in a motel room, from less flammable carpeting, as in a corridor. The intent is to prevent the spread of flame from a room fire to the corridor. The author has been told by an expert that this would be effective if the door were closed, but would have no effect if the door were open.

Flameproof the Material

Material can be painted with a fire-retardant coating. One difficulty with surface coatings applied after the hazard is discovered is that they do not protect the hidden surface of the material. Fire can burn unimpeded in the void. The surface treatment of the ceiling tile in the Hartford Hospital fire was defeated by the adhesive.

Fire-retardant surface coatings are effective only if applied as specified. The coverage is much less than that which is customary for paint. It is difficult to get painters to refrain from spreading the material too thin. There is an economic incentive to use as little as possible. One fire department requires that the requisite number of can labels be turned over to the inspector. This is not a foolproof method of assurance, but it at least demonstrates the fire department's interest that the job be done right. Another method is for the owner to buy the coating material, and to instruct the painter as to how many square feet of surface is to be covered by a gallon.

The Architect of the Capitol is responsible for a number of buildings on Capitol Hill, in addition to the Capitol itself. All materials used in construction or renovation are either non-combustible or treated with intumescent paints or solutions. Even structures that are con-

Fig. 9-10. Glass fiber-reinforced polyester resin plastic burns readily. It leaves a mat of glass fibers which should not be disturbed while doing an overhaul.

structed outdoors, such as the stand for an inauguration, are treated with intumescent paints when their loss may pose an inordinate risk to the people or operations of the Congress and the United States. An electrical fire erupted in the podium at the inauguration of John F. Kennedy. It was quickly extinguished by the D.C. Fire Marshal stationed there for just such an emergency.

Pressure-treated, fire-retardant wood (discussed in Chapter 3, Wood Construction) is often used where a wood surface is desired and flame-spread requirements must be met.

Changes in the formulations of plastics can reduce their ignitability. Materials can also be formulated to be flame-resistant. One manufacturer, for instance, offers a special hardboard that has a very low flame-spread rating and zero fuel contribution and smoke density. Such a board would be an excellent substitute for the common hardboard and pegboard that have high flame-spread characteristics.

One manufacturer of structural glass-fiber-reinforced plastics has produced panels which Factory Mutual will permit to be installed without sprinkler protection.

In 1988, the NBS investigated the fire hazard of a wide array of fire-retardant products, and nonfire-retardant but otherwise substantially identical products. The tests were conducted to determine the effectiveness of fire-retardant products, and to determine if quick and simple tests could be developed for them.[50]

Testing and Rating Materials

The first attempts to deal with flame spread by codes failed due to inexact, legally-unenforceable language, such as "flame spread no greater than wallpaper."

Fig. 9-11. A fire test in an anechoic chamber. Such chambers are lined with foamed plastic in various shapes. (Courtesy Factory Mutual)

Developing adequate tests is very costly and time consuming. Tests must be consistent, that is, tests performed in the same apparatus on the same material should produce the same results. Tests should be reproducible, that is, others using similar equipment and procedures should obtain the same results. Laboratories doing testing sometimes arrange round robin tests on samples cut from the same unit to help achieve uniformity.

The Tunnel Test

The basis for regulation of flame spread is found in NFPA 255, Method of Test of Surface Burning Characteristics of Building Materials, commonly referred to as the Steiner Tunnel Test. The test was developed by the late A.J. Steiner at Underwriters Laboratories Inc., and is also known as ASTM E84 and UL 723. The test is described in the NFPA *Fire Protection Handbook*.[51] Essentially, a test sample 25 feet long and 2 feet wide forms the top of a funnel or long box. A gas fire is lighted at one end, and fire progresses along the underside of the top of the box or test panel.

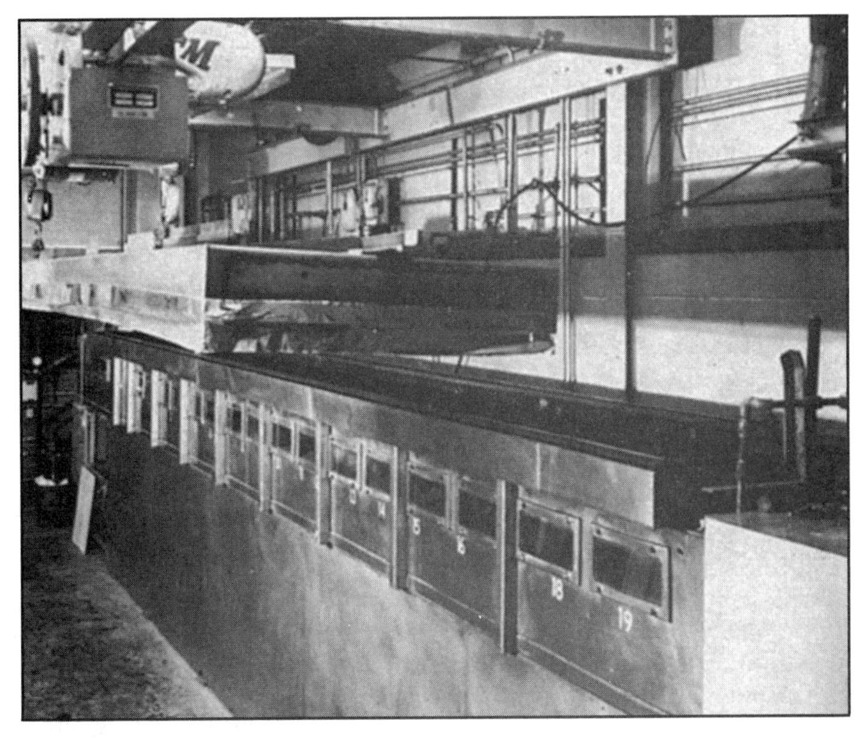

Fig. 9-12. The Steiner Tunnel (Courtesy Underwriters Laboratories Inc.)

There are two comparison points in evaluating test results. The flame spread over asbestos cement board is set at 0. The flame spread over red oak is set at 100. Tests have shown that fire can reach the end of a red oak test panel in 10 minutes. Other materials tested are rated on a scale determined by these points. The formula by which the flame spread index is calculated can be found in NFPA 255, *Surface Burning Characteristics of Building Materials.*

In NFPA 101®, Life Safety Code, surface materials are classified as follows:

Flame Spread

- Class A 0 - 25
- Class B 26 - 75
- Class C 76 - 200

In past years, Classes D and E were used for materials of very high flame spread. These have now been eliminated.

Most building codes have requirements for ceiling and wall surfaces based on the tunnel test standard. A typical requirement might be for a Class A flame-spread rating for corridors and exit ways, and less restrictive requirements for offices. Sometimes higher flame spreads are permitted for sprinklered buildings, which is another reason for fire department action when sprinklers are shut off. The Life Safety Code® allows a reduction in class given sprinkler protection, but in no case less than Class C, i.e., A→B, B→C, C.

When the tunnel test is run at UL, "smoke developed" is also measured and indexed.

Smoke developed is calculated by measuring the obscuration as the smoke passes a photoelectric cell placed in the stack from the test tunnel. Again, the loss of light from asbestos cement board and red oak are used as benchmarks.

It appears that materials with smoke-developed ratings of 300 or more can be expected to generate substantial amounts of smoke. Note that only light obscuration is measured, not any of the other effects of smoke or gases, such as toxicity or irritation.

The Life Safety Code makes use of the smoke-developed rating by including a requirement that surface finishes be classified at 450 or less on the smoke-developed rating of NFPA 255. In the author's opinion, this should be more restrictive.

The flame-spread testing standard is widely accepted. It is almost impossible to find literature for commercially-installed ceiling tiles without finding reference to the flame-spread rating. This is not wholly the case in the home market. Homeowners generally are unaware of the flame-spread testing standard or how it is applied to the tiles they buy.

In some cases, manufacturers have stressed low flame-spread ratings, neglecting to point out that the material advertised received a very high smoke-developed rating. All materials listed by UL are tested for both characteristics.

Caution About ASTM E84

Writing in the SFPE Bulletin of January 1981, David Breen, a well-known fire protection engineer, says in part:

> "Is the E-84 test a true to life test? No, because this test cannot be relied on to predict flame spread in the real world. Although the test may be very useful in comparing different families of materials, it is inherently unrealistic because it does not simulate real fires or end use applications. The spacial relationships between combustible objects becomes progressively more important in a developing fire. The ease with which an object catches on fire depends, in part, on the heat it has received from other burning objects, and the rate at which it burns depends in part on the flux of radiant heat it receives from other burning objects. These rates are compelled to change with time. Even when the physical size, chemical composition and spacial relationships of room furnishings and finish are well defined, a developing fire (one past the smoldering stage and on its way to flashover) in, say, a bedroom, is still difficult to chart with precision.

Howsoever the widespread acceptance of such tests over the years, together with their irrelevance to the true-to-life fire, suggests that the number of hotels, night clubs and restaurants throughout the United States which are not fully protected with automatic sprinklers, and which are therefore potential fire disasters, must be legion."

Fire-Rated

There is confusion in the building trades exemplified by the term "fire-rated." It is a commonly used, but undefined term, without a specific meaning.

Ceiling tiles may be part of a listed floor and ceiling assembly, fire resistance system or merely required to meet flame spread requirements. If replaced, it should be done with equivalent tiles observing the original installation practices.

Tiles that merely meet flame-spread requirements are not adequate for installation in required fire-resistive assemblies. Thus, the installation of a ceiling that meets flame-spread requirements does not provide any specific fire resistance.

A number of tradespeople and officials with whom the author has spoken seem to have no concept of the difference between installing a true fire-rated floor-ceiling assembly and simply putting up flame-retardant tiles under an existing floor.

The Radiant Panel Test

The National Bureau of Standards developed the ASTM E162, Radiant Panel Flame Spread Test. Samples for this test measure only 6 by 18 inches. For some materials, results from this test can be correlated with the tunnel test, but no direct relationship should be assumed. The radiant panel test has been used to develop information after serious fires, but would probably not be valid evidence in a prosecution based on noncompliance with a code. This flame-spread test is more fully described in the NFPA *Fire Protection Handbook*.[52]

Factory Mutual Corner Test

This corner test by Factory Mutual is designed to simulate an actual fire within the corner of a building. In the test the walls are up to 25 high; the east wall is 50 feet long; the south wall is over 37 feet long. The exposure fire is a pile of pallets arranged to simulate the NFPA 251 (ASTM E119) standard time/temperature exposure test for 15 minutes, or less. For further information see the NFPA *Fire Protection Handbook*.[53]

Carpet Tests

Floor covering, specifically carpeting, has been an important contributing factor in a number of serious fires. In recent years, many tests and standards have been developed for them. Commercial carpeting presently being manufactured is required to meet a test that measures ignitability of carpeting from a small source such as a dropped match or cigarette. It is a screening test (popularly known as the pill test), and seven out of eight samples must pass the test.

There is also a test to rate carpeting for its ability to spread flame when attacked with a greater ignition source than a cigarette. The test is NFPA 253, Standard Method of Test for Critical Radiant Flux of Floor Covering Systems Using a Radiant Heat Energy Source (also known as ASTM E648).

Radiant Flux Test

This Radiant Flux Test (NFPA 253) measures the ability to resist flame spread. The result derived from the test is the Critical Radiant Flux (CRF) of the sample. This is the amount of external radiant heat energy (measured in watts per square centimeter) below which a flame front will cease to propagate. The higher the CRF number, the less flammable the carpeting.

The test uses the radiant panel set at an angle. The carpeting sample is set flat in its normal position. The sample should be tested with the intended underlayment. If a carpeting is flammable, the flame spread will probably increase, that is, the critical radiant flux will decrease if an underlayment is present. The underlayment acts as an insulator, keeping the heat from the carpeting from being dissipated to the floor.

For a comprehensive and easily understood discussion of this important subject, study *Flammability Testing for Carpeting* by Irwin Benjamin and Sandy Davis of the National Bureau of Standards.[54]

Where regulated, the Life Safety Code considers two classes of interior floor finish:[55]

- Class I ≥ CRF minimum of 0.45 watt/sq cm
- Class II ≥ CRF minimum of 0.22 watt/sq cm

A reduction in class is permitted in spaces protected by an approved automatic sprinkler system.

Industrial Risk Insurers uses these same values to recommend protection for unsprinklered and sprinklered areas respectively.[56]

It is important to determine to what extent this standard has been adopted by the local code.

Critical radiant fluxes of carpeting involved in several serious fires have been determined. Thus, when regulations are adopted specifying a minimum CRF, there can be a relationship to specific fire danger which is not so apparent in other standards.[57]

Faked Tests

In one case, a carpet manufacturer took over another manufacturer. Shortly thereafter, the new owners learned that employees of the purchased company had faked the testing of millions of feet of carpet which had been produced and shipped, with respect to radiant panel and smoke density tests. Faked tests can bring about devastating results.

Don't Be Mousetrapped

Fire departments should be very wary of conducting their own tests. Tests which prove that a particular product is the answer to all problems should be suspect. Testing is a difficult and exacting profession. Leave it to the experts.

Conclusion

Research work is underway which might make it possible to write improved building codes "offering greater product differentiation and protection for life and property by including laboratory tests for product requirements with acceptance criteria based on the product's performance in real fires, as determined by validated computer models."[58] This research will be very beneficial to all concerned with fires in buildings.

Nevertheless, fire will always present hazards to responding fire fighters. It is difficult and sometimes impossible to determine the flame-spread characteristics of materials in place. Both low and high flame-spread materials can have the same appearance. Field testing can be dangerous. In one case in an eastern city, the fire department was conducting a campaign against flammable decorations. An inspector examined a display for a popular brand of blended whiskey. He decided to test the flammability with his lighter. The display was indeed flammable. The fire went to multiple alarm. The city paid a heavy judgment and two fire fighters were seriously injured.

If samples can be removed, a simple test may help determine the extent of the fire potential even if it is not adequate for legal purposes.

The ever present problems of fire growth must be carefully addressed by fire protection personnel. If not, the results may be tragic.

References

The addresses of all references are listed in the Appendix.

1. Grant, C. C. "Last Dance at the Cocoanut Grove." *NFPA Journal*, May/June 1991, p. 74.
2. Moulton, Robert. "Combustible Fibreboards." *NFPA Quarterly*, October 1949.
3. Gelbhaus, R. "Recalling Boulder." *Firehouse*, January 1988, p. 31.
4. Bell, J. "24 Die in Los Angeles Apartment Building Fire." *Fire Journal*, May 1983, p. 43.
5. Cote, R., and Timony, T. "Boarding House Fire Causes Fifteen Deaths." *Fire Journal*, January 1985, p. 44.
6. Burns, R. "Flashover Kills Reading Fire Fighter, Three Civilians." *Firehouse*, May 1985, p. 44.
7. Best, R. "Fire in Community Home Causes Seven Deaths." *Fire Journal*, March 1984, Vol. 78, No. 2, p. 19.
8. Timony, T. "Eight Mentally Handicapped Occupants Die in Georgia Fire." *Fire Journal*, May 1984, p. 91
9. Muir, J. "Exposed Insulation Facings Still Present a Fire Hazard." *Fire Journal*, September/October 1989, p. 15.
10. "Fire Record." *Fire Journal*, March 1985, p. 10.
11. Shaw, Gaylord, and Gillette, Robert. "Urethane, a Deadly and Pervasive Peril." *Los Angeles Times*, January 21, 1979.
12. Sawyer, R., and Eisner, J. "Cable Fire At Brown's Ferry Nuclear Power Plant." *Fire Journal*, July 1976.
13. Pryor, A. "Brown's Ferry Revisited." *Fire Journal*, May 1977.
14. Lathrop, J. "Telephone Exchange Fire NY, NY." *Fire Journal*, July 1975.
15. Kaufman, Stanley. "Fire Tests and Fire Resistant Communications Cables." *Fire Journal*, November 1985, p. 33.
16. *Factory Mutual Record*, "A Measure of Cable Flammability." September/October 1988, Vol. 65, No. 5, p. 3.
17. Prasso, C. "Cable Trays, Confining Cables and Losses." IRI *Sentinel*, April/June 1988.
18. McElroy, J. *NFPA Quarterly*, April 1989; Moulton, R. *NFPA Quarterly*, October 1989.
19. Bugbee, P. *Men Against Fire*. National Fire Protection Association, Quincy, Mass., 1971, p. 120.
20. *Fire Command*. "The Dundalk Deaths." March 1985, p. 48; Also April 1985, p. 18.
21. Comer, W. "Paterson N.J.'s Fatal Hotel Fire." *Fire Engineering*, January 1985, p. 24.

22. Isner, M., and Cahanin, G., "Fire Investigation Report Elderly Housing Fire, Johnson City, Tennessee." December 24, 1989, National Fire Protection Association, Quincy, Mass.

23. Brannigan, Vincent. (JD) "Record of Appellate Court Cases on Retrospective Fire Codes." *Fire Journal*, November 1981, p. 62.

24. Juillerat, E. "The Hartford Hospital Fire." *NFPA Quarterly*, January 1962, National Fire Protection Association, Quincy, Mass.

25. "Report of the MGM Grand Hotel Fire." Clark County Fire Department, Las Vegas, Nevada, 1981, p. v-16.

26. Stronach, I. "Multimillion Dollar Convention Center Saved in Fire (Montreal)." *Firehouse*, March 1986, p. 69.

27. *Fire Journal*. "White Grotto Becomes Black Tomb." Vol. 65, No. 3, May 1971, p. 91.

28. Rasbash, D. "Investigation Into the Stardust Disco Fire, Dublin." *Fire International*, April/May 1984, pp. 34-35, 38-39.

29. *Architectural Forum*. "Bucky's Biggest Bubble." June 1966, p. 74 (Bucky refers to famed designer Buckminster Fuller.)

30. Forgey, B. "Cityscape." Washington Post, June 29, 1991, p. 1, Style Section.

31. Lathrop, James K. "The Summerland Fire." *Fire Journal*, Vol. 69, No. 2, March 1975, p. 5.

32. "Investigation Report on the Las Vegas Hilton Fire." *Fire Journal*, Vol. 76, No. 1, January 1982, p. 52.

33. Sharry, J., and Stone, W. "School's Haunted House Burns." *Fire Journal*, March 1974, p. 14.

34. Demers, D. "Ten Students Die in Providence College Dormitory Fire." *Fire Journal*, July 1978.

35. Bouchard, John. "Fire in Haunted Castle Kills Eight." *Fire Journal*, September 1985, p. 45.

36. Saly, Alan. "Testimony on Trial." *Firehouse*, January 1986, p. 48.

37. *Firehouse*. "Forum." April 1986.

38. *Sprinkler Quarterly*, Fall/Winter 1986, The Journal of the National Fire Sprinkler Association, Patterson, N.Y.

39. Chapman, E. "Letter." *Firehouse*, May 1986, p. 6.

40. Nelson, H. "An Engineering View of the Fire on May 4, 1988 in the First Interstate Bank Building, Los Angeles, California." NISTIR 89 4061, NIST, Center for Fire Research, National Institute for Standards and Technology, Washington, DC., 1989.

41. Nelson, H. "Fire Growth, First Interstate Fire." *Fire Journal*, July 1989, p. 25.

42. Klem, T. "97 Die In Arson Fire at Dupont Plaza Hotel." *Fire Journal*, May/June 1987, p. 74. (Also in same issue, Willey, E. "The Lesson Taught by Dupont Plaza." p. 82.

43. Nelson, H. "An Engineering Analysis of the Early Stages of Fire Development in Fire at Dupont Plaza Hotel and Casino December 31, 1986." NBSIR 87-1560, May 1987, Center for Fire Research, National Institute for Standards and Technology, Washington, D.C.

44. DeCicco, P. *ed.* "Editorial." *Journal of Applied Fire Science*. Vol. 1., No. 2, 1990-91.

45. Fang, J., and Breese, J. N. "Fire Development in Residential Basement Rooms." NBSIR 80 2120, Center for Fire Research, National Institute for Standards and Technology, Washington, D.C., 1980.

46. Jervis, R. "Fire Resistant Furnishings." *Fire Journal*, November 1987, p. 56. This detailed article is up to date on work done to make upholstered furniture less combustible.

47. "Fire, Countdown to Disaster" and "Fire Power" are both available from NFPA, as 16mm film and videotape. Also, "A Rare Look at Flashover." *Fire Command*, August 1986 provides a series of spectacular pictures taken during "Fire Countdown to Disaster." Also Custer, R. "Fire Power, Making the Movie." *Fire Command*, November 1986.

48. Abbott, J. C. "Fire Involving Upholstery Materials." *Fire Journal*, Vol. 65, No. 4, July 1971, p. 88.

49. NFPA 70, National Electrical Code, National Fire Protection Association, Quincy, Mass.

50. Babrausks, V., *et al.* "Fire Hazard Comparison of Fire Retarded and Non Fire Retarded Products." NBS Special Publication 749, July 1988.

51. Cote, Arthur E., *ed. Fire Protection Handbook*, 17th edition, Section 6, Chapter 4, page 6-37, National Fire Protection Association, Quincy, Mass., 1991.

52. Ibid, *Fire Protection Handbook*.

53. Ibid, *Fire Protection Handbook*.

54. Benjamin, I., and Davis, S. "Flammability Testing for Carpeting." NBSIR 78 1436, Center for Fire Research, National Institute of Standards and Testing, Washington, D.C., 1978.

55. NFPA 101®, *Life Safety Code*, 1988, Sections 6-5.4 and 6-5.7.2.

56. *The Sentinel*. "Carpeting." April 1982, p. 8, Industrial Risk Insurers, Hartford, Conn.

57. Schnipper, S. "Carpet May Not Meet Fire Codes." *Engineering News Record*, May 14, 1987, p. 15.

58. Todd, N., and Ryan, J. "Improving Codes by Prediction Product Performance in Real Fires." *Fire Journal*, March/April 1990, p. 58.

Additional Readings

Lyons, J. *et al.* "Fire Research On Cellular Plastics: The Final Report of the Products Research Committee." 1980.

Parker, W., King-Mon, Tu, Nurbakhsh, S. and Damant, G. "Furniture Flammability: An Investigation of the California Technical Bulletin 133 Test. Part III: Full Scale Chair Burns." NISTIR 4375, NIST.

10

Smoke and Fire Containment

Fires generate a number of products of combustion. This chapter discusses what the average citizen, and some researchers, call "smoke." In fact, there should be distinctions made among smoke, particles, and fire gases. Many years ago, it was considered helpful to explain to the survivors of fire victims that they did not suffer because they suffocated before they were burned! Over the years there has been a gradual change to recognition that smoke and toxic gases are more significant as fire killers than thermal exposure.

Smoke

Smoke is made up of solid and liquid particles and is the usual visible product of combustion. Smoke often provides the first warning of most fires. It can do more damage to property than the fire or water, and may reduce visibility to zero. Its irritating particles may cause retching and immobility, and those irritants remaining in the respiratory system also may be toxic agents for a surviving fire victim.

Gases

Fire gases can cause injury or death when inhaled or in some cases, absorbed. Some fire gases can paralyze or slow human ability to function or escape. Carbon monoxide (CO), the most prevalent of the toxic fire gases, has this effect.

For most toxic materials the toxic effect is a product of concentration and exposure time. Habel's Rule states that any exposure in which the concentration (parts per million) x (minutes) equals 33,000 is likely to be dangerous. Carbon monoxide is probably the most common toxic fire gas. According to the NFPA *Fire Protection Handbook*, a 10-minute exposure to 3500 ppm of CO would be hazardous, possibly incapacitating. Higher exposures with greater concentrations are even more dangerous — 12,500 ppm may be fatal after only a few breaths. The toxic levels for other materials can be quite different.

Many other fire gases (such as nitric acid and hydrochloric acid, to name only two, for example) are toxic, and, in sufficient dosage, lethal. In 1929, the Cleveland (Ohio) Clinic, like many other hospitals, used nitrocellulose X-ray film even though safety film was available. A fire in film storage facilities sent clouds of deadly nitrous oxide gas (one breath can kill) up through the building and 125 doctors, nurses, and patients died. Some hospitals today use nitrocellulose test tubes for blood tests. There is very likely a substantial storage of these in many supply rooms. A fire in one may tragically duplicate the Cleveland Clinic fire results.

Until the end of World War II, a nitrocellulose base film was used for commercial motion pictures. Fires involving this film have caused many disasters. There still are places where old, deteriorating nitrate film is stored, such as in libraries and museums, and maybe even in some hospitals.

There are many other materials which can generate gases in a fire which are hazardous in quite small doses. Many people, including some we might think would know better, are unaware of the hazards. A retired dean of Massachusetts Institute of Technology included nitrocellulose in a list of plastics which might be useful in construction.

Polyvinyl chloride (PVC) is a very effective electrical insulator and has made itself indispensable as a contemporary material. When it burns, however, toxic gases are emitted.

It is this author's opinion that toxic gases from a burning combustible metal deck roof assembly were the chief cause of the loss of over 160 lives in the Beverly Hills Supper Club fire rather than the commonly accepted theory of gases from electrical insulation.

Tactical Considerations

Ask what your local hospitals use for disposable tubes. Radiologists may be aware of the Cleveland Clinic disaster, but other laboratory staff may be completely unaware of the hazard.

Tactical Considerations, continued

It usually falls to the fire service to maintain a memory of unusual fire hazards. There is very little "out-of-date" experience in the fire service — perhaps included in this category is how to shoe a fire horse.

Any location where nitrocellulose or any other especially toxic material is stored demands intensive preplanning with emphasis on defensive operations.

In any given situation one gas or another might be more toxic. Nothing generated by a fire is good to breathe. While most fire departments have accepted the concept of wearing SCBA for interior attack, the practice of removing SCBA for overhauling needs examination. It was noted earlier that moisture and vermin-resistant wood emit toxic products which indicate the need for SCBA on some exterior fires.

Flammability

Carbon monoxide is a flammable gas. When gases ignite there can be a wide range of overpressures ranging from the shattering of glass windows to the shearing of brick masonry walls. For a technical discussion of explosions see the *Fire Protection Handbook* 17th Edition, Section 1, Chapter 5. Most of the technical studies are made in connection with the design of containers. This author does not know of any technical investigation of what happens in a typical building fire.

By observation fire fighters note that in some cases CO trapped in a void or pocket simply "lights up," that is, the gas ignites when sufficient oxygen is available. This is sometimes seen as flames appearing in clouds of black smoke pouring from windows. This is also a sign that **flashover** is imminent.

Flashover is described as the sudden ignition of the gases distilled from the heating of the contents and interior surfaces of a structure. Much of the gas generated is carbon monoxide.

When flashover is imminent, fire fighters should be prepared to withdraw. In a tape film, "Flashover," FDNY Deputy Chief Vincent Dunn sets a limit of five feet for fire fighter penetration into a room where flashover is imminent.[3]

In other cases the ignition is more violent, with sufficient energy to break glass windows. This is sometimes observed by neighbors adjacent to a fire in a closed building. Newspaper reports often describe an **explosion** which called attention to the building.

In a few cases the gas-air mixture is just right for maximum energy release. In this case brick walls have been blown down.

Fig. 10-1. A carbon monoxide explosion blew the front off this Ft. Collins, Colorado building, killing one fire fighter.

Any or all of the above have been described as a **backdraft**. Fire fighters in the vicinity are not in a position to make detailed scientific observations. The many variables involved make a complete scientific study unfeasible. You must understand the subject as best you can with anecdotal information from survivors. See Chapter 2 for definitions.

The shock wave of a CO detonation has blown entire buildings apart. Carbon monoxide is not the only gas encountered under explosive conditions. For example, gases from burning polyurethane exploded and caused the collapse of a wall on the sound stage of a Hollywood studio in 1974.

Gases can accumulate in any enclosed area. Research at NIST has pointed out that CO can be generated up to 50 times as much in enclosed voids as in the open.[1]

Some 47 people died in a 1990 discotheque fire in Zaragoza, Spain. A small fire in the large void above the ceiling of the first floor produced huge quantities of CO and smoke, which poured down the stairs into the basement where the victims died, many of whom were sitting in their chairs.[2]

This tragedy is reminiscent of the 97-fatality fire in Our Lady of Angels School in Chicago in 1958, which this writer observed. Some fire professionals share the opinion that the fire had reached the attic overhead and blew down on the classrooms, possibly when a ceiling vent was opened to reduce the temperature.

Fig. 10-2 **(A)** *One moment the row store was visible.* **(B)** *Suddenly the wind shifted, everything turned black, and a carbon monoxide explosion felled several fire fighters. (Bottom photo: Wheaton, Maryland Rescue Squad)*

Ventilation of the building does not prevent CO explosions. The gas-air mixture can pocket in a fully or partially enclosed location. Several serious CO explosions are described in Chapter 4, Void Spaces.

Smoke vs. Gases

Smoke and gas have different physical effects on people. At one time, filter masks similar to those developed for use in World War I, were widely used in the fire service. They are very dangerous. The filter in that poison gas mask removed solid particles but allowed the odorless CO through. The fire service version contained a chemical which converted small amounts of CO to harmless carbon dioxide (CO_2), but in a number of cases the CO concentration was too high and fatalities resulted.

This same characteristic may be significant in high-rise fires. The smoke particles and CO moving up through the building cool off. Visible smoke often stratifies. Smoke particles may become deposited on surfaces. Yet, the colorless, odorless and tasteless CO can stratify well above the fire, on floors where there is little or no smoke, heat, or air movement. Fire fighters can walk into such death clouds. In 1961, a sub-basement fire occurred in New York City. Two fire fighters, who had left SCBA behind, died of CO poisoning on the 20th floor.

Smoke particles can plug up screens. This is a problem in prisons where screens have been used on windows to prevent passage of contraband. In one case, smoke from a mattress fire plugged the screen and inmates died. This same plugging effect should be anticipated in any location where the outgoing air is filtered, such as where toxic materials are handled.

In 1973 the National Bureau of Standards tested polyurethane and neoprene mattresses as reported in NBSIR 73-177, Parker, W. "Comparison of the Fire Performance of Neoprene and Flame Retardant Polyurethane Mattresses." Conclusion 10 reads:

> "In overall rating of fire hazard the neoprene mattress would appear to present the least contribution to the fire hazard under an intermediate level fire."

Tactical Considerations

Wear SCBA in any location where it is possible that toxic gases are present even though there is no apparent indication of them.

The fire department should be aware of the type of mattresses and cushions used in any place of confinement.

Smoke Damage

Water damage is often considered the most expensive byproduct of fire suppression. It is mentioned in most press accounts and, at time, is regarded as the sole responsibility of the fire department. To varying degrees, fire departments attempt to mitigate this damage. Some departments do extensive salvage work; others ignore it. Ironically there is never any mention of water damage at a total loss. As has often been noted, a good answer to those who complained of water damage is, "there wouldn't have been any damage if YOU hadn't had the fire." In fact, smoke damage may well be the most expensive byproduct of a fire, particularly where delicate equipment is involved or health-menacing contamination, such as food spoiled by the smoke, occurs.

Contaminated Smoke

It is costly enough to clean up the dirty, oily smoke which comes from such materials as plastics, but if the smoke is contaminated with another health hazard substance, such as radioactive material or PCB, the cost of cleanup can be astronomical. The relationship of health hazard to the fire loss is not nearly well enough understood though the problem has been around for some time.

The fire problem of radioactive material grows out of the simple fact that radioactive materials emit energy which can damage living tissue, thus the safety of personnel might demand abandonment of the property until it is cleaned up, or perhaps permanently.

In the Rocky Flats Nuclear Weapons Plant a fire did about $ 1 million in conventional damage. The cost of cleanup was 25 times greater.

Whether the fire fuel is radioactive, or a fire in another fuel disperses the radioactive material, the results can be similar. In the Hanford, Washington AEC Plant, a fire spread contamination more efficiently than a nuclear accident. A nuclear experiment went critical (when a nuclear reaction becomes self sustaining it is said to "go critical") and spread contamination around the building. It was successfully cleaned up with paper diapers, acetone, and nitric acid. The waste was then packaged to await disposal. The nuclear experts had forgotten their chemistry. The sealed packages ignited spontaneously. The fire spread the same contaminants so effectively that the building could not be cleaned up economically. It was closed off and years later dismantled and buried.

It is easy to control the health hazard of contamination — keep people away. Because the loss of use of a property can be very costly, however, and clean up can be prohibitive, it is sometimes less expensive to abandon the property.

A fire at another laboratory was stopped at a plywood wall. If it had passed the wall, it would probably have spread enough contamination to cause the permanent shut down of the laboratory.

High Efficiency Fire Resistive (HEFR) filters were developed by the AEC Safety and Fire Protection Branch. Filters used in the atomic energy program are individually tested. Such filters are listed by UL. HEPA (High Efficiency Particulate Air) filters used in industry may not be fire resistive. Massive filter banks require specially designed fire protection. A fire in the filters might disperse the contaminants collected on them. While these incidents involved radioactive material, the problem is the same for other contaminants such as PCBs, asbestos, or hazardous chemicals.

Work by H. Gilbert of the AECs' Safety and Fire Protection Branch led to the development of fire-resistant filters. If filters are used in a plant in your community, make sure your fire department is aware of their fire characteristics.

Fig. 10-3. This demonstration at the U.S. Navy Structural Fire Protection School involved four combustible HEPA filters which were mounted in a plenum built for the purpose. An air mover was drawing air through the filters and exhausting it. The four filters were ignited simultaneously. Carbon monoxide accumulated in the plenum behind the filters and ignited violently as seen here. This is a reliable way to demonstrate a backdraft.

Tactical Considerations

Prefire planning for potential contamination situations should embrace the entire problem. Particular study should be given to procedures which will minimize residual contamination.

Systems for contamination control should be examined for potential failures that could be caused by fire. These should include the electrical supply to fans and the type of filters used.

There have been a number of serious contamination incidents from polychlorinated biphenyls (PCBs) released from electrical transformers during fires. PCBs were used as non-flammable coolants in transformers. There are rules requiring notification of the presence of PCBs, but they are not infallible.

During a relatively small fire in a Binghamton, New York office building in 1982, a transformer cracked. PCBs that were used in the transformer coolant heated up and released toxic products into the atmosphere.

Clean-up costs have exceeded the value of the building and as of August 1991 the building is still unoccupied. The cleanup is very slow and costly because of the precautions which had to be taken to protect the workers. Fire fighters should be aware of the extremely toxic nature of PCB and the fact that it can be released from any transformer by fire or accident or even in routine maintenance.[4]

The telephone system is the nervous system of the community. Life is entirely disrupted when lines are out. Seriuos financial losses can also occur when the telephones are out. Fires in telephone company facilities are not ordinary fires. In 1975, a massive disruption of communications resulted from a fire in a basement cable vault of New York Telephone's main switching center. On May 8, 1988 a serious fire occurred in an Illinois Bell Telephone Central Office in Hinsdale, Illinois. Much of the multi-million dollar loss was due to smoke damage.[5]

Tactical Considerations

Fire Departments which have a telephone facility in their area should obtain and study such fire records. Possibly more important than the dollar loss of the telephone system is the effect on the community. There is conflicting information on telephone company opposition to sprinklers. I have a letter indicating that sprinklers are not only acceptable but desirable. Others have informed me otherwise.

Most important is realistic prefire planning involving both the fire department and telephone company personnel.

Fig. 10-4. A cable fire in the basement vaults of this telephone building caused serious health problems to a number of fire fighters. Note the tower ladder in operation. Those who oppose sprinklers should be reminded that this drenching is the alternative to much smaller quantities of water that would be delivered directly to the fire by sprinklers.

In another fire, a $30 million loss was suffered when a fire confined to burning PVC insulated cables and the roof above them generated smoke which corroded precision machinery.

In 1989, a Safeway grocery store warehouse burned for several days in Richmond, California. A single lawsuit presented the claims of 5,500 claimants, alleging failure to segregate hazardous materials from flammable materials. Eighteen people died allegedly from fumes. Newspaper reports indicated that Safeway agreed to pay $32 million to claimants. This case, or others like it, may be helpful in getting cooperation from warehouse management.

Corrosion

Delicate equipment as well as brickwork and concrete can be damaged or destroyed by corrosive products of combustion. Plastics containing halogens such as chlorine, fluorine, bromine, or iodine form corrosive acids when combined with hydrogen and oxygen, or moisture in the air.[6]

Fig. 10-5. The damage to these bricks was caused by corrosive smoke. (Industrial Risk Insurers)

Smoke Movement

The chief mover of smoke in a fire is gravity. We are all familiar with the fact that gravity causes things to fall down, but few of us know that gravity pulls on the surrounding heavier, colder air, causing the lighted heated air to rise upward.

When elevators were first introduced into buildings, designers saw the need for surrounding the shaft with a cage to prevent people from falling into it. There was no recognition of the fact that the open shaft was a perfect path for the extension of fire upward. All over the world, even in quite primitive societies, buildings are built which successfully defy the law of gravity as to collapse. Unfortunately, only where sad experience has caused adequate building codes to be developed are precautions taken to prevent the upward spread of fire and smoke.

Note that in buildings where dangerous gases, such as gasoline, which is heavier than air, are handled, vents are located at the floor line. By contrast, where hydrogen, which is lighter than air, is handled, vents are located at the ridge line. Such vents are called **gravity vents**, to differentiate them from **mechanical vents** (fans).

Figure 10-6. The designer of this open elevator shaft understood well that people might fall down it, and provided the wire guard. Unfortunately, there was no understanding that deadly heated smoke and gases would rise up the shaft.

There are many other factors which affect the movement of smoke from a fire. They are discussed in Chapter 11, High Rise Buildings.

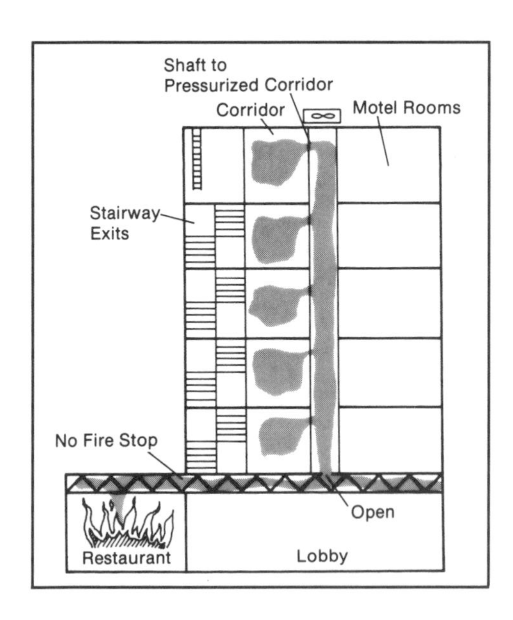

Fig. 10-7. Smoke originating from a fire in one portion of a building will travel upward through vertical voids, polluting the entire building. Deaths in many hotel fires have resulted from such movement of smoke.

Containment of Fire

When it was learned from such fires as the 1908 Parker Building fire in New York that fire could extend up open stairways in "fireproof buildings," the concept of compartmentation or creating fire areas no larger than one floor was developed.

All connections between floors were designed to stop the spread of fires. Stairways in new buildings were enclosed and provided with self-closing doors of rated fire resistance. Where there was a life safety concern, outside metal fire escapes or true fire towers with atmospheric breaks were installed.

Self-closing doors are considered a nuisance to many of those who must move from floor to floor. Many times open interior accommodation stairs are installed. These make a single fire area out of two or more floors, as was the case in the One Meridian Plaza fire in Philadelphia. (See Fig. 11-19 in Chapter 11, High Rise Construction.)

Some building departments require sprinklers above the stairway openings. There is no proof that this will prevent the extension of fire in an otherwise unsprinklered building. It certainly will not stop smoke and gases! A distant fire can fuse the sprinklers, and scalding water could pour down on fire fighters.

Tactical Considerations

Fire departments should know where the shut-off valves for such partial sprinkler systems are located.

Stairway doors are often blocked open with wooden wedges. These are sarcastically called "four-hour fusible links" by inspectors. A few years ago the United Kingdom had a campaign to enclose stairways and install fire extinguishers. Ironically, in some cases the extinguishers were used to block open the stairway doors.

In the United States the stairway doors at the office of a well known consumer safety advocate were blocked open with bundles of safety pamphlets. To counter this blocking problem, fusible links or door checks with integral 165°F fusible links have been used. The concept was that the fusible link would fail at that temperature and close the door. However, the links are not sensitive to smoke. The stairway can be totally polluted before the link operates. In one case in Austin, Texas the fusible links were at the bottom of the doors.

Enclosing of stairways does not often come naturally or easily. The open grand staircase is a crowning architectural feature of a monumental building. The most famous is in the Paris Opera. It appears that the volume of the stairway is greater than the auditorium. An architect who worked in a city hall admitted that the open stairways would

Fig. 10-8. This building at the University of Florida only has a single sprinkler over the doorway. A distant fire on the floor could cause fire fighters coming from the stairway to be showered with scalding water.

cause a loss of life, but felt that closing up the stairways would destroy the beauty of the building.

After several multi-fatality hotel fires in the 1940s there was an effort made to enclose open stairways. In many cases, the corrective measures were defective in that often the stairway was left open at the lower floors.

In some building restorations a "monumental" stairway was left unenclosed. The operative, yet strange, argument is that sufficient exits are already provided by protected stairways. This begs the question of the open stairway as a transmitter of smoke and heat.

Tactical Considerations

The fire department should bear down on the management of buildings to keep such stairway doors closed. Signs should be provided, employees should be educated, and citations should be issued when necessary. It is foolish to have a law requiring enclosed stairways and then take no steps to ensure it.

Recognize that interior stairways can deliver toxic smoke and gases throughout the building. This requires an early, painstaking search of all floors above the fire.

At a fire, command must designate which stairways are to be used for fire suppression and which for evacuation.

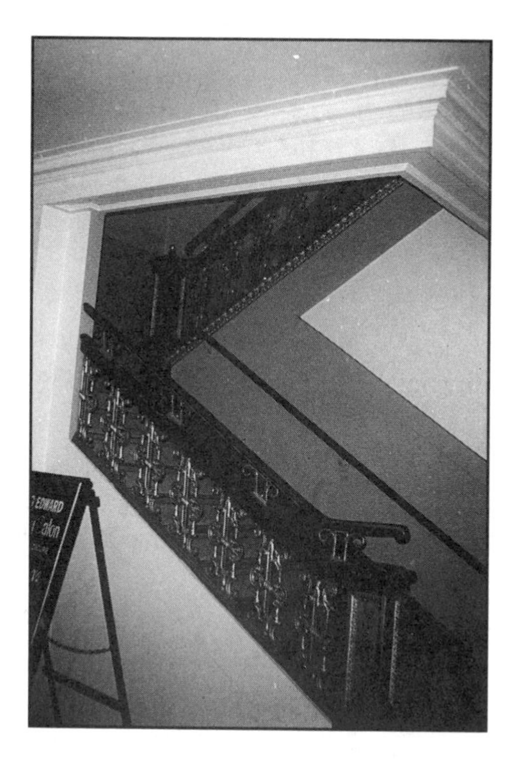

Fig. 10-9. The open stairway was closed up and equipped with doors on the upper floors, but the ground floor, where a great amount of fuel was stored, was left open. The stairway could quickly become completely smoke polluted.

Fire Door Closure Devices

There are two types of closure devices: self-closing and automatic.

A typical example of the self-closing type is the room or stairway door equipped with an automatic door check which closes the door after it has been opened. Earlier, the life safety deficiencies were noted of stairway door checks equipped with fusible links.

Fire walls with automatic fire doors intended to close only in case of fire are equipped with fusible links. When the link melts, the overhead rolling door comes down, or the sliding fire door closes horizontally. Sliding fire doors may roll down a track by gravity or be pulled shut by a weight. In either case, they should be checked by lifting up on the weight which holds the door open. The door should close properly without any assistance.

For a full discussion of protection of openings, see the discussion on Protection of Openings in the NFPA *Fire Protection Handbook*, 17th edition on page 6-85, and Protection of Conveyer Openings on page 8-214. More detailed information is found in the *NFPA Inspection Manual*, 6th edition, page 32.

Industrial Risk Insurers, a leading insurer of Highly Protected Risks (HPR), recommends the following, reprinted with permission from "Firewalls and Firedoors: Your Last Line of Defense" by Anson Smith in The Sentinel, Third Quarter 1991. The entire article is well worth reading.

Inspections should include:

- Operating fire doors and shutters.
- Raising the counterweights of automatic sliding and counterbalanced doors to be sure they close.
- Inspecting doors for damage.
- Inspecting all hardware including latches, guides, and thresholds for proper function.
- Checking the fusible links of automatic closing devices to ensure they are free of paint and other foreign matter that might impair their operation. (Note that the fusible links of sprinklers did not operate properly in a paper plant because they were covered with oily paper dust.)

Tactical Considerations

The fire department should not operate any of the equipment, but should supervise its operation by an industry employee. This might prevent involvement in a future lawsuit.

As with many devices or materials such closure devices may be completely inadequate to control smoke movement. They may not even control spread of fire. Unfortunately, it is a common practice to block open self-closing doors.

The fact that stairway and corridor smoke and fire barrier doors are often blocked open, however, should not prevent fire departments from working for self-closing doors in apartment and hotel occupancies. Many fire reports note that the spread of fire and smoke in such occupancies was due to doors left open — and not only in the unit of origin.

Non-fire resistive buildings, and older fire resistive buildings, were usually built without any consideration of the control of fire and smoke movement. As buildings are altered, there may be code requirements to upgrade the smoke and fire control. Some such alterations are more effective than others.

Smoke-Sensitive Releases

The development of the smoke detector has provided equipment which is sensitive to smoke and which may help to eliminate the problem

Fig. 10-10. The firesafety of this Canadian motel was upgraded by installing fire doors. But the staff promptly negated the improvement by placing wooden wedges to prevent the doors from closing.

of blocked open doors. In one method, the doors are held open by electromagnets. The door latch system can be triggered automatically by any fire or smoke alarm system, sprinkler water flow alarm, or manually. These doors should be designed to permit manual releasing. An alarm cuts the current to the magnets and the doors release. The door then becomes a typical self-closing door. Such a system is easy to test if the door contacts the magnetic latch when pushed to the wall. A recommended procedure for a patrolled building is for the test switch to be thrown after the cleaners are through for the night. All doors should close. A guard making rounds can report any non-functioning door. In the morning the current is restored. As doors are pushed open, they engage the magnet and stand open thereafter. Some doors are too heavy for magnetic latches. They are held by mechanical latches that are released the same way. Two-leaf doors should be checked to see that the doors close in the designed sequence. The roller that accomplishes this task is frequently damaged.

In other cases, the doors are controlled individually. One or two detectors (one on each side) trip a door as smoke is detected. In some cases, the detector is integral with the door check. If the door is open and the detector senses smoke, the door is released. Such doors should not have latches to hold them open.

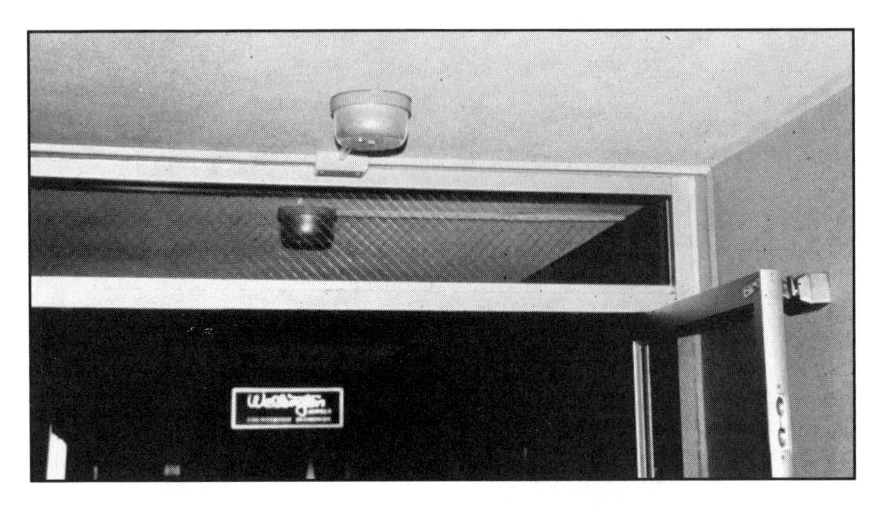

Fig. 10-11. Note the smoke detectors on each side of this smoke barrier door. The operation of the detector will cut the current to the magnet that holds the door open, and the door check will close the door.

Tactical Considerations

Fire fighters should carry wedges to hold electrically-released fire doors open when operations make this necessary.

Since many fire doors do not operate properly, units should be assigned to check all fire doors and close those which should be closed.

Fire fighters should never advance through a doorway protected by a fusible-link tripped fire door without blocking it to prevent its closing or dropping behind them.

Horizontal Barriers

In past years the concept of each floor being a single fire area was adequate. Today, however, floor areas in buildings have increased tremendously. This has necessitated horizontal barriers, either smoke or fire barriers. These are encouraged in health care facilities because the occupants lack the ability of self-preservation. The horizontal barrier is often resisted because closing down the opening to door size can damage architectural concepts.

A fire door was installed in the American History Museum over the objections of the Smithsonian Institution staff. It stopped a raging fire from destroying the Star Spangled Banner. The official who had fought the door later said, "I learned one thing in this investigation. If you are building a building to last 75 years, it will have a major fire, so fire is a design criteria."

Fire losses would be greatly reduced if designers and developers generally adopted this point of view. Shopping malls have minimum horizontal barriers. This allows smoke to pollute an entire mall. One Virginia mall has had two transformer fires with multi million dollar losses.

The huge Sawgrass Mall in Sunrise, Florida covers 2.2 million square feet. A central feature of life safety is a mechanical smoke removal system.[7]

In one case, a 1000-gallon propane tank fire seriously exposed a shopping mall. Flames broke windows in a Montgomery Ward store. To prevent smoke damage to the rest of the mall the doors were closed and taped.[8]

Tactical Considerations

The traditional interior attack may often result in smoke pollution which could be reduced by a change in tactics. In several cases, fires in basement storage and laundry rooms were attacked from the interior stairway, resulting in smoke pollution which would have been lessened by an alternate attack path, such as through a window. Of course, a standby line should be laid in the stairway to cover the storeroom door.

Fire fighters should be trained to control smoke damage to the extent possible. In malls this would include closing any available doors, particularly the sliding doors installed on major anchor stores. At times these stores have been "protected" only by a line of sprinklers across the opening. Store management should be advised that sprinklers will not prevent smoke from polluting the store contents. To reduce smoke pollution, where possible, fires on the lower floors of high-rise buildings should be attacked from aerial equipment while opening the interior of the building is held to a minimum.

In general, tactics should be reviewed to determine what changes might reduce smoke pollution.

Escalators

Escalators are widely used in department stores and public occupancies. There is strong resistance to enclosing escalators. A number of methods of "protecting" the openings with various types of sprinkler or spray systems have been adopted in various codes.

Fig. 10-12. There is an open sprinkler deluge system and conventional sprinklers above this escalator. The owner chose this combined system to avoid enclosing the escalator. Such a system could cause heavy water damage. The fire department should be alert to get the responsibility for the water damage properly placed. It also should know exactly how to cut off the deluge system.

One method involves the use of water spray nozzles directed downward through the opening. The fire department should be aware of this type of system because it can cause considerable water damage. Blame for this damage may be put on the fire department.

In some systems, a line of sprinklers is located around the escalators. The sprinklers are shielded from one another to prevent one sprinkler from cooling the other. While these sprinklers may extinguish a nearby fire, they would have no effect on smoke or gases moving from a fire not in the immediate vicinity.

Some escalator protection systems involve mechanical exhaust systems. In other cases, motor-driven shutters are used to close the escalator opening. These should be tested periodically.

In short, life safety in such occupancies depends on the full sprinkler protection of the building and not partial protection at floor openings. This is another reinforcement for the concept expressed elsewhere in this text: "If sprinklers provided for life safety are shut off, the occupancy should be closed."

Public Education

There is an ongoing problem of educating the public about the real situation in a fire. The public is bombarded with misinformation by movies and TV. A current hit movie "Backdraft" shows fire fighters running through flames without SCBA. The movie "Defend Your Life" shows Meryl Streep being commended for going back into a flaming house to rescue her cat. TV shows often depict the hero or heroine dashing through the flames. There is little or no smoke. If movies showed the real fire and smoke conditions, the audience wouldn't see anything.

In one tragic fire a 15 year-old girl who smelled smoke took four children to the second floor instead of getting them out of the house. Six persons died in the fire. Flammable paneling contributed to this disaster.

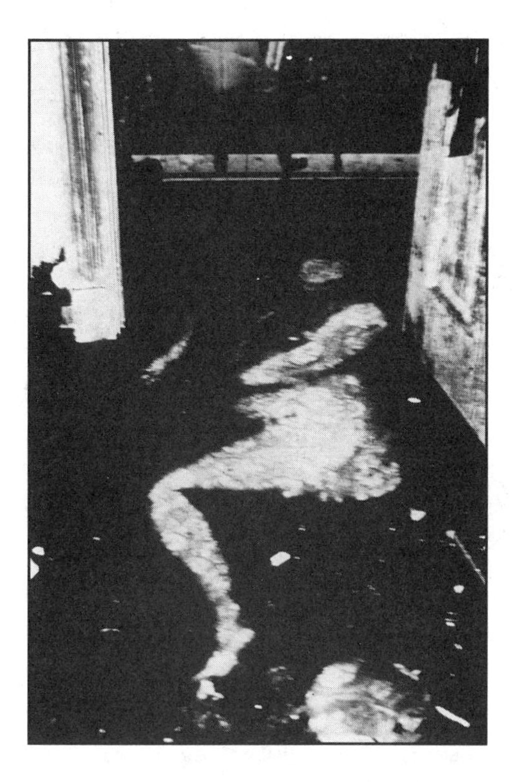

Fig. 10-13. Some tragedies could be prevented. This victim tried to change her clothes before leaving the building.

In her newspaper column, Heloise commended a contributor who included showing the baby sitter the location of the fire extinguisher, as one item in an otherwise excellent list of instructions. My letter pointing out the hazard of the baby sitter attempting to fight the fire went unacknowledged.

Are baby sitters in your community properly instructed to get the children out immediately and call help from a neighbor's house?[9]

A one-room fire in a fire-resistive hospice with a high level of fire protection, but without sprinklers in patient rooms, killed eight patients. The fire originated in a reclining chair with foam padding. A nurse rescued one of two patients from the room of origin. Smoke prevented the rescue of the second patient. The report reads: "Comments from several nurses indicate that they did not expect the rapid environmental changes experienced during the fire."[10] More likely they said "I didn't know that the fire could get that big and that smoky so fast." The only training most hospital personnel have is extinguisher training on small outside fires.

"Fire, Countdown to Disaster" an NFPA film/video, graphically shows how a chair fire in a bedroom goes to flashover in two minutes. It should be in the public education arsenal of every fire department.

Fig. 10-14. Another unnecessary tragedy. This victim thought he could use a fire extinguisher on a confined fire. Fire extinguisher training should also include when NOT to use them. With fire fighters wearing thousands of dollars of safety equipment, it doesn't make sense to install fire extinguishers and invite untrained, unprotected citizens to attack a fire with today's heavy fuel loads. The use of fire extinguishers by untrained persons not only can be tragic, but it often delays the alarm as well.

Fire Department Education- "Light Smoke Showing"

The on-arrival radio report, "light smoke showing," is commonplace. In most cases it indicates a trifling, easily controlled fire. In a significant number of cases, however, there have been serious fires or even disasters at "light-smoke showing" fires.

A sampling of Case Studies which highlight examples of light-smoke on-arrival conditions is included at the end of this chapter.

Smoke Detectors

Smoke detectors have saved many lives in residential fires. Unfortunately, too often the detectors are found to be missing or out of service. The educational effort stressing the importance of continuously operative smoke detectors should be intensified, particularly among low income and non-English speaking residents.

Often the detector is disconnected because of kitchen smoke. There are some detectors with a switch which will disable the detector for 15 minutes. Unfortunately those this author has tried have been unreliable. I keep my kitchen detector atop a cabinet. When it sounds off from cooking smoke I place it in the refrigerator, where it will be found after the meal and restored to its proper place.

There are two types of smoke detectors: ionization and photoelectric. They detect smoke differently. After reassuring the reader about the negligible hazard of ionizing radiation from a smoke detector, Consumers Union strongly recommends photoelectric detectors which are best at detecting smoldering fires, the most common type of home fire, and almost as good at detecting free-burning fires. Consumers Union reported that when they last tested smoke detectors the best performers were photoelectric detectors or combination units with both types of detection.[11] This author's personal preference is for a detector which incorporates both types of detection.

Earlier it was noted that a number of fires gained headway in attics without the knowledge of the occupants. It is worthwhile to repeat the recommendation that detectors should be installed in the attic of a residence.

In some cases, smoke detectors are relied on to the exclusion of automatic sprinklers.

One fire expert[12] has repeatedly warned that fire protection and fire detection are not synonymous. Only if the fire alarm is received where there is someone *ready, willing and able* to respond and control the fire can detection begin to be considered protection.

In some cases, the fire forces are not available, in others the fuel is of such a nature or the fire is so inaccessible, that the fire is out of control when the fire suppression forces arrive. In the case of a high-rise building, a realistic estimate of the time between the alarm and the start of extinguishment is at least 20 minutes.

Unwarranted Alarms

Unwarranted Alarms is a term used by New York City for alarms transmitted erroneously, except for sprinkler system emergency alarms from buildings required to be covered by an alarm system. Many unwarranted alarms received from the growing number of residential alarm systems are not included in the total below.

In 1987, the Fire Department of the City of New York responded to 20,400 such alarms. There are many problems with the regulation of private fire alarm systems. One article details New York's experience and recommendations.[13]

Some cities have instituted penalties for repeated, unwarranted alarms. However, such programs should have effective means to prohibit building managers from failing to transmit alarms. In several large-loss fires such as Los Angeles' First Interstate, Philadelphia's Meridian Place and the multi-fatality at Houston's Westchase Hilton, delay or failure to transmit the alarm received from detectors was a factor.

The City of New York is suing a building management company to recover the cost of fighting a fire, including subsequent medical costs, which sent 20 fire fighters to the hospital. The city charges that the building management delayed the alarm.[13]

Unwarranted alarms will probably increase as cable companies join alarm companies in offering fire and intrusion alarm service to households.

Ventilation

Fires produce huge volumes of heat, smoke, and gases. Starving the fire is one of the reasons sometimes advanced for compartmentation. Occasionally a fire has been cut off completely by compartmentation, but such cases are rare. The successes are offset by tragedies which occurred when openings were made into otherwise unventilated compartments. In 1932, eight fire fighters died when they opened the door on a smoldering fire in an unventilated storage room in the sub-basement of New York's Ritz Tower hotel. There have been a number of similar disasters.

Venting a fire, even at the risk of increasing its volume, is often necessary to make the conditions inside the fire building somewhat habitable for occupants and fire fighters. This was recognized over a hundred years ago in the case of multi-story fires in tenements in New York. Packed with hundreds of immigrants, these buildings were deathtraps. A single wooden stairway, with wood and glass doors to tenant apartments, rose through the center of a building. Fire fighters learned early on that opening the roof over the stairway relieved the pressure on the upper floors, where potential victims were often at the windows. In 1903, the New Law Tenement House Act required, among many improvements, that all stairways be masonry enclosed and noncombustible, and have a skylight, so that fire fighters could easily vent the stairway.

In 1930, Assistant Chief Thomas Dougherty, a man far ahead of this time, wrote "Fire."[14] In it, he not only explained the necessity of ventilation, he offered a design for automatic vents tripped by fusible links, to be installed atop stairways.

Some readers may object: "Automatic vents violate the cardinal rule, 'Don't vent till you have water'." Like all rules, there are exceptions. Theaters where live actors perform on stage with scenery are the first buildings ever designed so that a fire could occur and the occupants would be protected from the combustion products. Automatic vents are provided above the stage to reduce the pressure on the proscenium fire curtain. The stage may burn out, but the audience has a greatly improved chance of escape. This concept evolved from the hundreds of deaths which occurred in theatre fires.

However, in Chicago's 1903 Iroquois Theatre fire where 603 died, the stage vents were fixed closed. A gas light fixture blocked the curtain from closing. The program had boasted "absolutely fireproof." The building was, but the occupants weren't.

In 1919, the Holland Tunnel was being planned to connect New York and New Jersey under the Hudson River. It was designed to provide enough airflow through the tunnel to reduce the maximum amount of CO which could be generated by auto exhausts, to an acceptable level.

The permissible level for CO was determined by having volunteers drive cars, until affected, around a "racetrack" cut into the coal mine at the Bureau of Mines Experiment Station near Bruceton, Pennsylvania. This was the basis for the permissible level for eight-hour exposure. In the 1950's the permissible level was cut in half.

Tests showed that the airflow required would be about 50 mph. Ole Singstadt, the design engineer, objected. He claimed that if a fire occurred the occupants would be incinerated. He argued for a triple tunnel arrangement, an air supply tunnel below the car tunnel, and an exhaust tunnel above, so that toxic gases would be vented vertically, directly above the fire. The Port Authority management balked, citing the cost. Singstadt, a tough world class engineer, pointed out that he was the only person in the world who could build the tunnel, and it would be his way or no way. It was built his way. In 1951, a truckload of carbon disulfide burned in the tunnel. Fire showed out the New Jersey side vent tower, and it was necessary to use hose streams to cool the fans. Everyone in the tunnel escaped. Unfortunately, one fire officer died.

At this writing the Eurotunnel under the English Channel from England to France is under construction. Cars and trucks will be carried on trains. There are two separate train tunnels. A service tunnel between the two train tunnels is kept at a higher pressure than the train tunnels to exclude smoke. Access is provided through "smoke proof" doors at 375 m (415 ft.) intervals. Reports estimate that 1500 passengers will be evacuated from a burning train in 25 minutes.[15] Whether or not the planning is adequate remains to be seen.

Fig. 10-15. Ole Singstadt stood his ground against his bosses, and unprecedented loss of life has been avoided in the Holland Tunnel and many other tunnels because of his foresight.

Fire Department Ventilation

Fire department ventilation has consisted of opening and breaking windows, and cutting holes in the roof to allow smoke and heat to escape by gravity. A number of years ago smoke ejectors were developed to increase the volume of airflow. Later large fans were used to exhaust smoke, generally during overhaul. When this technique is used it is necessary to prevent outside air from entering the airstream, thus diluting the flow for combustion products.

Fig. 10-16. Los Angeles fire fighters used these positive pressure fans to ventilate this store. (Courtesy Chief J. Mittendorf)

Positive Pressure Ventilation

In recent years a new and promising method has been developed for use in ventilation.

Battalion Chief John Mittendorf of the LAFD authored an excellent text.[16] In it he provides good insights and strategies on effective ventilating. The following is abstracted from the text.

> Natural ventilation is effective with certain limitations. However, by utilizing portable smoke blowers, the natural ventilation process can be dramatically assisted or replaced by FORCING the movement of contaminants to:

- Exit through pre-selected and/or controlled openings.
- Allow the use of ventilation openings that are remote from the contaminants to be removed.
- Overcome the effects of humidity.
- Overcome interior-exterior temperature differentials.
- Move contaminants to the exterior of a building through areas or openings not normally utilized by natural ventilation.
- Reduce the time necessary to ventilate a building when compared to natural ventilation operations.

The term "blowers" refer to smoke ejectors and fans. These can be used to provide positive pressure ventilation (PPV).

To ventilate a typical one-story residential building, the door is opened and a blower is positioned OUTSIDE the building. This method will force clean, fresh air inside the building and create a positive pressure. The positive pressure will be equal at the top, bottom, and corners of the building. It is most important that one window be opened to permit the contaminants from ALL parts of the building to exhaust to the exterior.

The effective implementation of positive pressure is dependent of the following four considerations:

- Blowers must be positioned so the cone of air issued from a blower completely covers the entrance opening to the building to eliminate contaminants being forced back through the "unsealed portions" of the entrance opening and being reintroduced into the building. This is accomplished by varying the distance from the blower to the entrance opening.
- It is imperative that the flow or path of air between the entrance opening and the exhaust opening be controlled and directed to achieve effective ventilation. If air is directed from an entrance opening to an exhaust opening without being diverted to other openings, contaminants will be removed with the pressurized air in a minimal amount of time. Simultaneously opening multiple windows

and/or doors will not facilitate a successful positive (or negative pressure) ventilation operation.

- Exhaust openings can be selected to provide horizontal or vertical ventilation of contaminants. The size of the exhaust opening is dependent on the capacity and number of blowers being utilized.
- The key to effective positive pressure ventilation depends on controlling the entrance opening, path of the interior air flow, and the exhaust opening. The proper working relationship of all these elements of the ventilation operation requires training and an awareness of the ventilation goals.

Wind can have an adverse effect on positive pressure ventilation. Its effects are dependent on direction and velocity. As in any ventilation operation, maximum efficiency can be obtained by utilizing the prevailing wind direction to an advantage. It is most advantageous to pressurize the structure on the windward side and exhaust contaminants on the leeward side of the building. If it is not possible to use the prevailing wind advantageously, positive pressure has proven effective AGAINST winds of up to 25 mph. As winds exceed 25 mph, however, efficiency will be reduced.

Positive pressure has proven effective in all types of buildings including buildings with basements, multi-story office and industrial buildings, and high-rise buildings. The effectiveness of positive pressure (and other types of mechanical ventilation) have been dramatically increased by the recent availability of high-powered portable blowers.

Positive pressure can be utilized during overhaul operations to provide a flow of fresh cool air through the overhaul area. This can significantly improve heat, smoke, humidity and toxic gas (CO) conditions.

Although positive pressure can be effectively used with an attack line, it cannot be applied to every structure fire. It will achieve satisfactory results only through a combination of proper training and judicious implementation.

Tactical Considerations

Fire departments should keep up with current literature and training seminars on PPV. It would be wise to have an officer designated as the department's expert on the subject. Critiques of the PPV operation should be held to develop a body of recorded experience with the techniques.

Many fully enclosed air conditioned buildings are under positive pressure. Opening a single window on the leeward side of the building in the fire area will permit the excess pressure to vent, taking much of the smoke with it. Opening several windows can dilute the effect.

Tactical Considerations, continued

The use of PPV on initial attack should be carefully monitored. Bear in mind the fact that 537 Btu are generated for every cubic foot of oxygen delivered to the fire.

An attorney might convince a jury that negligence was involved in deliberately providing air to the fire, "in a new and untried manner," thus doing culpable injury to the client.

Summary

The nature of smoke, its paths of travel and its effects both on people and things should be understood by all fire fighters. Entering a toxic atmosphere is as dangerous as SCUBA diving; however, it is often done far too casually. In particular, many fire departments should improve the care given to SCBA.

Case Studies

Birr, T. "On The Job, Eugene, Oregon." *Firehouse*, September 1986, p. 40. "Lazy smoke" was reported on arrival at a four-story wood-framed commercial building. The basement and corridors were sprinklered. Fire spread through unfirestopped voids. After three hours a defensive mode was ordered.

Eisner, H. "Trapped." *Firehouse*, June 1987, p. 30. One fire fighter was killed in a fire in a dilapidated multi-story brick and timber building used for baled rag storage. Two others were killed in the collapse of the roof of an exposure. "Light smoke showing" was the initial report. Fire suddenly flashed over, apparently from another fire in an area beyond a fire wall. One officer fell from a window to his death, another was injured.

Fetete, A. "We Have a Roof Cave In." *Fire Command*, January 1984, p. 36. Air conditioning units were located directly above the stairway in garden apartments. Five minutes after the first alarm, the units collapsed onto the first-due engine company. Later, the truss roof collapsed on fire fighters operating in an apartment where there was only "light smoke showing."

Harold, Harvey. "Working Fire." *Firefighting in Canada*, June 1991, p. 15. Director of the St. Pierre Fire Department in Quebec, Canada provides an excellent discussion of the necessity for opening up voids.

Hart, Roger. "Attic Blaze Rips Historic Adrian Church." *Firehouse*, July 1985, p. 25. Fire involved the attic of a wood frame brick veneer church. Note that little smoke was showing on arrival, so no hydrant hookup was made. Intense heat was found when the door was opened. Fire fighters got out just in time. A backdraft blew out a portion of the brick gable of the attic.

Kyte, G. "Tragedy Strikes Biloxi." *Fire Command*, May 1987, p. 18. Two fire fighters died in an old house converted into seven small apartments. Fire was in concealed interconnected voids. "No smoke showing" was the first report. The fire was reported at 2:55 AM. It spread into the habitable spaces at about 3:38 AM. The victims were not immediately missed. Fire lurking in void spaces is like a tiger perched in a tree above a trail. It is hard for many fire fighters to realize the amount of fire which may be concealed, or the speed with which the fire can grow when a void is opened up, either accidentally or by the fire department. As an old timer told me years ago, "What you see on arrival ain't necessarily what you've got."

Stronach, I. "Death Knell Tolls for Two Fire Fighters at Church Blaze." *Firehouse*, November 1987, p. 42. Walls and roof collapsed on an aerial ladder. Other fire fighters had been ordered out only minutes before. Again, "light smoke showing" was reported on arrival. Fire originating in the crawl space spread upward through the walls. This article is a model for those who write for publication. The description of the structure is excellent.

Taafe, M. "Brooklyn Fire Factory." *WNYF*, 4th issue, 1985, p. 5. A "light smoke" on arrival ends as a multi-alarm fire involving several buildings. Standpipes and sprinklers were frozen. All the eggs were in the sprinkler basket and the basket failed. A delayed alarm was very significant, the elevator operator had "extinguished" a fire earlier and had not reported it.

"The Dundalk Deaths." *Fire Command*, March 1985, p. 48. Three fire fighters died in a furniture store fire. "Heavy smoke" was reported by a responding unit a mile from the fire. Only light smoke was seen inside. Apparently the presence of combustible acoustical tile ceilings was not recognized as a warning that fire hidden in combustible voids could burst out furiously.

References

The addresses of all referenced publications are contained in the Appendix.

1. Steckler, K., Quintiere, J. and Klote, J. "Johnson City Fire Key Factors." letter report, NIST, 1990.

2. Roson, A. "Spanish Disco Fire." *Fire Engineering*, Sept. 1990, p. 29.

3. Flashover, VHS videotape, *Fire Engineering*, 1989.

4. Connolly, J. "Fires Involving PCB: What Is The Risk?" SFPE Bulletin, July/August, 1988.

5. Isner, M. "Fire Investigation Report: Telephone Central Office, Hinsdale, Illinois," May 8, 1988. NFPA; Also Hinsdale Central Office Fire Report, Executive Summary: a joint report of The Office of State (Illinois) Fire Marshal and Illinois Commerce Commission Staff; Also a full two-volume report is available from Forensic Technologies International Corporation, 2021 Research Dr., Annapolis, Maryland 21401.

6. Linder, K. "Contamination and Corrosion, Double Trouble after a Fire." IRI *Sentinel*, 2nd issue 1987. Reprinted in SFPE *Bulletin*, November 1987, p. 1. Additional information can be found in "Non Thermal Fire Damage," *Factory Mutual Record*, Jul./Aug., 1989, p. 15.

7. Howard, H. "Megamalls: A Fire Protection and Suppression Challenge." *Fire Engineering*, March 1991, p. 94.

8. Gallagher, John. "Volunteer Firefighters Avert Disaster at Belair Mall," *Firehouse*, September 1984, p. 40.

9. Huether, L. "Six Die in Elgin Street Fire." *Fire Command*, February 1987, p. 23.

10. Isner, M. "Fire in Michigan Hospice Kills Eight Patients." *Fire Journal*, January 1987, p. 56.

11. Copyright 1990 by Consumers Union of United States, Inc., Yonkers, NY 10703. Reprinted by permission from CONSUMER REPORTS, November 1990.

12. Patrick, E. Phillips, SFPE Fellow, Automatic Fire Alarm Association "Man of the Year" 1975, is Chairman of NFPA's Signalling Systems Correlating Committee (NFPA 72E) and immediate past Chairman of NFPA's Committee on Fire Detection Devices from 1974 - 1988.

13. Miale, F. #2. "Cracking Down on Unwarranted Alarms." WNYF 1st issue, 1991, p. 19. This article gives insight into the tremendous problem of the volume of private alarms in New York City.

14. Dougherty, T. and Kearney, P. "Fire." G.P. Putnam's Sons, New York, New York, 1931. Unfortunately this book is out of print. When I was running a fire department I lent it to every military superior as a readable short course in many aspects of fire hazards and fire department operations.

15. Briefing Paper, Safety In The Channel Tunnel, 8th Dec 1986, The Channel Tunnel Group Limited, Portland House, Stage Place, London, England, SW1E5BT.

16. Mittendorf, J. *Ventilation Methods and Techniques*. Fire Technology Services, Inc., (permission given for quoted material).

11

High-Rise Construction

There are many definitions of high-rise buildings. At the International Conference on Fire Safety in High-Rise Buildings convened in 1971 by the United States General Services Administration, the generally accepted definition of a high-rise was a building beyond the reach of aerial ladder equipment. This author protested that this definition was inadequate and repeated a life safety mistake made 60 years earlier. In the aftermath of New York's infamous Triangle Shirtwaist Factory fire of 1911, in which 145 workers died, laws were enacted to provide factory buildings over six stories in height with sprinklers. This law was changed after a 1958 fire killed over 40 workers in a building less than six stories high in the same neighborhood. The aerial ladder is a limited tool. It is no guarantor of life safety.

The principal life safety question in any building is not its height but the time it takes occupants to reach a safe environment. Only a few feet from safety, 164 people died in the one-story 1977 Beverly Hills Supper Club fire. Twelve people died in 1982 on the fourth floor of the newly opened West Chase Hilton Hotel in Houston. That fire was confined to one room. In December 1980, 26 persons died on the third floor of the Stouffer's Inn in Westchester County, New York. In June 1989, five people died on the sixth floor of a ten-story office building in Atlanta, Georgia. (See Case Studies.)

This chapter is not intended to present a complete discussion of all the problems of fires in high-rise buildings. It focuses on those problems which arise from the construction of the building. The type of construction used, and the era in which the building was built, are very significant in planning for, or operating at, a high-rise fire. The height of the building is only one element of the problem. Material from Chapters 6 through 10 and 12 is particularly pertinent to this chapter on high rises.

When applied to fire department tactics, the definition of high-rise buildings is acceptable and valid.

Buildings of any height can present many of the problems associated with high rises. There are lessons to be learned from the study of high-rise buildings. Do not skip this chapter just because there are no buildings of more than a few stories in your area.

Airport terminals and large shopping malls also present many of the problems found in high-rise buildings.

It is a serious error to consider all high-rise buildings as a single problem. There are fire-significant construction differences among high rises. The particular buildings in your area must be studied in detail to determine the particular potential modes of building failure.

With a few specialized exceptions, such as the huge, unprotected steel vertical assembly building (VAB) at the Kennedy Space Center, all high-rise buildings are designed to resist the effects of fire on the structural frame of the building and the floors. Whether the design concepts used are adequate to cope with all these possible effects is quite another matter.

Early fire-resistive buildings were said to be "fireproof" and the designation persists in some codes. Fire resistive is a more accurate term.

It is important to understand the limitations of the term fire resistive. Fire resistance is intended to provide, within limits, resistance to collapse by structural members and floors, and resistance to the passage of fire through floors and horizontal barriers.

Note, however, that fire resistance standards are not directly concerned with life safety, the control of the movement of toxic combustion products, or with the limitation of dollar loss.

Historical Notes

Three fires of great significance occurred in New York City in the early 1900s. The Parker Building was a 20-story mercantile building considered to be the best type of "fireproof" building available. It was destroyed by a fire that started within the building and spread up the open stairways. This was of great concern to fire protection engineers because it demonstrated that even the best buildings could be destroyed, despite the efforts of the nation's largest fire department.

The Equitable Building was also believed to be fireproof. In fact, it was not. Its contents included millions of dollars of negotiable securities which were temporarily lost in the basement due to tons of ice and debris. The financial community was responsible for code changes in the wake of this fire. The National Bureau of Standards, Underwriters Laboratories, The National Board of Fire Underwriters, the Factory Insurance Association (now Industrial Risk Insurers) and Factory Mutual developed the standard for fire-resistive construction which exists today—almost unchanged as ASTM E119. (See Chapter 6, Principles of Fire Resistance.)

The Triangle Shirtwaist fire had tremendous political and sociological impacts. Of principal interest here is the resultant construction of scores of high-rise sprinklered factory buildings in New York City. These provide a vast body of successful sprinkler experience in high-rise buildings that effectively belies the argument that sprinklers are unproven or unwarranted.[38]

While some disastrous fires pass into history without making a ripple, others have a tremendous impact. When impoverished workers died in the Triangle Shirtwaist fire, and left heirs destitute, New York State passed the first workmen's compensation law. The law overturned legal principles which had required the injured employee to sue and prove negligence on the part of the employer. It also supplanted the infamous "fellow servant" rule, which held the employer free of liability if the employee was injured by the act of another employee.

Senator Robert Wagner, Alfred E. Smith, four-time governor of New York and presidential candidate in 1928; and Frances Perkins, the first woman cabinet member and Secretary of Labor in the Franklin Roosevelt administration, were three of the most prominent political figures whose careers began with the effort to correct such legal deficiencies in the aftermath of this fire. At this time too, the New York Fire Department organized the first Fire Prevention division in any U.S. fire department.

This section is not important just from a historical perspective. Many of these early buildings are still in use. Some deficiencies of high-rise type construction were noted in fires and corrected in local codes. Many of the same defects went unchanged in buildings erected in other cities.

Some buildings which are not high rises, such as courthouses, churches, libraries, and college and school buildings, were built to be fireproof. They contain some or all of the defects cited here, and some that are not noted here. In some cases, portions of ordinary construction buildings damaged by fire were rebuilt to be fireproof.

Some buildings of this era, like the famous Rookery in Chicago, have been rebuilt to a high degree of fire safety—with full sprinkler protection. Others, such as a Texas County Courthouse, have been restored to 19th century elegance, with little or no attention paid to a hundred years of tragic fire experience.

General Classifications of High-Rise Buildings

Fire-resistive, high-rise buildings have evolved over the last century. The following classifications are broad and intended only for general guidance. The dates given are approximate.

Early Fire-Resistive Buildings, 1870-1930

There were many fire safety defects in buildings of this era. Some of the major defects of this high-rise construction are:

- There were no standards for the protection of steel.
- Cast iron columns were often unprotected.
- Terra Cotta fireproofing was compromised by concealed light-weight conduit, which expanded and tore the tile off columns.
- Segmental (curved) brick or tile arch floors were tied with exposed steel ties.
- Wooden floor beams were placed on piers creating a void under the floor, connected to all other voids by the hollow columns. In some cases, floors were leveled with cinders, often overloading the structure.

Structures in this era were built without any appreciation of the impact of vertical openings. Ornate open stairways and light wells were a standard feature, and in many cases, still remain. Open elevator shafts, a common feature in these buildings, have been replaced in many cases, since new elevators must have shaft doors.

The District Building, the city hall of the nation's capital, is a typical fireproof public building of its period with a wide open ornate stairway. Its fire safety problem has been discussed from time to time. In an article in the *Washington Star*, the building manager is quoted as saying that "it would be difficult for a fire to spread from floor to floor."

The fire service should consider being more aggressive in attacking such unqualified opinions. The spread of fire and toxic gases is a serious life safety problem.

In the same article, an architect who works in the building was quoted:

> "We have never had a fire drill in this building, and there is no need for one. If a serious fire ever breaks out on one of the lower floors, there will be no way out for us."
>
> Then after a pause, the architect continued, "Do not write that. If they brought this building up to code, it would mean the rape of the District Building. The long winding stairways would be ruined."

Fig. 11-1. An open stairway was an architectural feature of older high-rise buildings. It is a natural path for smoke and gases to travel.

Underlying this statement is the arrogance of those who place artistic merit above life safety. It is very likely that the majority of those in the District Building might well vote for the "rape" of the stairways and a chance to go home to their families if a fire occurs.

The Southern Building in Washington, D.C. is a typical early era fireproof building. It took about 70 years for an inherent defect to be made evident by a fire. A florist shop occupied the first floor. An access stairway had been cut through to the basement. A basement fire completely polluted the building with smoke via the unenclosed stairway. Hundreds of occupants had to escape down the outside fire escapes. It was feared that the fire escapes would fall off. Fire department ladders were used to brace the platforms.

Many other hazards are present in older high-rise constructions. In some cases, segmental brick and tile arches were supplanted by flat terra cotta tile arches. In early construction, no protection was provided for the underside of the steel beams. Many such buildings still stand. Later, tile soffit blocks or "skew backs" were developed to provide protection to beams. These are often removed by workers, and the steel is unprotected.

Tile arch floors were delivered on site, already manufactured to specifications. Often the specifications for the floor units were wrong and the floor had to be laid in an improvised manner, lacking adequate protection.

Fig. 11-2. Hundreds crowded the fire escapes of the Southern Building. The interior was fully polluted by smoke. (Washington D.C. Fire Department)

Older buildings were built with high ceilings to provide summer ventilation. Subsequent to original construction, the ceiling may have been lowered one or more times. Each time this may have created a combustible void. The tiles used may have a high flame spread.

In many cases, the most recent alteration to older buildings may have been done with relatively non-combustible construction; however, this may often conceal old combustible materials. In 1963 in Jacksonville, Florida, 22 people died in the Roosevelt Hotel. A new fire-safe ceiling had been installed in the ballroom. The old combustible ceiling had been left in place. Sixteen died in 1989 in the John Sevier Retirement Home in Johnson City, Tennessee. An old combustible tile ceiling was left in place above a new low flame-spread grid ceiling. (See Case Studies and Chapter 9, Flame Spread.)

While most codes require newly installed ceiling tiles to meet flame spread requirements, it is doubtful that any code requires the removal of the old tile, despite the demonstrated hazard. One of the most common alterations of older buildings is to install new ceilings, often with void spaces to hide new utility lines. Very often the ceiling is of combustible tile. Pre-fire planners should look above dropped ceilings.

Fig. 11-3. This flat arch "fireproof floor" of terra cotta tile is seriously deficient. Note the unprotected steel beam.

Standpipes in older buildings may be inadequate in size. It is possible that the standpipe has not been tested and could fail in a fire. It is also possible that the standpipe outlets do not match the fire department's hose. The standpipe is a fire department tool as much as the pumper that supplies it.

Almost all older buildings have been altered at some time. Walls are opened up and doorways are closed. The closure is often not as fire resistive as original masonry walls. Stairways are cut through and others cut off. High-bay rooms may be made into two stories. Sometimes a wooden mezzanine is built in. In other cases, a steel bar-joist floor is installed to make an intermediate story.

Live loads in older buildings may have increased over the years to a point where the building is distressed. Safes to protect computer tapes from fire, for instance, are much heavier than other safes. In some cases, unprotected steel supports may have been inserted, as might be the case in a high-rise alteration to accommodate a mechanized filing system which necessitated strengthening of the floor.[1]

Fire loads also may have grown to high levels. This may be particularly so in the case of public buildings. Records and files accumulate. Many public buildings are domed. Usually there is an inner and outer dome. The space between makes a good void in which to store tons of paper. Basement rooms and attics are often similarly loaded with documents.

Tactical Considerations

Tile arch floors pass water easily and a serious salvage problem should be anticipated with this type of construction.

Most importantly, these floors should not be cut through without taking precautions. The floor units are arches, and cutting out a tile may drop the arch. This was reported to have happened when a sprinkler riser was being installed in the old Post Office building in Washington D.C. It is possible that the entire bay may collapse if the floor is disturbed. Even if personnel are protected against collapse, the collapse of a large floor section may cause a burst of fire upward.

The fire department should assure that the owner is keeping the building standpipe in proper operating condition. Make a strong effort to get proper closures provided for vertical openings and for the proper maintenance of those which are installed.

Outside fire escapes that may have been exposed to the weather for a long time should be carefully inspected for the safety of occupants and fire fighters.

Public buildings should be examined for excessive fire loads. Since these structures often have many defects, control of hazards such as these cited is of utmost importance. Jurisdictional conflicts or lack of legal authority in public buildings may prevent effective code enforcement inspections. However, prefire planning is within the right of fire departments expecting to fight fires. Prefire plans which disclose unmanageable situations should be brought to the attention of the responsible authorities—if only for defense against those who will blame the fire department for a disaster. A fire department, prevented from inspecting and preplanning a building, might announce that for the safety of fire fighters it would delay advancing into a serious fire until absolutely sure of the safety of personnel.

Later High-Rise Building Construction 1920-1940

The high-rise buildings constructed after these early developments and before World War II are excellent buildings. Those built after World War II, in this author's opinion, have poorer fire protection features.

Pre-World War II buildings were universally of steel-framed construction. Floor construction and fireproofing of steel were often of concrete or tile, both good heat sinks and slow to transmit heat to the floor above. The construction was heavy but no feasible alternative existed.

Relatively small floor areas were dictated by the need for natural light and air. Ads for the RCA Building in New York proclaimed, "no desk any farther than 35 feet from a window." This limited both the fire load and the number of occupants.

Each floor was a well-segregated fire area in these buildings. Wall construction was frequently of wet masonry, joined to the floor so that there was an inherent firestop at the floor line. Masonry in the spandrel area (the space between the top of one window and the bottom of another), was adequate to restrict outside extension. However, this author recalls such a fire extension which was caused by added combustible acoustical tile ceiling which provided for heavy fire out the window, and easily ignited fuel on the ceiling above.

In these buildings vertical shafts were enclosed in solid masonry with openings protected with proper closures. Fire department standpipes of adequate capacity were usually provided. These were wet and immediately pressurized by gravity from a tank in the building.

Exterior fire tower stairways with an atmospheric break between the building and stairway, the finest escape device available, were provided in many of the buildings. Such a stairway can be compared to an enclosed tower located away from the building which is reached by a bridge open to the weather, so that smoke cannot pollute the tower.

In the Empire State Building, the fire tower was built inside the building with an adjacent smoke shaft. The inadequacy of this construction was demonstrated in a fire in 1990. The interior smoke shaft became a chimney because it is impossible to separate smoke from heat. The fire was confined to one office suite, but five alarms were required for fire suppression and evacuation.[2]

The typical office was quite spartan, though executive suites and eating clubs often were paneled with huge quantities of wood. Nevertheless, most fire loads were low.

Windows could be opened in buildings of this era. This provided local ventilation and relief from smoke migrating from the fire.

The windows leaked, often like sieves, therefore there was no substantial stack effect. The author observed many fires in such buildings. They usually presented very little problem to the fire department. Some serious fires occurred in sprinklered garment factory lofts where accumulations of cuttings under the wide cutting tables could burn unaffected by water from the sprinklers. The occupants of the affected loft would evacuate, but thousands of others would continue to work.

Exposure Problems: Such buildings are almost immune to any serious fire attack except from exposures. In 1973, a six-story brick and

Fig. 11-4. This building contains a "Philadelphia Fire Tower." Occupants from each floor exit to an open balcony, then enter another interior stair shaft that is completely separated by a fire wall from the rest of the building. This provides a smoke-free exit path.

wood-joisted building was being demolished in downtown Indianapolis.[3] Only the floors remained. A violent fire erupted and ignited surrounding fire-resistive buildings. In 1982, ten floors of a fire-resistive high-rise bank building in Minneapolis were destroyed by a fire which originated in an adjacent wood-joisted building being demolished.[4]

In 1989, the largest commitment of fire fighters to a structural fire in the history of Los Angeles was required to suppress a fire in an occupied high-rise retirement residence which was heavily damaged by fire in an adjacent four-story wooden building under construction.[5] The NFPA produced an excellent study of an exposure problem in its 1982 Investigation Report entitled "Demolition Operations Exposure Fire. It is a worthwhile reference material for interested fire professionals.

Flame Spread: Like the earlier generation of high-rise buildings, some buildings of this era were modernized by installing combustible acoustical tile. This tile, whether installed on furring strips or glued,

Fig. 11-5. This Indianapolis high-rise building was ignited by a fire in a rear brick-and-wood building that was being demolished. The fire load was made up of the wooden floors.

presents the serious hazard of a fire spread through the building. Old tile should be removed. It cannot be made safe by painting it with an intumescent coating. As has been noted, it is most certainly asking for disaster to leave it in place and then install a code-approved ceiling below it.

Tactical Considerations

Most of these buildings have stood for many years without posing a serious problem to the fire department. This good experience is no guarantor of fire safety, rather it may encourage complacency.

Study high-rise buildings for special problems which may not violate any code. These include high fire-load areas, high flame spread potential, potential generation of toxic gases such as from wiring, potential for spread of gases and smoke pollution through penetrations, and via stairways habitually blocked open.

Modern High-Rise Buildings

After World War II, a number of significant developments occurred. Fire departments apparently bemused by the lack of problems in the previous generation of structures, generally failed to realize the dimensions of the new problem. City administrations, eager to get the spectacular new buildings onto the tax rolls, didn't question the hazard of cramming thousands of people into totally enclosed megastructures which created a potential for catastrophe.

In a unique book that was the basis of a public television series, one author followed day-to-day construction of a New York skyscraper.[6] Commenting on the subject of construction safety, the technical coordinator of the skyscraper project said, "New York has probably more codes and regulations to follow than any other locale in North America. Whether that's good or bad I'm not quite sure, but because there have been places in other municipalities where they have had a terrible loss of life and so forth, they have tried to avoid that. With the amount of construction going on in this city, and how few very serious fatal accidents on high-rise commercial construction there have been, it reflects the strength of the code we have to follow."* It appears to this author that the quote indicates that the technical coordinator has confused safety during construction with the safety of the occupants in the completed building.

The development of fluorescent lights and air conditioning helped to remove limits to the floor area. Thus, building populations could be enormously increased. As a result, many floors have substantial areas beyond the reach of hand hose streams.

Significant Construction Characteristics: Since this time there has been a definite push to lighten and thus reduce the costs of buildings. The Empire State Building weighs about 23 pounds per cubic foot. A typical modern high rise weighs approximately 8 pounds per cubic foot.

Because of the development of better reinforcing steel and new techniques, reinforced concrete became a serious competitor to steel as a construction material. No longer could the building industry be indifferent to the weight penalty caused by concrete fireproofing of steel. In addition, the necessity for fireproofing is an apparent cost disadvantage to steel. Fireproofing of a steel building is a separate item in the cost estimates; making concrete fire resistive is included in the cost of concrete. The spray-on materials and the floor and ceiling assemblies developed to provide lighter construction are discussed in Chapter 7, Steel Construction.

---

* From Skyscraper, by Karl Sabbagh, copyright (C) by Karl Sabbagh. Used by permission of Viking Penguin, a division of Penguin Books USA, Inc.

One of the results of reducing the concrete in the building is the loss of a valuable heat sink. Every Btu that is absorbed into the concrete is one less available to keep the fire extending. This retained heat can be very distressing to fire fighters, and personnel may require frequent rotation. On the positive side, though, the fire's growth has been slowed by loss of heat.

There has been an extensive increase in the requirements for electrical service and communications systems in modern high rises. Insulations are flammable, and the products of decomposition and combustion are toxic. Communications systems can breach fire and smoke barriers from floor to floor, and from vertical shafts to horizontal voids.

Steel-truss floor and ceiling assemblies provide useful voids to carry utilities and communications or as a plenum for conditioned air, and unintentionally for fire and smoke.

A most serious deficiency is the omission of fireproofing on columns within a rated floor/ceiling assembly. This very hazardous condition is discussed fully in Chapter 7, Steel Construction.

Slab floor concrete buildings do not have an inherent void. By dropping ceilings, useful voids are created and then connected via utility shafts. In a modern office building, possibly 25 percent or more of the floor volume is in the ceiling void.

Raceways for wiring are inserted in some concrete floors. Unless properly fireproofed below, heat can be transmitted to the floor above.

Gypsum rather than masonry is often used to enclose elevator and other shafts. In the One New York Plaza fire, gypsum enclosures were displaced leaving the shafts unprotected, thus hazardous to fire fighters.[7]

Fig. 11-6. The column and its connection to the beam are unprotected. The concept is that the floor/ceiling assembly will provide adequate fireproofing. The defects of this concept are discussed in Chapter 7, Steel Construction.

Fig. 11-7. Scissor stairs can be confusing. It would be helpful if buildings were required to use a consistent designation system. The standpipe outlets change from stairway to stairway on each floor. A prefire plan should provide specific information regarding them. One stairway should be designated for evacuation, the other for attack.

In earlier buildings, there was little concern about using the exterior of the building for fire towers. Modern construction discovered the value of exterior offices, particularly corner offices. Utilities are relegated to an interior "core" structure. The core often takes up the greater part of the wind load. Stairways are often located in the core, thereby eliminating the principle of remote exits.

Stairways are frequently "scissor" stairways—two stairways in the same shaft. The stairways are thus quite close together. Often there is no fireproofing on the underside of the stairways so heat or fire in one stairway can make the other untenable. If there is a standpipe, the outlets will be in alternate stairways, floor by floor.

Elevators are grouped together in the core; thus, it is possible that all will be affected by the same fire.

Not all high-rise buildings are of core construction. Some tall high-rise structures are of tube construction, that is, the wind load is taken principally on the exterior. However, this does not eliminate the use of a central core for utilities and services.

In an attempt to provide a smoke-proof fire tower in the interior of the building, a vent shaft is often provided adjacent to a stairway. Earlier, this text noted the Empire State Building fire in which the shaft became a chimney. Heat and fire reaching the shaft are directed upward. In the 299 Park Avenue fire in New York in 1980, Aluminum guard rails on such a shaft melted from the heat, exposing fire fighters to a potential fatal fall.[8]

Modern design often emphasizes an uncluttered exterior. Prefabricated panels, glass, aluminum, and other curtain wall developments make it difficult to provide a real barrier to the vertical extension of fire between the edge of the floor and the skin of the building. A glass wall system is described in detail in an engineering publication, but the article is utterly silent as to how this will stop a fire.

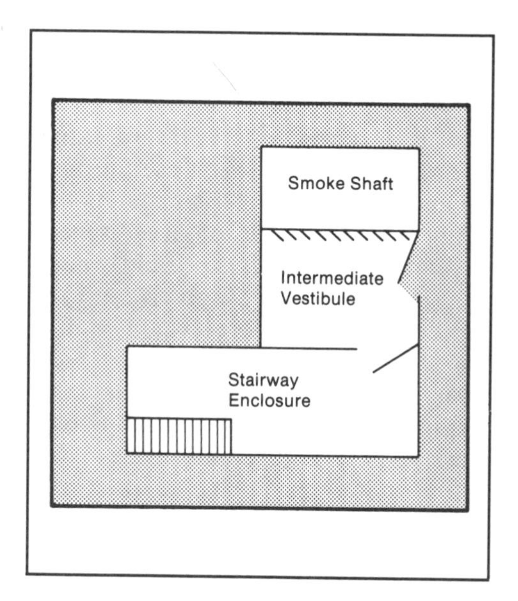

Fig. 11-8. The arrangement of an interior "smoke proof" stairway. The shaft is supposed to vent smoke, keeping the stairway clear. The problem is that smoke and heat travel together so the shaft can become a chimney.

Fire does not only spread upward; it can also spread downward. In December 1981, fire gutted the 23-story Barao de Maua Building in Rio de Janeiro, Brazil. Burning electrical wires in the 15th floor combustible tile ceiling void started the fire. It spread downward and upward through a six-inch gap between the floor slab and the alumi-

Fig. 11-9. These brick panel walls on a Denver building were not built in place. The panels were built on the ground, lifted into place, and attached to the steel frame. The fire department should be aware of the quality of the firestopping along the perimeter.

num and glass curtain wall. The enclosure of balconies by tenants or condominium owners may increase the potential for exterior vertical extension.

Deficient perimeter protection was also a factor in upward extension in the Los Angeles First Interstate Bank fire (See Case Studies).

As with buildings of many eras, methods and materials may not be consistent throughout the building. This author has observed several concrete buildings on which top floors or penthouses of light-weight steel are erected. The penthouse apartments of one high rise in Polk County, Florida had a light-weight wood truss and plywood roof.

Windows and Glass Exteriors: Many high rises have glass exteriors. It is often impossible to determine from the exterior which floor(s) are involved in a fire. Sudden massive glass failure may cause a serious change in conditions on the fire floor. It also presents other hazards. A fire fighter was severed injured by falling glass in the One Meridian Plaza fire in Philadelphia (see Case Studies). Falling glass had made many cuts in hoses feeding standpipe siamese connections. It appears reasonable that the building code should require an entry to the building and water supply connections which are safe from falling glass and furniture, and not leave this vital requirement to chance.

Fig. 11-10. Buildings may not be consistent in their construction. This concrete high rise has a light-weight steel top floor added to it.

Air conditioning requires fixed windows to control loss of treated air. Many buildings have no operable windows. Breaking glass is extremely hazardous because glass pieces can travel for several blocks. If windows are covered with a sun-screening plastic, the glass tends to hold together when broken and can be pulled into the building. More recently, some windows which can be opened are provided and keys are located in strategic but inconspicuous locations. In other buildings, windows which can be broken are marked with a designated symbol.

Energy efficient windows (EEWs) consist of double or sometimes triple panes of glass. They resist failure in a fire and eliminate the ventilation provided by the heat-caused failure of single pane sashes. They are more difficult to vent from the exterior. They can cause severe heat conditions to be retained in the fire area. They do resist auto exposure from fire below.[9] Auto exposure is exposure to other parts of the fire building as distinguished to exposure from fire in other buildings.

When Boston's John Hancock Tower was built, mysterious defects caused its large glass panes to fall out. They were replaced with plywood panels until the tower looked almost like a wooden structure. When the flame spread hazard was later realized, plywood panels were treated with fire retardant.

Atriums: The open light well is back. Today, it is frequently called by an ancient name, atrium. This huge void presents tremendous

Fig. 11-11. The John Hancock Tower was known as "Boston's pane in the glass." Note the huge plywood area. Fortunately, there was no fire.

difficulty in the control of smoke, as was experienced in a fire in 1979 in which the smoke removal system failed to operate.[10] A costly fire at the National Institutes of Health involved a modern atrium.[11]

Building codes have only recently come to grips with the atrium problem. Typical requirements are sprinklers throughout the building, smoke exhaust systems with standby power, and separation from floors by a fire barrier, or sprinkler protected glass as an alternative. Atriums are glass enclosed with closely spaced sprinklers in lieu of one-hour rated fire barriers. The number of floors that can be open to the atrium without a fire barrier is limited. Central to atrium protection is the limitation of the fire load on the floor of the atrium.

This provision may be violated by attempts to get the greatest revenue from the building.

Some codes allow wooden structures within the atrium. This author believes it is a mistake. I am aware of a multi-story wooden bar and lounge erected within the atrium of a motel. The structure is sprinklered. The interior sprinklers would not affect a fire on the exterior of the structure. There are sprinklers at the top of the atrium but it is possible that the wooden bar could be well involved on the exterior and from there into the voids before the sprinklers could control the fire. The exit from all the rooms is through the atrium. This is a potentially dangerous situation.

The fire marshal has assured me that if the sprinklers are out of service, the motel will be too. But how will he know?

Fig. 11-12. This wooden gazebo is built in an atrium. The gazebo is sprinklered and there are sprinklers in the overhead of the atrium. The exits from rooms are along the perimeter of the atrium. Sprinkler protection must be maintained at all times. If sprinklers are shut off, strong protective measures should be required.

There is a wide variety of technical opinion on fire protection of atriums. The NFPA convened a round table of experts in December 1988. The proceedings were published by NFPA in 1991.[12]

Parking Garages

The parking garage of these buildings may be partially or totally above grade and open to the atmosphere. In one high-rise office building, this led to permission to use plastic conduit on the ceiling of the garage. A rubbish fire involved the conduit and wiring. The high-rise building, fortunately still under construction, was heavily smoke-polluted. All garage areas under buildings should be sprinklered.

Fig. 11-13. Now completed, this Maryland office building houses thousands of government employees. ABS (acrylonitrile, butadiene, and styrene) plastic conduit and combustible wiring insulation were hung from the ceiling of the garage. A rubbish fire inside the garage ignited the plastic. Note the dense smoke. The entire building was heavily smoke damaged. Imagine what might have happened if the building had been occupied. All garages under buildings should be sprinklered, even though they are open to the atmosphere. (Montgomery County, Maryland Fire Marshal)

Fig. 11-14. This Revere, Massachusetts high-rise apartment building is unique. The doors seen on the right may lead to an apartment on the same floor, up one flight, or down one flight. There is one advantage to this design — a row of windows opposite the doors. The corridor could be vented in case of fire.

A Unique Design

A Revere, Massachusetts designer has found a unique way to increase rentable space in a high-rise residential structure. The design may be copied elsewhere. The potential problems appear to be very serious. One corridor serves three floors. Apartment doors along the corridor lead to an apartment on that floor, an interior stairway leads to an apartment on the floor above, and another interior stairway leads to an apartment below.[13]

The building is sprinklered, not because of the unique design but because of its size. If the sprinklers fail, a fire downstairs would be especially serious. A fog stream might endanger any occupants still in the apartment. An examination of the building indicated that an approach from the adjacent apartment through the gypsum board wall might be more productive.

Fire departments should be aware of this construction, especially if it is proposed for their area.

Tactical Considerations

Do not be misled by the "high-rise" title of this chapter. Many buildings up to six stories in height have the same construction defects, and present similar hazards, as high rises.

Core construction may make operation of hand lines from a safe location, i.e., the stairway, utterly inadequate. Prepare to use

Tactical Considerations, continued

monitor nozzles to knock down fire and reach distant points if the standpipe system can supply enough water. The Wilmington, Delaware fire department has developed the use of light-weight monitor nozzles fed by two 1-3/4" lines which can deliver 300 gpm.

In many cases 1 3/4" hose is adequate to handle the fire but each piece of equipment should carry some 2 1/2" hose already packed flat for use when needed. Hose rolled up from one end to the other will come out like a spring and often kink.

High-rise fire plans may propose cutting through floors to reach the fire. Note that it is not safe to cut through post-tensioned concrete floors. The fire department should be aware of all post-tensioned concrete buildings.

Because of changes in office decor and the elimination of interior partitions, in many cases heavy fire loads in huge unbroken areas will be found. Conversely, compartmented floor areas, with maze-like corridors, require long hose stretches and are extremely confusing when smoke-polluted.

Each high-rise building is a problem all its own. A high-rise with concrete floors and masonry panel walls built on each floor presents no problem of fire extending up the outside behind the wall panels. Another concrete-floored building, with precast concrete or imitation

Fig. 11-15. Members of the Wilmington, Delaware Fire Department carry their light-weight high-rise monitor nozzle kit into a building. (B. C. Huelsenback, Wilmington Fire Department)

Fig. 11-16. After being set up inside the building this stream is fed by a pair of 1¾" lines. (B. C. Huelsenback, Wilmington Fire Department)

Tactical Considerations, continued

concrete panel walls, may have a serious problem of perimeter extension.

For fire fighting operations, it is very difficult to determine exactly from the exterior, a specific floor of an all glass building. Yet, this is vital to communication of observed conditions to personnel inside. One method to make such determination, without costly waste of time, uses sheets of heavy paper, tape, and marking brushes. Fire fighters on a non-fire floor could mark the floor number on the paper and paste it on the window as a reference point for communication between exterior and exterior. Tinted windows or nighttime situations may require other methods. Binoculars should also be available.

Parking garages present a number of problems. The garage area may extend out beyond the building. The covering slab may be designed only to carry the weight of cars. Heavier fire apparatus may be restricted. One shopping mall in the midwest consists of many buildings on a light-weight slab over the garage. No horizontal standpipe was installed. Fire apparatus cannot drive on the slab. Extremely long, hand-hose line-stretches are required.

Fig. 11-17. It isn't always easy to identify a floor in a high-rise building. Can you readily identify the tenth floor in this Ft. Worth, Texas all-glass high rise? The building is a common all-glass type. A suggestion for a marking system is made in the text on page 472.

Tactical Considerations, continued

The huge populations of modern high-rise buildings may make evacuation an almost unmanageable problem. Where will the evacuees go, particularly if the weather is bad? People will be anxiously, perhaps even hysterically, searching for co-workers. A strong police presence may be required to provide the working space and order needed for fire department operations.

The need for search and rescue should never be overlooked. Do not be deceived by the "nine to five" aspect of office buildings. A building that is guarded may have sign-in and sign-out sheets, but many buildings may have no night security and cannot be assumed to be empty. Such persons as cleaners, computer operators, and attorneys, for example, may be working at any hour. Often they are alone and unknown to anyone else in the building.

Fire fighters' access to stairways may be impeded by the sheer numbers of people descending them. In residential occupancies, it may be possible to train the occupants to keep to the rail side, leaving the wall side for fire fighters.

Because of the falling glass hazard the preplan should include pertinent information on safe entry to the building, such as through an underground passage.

General Problems and Hazards with High Rises

Many of the hazards presented by high-rise buildings are not solely associated with buildings of one particular era. Due to their size and form these buildings share some common problems that fire fighters must address.

Exits

All exits should provide a clear path to the outside. In one high-rise building, the exits opened to a rear yard. The yard was then planted and enclosed with an unbroken 10-foot high masonry wall.

In another example the exit in a group of three interconnected buildings leased by the government was extremely poor. Several alterations were made to improve the situation. The added exits led through a small garage area to the open street. The next occupant decided to use the garage for the storage of cardboard box files—a huge smoke-producing fire load. They asked the building department for advice. The advice was to enclose the garage in two-hour fire-resistive walls. This advice created a serious situation. The pressure generated by a fire would be contained. Smoke and gases would be forced into the exit ways through the space under the exit doors and up the stairways

Fig. 11-18. A serious situation may not be apparent on arrival at a high-rise building. Lines should be stretched to both sprinkler and standpipe siamese connections so as to be instantly available if needed.

into the building. Though this author was without responsibility for the building, he prevailed on the management to install a dry-pipe sprinkler system in the area.

Occupancy

The occupancy of a particular building changes its problem. Offices, hotels, apartments, homes for the elderly, factories and showrooms are all different. Some buildings have mixed occupancies. One rigid code interpretation provided smoke detectors in occupied apartments but none in an architect's office which used two apartments and had the typical high fire load of exposed paper contents. Four lives were lost in a New York high-rise apartment building which housed several commercial occupancies on the first floor.

Accommodation or Access Stairs

Many tenants occupy more than one floor in a building. They find it convenient to have an accommodation or access stairway installed. This is usually done as an alteration and is rarely enclosed. The result is that two or more floors become one fire area, completely negating the concept of floor integrity. In the One Meridian Plaza fire in Philadelphia, fire was seen falling down through such an accommodation stairway from the 22nd to the 21st floor.[14]

In some cases, codes require sprinkler heads above the opening in otherwise unsprinklered buildings. There is no proof that this would be effective. In fact, sprinklers operating at a distance from the actual fire may deluge fire fighters with scalding water.

Duplex and triplex apartments, (two- or three-story "houses,") can be found in high-rise apartments. Often, there are no exits from the upper levels. If there is an elegant restaurant located at the top of a structure, it may be a multi-level with an open stairway connecting the levels.

All of these encroachments on good containment increase the fire area and can multiply the fire problem. In some codes, and among some designers, there is little concern about open stairways, provided the required exits are properly enclosed.

A Canadian fire department responded to an automatic fire alarm from the 28th floor of a 43-story building. They took the elevator to the 27th floor (which they assumed was the floor below the fire). The door opened on the fire. Fortunately, they were able to get the door closed and descend. One firm had occupied the two floors. There was an access stairway. The smoke detector on the 28th floor detected and sounded an alarm. Smoke detectors that are far removed from the fire may be the first to operate.

Fig. 11-19. This accommodation stairway connects two floors of the library of a major law firm, making the two floors a single fire area. Many libraries are interior areas without windows, and are impossible to vent.

Tactical Considerations

The "floor below the fire floor" operating base was adequate for the pre-World War II towers for which the concept was developed,—It is no longer safe.

Forcible Entry

Security against crime is a serious consideration in all occupancies. Heavy doors with multiple locks may delay entry. The fire department should be aware of how fire separation was accomplished. In most cases, the fire resistance may be achieved by gypsum wallboard on studs, or gypsum block or terra cotta tile, all easily penetrated. On the other hand, if the building is of reinforced masonry construction, the corridor walls are likely to be of reinforced masonry, which would be very difficult to breach.

Some stairways are locked against re-entry on all floors. Some codes require doors to be unlocked for re-entry on certain floors, typically no more than four intervening floors between re-entry floors.

Tactical Considerations

It may be much simpler to go through the wallboard and ignore the lock. This may also be possible from apartment to apartment and into the upper levels of duplex and triplex apartments.

Fire fighters forcing entry to a stairway door into the fire floor may face a burst of fire as the door springs wide open. Arrange for keys to stairway doors to be available at the security desk.

Consider having each team carry minimum forcible entry tools. A forcible entry team may not be available.

Elevators

The development of safe elevators which would automatically stop if the cables failed helped to make high-rise buildings practical. A complete discussion of the problems which elevators present are beyond the scope of this book. However, there are several facets to the problem which should be touched upon here.

Extrication of persons trapped in stalled elevators requires a detailed knowledge of elevator construction and operating systems.

The use of elevators by occupants for fire evacuation is dangerous and is universally warned against by signs placed in or near elevators.[15]

Legislation making discrimination against the handicapped in employment subject to severe penalties had created a serious evacuation problem for handicapped persons who cannot use stairways.

Guards and maintenance personnel routinely use elevators to reach any floor where there is a problem. If the problem is a fire, opening the car door on the fire floor can be fatal as it was for the maintenance person sent to check the alarm in Los Angeles First Interstate Bank fire.

It was almost fatal for the maintenance person who went to the 22nd floor of the One Meridian Plaza building in Philadelphia. He was greeted by dense smoke and heat and could not close the doors. By radio he was able to instruct the guard how she could override the controls and return the car to the first floor and safety.

In a hotel fire in Orlando, Florida a guard attempted to go to the alarm floor. Fortunately the elevator did not respond. The incendiary fire had been started in furniture in the elevator lobby.

Tactical Considerations

Fire departments should stress to building staffs the hazards of going to the alarm floor.

The use of elevators by the fire department in fire combat presents many hazards.

In February 1991, at Baltimore, Maryland, The American Society of Mechanical Engineers, The Council of American Building Officials, and the NFPA jointly sponsored *The Symposium on Elevators and Fire*. Experts discussed the various phases of the problem.[16]

Deputy Chief Elmer Chapman, FDNY (Ret.), made the following points regarding the use of elevators for evacuation.

> "In some locations it is the practice for security personnel to use the fire fighters' elevator key to evacuate handicapped persons. This procedure is potentially dangerous. No one should operate an elevator under "Phase Two Emergency In Car Operations" except trained emergency service personnel who are: (1) fully trained in fire fighting as prescribed by NFPA 1001; (2) fully protected by fire fighter protective clothing; and (3) following all safety precautions prescribed by the fire department for its own personnel."

A complete set of recommendations for the hardening of elevators to permit their use in evacuation of the occupants has been developed by Chief Chapman. The recommendations are summarized below:

- Fully-sprinklered building
- Pressurized elevator shaft
- All elevator lobbies enclosed
- Elevator lobbies pressurized
- All intakes for pressurization in smoke free location
- Smoke detectors in elevator lobbies
- Elevator systems resistant to water
- Power failure: elevators return to designated location
- Emergency power for all elevators
- Elevator lobbies have safe access to pressurized stairway
- Two-way communication from elevator car to fire command station
- Two-way communication from elevator lobbies to fire command station

Some Additional Points

All experience points to the possibility that at some fires, all elevators may be totally unavailable to the fire department. This happened in the One Meridian Plaza fire. All power for lights and elevators failed. In total darkness all equipment had to be hauled up 20 or more floors.

Just imagine the problem if the loss of power occurred in a fire on a floor such as the 60th. Plans should embrace this contingency.

The fire department should be fully familiar with the elevators in all high-rise buildings. They were probably installed under different codes and may have markedly different operating systems. It might be useful to appoint an officer as elevator information coordinator.

Very tall high-rise buildings have blind shafts in which elevators serving upper floors pass many floors without door openings. In some occupancies, commercial and residential elevators open directly onto the floor, without any vestibule.

Shaft doors are often left for last-minute installation due to possible damage during construction. Inadequate makeshift barriers may be used to close these openings. The opening of one high-rise Florida hotel was rushed to accommodate a big banquet. The decorator wall-papered over the open shaft doors!

Tactical Considerations

Some fire departments practically prohibit the use of elevators by fire fighters attacking a fire. Others consider it necessary but under strictly regulated conditions. These include approval of the incident commander; SCBA on all personnel; portable radio (through efficiency is doubtful, wired car telephones are better); forcible entry tools; stopping every five floors to check the shaft for smoke; and limiting the number of fire fighters in the car.[17]

When an elevator shaft fire-resistive enclosure is made of gypsum board, it is possible that hose streams have disintegrated the gypsum board, leaving an open shaft. Shaft doors may be open on the fire or other floors, without the car in place. Two Chicago fire fighters were killed in an open shaft and one Maryland fire fighter survived a multi-story fall down an open shaft.

Smoke Movement in High-Rise Buildings

There are a number of mechanisms which affect the movement of smoke in a high-rise building.

Thermal Energy

The principal smoke-moving mechanism is the thermal energy of fire (See chapter 10, Smoke and Fire Containment). The thermal energy of a fire can be massive. The burning rate of the fuel in the MGM

Grand Hotel fire at Las Vegas was estimated at three tons per minute.[18] The burning rate at the Los Angeles First Interstate Bank fire was comparable.

The other mechanisms which can greatly affect the movement of smoke, also may be of great significance and can add to the problem.

Atmospheric Conditions

When the atmospheric temperature is constantly decreasing as height increases, the condition is called **lapse**. Under lapse conditions, smoke will move up and away from the fire. If there is a layer of air warmer than air below, the condition is called **pause**, and the layer is called the **inversion layer**. It acts as a roof to rising smoke. A high-rise building may be tall enough to penetrate the inversion layer. This would cause substantial differences in the smoke conditions above and below the inversion layer.

Wind: Wind is very powerful and influences smoke movement. Note the tremendous bracing required in a high rise to overcome the force of the wind. The wind exerts a pressure on the windward side of the building and a suction on the leeward. If the windows are out and the

Fig. 11-20. This photo of the 1980 MGM Grand Hotel fire shows how smoke spreads horizontally as it meets an inversion layer.

fire is on the leeward side of the building, the suppression may be "a piece of cake." Given the same fire, the windows out, and the fire on the windward side of the building, it may be impossible to move into the fire floor.

Tactical Considerations

Note again that interior personnel should be notified immediately if the windows blow out. This can change the situation on the fire floor radically for better or worse. In the Empire State Building fire, the windows failed and the wind, blowing the windows, pushed the fire down the hallway at the fire fighters.

Where an interior attack is required, the use of positive pressure ventilation into the fire floor through the attack stairway should reduce smoke pollution in the stairway. The difficulty is the limited space in the stairway.

Closely-spaced high-rise buildings can create a canyon effect, where the wind increases velocity as it squeezes through a narrow opening. The wind can shift direction many times during the fire. Ground level observations are not valid. At upper levels, the wind can blow harder and from a different direction.

Wind blowing against a building seems to split about two-thirds of the way up the building. The upper portion flows up and over the roof. The lower portion flows downward forming a vortex next to the building. It increases in velocity as it flows downward. The effect of this wind on fire pouring out windows appears to be unevaluated in the design of high-rise structures.

The effect of wind on the structure is becoming better understood. As previously mentioned, modern high-rise buildings weigh about eight or nine pounds per cubic foot. The result is that the top stories of some such buildings sway noticeably in the wind. To counter this, some buildings have **tuned-mass dampers**—heavy weights installed high up in the building which are adjusted by computer to counter the wind-induced oscillations.

Stack Effect: Stack effect is the term used to describe the movement of air inside a tightly sealed building due to the difference in temperature between the air inside the building and outside the building. Stack effect is most significant in cold climates in the winter time because of the great difference between the inside and outside temperatures. Stack effect also occurs, with opposite airflow direction, when the outside temperature is greater than the inside temperature. This condition is, however, much less significant, because the amount of stack effect is proportional to the difference between the two temper-

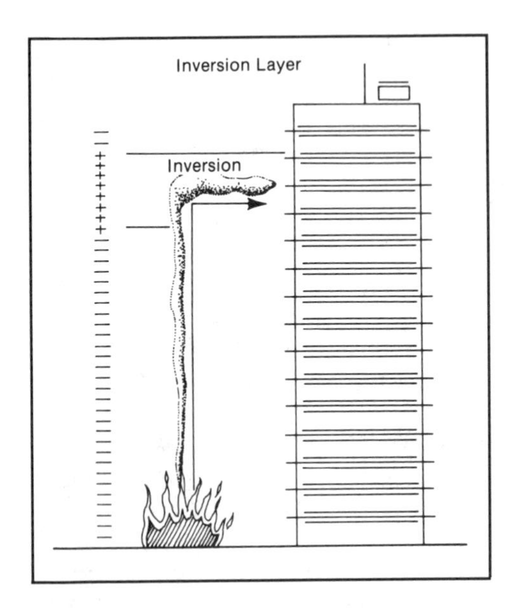

Fig. 11-21. Note how the top of a high rise may be above the inversion layer.

atures. There is a much greater difference between the inside temperature and outside temperature during the winter than during the summer. The winter differential lasts for longer periods of time. In a cold climate, a differential of about 70°F might exist for weeks on end, while in the heat of summer, the difference might be only about 15°F, mostly in the afternoon on days in which the building is occupied.[19]

Stack effect is not caused by a fire. The products of combustion ride the stack effect currents.

Fig. 11-22. In cold climates the winter stack effect is significant. People who live in mild climates do not always see an appreciable stack effect at fires because of the small difference in temperature. In very warm climates summer stack effect can be experienced but it is not as significant as winter stack effect because the temperature differential is much less.

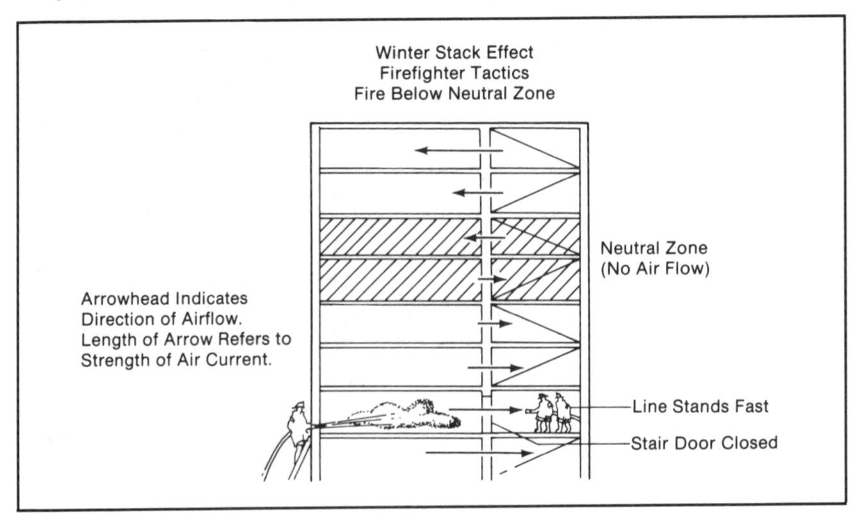

Winter Stack Effect: Under winter conditions, stack effect causes a movement of air from the floors into the vertical shafts, stairways, elevators, etc., in the lower portion of the building. The greatest flow will be at the first floor, with the flow gradually decreasing as the height of the floor above ground increases. At a point about one-third to one-half the height, the flow is reduced to zero; this is called the neutral zone. Above the neutral zone, the flow reverses and travels out to the floors from the shafts. The pressure increases floor by floor and is greatest at the top.

If floors were completely cut off from one another, the stack effect would exist separately on each floor. Air would flow in at openings at the bottom of each floor and out the top. There would be no overall stack effect since there would be no connection between the floors. The buildings in which fires are fought do have interconnected vertical openings, stairways, elevator shafts, and pipe openings, so the stack effect can be significant throughout the building.

Smoke which has lost much of its thermal energy can be delivered to upper floors by the stack effect.

A fire that vividly demonstrated winter stack effect occurred in a Federal office building under construction in Cleveland, Ohio. The fire was in a pile of rubbish on the 9th floor. Fumes from degraded and burning plastics traveled to the top floor by the stack effect. The workers started down the stairways. There were so affected by the fumes that they went into the 25th floor for refuge and started to break windows despite the hazard to those below. In another case, a winter fire on a loading dock was first reported from almost the top floor of a 40-story building.

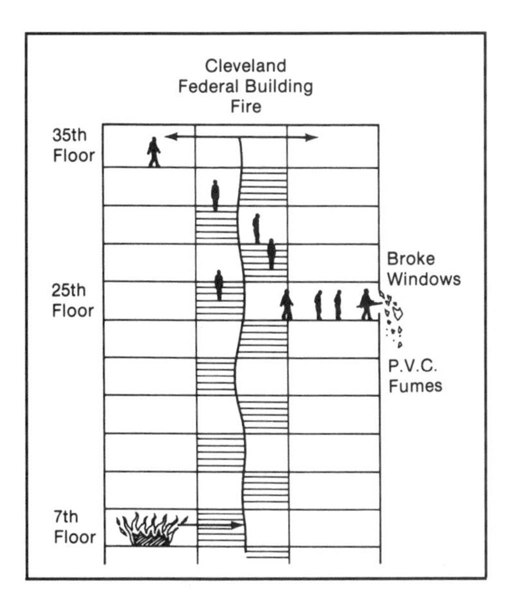

Fig. 11-23. The stack effect delivered gases to the thirty-fifth floor from a minor fire on the ninth floor of a building under construction. The fire led to improved fire protection standards in buildings constructed by the United States General Services Administration.

Summer Stack Effect: In a sealed air-conditioned building in the summertime, the stack effect is reversed. The flow is downward. The situation is not as extreme as the winter situation because the differential is usually much less, but it still can be serious, particularly in very warm climates. In a Puerto Rican hospital, in the summer, smoke pollution occurred several floors below the fire floor due to the summer stack effect.

At a very serious summer high-rise fire in New York, five floors below the fire floor were polluted by smoke so as to be unusable for staging the fire. Air bottles were seriously depleted before personnel reached the fire floor.

Fire fighters experienced in refrigerator plant fires have long been acquainted with **cold smoke** which falls downward. In 1946, Naval Base fire fighters, under this author's command, assisted the Norfolk, Virginia fire department in a refrigerator plant fire. The fire was in fungus-treated construction lumber in a cold box. The boxes surrounding the box on fire were at zero temperature. The box where the fire was located was at 30°F. By rigging salvage covers and setting up fans as preplanned for a similar building on the Naval Base, the smoke was pumped into the elevator shaft where it fell down the shaft to the street. At that time, the term "stack effect" was unknown to fire fighters, but this was a perfect example of extreme reverse stack effect conditions.

Stack effect is but one of the factors which can affect smoke movement. In a given situation, it may be dominant, as in a relatively small rubbish fire in the basement which causes acute distress to tenants on upper floors. In other cases, it may be subordinate or inconsequential.

Tactical Considerations

When stack effect is a significant factor due to weather conditions, all should be reminded of the situation as is done, for example, when the burn index is high. All should be aware that the first alarm might be from a person or a detector many floors removed from the scene of the actual fire. Fire ground commanders should avoid committing all forces until the location of the fire is determined.

Alarm operators should be trained to take particular note of callers who say ***fire or flames*** *when multiple calls about smoke are being received. This information should be passed to the fire ground commander rather than assume that the combat forces will find the fire location immediately upon arrival.*

It is a common practice to hook open the lobby doors to make it easy to handle or access equipment. This may lower the neutral zone several floors and increase the number of floors polluted by a fire below the neutral zone.

Tactical Considerations, continued

Stack effect might lead fire departments to change some operations. Consider a winter fire in a high-rise building. The fire is below the neutral plane and accessible from the outside of the building. Using conventional operations, the fire department would attack the fire from the interior, thereby freeing all the smoke to ride the available stack effect currents and pollute the upper floors. An alternate attack might be to protect the interior exposure with charged lines but not to open the fire floor to the stairways. The attack is made from the outside. Most of the smoke is vented out the windows by thermal pressure, fire department fans, and, with a little luck, by the wind blowing in the right direction. The fire loss would probably be the same, but the smoke damage and apprehension of occupants throughout the building could be substantially reduced.

Air Conditioning

Whether or not air conditinging is a significant factor in smoke movement depends on the type of system.

Individual Room Units

Individual room units which recirculate room air or draw air only from the outside, as in many recently built motels, will have no effect on the general situation.

On the other hand, if the individual units draw air from the corridor, as was the case in the MGM Grand Hotel fire in Las Vegas,[20] polluted air from the corridor will be drawn into the rooms. It appears that the opening of windows at this fire increased the flow from the corridors.

A similar system was used in the Du Pont Plaza Hotel fire in San Juan. The air supply was drawn from the corridor ceiling. The only saving factor was the balconies. Tenants who retreated to the balconies spent several uncomfortable hours but survived.

Single-Floor Systems

Single-floor systems are seen in many smaller office buildings and in retrofitted buildings built before air conditioning. In a typical system, there is a fan unit where the air is treated on each floor. Air flows from the corridor to the fan unit and is distributed through ducts which spill into the ceiling void. The air comes into the offices through openings in the ceiling tiles. It returns to the fan unit via the corridors. Smoke fron any fire on the floor will be distributed throughout the entire floor. There may be vents from the ducts feeding the floor above. If the building code has permitted the dangerous situation of assuming the floor and ceiling assembly protect the columns in the plenum space, then smoke can move along the re-entrant space of the column to the voids on other floors. Telephone service is often provided by cabinets located one above the other. Holes in the floors connect the cabinets. Openings from the cabinet to the ceiling void provide another path for smoke distribution.

In short, the fact that the building has single-floor separate systems will not prevent smoke migration throughout the building.

In one of New York City's World Trade Center fires, fire spread through many unprotected floor openings. The smoke removal system did not operate because it was shut down for the night.[21]

Multi-Floor Systems

When the entire building is air conditioned by one or more building systems, the problems become severely complicated.

In the 1950s, for instance, when total building systems were first coming into use, the treated air was regarded as a precious commodity. Only certain exhausts were passed to the outside; all other air was treated and returned to the occupants. In the case of the Energy Department Headquarters at Germantown, Maryland, this author was assured that in the case of fire, the fans could be put on "full exhaust" which was taken to mean that all air reaching the fan would be exhausted. In fact, of the air reaching the exhaust area only 14 percent of it would be exhausted. The balance would be recycled so that most of the smoke would be returned to the building. As technology progressed and energy costs rose, it was learned that there are many days when the outside air is quite satisfactory and could be pumped into the building. More modern systems often have a true full-exhaust capability.

Tactical Considerations

The fire department must develop officers who are familiar with all the many aspects of Heating Ventilation Air Conditioning (HVAC) systems. It is simply not adequate to accept somebody else's idea of what will be an adequate functioning of the system when there is a fire in the building.

Fig. 11-24. The outside of this building shows several floors called mechanical floors where equipment such as air conditioning is located. Each mechanical floor usually serves floors above and below it. Every fire department should know which mechanical floor serves the other floors.

Smoke Removal Systems

Complete knowledge of such systems is particularly important. Will the fire department operate it? Will the building engineer operate it?

In one high rise, the operating station for the engineer who allegedly will operate the system in an emergency is within the main plenum of the system. Operator instructions include provision for the engineer to leave immediately in the event of a fire. The control room for the air-conditioning smoke removal system must be invulnerable to the effects of the fire with which the system must cope.

There are those who have insisted that high-rise fires can be controlled simply by manipulating the air supply. If such a system is proposed as the primary defense, proponents should be required to provide proof of the system efficiency in a realistic fire—perhaps with the windows out and a 35-mile-per-hour-wind blowing into the building.

One such air supply advocate argues, "Let the fire burn, but let it burn clean"—presumably so that there will be no dangerous products of combustion. This is a fallacy. Heat is a dangerous product of combustion. There is no such thing as a clean-burning, hostile fire. Even if the fire itself is burning freely, above, below, and on all sides of the fire, materials will be in various stages of degradation, generating toxic and explosive gases.

The design of smoke removal systems is an extremely complicated task.[22] A well-designed and properly maintained smoke removal system can certainly supplement the primary defense—automatic sprinklers—but it is certainly no substitute for adequate protection.

This author recently looked at a smoke removal system in a five-story sprinklered office building with a huge atrium. The system is triggered by detectors in the air ducts, by the operation of a sprinkler, or by a manual fire alarm. The normal air conditioning is a ducted supply with an unducted return through the ceiling void. In the smoke removal mode, the supply is cut off to the fire floor, and all air returned from the floor from which the alarm was received is spilled outside the building. Air from other floors is drawn to the fire floor through the atrium. It is possible that a person looking across the atrium might see a fire on another floor and pull the nearby alarm. The effect of this contingency is unevaluated. One tenant in this office building installed an insulation barrier in the void to control sound from a typing pool. This completely negated the system. The void must remain unobstructed, and all ceiling tile units must be in place for proper operation.

In larger buildings, the smoke removal system can be complex and customized. The fire department must be aware of its installation and operation from the beginning. One of the possible deficiencies is that it may be necessary that the system remain running, even if not refrigerating or heating, so that smoke will reach the detectors when the building is closed. An unaware future energy-conscious manager may turn it off over weekends.

Shortly after New York's Citibank building opened, one of the stores in its atrium had a serious fire. To clear the smoke, the supply fans were shut down, doors were opened, and the exhaust fans started. The smoke did not clear even though the exhaust fans were running at full speed. The fan system was interlocked. If the exhaust fans received a signal from the supply fans that they were not passing air, the variable pitch blades automatically adjusted to no airflow, even though the fans kept running. After this experience, the system was modified to make it possible to bypass the automatic controls.

This demonstrates the importance of studying how the system will react to every imaginable set of circumstances.

Tactical Considerations

All the necessary information on the HVAC system should be developed as part of the fire department's retrievable information system. It is dangerous to rely on the building staff. When the fire occurs on a weekend, a temporary replacement may be on duty who knows nothing specific about the fire operations mode of the system. Similarly, the fire ground commander may be from another part of the city and may never have studied the fire building.

NFPA Standard 92A covers "Recommended Practice for Smoke Control Systems."

Deputy Chief Elmer Chapman FDNY (a former commander of the midtown high-rise area), a student of smoke control, and member of the NFPA 92A committee, has serious and basic reservations about the installation of automatic smoke detector-triggered, smoke control systems in *unsprinklered buildings*.

- "To advocate the installation of a zone smoke control system as a life safety system *in a non-sprinklered building* without any proof of its capability to perform may create liability problems for the owner, safety problems for the occupants, and fire control problems for the fire department.
- A basic principle of fire suppression tactics is to not ventilate a fire before the hose line is in place. A zone smoke control system removes smoke from the fire area, replacing it with fresh air. This supplies oxygen to the fire. Recall that 537 Btu are generated for each cubic foot of oxygen delivered to the fire.
- Fire fighters know well that in venting a building manually, the fire area must be defined, so that it is not spread horizontally. Automatic systems have spread fire laterally for considerable distances.
- The zone control system compromises the effectiveness of compartmentation by permitting the overriding of fire dampers in ducts which pass through the compartments.
- The location of return air shafts near the exit stairways in the core means that smoke will be drawn toward the exits.
- It is very possible that a smoke detector would operate on the floor above the fire. In this case the control system would supply air to the fire floor in the erroneous belief that positive pressure is needed to prevent smoke migration from the supposed fire floor.

Chief J. O'Hagan's book "High Rise/Fire and Life Safety" is an excellent discussion of the entire high-rise problem by a man the author knew and admired as a brilliant and innovative fire officer, O'Hagan was appointed the youngest Chief of the New York Fire Department. R. Mendes' "Fighting High Rise Fires" is another excellent work on this topic. This book, published by the NFPA, is out of print but is available in many fire service libraries.[23]

The following three fires in New York City are of interest because they illustrate and have focused interest on some of the deficiencies of the newer types of high-rise construction.

One New York Plaza—August 4, 1970.

The HVAC system was designed to shut down the supply fans but allow the exhaust fans to run in the belief that they would remove smoke. The quantity of smoke generated in the fire overwhelmed the fans, but it did draw the fire 200 feet toward the return air shaft. This caused extensive structural damage. It was fortunate that the return fans were drawing smoke from several floors, as the effect was diluted. Had the fans been drawing only from the fire floor the effect might have been even more devastating.

810 7th Avenue—January 17, 1974

This fire started in a trash cart in the elevator lobby on the fifth floor. The smoke detectors sensed the fire and exhausted the fire area through the plenum space above the dropped ceiling. The fire heated the non-combustible ceiling tiles to incandescence. As the tiles dropped they started numerous other fires. Fire fighters were able to extinguish the fires before the evidence of this system failure was destroyed.

777 Third Avenue—January 6, 1976

The HVAC system was completely shut down for the night. At 6 a.m. an employee found a smoke condition on the 40th and other upper floors. The fire department searched the building unsuccessfully for over an hour. Smoke conditions were present from the 17th to the 40th floors and were increasing.

The command officer ordered the return fans started to clear the smoke. Five windows on the 20th floor blew out and the fire accelerated. It was quickly extinguished. The fire had started shortly after midnight and had been limited by the compartmentation of the fire space.

In short, automatic mechanical exhaust of smoke in an *unsprinklered building* can cause disaster. If the plenum space is part of the smoke removal systems, sprinklers should be provided above and below the ceiling.

Note: In any particular fire, all the factors affecting smoke movement may be operative, assisting or canceling one another. The situation changes from minute to minute in many ways. How they will combine in a specific situation remains uncertain.

Compartmentation

Some people assume that fire-resistive buildings automatically provide compartmentation which limits the transmission of fire. This is often the case in older buildings with masonry panel walls built on concrete floors, without air conditioning and without the huge number of penetrations necessary for current communications requirements. Modern buildings often have poor perimeter fire stopping, and multiple penetrations for wiring. Air conditioning ducts are sometimes smoke paths. Steel bar-joist floor and ceiling assemblies can pass smoke. T.Z. Harmathy, a very competent Canadian fire researcher, provides a good technical discussion of problems of compartmentation.[24] (Also see discussion of compartmentation in Chapters 5 and 8.)

Tactical Considerations

The fire department should designate an officer to be its smoke control expert. This officer should stay abreast of all new developments.

The pre-plan for a high-rise fire should include a smoke control component.

Pressurized Stairways

One or more of the stairways in a building may be equipped to be pressurized when fire occurs. Fans are installed to pump outside air into the stairway so that the pressure differential will keep the stairways free of smoke. This requires a dependable source of power and intakes located so that they will not pick up smoke from the fire. Occupants must be trained to use the proper stairway, and should be drilled on this, since the noise of the fans may frighten people already upset and trying to escape. Since the temperature should be close to the outside temperature, there should be no stack effect. The overpressure may make opening doors difficult, and mechanical assistance may be required. The system does nothing to clear corridors.

A 21-story high rise in Liverpool, England was built with only one exit stairway because the stairway was pressurized. A fire started in rubbish in a passenger elevator. The smoke for the fire vented out at the roof level in proximity to the intakes for the pressurization fans. The entire stairway was polluted. Fortunately, there were only a few occupants in the building.

This demonstrates the necessity of careful attention to the location of intake fans for pressurized stairways and for ventilation systems generally.

Tactical Considerations

A pressurized stairway should be reserved for evacuation.

The high-volume fans used for positive pressure ventilation may be used to clear a stairway of smoke so that it could be used for evacuation, or be more habitable for back-up fire fighting teams thus preserving the SCBA air supply.

Installation of Special Equipment

Any equipment designed to function in case of fire is in effect fire department equipment. The fire department should approve its installation, be familiar with its operation, and supervise its testing and maintenance by the owner or owner's contractor.

The basis for this argument, opposed by some designers and building officials, is the fire chief's unquestioned authority over fire fighting operations. Of course, there is no guarantee that the concept will be upheld, but it is the position with which the fire department should start.

Over 100 years ago, when automatic sprinklers were first developed, the International Association of Fire Chiefs voted to have nothing to do with them. This decision was reversed at their next annual meeting.[25] Those who oppose fire department supervision of installed equipment are courting disaster.

An $80 million loss in a Montreal fire shows the consequences of insufficiently checking out installed equipment. See the Alexis Nihon Plaza Fire in the Case Studies section.

Fire Load and Flame Spread

Consideration of the fire problem of a high-rise building must go beyond the basic structural considerations to the question of interior trim and contents.

A developer who was a vociferous opponent of improved high-rise fire protection in New York City argued that his buildings could not be burned with a blowtorch when finished. He argued that the problems arose from what the tenants did—adding interior trim and contents.

In general, codes have adequate controls on interior finish with respect to flame spread. The difficulty is that the interior finish is often installed by the occupant without a building permit. Even when flame spread requirements are adequate to protect life safety during evacuation, fires can gain great headway in combustible trim in the interval before an effective attack can be mounted. Multiple layers of wall covering were a major factor in an Atlanta office building fire in which ten died.

Clark County, Nevada fire fighters faced a pillar of fire extending up the exterior of the Las Vegas Hilton Hotel. Eight guests died. The fire was started in furniture in an elevator lobby and spread up the exterior from the eighth to the 30th floor by way of carpeting on the ceilings of the elevator lobbies.[26]

Fig. 11-25. Furnishings in the elevator lobby of each floor of the Las Vegas Hilton high-rise fire consisted of a small wooden bench with a polyurethane foam plastic cushion. Draperies were on both sides of the window and carpeting covered the floor, walls, and ceiling.

The rapid development of a fire in one suite of offices in the Empire State Building was due to cork tiles glued to the walls in the reception area. Five alarms were sounded due to the difficulty of getting to the fire and the need to check the safety of several thousand persons.[27]

In his book, *From Bauhaus to Our House*,[28] Tom Wolfe uniquely describes the problem of interior trim and contents:

> "I have seen the carpenters and cabinetmakers and search-and-acquire girls hauling in more cornices, covings, pilasters, carved moldings, and recessed domes, more linenfold paneling, more (fireless) fireplaces with festoons of fruit carved in mahogany on the mantels, more chandeliers, sconces, girandoles, chestnut leather sofas, and chiming clocks than Wren, Inigo Jones, the brothers Adam, Lord Burlington, and the Dilettanti, working in concert, could have dreamed of."

The Casino of the MGM Grand Hotel was 68,000 square feet in area, equivalent to about one and one-half football fields. The fire load is not reported in terms of Btu per square foot but 12 tons of combustible adhesive held up the ceiling tiles.

In August 1991, a wood paneled dining room was destroyed in an early morning fire at the headquarters of the U.S. General Services Administration. The fire apparently started in materials used to refinish the wood. This has been cited as the cause of other high-rise fires. The dollar loss was estimated at about $100,000. It is estimated that the government paid over $200,000 to employees who were sent home for the day. Indirect losses may far exceed direct losses in office building fires. Lawsuits involving the Meridian Plaza fire are in the billions of dollars.

Contents

The wide open floor with scores of computers and associated equipment required in many of today's businesses has introduced a new flame spread problem. It was seen in the spectacular First Interstate Bank fire. The fire spread rapidly while the guards mistakenly reset the alarms. The Los Angeles Fire Department received the first alarm from an observer in a building several blocks away. At that time heavy fire was coming from many windows. NIST's Harold Nelson, an SFPE Fellow, studied the rate at which this fire grew.[29]

He concluded that the fire started about 10:27 p.m. and that by 10:35 the smoke temperature was 325°F, beyond the level of human endurance. In this author's opinion, had the building been in normal, daytime operation, there would have been a terrible disaster.

A fire chief ventured the opinion that occupants would have used the fire hose and extinguishers and put out the fire. This is doubtful.

Like scientists, computer operators are almost paranoid about water on their equipment. From experience, I can testify that it takes careful argument and demonstration to convince them that sprinklers are necessary and justified.

The fire was on a "trading floor" where fortunes can be made or lost in seconds. Consider the reaction of those attending the soccer match at the Bradford Stadium Fire.[30] They are seen in a dramatic video, torn between watching the fire and watching the game. It is unlikely that a trader would hang up on a big deal with a counterpart in Tokyo just because there was a fire across the room.

In 1966 a fire occurred in an architect's office on the second floor of a New York high rise. The occupants were seen trying to beat out the fire with seat cushions. Twelve died in the fire. Ironically, the lens grinding shop on the first floor was a factory occupancy under the strict New York State Labor law. The building should have been sprinklered. The fatalities were settled for about $250,000 apiece (1968 dollars).

Heavy fire loads may be found in special locations in high rises such as club rooms and restaurants. Wood paneling and imitation wood beams and heavy loads of plastic are common. The wood paneling is installed on furring strips leaving a void behind it where fire can burn untouched by hose streams. Five occupants and two fire fighters died in a New Orleans High Rise fire fueled by wood paneling in a restaurant. Inadequate hose streams contributed to the disaster.[31]

Storerooms for office supplies and telephone rooms are other high fire-load areas. Not only is the wiring insulation combustible, but a typical telephone room is usually the supply room and contains boxes of equipment packed in foamed plastic.

Fig. 11-26. Heavy fire loads may be found in conference rooms. This one is decorated with a virtual lumber yard of planks on the overhead. It is possible that the planks are treated to be flame resistant. In any case, fire involving the desks and chairs might cause failure of the hangers, ususally set in lead anchors, allowing the planks to fall.

A significant high fire-load fire occurred in an insurance company building in Philadelphia, built as a nine-story office building in 1914. A tenth floor was added in the 1920's. The ninth floor was used for the storage of old records, in a room within a room. The area was 60 x 160 feet. With 11-foot ceilings, the volume was over 100,000 cubic feet. It was completely unventilated. Sprinklers had been recommended but the value of the old records did not warrant protection. Two smoke detectors were provided. The fire department fought this difficult fire for 72 hours with two monitor nozzles. The water damage to computer installations on the floors below was enormous.

When full sprinkler protection is not required, some codes require sprinklers in high fire-load areas as designated by the fire department.

Tactical Considerations

When sprinkler recommendations are made, the fire department would be well advised to append an analysis of the effect of alternative fire suppression. Showing that building operations would be seriously affected and the potential loss might provide a convincing argument for the sprinklers.

Maintenance Operations

Maintenance operations can provide unexpectedly serious fire loads. If arranged for by a tenant, the building management may be completely unaware of this hazard. This happened at the Union Bank Building fire in Los Angeles on July 18, 1988.[32] A contractor employed by a tenant was refinishing book cases with flammable liquids. A worker brushed a solvent over an electrical outlet and the fire erupted. The fire was quickly contained by the fire department but the loss was estimated at $500,000. An almost identical fire occurred in Los Angeles in 1975. Flammable paint thinner was used in an occupied building. A light switch ignited vapors. The air conditioning system spread smoke through two connected floors.[33] The One Meridian Place fire started from spontaneous ignition of linseed rags used to restore wood paneling.

Rubbish

As rubbish is gathered in these buildings, it is often concentrated in one location, making another fire load problem. The condition of the material provides a high rate of heat release. The concentration takes place in halls, elevators and basements or lobbies, all particularly bad locations. Collecting rubbish on elevators tempts the potential vandal and provides the circumstances where otherwise inconsequential fire

Fig. 11-27. The concentration of rubbish as it is collected presents a major fire load problem in any high-rise building. This picture shows the daily load from a building that is only eight stories high.

can pollute an entire building. A rubbish fire in an elevator parked at the ninth floor contaminated the 26 floors above.[34]

Many apartment houses have huge compactor units for rubbish located in the basement where it is delivered through chutes. In one apartment house, the aluminum electrical conduit passed through the compactor room. A small rubbish fire destroyed the electrical system and forced all the tenants out of their homes for an extended period. Seven people died when the fire roared 35 stories up a compactor chute blocked with rubbish.[35] When high-rise apartments had incinerators, it was a common practice for tenants to set fire to rubbish piled up in the shaft. The shaft was built as a chimney and easily confined the fire. Unfortunately, this practice has continued where compactors are installed. The compactor shaft cannot contain the fire. Fire has broken out in apartments on several floors along the line of the compactor shaft. Sprinklers in the compactor shaft had been shut off.

A 4-foot by 15-foot, free-burning rubbish fire on a first-floor stair lobby of an 18-story apartment hotel in New York quickly became an inferno that almost cost the life of a fire fighter and spread 185 feet vertically and 100 feet horizontally in less than 10 minutes. The rubbish was found to contain cans which probably had held flammable liquids.

The rubbish handling system is important. *It should be properly planned with the anticipation that there will be fires.*

Tactical Considerations

Rubbish is usually accumulated in the service lobby on each floor. Fire departments should be wary of using the service elevator routinely to approach all fires.

Tactical Considerations, continued

Because of the prevalence of rubbish fires in high rises, New York City fire fighters are not permitted to use a service elevator until it is declared safe by command.

Particular attention should be paid to the hazard created when the regular removal of rubbish is interrupted by a strike or other causes.

Alterations to Occupied Buildings

A very serious hazard exists when a building is altered or rehabilitated while occupied.

In a university hospital being altered, an untreated plywood wall was erected to separate the construction area from the maternity ward and nursery. The construction area had the typical fire hazards of any construction job.

Hotels are renovated periodically. This usually involves replacing the furniture, particularly the beds. The new furniture is often stored wherever space is available, including the basement. Sometimes a floor is placed out of service and used for furniture storage. Mattresses removed from rooms may be stored in halls, awaiting disposal. Two serious fires occurred in Maryland suburban motels from such mattresses, which were inviting targets to an arsonist.

The disastrous Du Pont Plaza Hotel fire in San Juan, Puerto Rico was started in a large pile of new furniture temporarily stored in a ballroom.

Tactical Considerations

Watch for a sign like, "pardon our dust, we're remodeling." It may be the clue to a potential disaster.

There are no real standards for construction work in occupied buildings. The fire department should be alert for any preliminary signs that this hazardous condition is about to take place. Newspapers, including the real estate pages, should be read with this possibility in mind. All clues should be forwarded through channels to the proper authority. A pre-construction conference may eliminate many of the problems and all concerned will be on notice that the fire department intends to give the operation very close attention.

Fig. 11-28. The Fire Prevention Division of every fire department should stay on top of the situation when a hotel is being refurbished. Temporarily stored furniture or mattresses are tempting targets for arsonists.

Partial Occupancy of Buildings Under Construction

It is a most dangerous situation to have a building partially occupied and partially being finished. This will be the case sometimes. Fire protection systems are not complete; doors may not yet be installed on stairways and elevators. Temporary heating using LPG may be used in some areas and all the hazards of a building under construction exist. The strictest special precautions should be demanded if the building department permits partial occupancy.

The aforementioned book *Skyscraper*, which followed the planning and erection of a New York Skyscraper in 1988, gives an excellent picture of the advantages from the management point of view of moving tenants into a partially completed building—specifically $300,000 per month in rent.

The book particularly describes the strictness of the fire department inspections required to gain a Temporary Certificate of Occupancy.[36]

This author examined another high-rise building where the fire protection was the pride of the management. The building was voluntarily sprinklered. A floor under construction had a number of wood construction shanties erected on it, plus a large accumulation of combustibles. The doors had not yet been installed on the elevator shaft.

There were many possible fire causes, but this floor was not yet sprinklered. The sprinkler piping was in, but plugged off. There were occupied floors above this floor. The sprinkler should go in first in such a situation, not last, after the ceiling is finished.

At least two serious fires are on record as occurring in occupied buildings under construction.[37]

Fire in building materials on the uncompleted 14th floor of Boston's 52-story Prudential Building endangered 1500 occupants and required 60 percent of the Boston fire department for extinguishment. Twelve people were injured. The building is now being sprinklered. As of June 1991, 57% of the tower above the sprinklered basement is sprinklered. Retrofitting sprinklers takes a long time. The building was classified "protected non-combustible" construction. Most codes would describe it as fire resistive. This fire was alarmed by a smoke detector and was a raging inferno by the fire fighters reached it.

A New York City 45-story high-rise office building was partially occupied while still under construction. The contents of the third floor included many bottles of propane and acetylene gas. The sprinklers were not yet installed on the third (fire) floor except for the elevator lobby. Both scissor stairway doors were open on the fire floor. It was thus impossible to use one for attack and the other for evacuation, as

Fig. 11-29. This floor of Boston's Prudential Building was devastated by a fire in construction materials. At the time the building was already heavily occupied. (Bill Noonan, Boston Fire Department)

planned. Many distant sprinklers fused, shorting out the elevators, forcing the search teams to climb 45 stories. A disaster was narrowly averted because, by chance, the fire was reachable by ground attack. An exterior tower ladder attack was carefully coordinated with the withdrawal of interior attack personnel.

Automatic Sprinklers

In this author's opinion, an examination of the disaster potential inherent in the typical enclosed structure with its occupants, fire load and fuels, and construction materials and configuration, leads to one conclusion. There must be absolute assurance that the toxic gases released in a fire will be severely limited, and not dependent on control by a smoke control system. In theory, this could be accomplished by limiting the quantity of fuel present. In fact, the combustible materials are necessary to the functions performed in the building. The only practical solution is to provide a system which will limit the amount of gas produced.

The only such system available is automatic sprinklers. Fire department reaction time (from the time the alarm is received until water is put on the fire) is not the fabled 1½ minutes, but more nearly 20 minutes or more. Burn time between ignition and alarm can be any length. The relatively short reaction time of the sprinklers, plus the fact that they can automatically transmit the alarm, makes them infinitely superior to any other system. If the sprinkler system water supply is properly tied to the domestic water system, the system is almost foolproof.

This is not to say that other systems are not necessary to supplement the sprinkler protection, but sprinklers are the core of fire safety for the occupants of high-rise buildings.

It is very possible that the world's worst fire disaster will take place in a fully occupied unsprinklered high-rise office building. Study the reports of some of the high-rise fires which have occurred. Note that most have occurred after hours. Add to this fire scenario one or more of these very reasonable contingencies:

- Fire forces are committed heavily to another fire. For instance, the Los Angeles fire department regularly makes a huge commitment to wildland fires which involve populated areas.
- Heavy snow conditions.
- Traffic gridlock, such as occurs in New York when it is impossible to move apparatus through the streets.
- Gale wind conditions.
- Total elevator power failure.

This list of potential complicating factors is long and the danger of catastrophe is real. Those opposed to sprinklers sometimes advance the argument that a fire in an occupied building would be quickly detected, reported and attacked by the occupants. No doubt this has occurred in some circumstances. However, the history of fire disasters convinces this writer that this approach can be cynically called the "Falsenza Security System."

For You to Do

As an exercise, set out the commitments of resources to a recent major non-high-rise fire and, assuming the fires are simultaneous, superimpose the requirements of a recent serious high-rise fire. Estimate the time it would take to deliver the required units to the action scene. Consider the time delay in performing necessary operations. Realistically assess the probable outcome of the high-rise fire.

When the gigantic World Trade Center was being designed in New York City, staff members of the Port Authority of New York and New Jersey were very confident that fire problems would be minimal since they would have a well-trained building staff. The Center has had several serious fires. After a third-alarm fire which extended to several floors via wiring channels, a number of changes were made. Sprinklers were installed in high fire-load areas and lobby areas with additional capacity to cover future high fire-load developments. Many floors are fully sprinklered. A number of other improvements, which should have been installed when the gargantuan structures were built, were retrofitted.

Sprinkler Experience in High-Rise Buildings[38] by Robert Powers, then Superintendent of the New York Board of Fire Underwriters, is a most useful document and certainly contradicts those who allege that "sprinklers are untried in high-rise buildings."

This report shows that in 254 fires in office buildings, the sprinkler performance was "satisfactory" in 98.8 percent of the cases. Of 1371 fires in non-office buildings, 98.4 percent were satisfactory. In high-rise buildings (other than office buildings), 1394 fires were recorded in a 10-year period. Only 23 fires were not controlled by sprinklers. In 21 of the 23 cases not controlled by sprinklers, closed valves caused the loss.

Powers also mentions that the overwhelming number of fires occurred in rubbish, again pointing up the seriousness of this problem. He goes on to note that water supply reliability is achieved by taking the domestic supply from part of the sprinkler supply. When the domestic supply fails, corrective action is immediate.

Some of the worst high-rise buildings are in areas where the fire department is only of modest strength. The 293 Park Avenue fire in

New York was essentially confined to the 7500-square-foot compartmented area specified in the code as an alternate to sprinklers.[39] Yet, the task took more than 500 fire fighters of whom over 100 were casualties. In Los Angeles, it took more than 300 fire fighters to suppress the Occidental Building fire, and almost 400 to control the First Interstate Bank fire. How many fire departments have staffs approaching these numbers?

What gives the developer of a high-rise the right to erect or maintain a structure which places such a burden on the public purse and strips the city of normal fire protection? At one time during the February 1991 Meridian Building high-rise fire in Philadelphia, there were only 16 companies available to protect the rest of the city.

The argument against sprinklers is usually an economic one, buttressed at times with arguments from some construction material suppliers and devotees of managing fires by controlling smoke.

The basic argument for sprinkler protection is one of equity. The builder is creating the problem for profit. It is up to the builder to provide the solution. This parallels what is done in the case of parking facilities, sewage facilities, and other amenities.

The fire chief is often all alone in an argument before the city council. Enormous social, political, and financial pressures may be placed on the public officials. The fire chief is arguing the case on behalf of people who probably do not even know they are going to be in the high rise, and who in many cases would be oblivious to the hazard. The fires discussed in this chapter, as well as the Case Studies and Additional Readings, can be the raw material of an effective argument. The huge losses sustained by tenants in the Philadelphia high-rise fire who were not directly involved in the fire may help to make tenants prefer to be located in sprinklered buildings.

When the improvement of existing buildings is being discussed, many attorneys immediately dismiss any effort to improve the fire safety of an existing building as being unconstitutional. The United States Fire Administration asked Professor Vincent Brannigan, J.D., to research the subject. He found no U.S. appellate court case which accepted a constitutional barrier to the enforcement of a new code to an existing building.[40]

It is well worth repeating that when sprinklers are installed for life safety, they should be in service all the time the building is occupied. If evacuation of the building is not feasible, special precautions should be taken during the out-of-service period. Many code provisions are reduced when sprinklers are installed. *If all the fire safety eggs are placed in the sprinkler basket, the basket requires close watching.* Possible deficiencies of sprinkler systems are discussed in Chapter 13, Automatic Sprinklers.

Tactical Considerations

There are many good texts available on the problems of fighting high-rise fires. Only some observations will be presented here.

Note that many fires in high-rise buildings occur at the lower levels where ordinary tactics will suffice (except for the problem of smoke pollution of upper floors). One Sunday night, the entire block-long second floor of a New York office high-rise was ablaze. A well-placed heavy caliber deck pipe stream knocked the fire down before inside units could force the door. If the fire can be handled by units operating over ladders from the street, with the interior closed up, but guarded with units with charged hose lines, much smoke pollution of upper floors may be avoided.

Each fire-resistive building is different. The fire department should make a systematic evaluation of each building, studying the fire origin potential and the potential for the spread of smoke and fire. The cooperation of the building management should be actively sought, but lack of cooperation is no reason not to make the evaluation.

All the information the fire department gathers that would be useful at the fire scene should be reduced to a single record and retrieval system which should be immediately available to the fire ground commander at the scene of the fire. In simple situations, plastic protected pages in a loose-leaf book may be adequate. Where there are a number of buildings, one of the many information storage and retrieval systems available should be used. It is possible the storage function would be maintained at a central computer, and at the scene there would be units with display and hard copy printout capability.

Checklist: There are innumerable items of information which would be useful for the fire ground commander to have at the scene. The following is merely suggestive of useful building information.

Even though a sophisticated information retrieval system may not be available, this valuable building related information should be gathered. The day is past when fire departments can depend on information known only to "Captain Joe, the old-timer," who won't always be around. The information must be institutionalized—available to all. The building may be there for a century. It may outlast several generations of the fire department personnel.

Some Building Inventory Items

Structural

Frame

Cast iron—What is the estimated value of "fireproofing"? Are there connections from column voids to floor voids?

Steel—What type of fireproofing? Are columns unprotected in plenum space?

Concrete—Are there visible rods and cracks? Any fire experience in this building, which might have caused structural damage? Is corridor fire resistance achieved by gypsum on studs, making breaching easy?

Reinforced masonry—Are corridor walls reinforced? If so, breaching would be difficult.

Floor Assemblies

Bar-joist floor and ceiling assemblies—What is the integrity of the floor? Early collapse is possible. The floor above may be very hot.

Concrete—Are the floors of solid reinforced construction (good insulator and heat sink) or composite decks (Q-floors or cored slabs)? The floor above may be hot.

Post-tensioned: Don't drill or cut.

Tile arches—Dangerous to cut as entire panel may collapse.

Overloaded floors—potential for multistory collapse.

Floor Containment

Floor and ceiling assemblies—Ceiling tile failure may permit at least partial collapse and open fire and smoke passage.

Unprotected columns in plenum—Smoke and fire may pass to voids above, via reentrant space.

Firestopping—Is there effective firestopping in void spaces?

Masonry floors—Utility openings may pass fire.

Accommodation or access stairways—these create huge fire areas.

Perimeter Integrity

Joints—Are masonry wall to masonry floor joints solid (good inherent firestops)?

Panel walls—How are exterior panel walls joined to the floor? What is their value as firestops?

Glass walls—Is vertical extension of fire probable?

Exterior architectural panels—Voids between panels and walls may negate firestops.

Contents—High rate-of-heat release furnishings near windows or combustible tile ceilings increase potential for exterior extension.

Lower levels—What is the potential for exterior hose streams to block the extension of fire?

Horizontal Containment

Compartmentation

Integrity of compartmentation—Are there utility openings or underfloor openings such as for computer cables? Do barriers extend to underside of the floor above? Will fire doors function properly?

Gypsum board compartmentation—Consider penetration of relatively light-weight gypsum partitions as a substitute for forcible door entry.

Stairways—Are there deficiencies of stair enclosures, particularly on lower floors? Are there proper operating door closures on all floors? Are stairways locked to deny entry to floors? Do all units carry forcible entry tools?

Elevators and other vertical shafts—Gypsum board enclosures may be destroyed by fire fighting operations.

(See Chapter 13, Sprinklers, for further information on sprinkler protection.)

Case Studies

A number of fires have been referred to in this chapter. Each one contains lessons beyond those mentioned in the text. They should be studied carefully.

The following fires provide particularly valuable information. In many cases, NFPA Fire Investigation Reports are cited as information sources. These comprehensive, well-written and documented reports are a gold mine of information. They should be in the library of any fire department which realizes that the knowledge from the experience of others, both good and bad, is both inexpensive and priceless.

First Interstate Bank Building fire, Los Angeles, California, May 4, 1988[41]

At the time of the fire, the building was being retrofitted with complete automatic sprinkler protection. The fire started about 10:30 pm on the 12th floor. This floor was almost completely one big open area. The contents were combustible work stations with computers, associated equipment and other combustible fuel. Laboratory tests indicate that such arrangements can develop a 1 MW (Megawatt) intensity in ten minutes, but that this intensity would double in another minute.[42]

As alarm signals started coming in from the 12th floor, security personnel, in accordance with standing instructions, kept resetting the alarm. Even though many detectors were going off, a maintenance person was sent to the 12th floor to investigate. He died while opening the elevator door. The fire had flashed over, broken windows, and was extending vertically on arrival of the fire department.

Fire attack was further delayed because the standpipe system had been shut down to make connections to the sprinkler system.

By a massive commitment of resources and dogged determination by the fire fighters, the fire was confined to floors 12 through 16. The margin of victory was extremely thin.

The fire had spread vertically by outside extension and also through the space between the floor slabs and the exterior glass and aluminum panels. The interior of the panels were typical assemblies of gypsum board and insulation.

By contrast, a raging fire involving a floor of the Occidental Building is Los Angeles in 1977[43] did not extend along the perimeter due to the inherent fire stop provided by the masonry curtain wall erected on the masonry floor.

The spray-on fireproofing of the steel was cementatious rather than asbestos. The steel structure survived without structural damage.

In a post-fire interview Chief Engineer Donald Manning pointed out that few fire departments have the resources to control such a fire. He also stressed the need for operable sprinklers. Since 1974, California law has required that buildings over 75 feet in height be sprinklered. Los Angeles now requires retrofitting of sprinklers in pre-1974 buildings. The State of California rejected such retrofitting legislation for the rest of the state, reportedly because many state-owned buildings would be required to be sprinklered.

As a result of this fire the Los Angeles Fire Department developed a list of high-rise recommendations for their future high-rise operations.[44]

Postal Headquarters Building fire, Washington, D.C., 1984[45]

Heavy fire engulfed the 11th floor of an unsprinklered concrete high rise occupied by Postal Service headquarters. Heat and smoke made an inside attack impossible after only 20 minutes. Standpipe risers were located 140 feet from stairwells (a common but bad practice) and could not be found because of smoke.

The interior attack was suspended until the fire was knocked down by exterior master streams. Two sides were inaccessible. Falling glass endangered fire fighters. The building is now sprinklered.

The John Servier Retirement Home fire, Johnson City, Tennessee, December 24, 1989[46]

This reinforced concrete 11-story high rise was built as a hotel about 65 years ago, and later converted to elderly housing. During a 1950 renovation, a combustible tile ceiling was installed on a wooden frame about one foot below the original plaster-on-concrete ceiling.

In 1977, a non-combustible tile ceiling was installed below the combustible ceiling.

The fire started in an occupied first floor residence. The fire extended upward into the combustible tile above the non-combustible ceiling. The fire then spread to the combustible tile above the corridor ceiling and to combustible furniture and wall finishes. Fire spread was confined to the first floor and part of the second. Smoke, however, spread throughout the building. Smoke was pushed up through the building by natural buoyancy and by stack effect because of the great difference between the inside and outside temperature which was 17°F.

Two victims were found on the first floor. The others were found on six upper floors.

Stouffer's Inn, Harrison, New York, December 1980[47]

This was not a high-rise building, but the lessons learned are pertinent to the meeting facilities often located on the lower floors of high-rise hotels. It questions the adequacy of rules which require sprinkler protection only in buildings above six stories in height.

There were 95 occupants attending meetings in several conference rooms in the fire-resistive conference facility. The fire started on the third floor; 29 persons lost their lives and 24 were injured.

The fire was of incendiary origin. An accelerant was apparently used on the carpeting. The fire involved coats on a rack, tables for coffee service and other furnishings including polyurethane chair padding.

Outlets on a hydrant did not fit the local fire hose. About a half hour into the fire a backdraft explosion occurred in the roof area over the ballroom. It had sufficient force to lift the concrete slab roof, crack concrete encasement of steel beams, and dislodge brick. Note that this contradicts a common myth that backdrafts occur only early in a fire.

The key factors in this fire were the location of the fire at the intersection of exit corridors, rapid fire development, lack of remote egress from some rooms, and lack of automatic sprinkler protection.

Apartment High-Rise fire, New York, New York, January 11, 1988[48]

The building was a 60 year-old, ten-story fire-resistive apartment house, with commercial areas on the first two floors. A non-fire-resistive one-story addition was roofed by concrete slabs on unprotected steel. Some of this roof collapsed early in the fire. Flames extended five stories up the face of the building.

Alterations of the first floor occupancies had provided a large quantity of combustible partitions and trim. Upholstered chairs, shelves of books and paper goods added to the fire load.

Originally, each apartment had both a stairway egress and a fire escape. Alterations had eliminated the separate escape routes.

Arriving fire fighters were told that a "cigarette was burning on a couch." Fire fighters found smoke down to about a foot from the floor. Flames were overhead.

The doors to both interior stairways were blocked open making the stairways, where three of the four dead were found, impassible.

Though many people were at the front windows, they were judged to be in no immediate danger due to the wind direction. Aerial ladders were not raised because of the potential for jumping by occupants.

The following factors were deemed significant at this fire: building modifications had increased the fuel load; there was an absence of automatic detection or suppression; there were open stairway doors at the first level.

The report speaks well of fire department operations and the command and control system used.

Fatal Office Building fire, Atlanta, Georgia, June 1989[49]

An electrical arc started a violent fire in an exit corridor of a ten-story Atlanta office building. The electrician and four occupants of the building died. Twenty occupants and six fire fighters were injured. Multiple layers of wall covering contributed to the rapid fire growth in the exit corridor.

The fire was within reach of aerial ladders and 14 occupants were rescued over aerial ladders. An additional 14 were brought out by fire fighters via stairways. Hundreds of federal government employees evacuated successfully, following fire drill procedures.

A similar fire occurring on a much higher floor would undoubtedly have even more serious consequences due to the necessarily slower and more difficult fire department response to the scene of the fire.

The key factors were the rapid development of the fire, and the lack of automatic sprinkler protection.

Westchase Hilton fire, Houston, Texas, March 1982[50]

A fire which principally involved only one room on the fourth floor of this new 13-story high-rise hotel claimed 12 lives on the floor of origin. The occupant of the room of origin was dragged to safety by his returning roommate who attempted to extinguish the fire with a pillow. When they left the room the door was apparently left open. While there was some fire extension, the contents of one room generated sufficient carbon monoxide to kill twelve people.

The hotel alarm system had a number of deficiencies. In addition, the hotel personnel reset the alarm several times, indicating inadequate training.

As a result the fire department apparently was notified about 20 minutes after the fire was first reported to a hotel employee.

Key factors were serious delay in reporting the fire, lack of automatic sprinklers, failure of the fire room door to close completely, and lack of an evacuation alarm signal.

Tactical Considerations

Many hotel personnel have an instinctive aversion to calling the fire department. They fear bad publicity. Others are simply untrained and refuse to function independently, calling management or security, thus wasting time. The fire department should designate a hotel liaison officer. This officer should keep in touch with the hotel association on training, and instruct field personnel in random checks of desk personnel, particularly at night. These positions are often filled by untrained workers, particularly in lower-priced occupancies.

DuPont Plaza Hotel fire, San Juan, Puerto Rico, December 1986[51]

Some 97 people died in a mid-afternoon fire, nearly all in the lobby or casino.

There was little or no organized fire protection in the hotel either in physical facilities or training.

Disgruntled employees set fire to a large stack of newly delivered furniture in the south ballroom. The fire grew rapidly and vented violently into the lobby and casino.

The high-rise section was seriously smoke polluted, and one occupant died. Exterior balconies provided occupants with a safe, if uncomfortable, refuge for an extended time. If the fire had occurred at night it appears reasonable to assume that there would also have been massive casualties in the high rise. The manual evacuation alarm system was reportedly out of service.

Estimates of settlement claims for death and injury are reported to be approximately $240 million.

There is extensive literature on this fire, much of it published by the NFPA.

Alexis Nihon Plaza fire, Montreal, Canada, October 1986[52]

At 5:15 p.m. on a Sunday, guards discovered a fire on the tenth floor of a 15-story office tower at the Alexis Nihon Plaza. The fire lasted 13 hours.

The fire involved the 10th, 11th, and 12th floors, bypassed the 14 and 15th (no 13th floor) and totally burned out the 16th floor. A 30 x 40 foot section of the 11th floor collapsed.

The information provided to the owners and fire department about the standpipe system was seriously inadequate.

Aerial ladders were used to improvise an exterior standpipe system. It was fortunate that the fire was within reach of this hastily improvised arrangement.

The fire alarm system and the standpipe system were totally inadequate. As a result of this disaster, Montreal has passed a very strong sprinkler retrofit law.[53]

One Meridian Plaza fire, Philadelphia, Pennsylvania, February 23, 1991

This fire is very significant. It should be carefully studied. Three fire fighters died in this fire which raged uncontrolled through the unsprinklered 22nd to 29th floors. It was stopped by sprinklers installed on the 30th floor.

An excellent presentation of the defects of the building fire protection system and tremendous difficulties faced by the fire department is presented in "One Meridian Plaza Fire" and several sidebar articles in *Fire Engineering*.[54]

Read the detailed story of the system failures and apply these conditions to high-rise buildings in your jurisdiction. If building management does not address critical fire protection issues the fire department should seriously consider going public. Citizens occupying high-rise buildings have a right to be informed as to the hazards. Building management should be made aware of the tremendous liability exposure and the real possibility of personal criminal prosecutions.

Every fire department should have copies of the One Meridian Plaza fire articles and the NFPA Investigative Report[55] on hand to press the case for full automatic sprinkler protection as the only best prevention and protection for the almost inevitable high-rise disaster.

Fig. 11-30. The view looking down from a nearby building at Philadelphia's Meridian Place fire. In the photograph, note (A) hose lines being operated from another building to check outside extension of the fire, (B) showers of water from a hose line that has been cut by falling glass, and (C) plywood placed by fire fighters to protect standpipe supply lines from falling glass. (Philadelphia Fire Department)

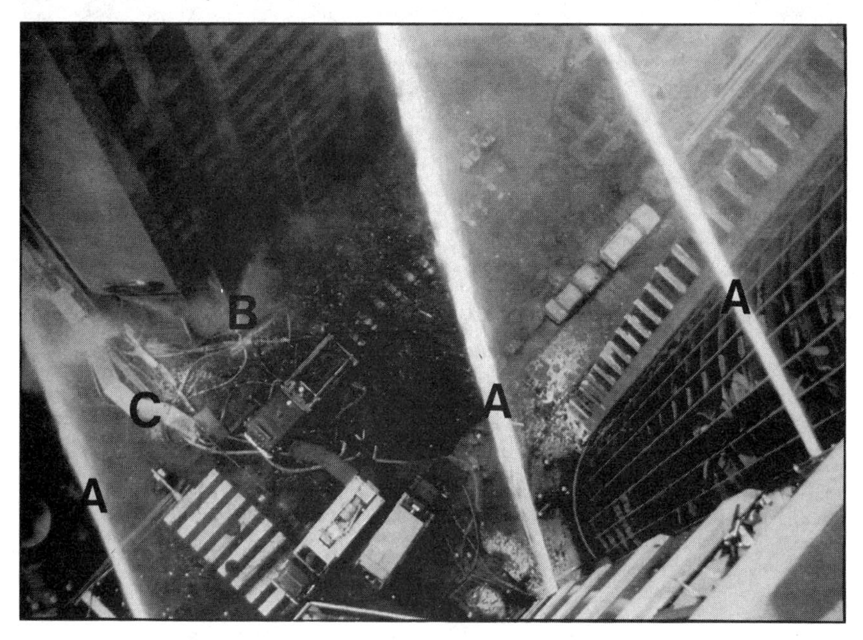

Fig. 11-31. The One Meridian Place fire as fire fighters were arriving. (Philadelphia Fire Department)

In this fire, as is the case all too frequently in high-rise fire reports, a smoke alarm was not transmitted to the fire department. Instead, a maintenance person went to the fire floor where he almost lost his life. The fire was reported from another building, an indication of the headway it must have had before the fire department responded.

All power failed in the building—regular and emergency. There were no lights or elevators. These problems may have been overcome if the pressure control valves installed on the standpipe outlets were not set at too low a pressure for fire department operations.

The valves could not be adjusted for several hours until a contractor familiar with the valves reset them. All fire departments should be aware of all the types of pressure-regulating valves used in buildings in their city. (For details see the NFPA special bulletin about these valves.[56])

The whole question of pressure-regulating valves could stand rethinking. The original purpose was to reduce the pressure to a level safe for citizens using standpipe hoses. Early valves had a simple break piece which permitted the fire department to open the valve fully. In many cases now there is no hose provided for citizens to use. In such cases it might be appropriate to consider having no regulating valve at all. The fire department would then bring their own regulating valve attached to the standpipe pack. When New York had a high-pressure system, any hand lines attached to a high-pressure hydrant were connected through such a valve.

The loss of all power in a building presents an extremely serious situation for fire fighters. Imagine the situation if the daytime "life load" was in the building. Hundreds of people would be trying to descend smoke-filled stairways in total darkness.

All heavily occupied buildings should be examined with the scenario that all power has failed. Dependence on a total building emergency system can mean total loss of lighting. Individual battery-operated lights which come on when power fails are not subject to total system failure. After 11 hours the One Meridian Plaza fire still raged. A structural engineer warned of possible collapse. All fire fighters were withdrawn. The 30th floor sprinklers were supplied by fire department pumpers. The pumpers increased the sprinklers' discharge density by as much as three to four times and stopped the fire.

Because of the duration and intensity of this and the First Interstate Bank fire, attention was focused on the possible loss of strength and stability of the structural frame of high rises. In both cases the columns of the structural steel frame endured the fire without significant damage.

Can we draw from this experience that there is no need to fear collapse even in a raging high-rise fire? I believe the answer is **NO**.

In these fires the steel was protected with a cementatious coating or by a mineral fiber material with a cement binder and some asbestos. Over the years many different methods of insulating steel have been used. Some are more reliable than others. Asbestos fireproofing has been inadvertently removed and/or washed off by hose streams. (The dubious reliability of columns which are unprotected in the plenum space and dependent solely on the integrity of a suspended ceiling is fully discussed in Chapter 7.)

The following is an abstract which describes a fire in a partially completed building from Factory Mutual Data Sheet 1-3.

In a 41-story high rise under construction, sprinklers were installed on the 19 floor but were not in service. The elevator lobby had combustible finish. A smoke detector operated on a suspicious fire in the elevator lobby. An employee turned on the sprinkler system three minutes after the smoke alarm. The fire department was on the fire floor in 13 minutes. Fifteen sprinklers operated because the system was turned off initially, resulting in almost a million dollars in property damage. The shut sprinkler valve was obviously responsible for a great part of the damage.

References

The addresses of all referenced publications are found in Reference Addresses after Chapter 14.

1. Sabbagh, K. *Skyscraper*. Viking, Penguin, New York, N.Y., 1988, p. 361.
2. Hassett, B. "Fire In The Empire State Building." *Fire Engineering*, November 1990, p. 50.
3. Sharry, J. "Group Fire Indianapolis, Indiana." *Fire Journal*, July 1977, p. 13.

4. Best R. "Demolition Exposure Fire Causes over $90 Million Loss to Minneapolis Bank." *Fire Journal*, July 1983, p. 60.

5. Meadows, M. "On the Job California." *Firehouse*, June 1990, p. 28.

6. Op Cit - Reference 1. See Sabbagh, K.

7. O'Hagan, J. *High Rise Fire and Life Safety*. Fire Engineering Publications, Saddle Brook, N.J., 1977, p. 143.

8. Bell, James R. "137 Injured in New York High-Rise Building Fire." *Fire Journal*, March 1981, p. 38.

9. Kennedy, T. Training Notebook: "Energy Efficient Windows." *Fire Engineering*, December 1990, p. 14.

10. Lathrop, James K. "Atrium Fire Proves Difficult to Ventilate." *Fire Journal*, Vol. 73, No. 1, January 1979, p. 30.

11. Morehart, J. "Sprinklers in the NIH Atrium: How Did They React During The Fire Last May?" January 1989, pp. 56-57.

12. Proceedings, "Roundtable On Fire Safety in Atriums. Are The Codes Meeting The Challenge?" NFPA, Quincy, Massachusetts, 1988. (Contact A. Cote, NFPA)

13. Moschella, J. "New dimensions in High Rise Multifloor Corridor Design." *Fire Engineering*, August 1990, p. 95.

14. Louderback, J. "On the Job, Pennsylvania." *Firehouse*, July 1991, p. 31.

15. "High Rise Evacuation." NFPA Video tape/Film, 1991.

16. Proceedings, American Society of Mechanical Engineers, New York, N.Y., 1991.

17. Chapman, E. "Elevator Use," *Fire Engineering*, November 1988, p. 47. Also, Letters: *Fire Engineering*, April 1989, p. 22.

18. Breen, D. "The MGM Fire and the Spread of Flames." *Bulletin of the Society of Fire Protection Engineers*, January 1981, p. 1.

19. Klote, John H. "Considerations of Stack Effect in Building Fires." NISTIR 89-4035-NIST [(Center for Fire Research) Gaithersburg, Maryland, 20899] 1989.

20. "Report of the MGM Grand Hotel Fire." Clark County Fire Department, Las Vegas, Nevada, 1981. Also see, "Fire at the MGM Grand." *Fire Journal*, January 1982, p. 19, based on NFPA Investigative Report by R. Best and D. Demers; and Bryan, J. "Human Behavior in the MGM Grand Hotel Fire." *Fire Journal*, Vol. 76, No. 2, March 1982, p. 37.

21. Lathrop, James K. "World Trade Center Fire." *Fire Journal*, July 1975, p. 19.

22. Boyd, Howard. "Smoke, Atrium, and Stairways." *Fire Journal*, January 1974, p. 9.

23. Op. Cit. See reference 7, O'Hagan, J. *High Rise/Fire and Life Safety*. Fire Engineering Publications, 1977, p. 28.

24. Harmathy, T. Z. "Design of Buildings Against Fire Spread." *Journal of Applied Fire Science*, Volume 1, No. 1, 1990.

25. O'Brien, D. "A Centennial History of The International Association of Fire Chiefs." IAFC, Washington, D.C., 1973, p. 41.

26. Demers, David P. *et al*, "Investigation Report on the Las Vegas Hilton Hotel Fire." *Fire Journal*, Vol. 76, No. 1, January 1982, p. 52.

27. Hassett, B. "Fire in the Empire State Building." *Fire Engineering*, November 1990, p. 50. Also Dunn, V. "High Rise Survival." WNYF 1st Issue, 1991.

28. Wolfe, T. *From Bauhaus to Our House*. Farrar, Strauss Giroux, New York, New York, 1977.

29. Nelson, H. "Science in Action." *Fire Journal*, July 1989, p. 25. Also, Nelson, H. "An Engineering View of the Fire of May 4, 1988 in the First Interstate Bank Building, Los Angeles, California." NSITR 89-4061, NIST (Center for Fire Research) Gaithersburg, Maryland 20899.

30. Klem, T. "56 Die in English Stadium Fire." *Fire Journal*, May 1986, p. 128.

31. Watrous, L. D. "High Rise Fire in New Orleans." *Fire Journal*, May 1973, p. 6.

32. Peterson, C. "Union Bank Building Fire Los Angeles California July 18, 1988." *Fire Investigation Report*, NFPA, Batterymarch Park, Quincy, Massachusetts, 02269. See also Anthony, D. and Brooks, C. "LAFD Battles Second Highrise Blaze In Ten Weeks." *American Fire Journal*, November 1988, p. 32.

33. "High Rise Office Building Fire. Los Angeles, California," *Fire Journal*, November 1975.

34. Courtney, N. "High Rise Office Building." (Fire Record) *Fire Journal*, September 1987, p. 29.

35. Eisner, H. "Death Leap." *Firehouse*, June 1989, p. 81.

36. Op Cit - See Reference 1, Sabbagh, K. "Finishing Off." pp. 362-4.

37. Klem T. and Kyte, G. "Fire at the Prudential Building." *Fire Command*, March 1986, p. 34. See also De Caprio, A. "Tower 49." *WNYF* Second Issue, 1986, p. 5.

38. Powers, R. "Sprinkler Experience in High Rise Buildings." Society of Fire Protection Engineers, *Technology Report* 79-1, 1979.

39. Mulrine, J. "Compartmentation vs. Sprinklers." *Fire Engineering*, December 1980.

40. Brannigan, V. "Record of Appellate Court Cases on Retrospective Fire Codes." *Fire Journal*, November 1981, p. 62. Also see Brannigan, V., JD, "Applying New Laws To Existing Buildings: Retrospective Fire Safety Codes." *Journal of Urban Law*, Vol. 60, Issue 3, Spring 1983.

41. Klem, T. *Fire Investigation Report: First Interstate Bank Building Fire, Los Angeles, California, May 4, 1988.* NFPA, Batterymarch Park, Quincy, Massachusetts.

42. Walton, W. D. and Budnick, E. K. "Quick Response Sprinklers in Office Configurations: Fire Test Results." Report NBSIR 88-3695, National Institute of Standards and Technology, Gaithersburg, Maryland, 1988.

43. Lathrop, J. "Building's Design, 300 Fire Fighters Save Los Angeles High Rise Office Building." *Fire Journal*, September 1977, p. 34.

44. "The LAFD's Highrise Recommendations." *American Fire Journal*, November 1988, p. 32.

45. Irvin, P. "Blaze Causes $100 Million Loss to Postal Headquarters." *Fire House*, February 1985, p. 35.

46. Isner, M. *Fire Investigation Report: Elderly Housing Fire, Johnson City, Tennessee, December 24, 1989*, NFPA, Batterymarch Park, Quincy, Massachusetts.

47. Bell, J. "*Fire Investigation Report: Stauffer's Inn of Westchester County, Harrison New York, 26 Fatalities, December 4, 1980.*" NFPA, Batterymarch Park, Quincy, Massachusetts.

48. Isner, M. *Fire Investigation Report: Apartment High Rise Fire Manhattan New York, January 11, 1988.* NFPA, Batterymarch Park, Quincy, Massachusetts.

49. Isner, M. *Fire Investigation Report: Fatal Office Building Fire Atlanta Georgia June 30, 1989*. NFPA, Batterymarch Park, Quincy, Massachusetts.

50. Bell, J. "*Fire Investigation Report:* Westchase Hilton Hotel Houston, Texas, March 6, 1982, 12 fatalities." NFPA, Batterymarch Park, Quincy, Massachusetts. Also, "Twelve Die in Fire at Westchase Hilton Hotel, Houston, Texas." *Fire Journal*, NFPA, January 1983, p. 11.

51. Klem, T. "*Fire Investigation Report: Dupont Plaza Hotel Fire, San Juan, Puerto Rico, December 31, 1986.*" 1987, NFPA, Batterymarch Park, Quincy, Massachusetts.

52. Isner, M. "*Fire Investigation Report: High-Rise Office Building*", Montreal, Canada, October 26, 1986, NFPA, Batterymarch Park, Quincy, Massachusetts. See also Isner M. "Montreal High-Rise Fire." *Fire Journal*, January 1987, p. 64.

53. Stronach I. "City Brings in Retroactive High-Rise Sprinkler Law." *Fire International*, February/March 1990, p. 23.

54. Eisner, H and Manning, B. "One Meridian Plaza Fire." *Fire Engineering*, August 1991, p. 51.

55. Klem, T. "NFPA Investigative Report: One Meridian Plaza Fire, Philadelphia, Pennsylvania." NFPA, December 1991.

56. NFPA *Alert Bulletin*, No. 91-3, May 1991.

Additional Readings

"Arson in Seattle Hotel." *Fire Journal*, November 1975, p. 5. A fire started on an open stairway mezzanine and smoke polluted the entire stairway.

Best, Richard. "High-Rise Apartment Fire in Chicago Leaves One Dead," *Fire Journal*, September 1975, p. 38. Chicago Fire Department uses master streams on high fire load of furniture fire in 17th-floor apartment. Fire confined to unit of origin by self-closing door despite unsprinklered tenant storage area.

Bimonthly Fire Record. *Fire Journal*, May 1973, p. 59. A Christmas tree fire in a Dallas high-rise did $340,000 in damage. Pipe chases were enclosed only to the ceiling and voids above were pierced for air conditioning. The building lacked integrity. A force of 85 fire fighters prevented a more serious outcome.

Birminghan, J. "Firestops Protect Property and Lives," *Fire Journal*, March 1986, p. 41. A good discussion of firestopping materials which can be used to restore the integrity of fire-rated floors and walls after they are penetrated for wiring.

Birr, Timothy. "Landmark Hotel Closed for Fire Code Violations," *Fire House*, June 1981. The Eugene, Oregon Fire Department is successful in its battle to close a hotel which failed to comply with retrospective fire code provisions requiring sprinklers.

Break, R. "High Rise Fire Handled Successfully," *Fire Command*, August 1987, p. 47. A basement fire polluted all 27 stories of an old St. Louis hotel housing scores of elderly people. All the residents were evacuated safely under difficult conditions. The operation required almost the entire 44-company department. Fire departments with limited personnel should study this story and determine whether their evacuation plans are realistic, and, if not, determine what is to be done to forestall a disaster.

Burns, R. "Philly Four Alarmer," *Firehouse*, Sept. 87, p. 73. An old 30-story fire-resistive hotel was being remodeled into apartments. Units found a raging fire on the fourth floor. Occupants were removed over ladders and from smoke-filled halls up as high as the 30th floor.

A maintenance man, investigating smoke, opened windows and then found a smoldering mattress in a bedroom. He went for an extinguisher and a house line. The tenant across the way opened her door and cross ventilated the fire. Maintenance personnel and other low-paid employees are usually the persons who find the fire.

Every fire department should educate such personnel to never open a smoky door, and to call the fire department before taking any other action, including using fire extinguishers.

The code should have provided for self-closing doors for all apartments in the remodeling. This was required in the New York City Tenement House Act of 1903.
It was extremely fortunate that this fire took place in the daytime and on a lower floor, accessible to ground operations.

"Dallas Fire Kills Two Fire Fighters." Fire Journal, July 1977, p. 108. Describes fire which started in an apartment that flashed along glued-up combustible tile in a corridor.

Fisher, Charles A. "Lessons in High-Rise Rescue," *Fire Service Today*, March 1982, p. 24. Outlines lessons from a fire in which eight persons died, and is recommended for departments which have never had a high-rise fire.

Gerard, John C. "An Incident Command System for High-Rise Fires," *International Fire Chief*, January 1981, p. 17, published for the IAFC, Washington, D.C. Presents a picture of the command staff required to control a high-rise fire manually. Helps in demonstrating why building codes in smaller cities must be even more stringent than in larger cities.

Isner, M. "Fire in Los Angeles Central Library," *Fire Journal*, March, 1987, p. 56. A fire-resistive, unsprinklered central library suffered tremendous damage. There is no solution but full automatic sprinkler protection. All the stack areas of the Library of Congress are now sprinklered. Even so, because of passages and stairways even narrower than those on a ship, and difficulty in venting, a fire would probably be a tough fight. Note also that older libraries often used marble for the flooring in stacks. If unsupported, such as in the Library of Congress's main building, it can look good after heat exposure, but fail under a fire fighter's weight.

Kravontka, Stanley, J. "Elevator Use During Fires in Megastructures," Technology Report 76-1, Society of Fire Protection Engineers, Boston, Mass.

Lathrop, James K. "Nineteenth-Floor Dormitory Fire Kills One Student," *Fire Journal*, March 1976, p. 77.

Lathrop, James K. "Two Die in High-Rise Senior Citizens' Home," *Fire Journal*, September 1975, p. 60. Intense fire in Albany, N.Y. Gypsum board over polystyrene disintegrated. One-quarter-inch plywood concealing a drain pipe provided a "pinhole" in the fire-resistive containment, allowing extension to the next apartment.

Lebel, R. "Preparing for High Rise Fires," *Fire Command*, August 1986, p. 40. An excellent program undertaken by a small city fire department facing high-rise construction.

Marshall, Steve. "Two Arlington High-Rises Struck by Major Fires," *Firehouse*, October 1985, p. 38. A fourth-floor office fire in an unsprinklered 11-story office building required a third alarm assignment. Again the deficiency of locating standpipe outlets outside of stair towers was evident.

Mulrine, Joseph. "Compartmentation vs Sprinklers," *Fire Engineering*, December 1980.

Murray, R. and Lathrop, J. "Fire Barrier Penetration Must Be Firestopped." *Fire Journal*, November 1987, p. 66. A four-story Waltham, Mass. office building suffered a multi-million dollar loss because fire extended through openings in the floors for computer cables. The article goes on to discuss methods of firestopping around cable openings.

Paul, G. H., Clougherty, E. W. and Lathrop, J. K. "Federal Reserve Bank," *Fire Journal*, Vol. 71, No. 4, July 1977, p. 33. An excellent source of information on what can go wrong when foamed plastics are being installed. Exterior panels are aluminum with foamed insulation. There is no statement of any test to validate the system. The Boston Fire Department required a number of corrective actions.

Sharry, John A. "A High-Rise Hotel Fire, Virginia Beach, Virginia," *Fire Journal*, January 1975, p. 20. A useful article when instructing hotel personnel. The assistant manager died trying to fight the fire. With the standpipe out of service, the fire department used a ladder pipe as an auxiliary standpipe.

Sharry, John A. "South America Burning," *Fire Journal*, July 1974, p. 23. Recounts three serious high-rise fires: the Joelma Building, Sao Paulo, Brazil, in which 179 lives were lost; Avianca Building, Bogota, Columbia; Caixa Economica Building, Rio de Janeiro, Brazil. Exterior extension, open stairways and combustible interior finishes were principal factors.

Willey, A. Elwood. "Baptist Towers Housing for the Elderly," *Fire Journal*, March 1973. In Atlanta, Georgia, ten died in an 11-story fire-resistive home for the elderly. The door to the fire apartment was left open (no self-closer). The windows were out. High flame spread carpeting extended fire into the corridor.

12
Trusses

The current practice of designing buildings of lighter and lighter materials and substituting geometry for mass has brought about significant strides in the architectural field. It has also brought about significant problems in the field of fire fighting.

Knowledge of the hazard of trusses has developed gradually in the fire service. In 1948, this author wrote the first article to appear in the fire service press[1] on the then new concept of pre-fire planning. The article described a Navy drill hall which was converted into a warehouse. The pictures clearly show 100-foot timber trusses. The preplan doesn't mention them as a hazard. Any fire department preplanning such a structure today would give serious attention to such trusses.

When the first settlers arrived in this country, huge virgin forests yielded countless inexpensive beams and columns of every desired dimension. Structures were massively built of wood and masonry, without regard to their weight. Times have changed.

Today, the emphasis is on lighter-weight structures. For example, light-weight gypsum panels are advertised as suitable replacements for heavier masonry shaft enclosures. The steel industry seeks lighter fireproofing for steel to replace the concrete that was commonly used.

The weight of materials and the weight of the structure required to support heavy materials are now important cost considerations. Suppliers of materials lighter than those previously used proudly brag about weight savings. The Empire State Building weighs 23 pounds per cubic foot. Modern high-rises weigh as little as 8 pounds per cubic foot.

The largest practical clear areas and the maximum amount of off-site prefabrication are other important considerations in today's construction.

Many of the techniques for making buildings lighter and providing large open areas without columns have been available for centuries. The hammerbeam truss, for instance, used in churches and other monumental gothic structures, was invented in the 13th century and used for Westminster Hall, (not Abbey) which still stands in London.

In recent years, increased material and labor costs have forced builders to maximize use of more economical construction techniques.

Trusses are most often used as beams, but there are many trussed columns, especially in huge buildings and special-purpose structures, such as high-energy transmission lines and tall radio towers.

The Truss

The truss satisfies many building requirements.

- It is lighter in weight than solid construction, thus reducing the weight of supporting walls or columns.
- It provides long clear spans, thus giving the maximum flexibility in the use of space.
- Many trusses, such as light-weight wood trusses and steel bar joists, can be delivered prefabricated. Huge steel trusses can be delivered partially prefabricated, with the final connections made in the field.

What Is a Truss?

A truss is defined as a framed structure consisting of a triangle or group of triangles arranged in a single plane in such a manner that loads applied at the points of intersections of the members will cause only direct stresses (tension or compression) in the members. Loads applied between these points cause flexural (bending) stresses.

The Triangle Principle

The rigidity of the truss rests in the geometric principle that only one triangle can be formed from any three lines. Thus, the triangle is inherently stable. An infinite number of quadrilaterals can be formed from four lines, so the rectangle is inherently unstable.

The economy of a truss is derived from the separation of compressive and tensile stresses so that a minimum of material can be used. There are many designs of trusses, each with its own name.

Sketches of many types can be found in any good construction book such as Huntington's Building Construction.[2]

Components

The top and bottom members of the truss are called **chords**. The compressive connecting members are called **struts**. The tensile connecting members are called **ties**. Connections are called **panel points**. As a group, the struts, ties, and panel points are called the **web**.

Trusses can be built of wood, wood and steel combined, or steel. Cast iron was used for compression members in early metal trusses, and may be found in 19th century structures.

Concrete trusses are not common, though hugh concrete trusses are seen in the Tampa, Florida, and Dallas/Ft. Worth, Texas Airports. An American Airlines hanger at Dallas Ft. Worth has a 1280-foot × 280-foot clear span, with doorways 74 feet high. Concrete towers 80 feet high support concrete trusses 76 feet high. Each pair of concrete trusses supports steel box trusses 20 feet wide and 280 feet long.

Principal Types

The principal types of trusses of interest to fire protection professionals can be described by their overall appearance.

Bowstring Truss: The bowstring truss gets its name from the curved shape of the top chord. Sometimes bowstring trusses are known as **arched trusses**. This can be confusing. The thrust of a truss is downward. The thrust of an arch is outward, and usually resisted by a mass of masonry. Some arches are **tied arches** in which a steel tension rod ties the ends of the arch together to eliminate the need for the masonry. Early Arby's restaurants featured arched roofs. Some of the arches are tied arches.

Fig. 12-1. There are several kinds of trusses. See how many you can find in your community.

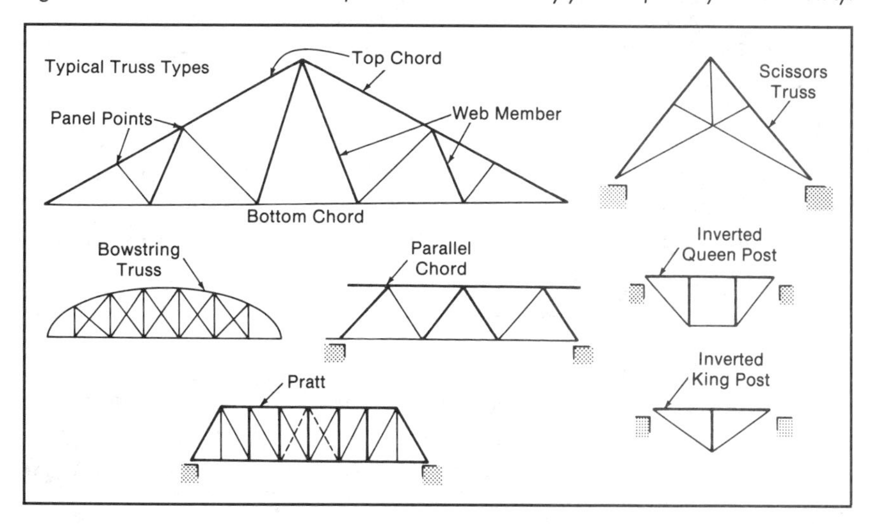

However, there are also true **trussed arches**. The arch of a steel arch bridge is often a truss. The arch is a compression structure. In this unique case the top and bottom chords are both under compression. The Cascades area of the Opryland Hotel in Nashville, Tennessee is covered with a skylight. The building specification called for no deflection. Tied arches, which actually look like trusses, were used for the 165-foot span. The arches weigh 95 pounds per foot. A simple span truss would have weighed 200 pounds per foot.[3]

Parallel-Chord Truss: In this type of truss the bottom and top chords are parallel. The steel bar joist (a parallel-chord truss) used both for roofs and floors has been around for some time. Wire trusses used horizontally, to tie brick and block together to produce a composite masonry wall, are also parallel-chord trusses.

Very long-span parallel-chord roof trusses may have a slight upward pitch to the center to facilitate water runoff.

Very deep-parallel-chord trusses have been used as floor beams in hospitals. Engineers borrowed a medical term and called the space produced by such trusses, **interstitial space**. The hazard of such huge spaces is the temptation to use it for storage, maintenance shops, etc. Such voids are dangerous. Automatic sprinklers should be required in them.

Heavy parallel-chord trusses have been used as **transfer beams**. A transfer beam is used to laterally relocate the vertical load of columns to clear an area, such as in a hotel ballroom. The trusses are often hidden in partition walls.

Truss floors have been designed in which the trusses were buried in the walls of apartments, so that the living space is within the truss.

Wooden parallel-chord trusses are being used for floors and roofs in single-family homes, row houses, apartment houses, and smaller office buildings.

Triangular Truss: Many roof trusses are triangular in shape in order to provide a peaked roof. Light-weight triangular trusses must be closely spaced. As a result they leave no clear space.

Recently, builders have been using wooden parallel-chord trusses as roof rafters. This provides a peaked roof, with usable attic space. Dormer windows may be a clue to this type of roof.

Vierendeel Truss: This truss is composed of rectangles. Extremely stiff corners are used to provide the rigidity. The entire exterior of New York's World Trade Center consists of steel Vierendeel trusses.

A **space frame** is a three-dimensional truss, thus contradicting the standard truss definition of "in a single plane."

A serious and continuous problem in truss design has been **connectors**. In recent years, there have been tremendous strides in the development of connectors capable of transmitting heavy loads. Trusses can provide huge clear spans at a deadweight considerably less than that of a corresponding ordinary beam. In some cases, no ordinary beam can do the job.

Exterior Trusses

Trusses are most often located within the building. They can also be located above the building to support the roof and ceiling of the top floor so that a large area below can be free of columns.

Compression vs. Tension

The top chord of a truss is in compression. The bottom chord is in tension. When a truss is **cantilevered** out, the situation reverses. Beyond the point of support, the tension is in the top chord and the compression is in the bottom chord.

In sketches of trusses, compression members are often shown as thick lines, while tension members are shown as thin lines. This takes graphic account of the fact that compressive loads are best resisted by columns whose material is as far as possible from the center (e.g., a cylinder). Tensile loads need only the required strength of material. Shape is unimportant.

Truss Principles

Consider a building span of 20 feet. Two 10-foot beams extend from the walls on opposite sides to a column support in the center. Assume that each beam can carry 1000 pounds. If the column is removed, the beam would have to be 20 feet long. According to the principle — doubling the length of a beam decreases its carrying capacity by half — a beam 20 feet long, of the same material and cross-section area, can only carry 500 pounds.

Suppose the column is cut off and a stub remains at the junction of the beams. If a tension member is tied from the stub to the beam ends at the wall, a triangle (truss) will be formed in effect, restoring the load-carrying capacity of the beam, yet removing the obstruction of the column from the floor below. This truss is called an **inverted king post truss**. It is called king post because of the single compression member, and inverted because the compression member extends

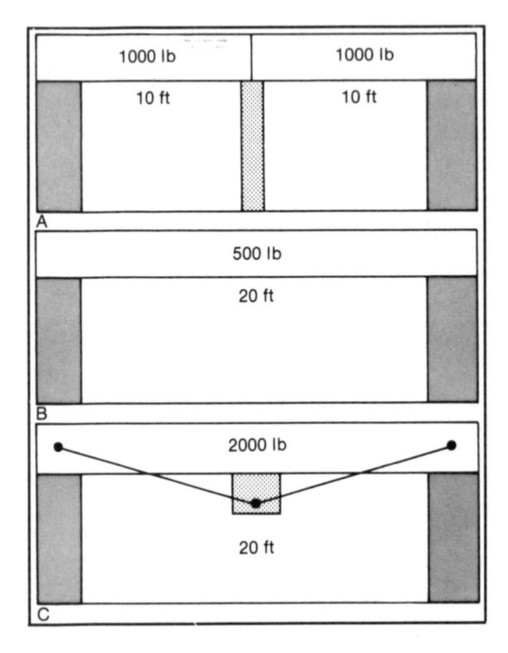

Fig. 12-2. **(A)** *Two short beams supported in the center by a column will carry a 20-foot span. Each side of the roof will support 1000 pounds.* **(B)** *If there is to be one large room in the same space, the two short beams can be replaced with a long one. It drops the load-carrying capacity to 500 pounds.* **(C)** *A better alternative is to create a truss. The load-carrying capacity is restored to 2000 pounds.*

downward. A truss with two compression members is called a **queen post truss.**

This simple illustration explains the basic theory of trusses.

Problems with Trusses

The truss can be designed and constructed as the minimum structure which will carry the designed load. This is emphasized in some engineering schools. Competitions are held in which the winner produces a truss with the highest ratio of superimposed load to weight of the truss.

"Our Trusses are Engineered," is sometimes presented as an unquestioned virtue. The question is, "engineered to what objective?" Unfortunately, the answer is often, "to meet the lowest price in a competitive market."

Trusses can fail in a variety of ways. All parts and connections of a truss are vital to its stability. The failure of one element of a truss may cause the entire truss to fail.

Multiple truss failures can occur. The failure of one truss can have serious impacts on other parts of the structure, even parts far from the initial failure point. For example, steel trusses are tied together for stability to increase wind resistance. When one truss fails, undesigned torsional stresses can be passed to other trusses by way of the ties, and may cause multiple failures. In one case, a 1400-foot TV tower collapsed as an antenna was being installed. Two diagonals had been removed to permit the antenna to be hoisted up the interior of the tower.

One major insurer was careful and warned its inspectors about trusses as follows:

> "Problem: Missing Truss Members. Steel trusses are designed in such a manner that all of the structural members are needed in order that stresses may be distributed properly. In some cases there are several members originating at one point in the truss in such a manner that it may appear to the layman that some of them are not needed. As a result members have sometimes been removed to allow space for ducts and other apparatus to pass through. This practice can greatly reduce the strength of the truss and result in future collapse."

Some trusses come with labels warning carpenters not to cut any part of the truss. Despite this, trusses are often cut to accommodate plumbing or other fixtures. Kirby Keifer of the Minnesota Technical Institute System told me of the case where a builder complained to the supplier that the trusses he ordered were too long. He had to cut off the "excess," double vertical end members and part of the web.

Truss Frame structures are discussed in Chapter 3.

Fig. 12-3. This truss collapsed because one of its members was removed.

Rising roofs present a serious wood and metal plate truss problem. For example, some trusses bow upward, causing ceilings to separate from walls. The *Journal of Forestry* blames this defect on early harvested tree species. Juvenile wood is more prone to shrinkage and warping. From a fire fighting point of view, the most serious effect of rising roofs is the weakened joints which can result from shrinking wood pulling away from gusset plates. Gas and electric lines located in such attics may be dangerously affected as well.[4]

Designers rely on standard tables of allowable design stresses in building structural components. "Juvenile wood, often well below normal average strength, is not adequately represented in average strength values used to derive allowable design stresses."[5]

Timber trusses can have many defects which may cause collapse under non-fire conditions such as high wind or snow loads. These same defects may precipitate early collapse under fire conditions. Shrinkage of wood may cause undesigned stresses. Chemical action, wet and dry rot, enzyme decay, and insect infestation can destroy the timber. Excessive loads may be carried on the trusses.

Defective design may lead to the truss not acting as a truss. Poor maintenance also affects trusses. It may be responsible for a failure to tighten loose tension rods. Light-weight gusset plate trusses are sometimes poorly made or mishandled in the field.

A typical structure may have many structural defects. Some common defects were detailed in a recent *Progressive Architecture* article.[6] In a collapse of a truss roof under construction, some of the following defects were noted.

- Use of inadequate dimension lumber at joints where forces were high.
- Knots located in gusset plate contact areas.
- Gusset plates not centered on several joints.
- Gusset plate lugs not adequately embedded in the lumber.
- Defective lumber used.
- Attempt to repair split lumber with a plate.
- Lack of fit-up at truss end.
- Poor fit-up of joint.
- Inadequate connector sizes.
- Reduced lumber section at joint due to improper finishing.

These defects referred only to one specific case. There is intense competition in the building materials supply business. There is incentive for the builder to purchase from the least expensive supplier. Obviously, there is the potential for many more such cases. Trusses and any of their potential defects are unseen in a completed structure. With luck the defects may not cause failure for some time, if ever.

Fig. 12-4. Poor workmanship is evident in this defective connection in Los Angeles County. Once completed, the bad connection will be out of sight.

Unique Alterations

Lieutenant A. Jesionowski of Pleasanton, Illinois wrote and described a unique truss hazard. A building had several long-span bow string trusses supported on masonry bearing walls. To enlarge the building, one bearing wall was removed. A steel beam was erected above the ends

Figure 12-5. Another truss roof, this one in Los Angeles, collapsed while under construction. (Courtesy John Mittendorf, LAFD)

of the trusses. Tension rods suspended from the beam supported the ends of the trusses.

In another case, wooden I-beams were used in a Texas apartment house. The roof remained open to the weather for several months after the fire. The plywood web of ground floor I-beams separated due to moisture. The beams were repaired by nailing pieces of 1 × 6 wood diagonally from the top to bottom flange to create a homemade truss.

Tactical Considerations

Be alert to any alterations of trusses. All alterations are questionable, most likely they are hazardous.

The Bar Joist Gap

A fire in a store surprised Bethesda, Maryland fire fighters by extending to adjacent stores through the gap between the top of the wall and the roof formed by the bar joist resting on the top of the wall. Thereafter, the county building department required this gap to be blocked.

Fig. 12-6. These wooden I-beams were damaged by rain after a fire in a Texas apartment building. The plywood web was disintegrating. The builder converted them into trusses. Know your buildings!

This gap also existed in the interior bearing walls of the Dardanelle Nursing Home in Arkansas in which four patients died in March 1990. The walls were also the corridor walls. The NFPA reports of the fire concluded, "the corridor walls were neither an effective smoke barrier nor fire barrier."[7]

Excessive Loads

Often the truss is designed simply to support the roof and provide the desired clear span. Occupants often use the space between trusses for equipment or storage. A heavy suspended ceiling might be installed. Such undesigned loads were a factor in two fatal collapses. In theory, the top chord of a light-weight truss is only in compression. A load, such as an air conditioning compressor hung from a top chord, may introduce undesigned-for bending (compression and tension) stresses. An undesigned load can shorten the time to failure in a fire. Sometimes trusses are set in multiples to cope with a concentrated load. This does not suffice in a fire because all the trusses can be exposed to the same fire conditions.

Fig. 12-7. Near Halifax, Nova Scotia, these heavy gusset plate trusses are spliced at midpoint. All the joints are in the same place. There is no overlap. A failure at this point would collapse a huge section of the roof, since this truss supports many others.

Arena Collapses

Arenas which often incorporate long-span roof trusses have suffered major collapses. (See Chapter 7.) Providentially, none of the collapses occurred when the structures were occupied. The American Institute of Architects commissioned a report on long-span roof collapses. The report noted a number of problems with such structures which posed a threat to human life in halls designed to house large groups.

- Lack of Redundancy where failure of a member or joint tends not to remain localized.
- Lessened resistance to a member buckling.
- Over-reliance on standard practice and codes which are derived from shorter spans.
- Excessive concentrated loads, such as snow drifts, should receive more attention.[8]

In a Letter to the Editor of *Engineering News Record* in April 20, 1978, a chief structural engineer, J. E. Varga, noted the fact that thousands of hockey fans were in the Hartford Arena only hours before it collapsed. He argued that the life hazard should be recognized by an **importance factor** incorporated into codes to mitigate potential catastrophes.

Truss Failure in Fires

The economy of the truss is found in the fact that it separates compression and tension forces. However, this economy can also cause disaster.

The bottom chord of a truss is under tension. Recall that a tension member is like a rope. One break precipitates failure.

Conversely, the top chord of a truss is under compression, therefore it responds exactly like a **column**. Chapter 2 showed how the load-carrying capacity of a column decreases as its length is increased. Now consider a truss with panel points five feet apart. This is loaded like a column tied off every five feet; a series of five-foot columns one atop the other. If a panel point or connection fails, two five-foot segments join to become a ten-foot segment. This segment can carry only 25 percent of the design load. Its failure can precipitate the failure of other trusses.

Compare a light-weight wooden parallel-chord truss with a sawn beam.

As you also learned in chapter 2, in an ordinary beam the top of the beam is in compression, and the bottom is in tension. The greatest compression is at the top, gradually lessening as the neutral plane is approached. The greatest tension is in the bottom, gradually lessening as the neutral plane is approached. There is a continuum in the loading of the beam. The loss of some material therefore is not necessarily fatal.

In the truss, all the compressive load is carried on the top chord, typically a light member such as 2 × 4 or 2 × 6. Loss of wood can precipitate crushing failure in compression.

All the tensile load is carried on the bottom chord. Except for quite short trusses, the bottom chord is made up of two or more pieces of wood, joined end-to-end with metal gusset plates. The loss of a gusset plate would cause the tensile load to pull the truss apart.

Web members may be connected by gusset plates, or nailing. In one design, the **ties** or tension members are stamped from one piece of metal, and flattened on the end to make the gusset plate. The steel tie could act as a heat collector, delivering heat to the gusset plate teeth, which can pyrolytically destroy the tensioned wood fibers which had been gripping the 3/8-inch deep metal teeth.

Tubular steel and wood trusses can burn through at panel points, where there is little wood. A steel pin which carries heat into the wood can cause pyrolytic decomposition.

Large triangular trusses such as those used in apartment house roofs have bottom chords made up of two or more pieces of wood butted together and held by gusset plates. In the case of such trusses used in

Fig. 12-8. **(A)** *The tubular steel and wood trusses failed early in a fire. (Phoenix, Arizona Fire Dept.)*

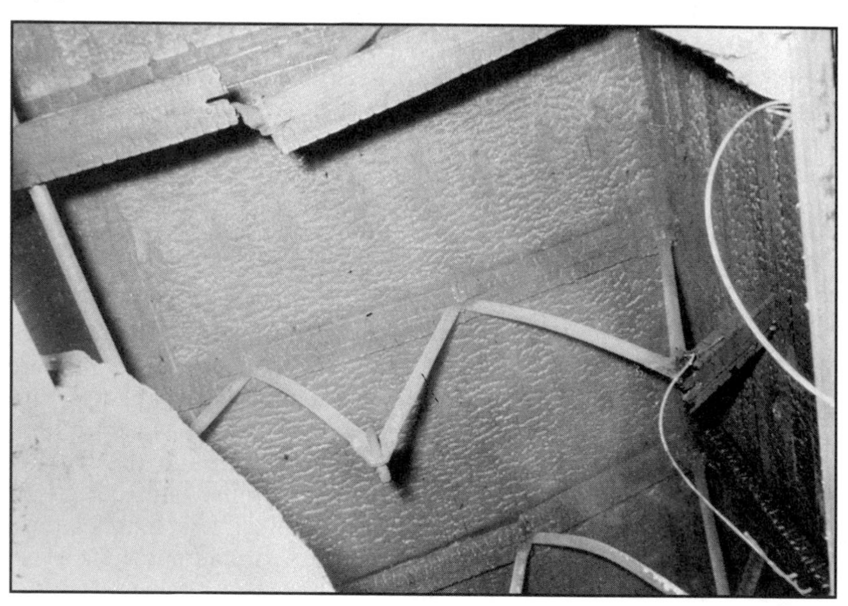

Figure 12-8. **(B)** *Note how the wood burned away from the steel pin of a tubular truss. (John Mittendorf)*

apartment house roofs, disaster to fire fighters operating below may be averted by the fact that room partitions support the failing trusses, so that only parts of the truss fall into the top floor. The same truss, used in a commercial building to provide a wide clear space, can and has failed catastrophically.

In February 1989, two Orange County, Florida fire fighters died in the collapse of a triangular truss roof of a store. The trusses spanned 50 feet front to back. The roof collapsed in less than 12 minutes. The fire was above the ceiling and did not appear to be threatening from the inside. The first-due engine was the only unit on the scene when the collapse occurred. The Orange County Fire Department produced a searching, detailed and well-illustrated report.[9]

In 1991, a neon light started a fire in the gable end of a truss roof restaurant, also in Orange County, Florida, not far from the fatal fire described above. The staff and patrons of the restaurant were unaware of the fire. Patrons of a nearby restaurant gave the alarm. The first unit on the scene raised a ladder to the attic scuttle hole. The acting lieutenant noted that the fire was moving rapidly through the trusses, pushed by a high wind, so he ordered his unit out. The acting battalion chief ordered the building cleared. About five minutes later a large section of the roof collapsed. Pictures of the restaurant show heavy

Fig. 12-9. **(A)** *This large triangular truss is on a delivery truck. Turn the picture upside down. Now note that the bottom chord, which is under tension, consists of three pieces of wood joined together by gusset plates. There are four failure points. When a truss like this fails in a building that has a long clear span, everything falls to the floor. Fire fighters below are in serious danger.* **(B)** *By contrast, when the trusses collapse in an apartment house fire, often they are caught on interior partitions and the situation may be less catastrophic.*

Fig. 12-10. The truss roof on this commercial building collapsed moments after Louisville, Kentucky fire fighters were ordered out. Note the total collapse when a truss spans a long clear area. (Louisville Fire Dept.)

equipment on the roof. There was a gypsum board ceiling on the bottom of the trusses, probably intended to protect the trusses for a few minutes, from fire originating below. The duct work was suspended below it. A grid ceiling concealed the duct work. The construction made the fire impervious to an interior attack.

These fire officers are to be commended. I am aware of at least two tragic collapses in which acting officers hesitated to take necessary decisive action.

Tactical Considerations

It cannot be stressed too often that heavy fire conditions can exist in concealed spaces, particularly overhead, and not be evident in the space below.

The roof of a Phoenix, Arizona house was built of light-weight trusses with a plywood roof covered with built-up roofing. There was no evidence on the top of the roof that a fire, spreading fast on the trusses and plywood, had destroyed the trusses. A fire fighter plunged to his death in 1979. More recently three Phoenix fire fighters went through a trussed roof but managed to extricate themselves. (See Case Studies at the end of the Chapter.)

Fig. 12-11. Alert Orange County, Florida fire officers evacuated this trussed roof restaurant moments before a general collapse.

Trusses built in the field by carpenters often have plywood gusset plates. The relative inherent fire resistance of such plates, as compared to metal connectors, is uncertain. A fire in La Plata, Maryland involved carpenter-built trusses with a 40-foot span and wooden gusset plates. Plywood gusset plates of the bottom chord burned away. The roof fell, injuring eight fire fighters.

In another case replacement trusses were built in the field. A sheet of metal was used on one side of a joint. On the other side was a piece of heavy cardboard to catch the nail points.

Very large and heavy gusset-plate trusses have been used as a girder to support other trusses. In a least one instance near Halifax, Nova Scotia the bottom chord consists of three 2 × 10 planks, side by side. All the gusset plate connections were located at the same point. There is no overlap and apparently no serious attempt to tie the several bottom chord members together. A failure at that critical point would drop the entire structure.

Trusses supported by trusses are not uncommon. They are a characteristic of one chain of family restaurants.

Heavy Trusses

Heavy trusses of combined timber and steel, and light-but-strong all steel trusses, have long been used for commercial and wide-span public buildings. They can be bowstring, parallel-chord, or triangular in shape.

For many years, almost all wood trusses were composed of heavy planks or timbers bolted together. Steel rods commonly were used as tension members.

Some light-weight **lattice trusses** were also used. The web in these trusses consists of closely spaced boards typically 1 inch × 6 inch. Such a truss would fail rapidly.

There are some heavy trusses which appear to be able to resist substantial fire exposure. However, one problem is that these trusses are stronger than the thin roof boards. A Yonkers, New York fire officer died when he stepped onto the roof boards.[10]

Another problem is the way some trusses are assembled. If the members making up a chord are designed to be spaced apart, rather than being solidly together, the surface area is increased and such a truss is liable to earlier failure. While some truss roofs may last longer than others, all trusses are inherently hazardous. If the truss design has been examined during prefire planning, it may be determined that it is possible to operate on or under the truss, as long as the condition of the trusses can be observed.

As has been noted, a failure of any part of a truss is a failure of the entire truss. Heavy timber trusses have metal connections that can fail early. A group of heavy timber trusses rest on an unprotected steel girder in a completely wooden unsprinklered Florida country club. The girder, upon being heated, could elongate, and overturn and drop all of the trusses. At higher temperatures the girder would fail and drop the trusses.

Some timber bowstring truss collapses have caused fire fighter deaths and injuries. In 1988, five Hackensack, New Jersey fire fighters died in a bowstring truss roof collapse. Subsequently there was immediate attention to the hazards of the Bowstring Truss. This emphasis on only one type of truss is inadequate.

Retired Deputy Fire Commissioner, William Foley, described an incident some years ago in Chicago where three fire fighters died in the collapse of a bowstring truss. Officers were warned to become aware of truss hazards. The general belief was that the hump denoted a truss roof. At a later fire, the command officer, noting the clear span, asked the roof officer if there was a trussed roof. Seeing no hump he answered, "Negative." Three fire fighters died in the collapse of its parallel-chord truss.

Heavy trusses are located so far apart that to some there is apparently useful space left between them. The space can be used for storage or to erect a sort of mezzanine floor for storage or office space. It any such use is noted the building department should be queried to learn if the loads were calculated in the initial design. In a supermarket collapse where six New York fire fighters died, the fire involved the mezzanine offices between the bowstring trusses. Only one truss failed.[11]

In the case of the Hackensack, New Jersey collapse, the truss space had been used for heavy storage. In addition, a heavy wire lath and plaster ceiling added to the weight and concealed the fire from the fire fighters below.[12]

Los Angeles City Fire Department Battalion Chief John Mittendorf has a different view of heavy trusses than this author. Chief Mittendorf holds the position that some timber truss roof fires can be fought successfully with an interior attack under certain well defined conditions. This author has seem some such "safe" trusses — but not many. A joint article on this subject was recently published in *Fire Engineering*.[13] Despite divergent opinions, both of us are dedicated to fire fighter safety.

Steel Trusses

Some long-span steel trusses are being designed with a **steel cable** as the bottom chord. Such cold-drawn cables totally fail at 800°F—a low fire temperature. A failure would lead to the total collapse of the truss.[14] Steel is such a strong material that very light-weight steel members can carry substantial loads. Such light-weight members have little inherent fire resistance because they heat up rapidly.

A Wichita, Kansas automobile dealership was located in a building with typical long-span steel bowstring trusses. As fire fighters turned out, a second alarm was ordered because of the visible fire and smoke. Upon arrival, a line was stretched into the building. The steel bowstring trusses collapsed and four fire fighters died. The trusses were loaded with automobile parts.

The well documented McCormack Place fire which destroyed the hall in less than one hour attacked massive unprotected steel trusses 30 feet above the exhibit floor. (See Chapter 7 for details about the fire.)

The steel **bar joist** (a parallel-chord truss) is used both for roofs and floors. In standard fire resistance tests conducted in 1954 at the National Bureau of Standards (NBS is now NIST), unprotected steel open-web bar joists reached 1200°F in 6 to 8 minutes.[15] The steel is so light that it reaches elongation temperatures of over 1000°F very rapidly. At this temperature, bar joists exert sufficient lateral thrust to push a masonry wall to the point of collapse. At higher temperatures the steel fails. While wood trusses fail suddenly, steel trusses are inclined to sag, thus giving some warning to fire forces. However, a sagging roof may be hot enough to liquefy tar — fire fighters may slide down such a slippery surface and into the fire. (See Fig. 7-10.)

Insulated Metal Deck Roofs

Steel bar joists are often used in conjunction with **insulated metal deck roofs**. Looking up from the underside, the construction appears to be entirely non-combustible. The hidden fuel is the tar adhesive-vapor seal on top of the steel decking. When heated, gases cannot escape upward so they blow down through the openings in the decking, providing a rapidly moving, self-sustained roof fire, independent of the fire in the contents.

Such a metal deck roof fire can heat up bar joists, elongate them, and push down a masonry wall. In a Dallas, Texas metal deck roof fire, six fire fighters were on the roof some distance from the visible fire. Heat elongated bar joists which apparently pushed down a masonry block wall, causing the roof to collapse. The six fire fighters were dropped into the building. (Chapter 7 contains information on such metal deck roofs.)

Rated Steel Floor-Ceiling Assemblies

Steel bar joists are used in some rated fire-resistive floor-ceiling assemblies. To provide the fire resistance, such assemblies must be installed exactly as they are in the test laboratory—a requirement difficult to meet. In addition, the removal of one ceiling tile can subject the entire assembly to destructive fire temperatures. (See Chapter 7.)

Some building departments permit fireproofing to be omitted from columns passing through the plenum space or truss void. A ceiling failure in such a building might well mean a column failure and a catastrophic loss.

Such rated floors are tested for passage of fire **upward** through the floor, and for collapse. They are not tested for the problem presented by a trussloft in every floor of the building. This void can and has extended fire over partitions into other occupancies on the fire floor.

A high-rise building's ceiling void was used as the return side of the air conditioning system (plenum). The system provided automatic exhaust when alarmed. In one case it pulled the fire 200 feet through the plenum. The mineral tiles which lined the ceilings became incandescent and started new fires as they dropped.

Wood Truss Floors

The now common wood truss floors are a hazard to fire fighters. They are built of small cross-section wood, with the high surface-to-mass ratio characteristic of kindling. Metal gusset plates dig into the wood approximately $\frac{3}{8}$ of an inch. Heating of the gusset plate will decompose tensioned wood fibers holding the teeth. Not only is there the hazard of early collapse, but the void is a reservoir for explosive carbon monoxide gas waiting for a fire fighter to pull the ceiling. Tactics based on sawn joist floors will kill fire fighters if used on buildings with truss floors.

The truss floor building ordinarily gives no outward indication of its presence. The only solution is for the fire department to pre-plan, record and retrieve, on the fireground, the construction information. During a fire the presence of truss floors may be disclosed by smoke showing through the wall at the floor line. This is less likely in a brick-veneered building. In a San Antonio, Texas fire, fire fighters escaped just moments before a multiple truss floor collapsed.[16]

Fig. 12-12. An "undressed" truss floor multiple-occupancy building. The truss voids are connected by pipe chases. There is no effective firestopping. Any fire in the truss void has access to the entire building.

Fire-Resistance Ratings and the ASTM E119 Fire Endurance Test

Fire Journal[17] published an article by Erwin Shaffer, an official of the Forest Products Laboratory. The article sought to allay the fears of fire fighters as to the collapse potential of wood truss floors during a fire. It raised much protest from the fire service.

Mr. Shaffer relied principally on tests conducted in accordance with ASTM E119—the standard fire-resistance test. UL rates truss assemblies tested to this standard as **one-hour fire-resistive**.

When this author first found that ordinary apartment house fires were burning down into the floor void early into the fire and reported this as part of a study of Combustible Multiple Dwellings for the National Bureau of Standards (now NIST), one researcher exclaimed, "The fire is not supposed to burn that way." My reply was, "Who put you in charge of how the fire burns?" The point is: test procedures should be realistic representations of actual fire conditions, if the results are to be cited to reassure fire fighters. Often, they are not.

Does ASTM E119 show this characteristic in the case of wooden floors of any kind? I think not, certainly not to such a degree as to reassure or add safety for fire fighters. The following are noted in no particular order.

- The 70 year-old standard time-temperature curve is not a valid measure of today's fires.

  In a 1975 Bulletin, the Society of Fire Protection Engineers in discussing the fires in today's contents said "... the rate of fire development can create a condition that may tax or overpower traditional fire defenses."

  In 1985, the NFPA produced the remarkable film, "Fire, Countdown to Disaster." It shows the rapid growth of a typical very ordinary bedroom fire. Note that in TWO MINUTES AND TWELVE SECONDS flashover had occurred, and floor temperatures were 450°F. The carpeting can be seen burning. THIRTY SECONDS LATER the temperature exceeds 1230°F all over the room.[18]

  In 1980, The National Bureau of Standards published "Fire Development in Residential Basement Rooms." (NBSIR 80-2120) Don't be

---

Author's Note: Chapter 6 discusses fire-resistance testing and the fact that the one-hour rating does not mean one hour in a real fire. Chapter 5 discusses combustible floor assemblies and the fallacy of assuming that nailing up gypsum board necessarily provides rated fire resistance. This is particularly noteworthy in the case of gypsum board top floor ceilings nailed to roof trusses. There are **NO** rated triangular truss-gypsum board assemblies. This author believes that there is a serious question whether the ASTM E119 Test is adequate when used to determine the fire resistance of wooden structures.

misled by the title. The rooms were typical living rooms, or office reception rooms. The fires were ignited in newspapers spread over the back of the couch. One Figure revealed heat fluxes at the floor level in the range of 100 to 160 kW/m2, in the time period from 6 minutes until 12 minutes. This energy level is sufficient to ignite a wooden floor. Another figure shows a comparison of the fire exposure curve with the ASTM standard curve. The test curve peaks at over 1000°C (over 1800°F) in 10 minutes.

Conclusions of the report are noteworthy.

"The rate of development and intensity of real fires involving the burning of typical furniture and interior linings in a room during the first 20 minutes may be significantly greater than those defined by the ASTM E119 standard time-temperature curve. A more realistic time-temperature curve for residential occupancies is presented in this report. This curve is considered suitable for testing exposed floor construction, floor-ceiling assemblies, wall assemblies, columns or doors."

A dramatic demonstration of the amount of fire the contents of an apartment can generate was seen on NBC news in Sept., 1991. A two-fatality fire belched flames 20 feet out the windows of the 19th floor of a Los Angeles high-rise building. Tom Brokaw noted, "There were no sprinklers."

- The ASTM E119 test is conducted on a tightly firestopped structure 14 × 17 feet. This cannot be regarded as representative of structures as built in the United States.

The ASTM test structure is built by the organization sponsoring the test. It is obvious that the test structure is the best that can be built, and is not the typical or average structure. The typical structure may have many structural defects. Some were detailed earlier in this chapter.

Fire fighters fight fires in buildings as built, not in a world of perfect buildings and construction methods.

- The size of the test structure is not realistic. The ASTM void might contain 238 cubic feet of air which would contain about 50 cubic feet of oxygen. One cubic foot of oxygen provides 537 Btu. The 50 cubic feet would generate a maximum of 27,000 Btu which would be provided by about 3.5 pounds of wood. The burning of the wood in the trusses is therefore self limiting until the ceiling opens up to admit more air.

A realistic structure would be about 1200 square feet. A greater amount of oxygen would then be available to support combustion. A realistic structure would have provision for additional air, representing the air that would be available through the alleged firestopping which, as observed and photographed, is often ineffective.

- The test assumes, incorrectly, that the fire will only burn upward. The test does not consider that the fire can burn downward. As noted, today's fires can burn down into the floor void in as little as five minutes.
- The test does not simulate fire entering the truss void through the floor above.
- The test does not simulate fire entering the truss void laterally, such as by an exterior fire burning through the combustible exterior into the truss void.

In Lake Benbrook, Texas, an exterior fire burned laterally into the void through the fascia board. The trusses were destroyed by fire passing through or around the firestopping.

- The test does not provide for any penetrations of the gypsum sheath. Light fixtures, HVAC outlets, and ventilating duct inlets are typical of pinholes or other penetrations of the gypsum sheath.
- The test does not simulate fires which initiate in the void, such as from wiring or metal chimney ducts. Fire can be directly delivered to the void by ventilation ducts from interior kitchens or bathrooms. A number of the fires investigated for the NBS started above or behind the gypsum board sheath.

Fig. 12-13. This Texas town house had a truss floor. The building was destroyed by an outside fire that entered the truss void through the combustible exterior wall. This can happen in any truss-floor structure with combustible walls.

- The test was conducted with a static live load of 30 psf. If the test results are to be used to reassure fire fighters, it is reasonable that in addition to the static live load, there should be a moving live load with some impact component representing two fire fighters making a primary search for victims. Two fire fighters, with modern light-weight SCBA and the minimum tools used for a search, weigh about 450 pounds.
- The test is conducted under negative pressure. This exhausts toxic and explosive carbon monoxide which would be present in any real-life void space fire.
- There is no evaluation of the extension potential presented by the vertical voids (such as for piping) open to the floor void.

Dick Sylvia, a fire officer, fire science instructor, and, at the time, Editor of *Fire Engineering*, wrote an article on the hazards of truss floors.[19] A representative of the National Forest Products Association wrote to argue that firestopping and education of architects could eliminate the hazard. The deficiencies of firestopping have long been noted and discussed.

Education of architects has been offered as a solution to many fire service problems. Imagine an architect, well aware of the problem of wood truss construction, who designs fire fighter-safe, yet necessarily more costly, buildings. The architect would most likely have few commissions.

The fire service must take care of itself.

Non-Standard Tests of Wood Assemblies

Chief John Mittendorf of the Los Angeles City Fire Department[19] reported on some nonstandard tests on unprotected light-weight truss and wooden I-beam roofs. The test fires were cut-up pallets with four gallons of thinner. The test structures carried no superimposed load, which is a bias in favor of the structure. The trusses failed in 90 seconds. The pictures show no sign of protection of the wood by the gusset plate as reported by Schaffer.

The Illinois Fire Service Institute ran tests on a number of truss floors and a wood-joisted floor. All were loaded to an evenly distributed 31 psf. The fires were in cut-off 55 gallon drums containing four gallons of diesel fuel and one gallon of gasoline involving five gallons of water. The solid sawn joists gave ample warning of failure and survived until the fire was extinguished in 13 minutes. The wooden I-beams failed completely in 4 minutes and 40 seconds without warning. Open web

Fig. 12-14. One of a number of tests run in Los Angeles. The truss failed. (John Mittendorf)

gusset-plate trusses sagged and gave some warning of pending failure. Trusses with metal webs failed in about six minutes.

As is the case with the standard ASTM E119 test, the applied live load may be representative of an evenly distributed furniture load. The effect of a concentrated load, such as several fire fighters, is unevaluated. Note that the tests were just comparative. The test structures would have collapsed much earlier had they been loaded to represent searching fire fighters.[21] The *Fire Command* article describing the test cautions that the times indicated do not represent the time to collapse in a real fire.

The Truss Void

The wood truss floor introduces a catastrophic new dimension into fire suppression in combustible buildings, a **truss void** or **trussloft**.

The hazards of the cockloft or void between the top-floor ceiling and roof, which are not high enough to be called an attic, have long been familiar. This floor-ceiling truss void can properly be called a trussloft.

All comparative strength controversy aside, the joisted floor has one tremendous advantage over the truss. Each joist acts as a fire stop. The area of the void is limited. The truss void, on the other hand, represents a large volume in which explosive carbon monoxide can accumulate. The use of trusses provides a huge increase in the volume of concealed voids in the structure. Voids are almost invariably interconnected horizontally and vertically by utility pathways. The generation and accumulation of carbon monoxide in void spaces is not as well recognized by the fire service as it should be. It is both toxic and explosive.

The hazard of hidden fire is as important to the fire service as collapse. See Chapter 2 for a discussion of the hazard of hidden fire.

The NFPA *Fire Protection Handbook* provides good information on carbon monoxide. A rule of thumb is that any exposure in which the product of concentration (ppm) × duration (minutes) = 35,000 is likely to be dangerous. A ten-minute exposure to 3500 ppm (parts per million) of CO would be hazardous and possibly incapacitating. Higher ppm values are even more dangerous—12,500 ppm can be fatal after a few breaths. The potential for loss of life or incapacitation of occupants from CO generated by a fire in the void even before the alarm is sounded, cannot be dismissed.

Firestopping, even if installed, is often made ineffective by maintenance work or new installations, for example, cable TV. No tests exist which demonstrate the effectiveness of firestopping or draftstopping for its intended purpose. (See Chapter 3 for a discussion of firestopping.)

The flammability range of carbon monoxide is from 12.5 to 74 percent. Its ignition temperature is 1128°F. When an ideal mixture of CO in air is present, a detonation can result which is sufficient to blow a building apart. Generally a deflagration—a huge sudden burst of flame—appears.

In December 1989, a loss of 16 lives occurred in the fire-resistive John Sevier Retirement Home in Johnson City, Tennessee. Researchers examined the circumstances and determined that the deaths were caused by CO generated in a ceiling void. The old original combustible tile ceiling was left in place when the newer ceiling of flame spread-resistant tiles was installed, suspended below. The researchers stated, "The rapid involvement of a large combustible surface combined with a restricted air supply alters the fire chemistry to increase CO production as much as 50 fold over burning in the open."[22]

In October 1989, NIST presented a proposal for a five-year research project into the production of CO. The thrust was on toxicity, not explosiveness. Plywood or chipboard floor, the paper on the gypsum board, and the wood of the trusses will all generate CO in the confined atmosphere of the truss void. This is in addition to CO generated from contents which becomes trapped in the void.

Catastrophe Potential

A most serious threat to life safety by lateral extension exists when the trusses are extended out, either as cantilevers, or continuous beams, to support the balcony which is the only exit for the occupants and access for the fire department. Where the truss passes over the exterior wall, firestopping or draftstopping in the form of a sheet of gypsum board is "buttered into place" with cement. This "firestopping" is penetrated for the lighting circuits of the balcony. In any event, the use of such an unreliable barrier to the extension of fire from the void is inexcusable. In some cases, the ceiling of such a balcony is plywood, permitting early access to the void for fire originating on the balcony, or coming out the window.

It is a credible scenario that the stairway exit can be involved in fire and collapse before the occupants have escaped. In many suburban areas a typical response might be two engine companies and a truck, each with three or four fire fighters. It is impossible for this response to remove, via ground ladders, all the occupants of a three- or four-story truss-floored multiple dwelling. Even with greater response it is unlikely that all the occupants could be located and removed before interior collapse forced withdrawal. The potential for a disaster should force a code requirement that the exit facilities be totally separated from the truss void.

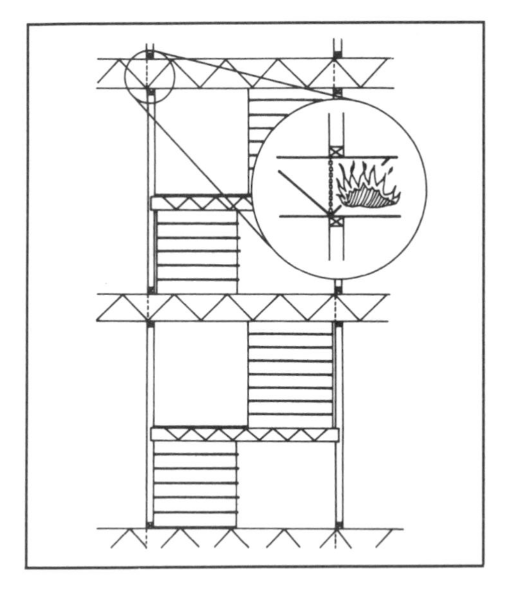

Fig. 12-15. Note how the stairway platform is supported on trusses which extend out from the truss void. Any fire in the truss void can involve the platform, cause the stairs to fall, and leave the occupants without an exit. Perhaps this is the most dangerous use of trusses. There is no proof that the typical firestopping at the building wall will stop a fire.

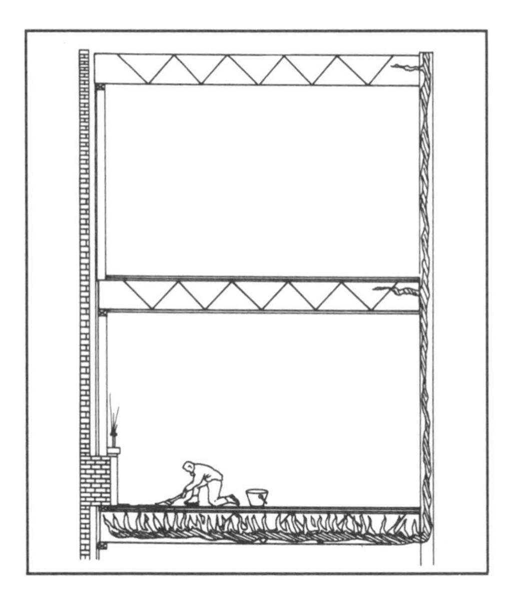

Fig. 12-16. The worker used a flammable solvent to clean a fireplace. The solvent leaked down into the truss. It was ignited when the fireplace was lit. The fire spread through the truss voids. San Antonio, Texas fire fighters escaped moments before the building collapsed.

Fig. 12-17. Many local codes require gypsum sheathing only on the interior of the building. The plywood ceiling of a porch would allow a fire on the porch to enter the truss void.

Tactical Considerations

Both in writing and orally, this writer has urged the truss industry to refuse to supply trusses for such designs. The potential for a "Sindell Type" action is very real. With this type of action the entire industry is sued, because it may be impossible to determine precisely who made the trusses. With some notable exceptions the response of the industry to this hazard has been characterized by denials and personal attacks. This is reminiscent of the attitude of other industries when hazards have been pointed out. Percy Bugbee's 1971 NFPA book "Men Against Fire," noted bitter opposition of wood shingle manufacturers to limitations on their products. The same situation existed with combustible fiberboard manufacturers.

The philosophy that a problem can be eliminated by killing the messenger, unfortunately, is alive and well.

Building department records may be lost or destroyed because of lack of storage space. The fire department also should keep records of the builder, the architect and the truss supplier for possible use in future litigation. If the trusses are truly safe there should be no objection to storing this information for the benefit of citizens who may be victims of fires. Such records may help the fire department be a competent witness instead of a defendant. When this author was working in government fire protection, the notation, "cc grand jury file" on the bottom of a memo got immediate attention and often action.

Press the building department on the problem of trusses extending into the stairway. If the defense is that the "state will not permit us to amend the code," make strong representation to your local state legislators. If they are indifferent, press the city or county for a full-strength ladder company for the area.

Get as many of those responsible onto the record as possible. It is reasonable to assume that efforts will be made to blame the fire department in the event of a catastrophe.

Miscellaneous Steel

Steel I-beams are being used with increasing frequency as main girders to support wood truss floors or roofs. Apparently building officials believe that the gypsum sheathing also protects the steel. This is unevaluated. The steel is exposed to heavy fire in the void. As frequently noted, the elongating steel, usually restrained, will rotate, overturn, and dump the wood trusses. At higher temperatures it will fail.

Fig. 12-18. Cold-drawn steel cables which fail at 800°F are often used to hold failing buildings together.

Trusses are beams and, like all beams, exert a vertical load. This load is delivered to the two opposite walls. Unless the walls are perfectly vertical, one or both might overturn. In some cases, this problem is cured by tying the bearing walls together with cold-drawn steel cables, which was noted as failing completely at 800°F.

A Colorado building-supply store has big parallel-chord trusses extending from front to rear. Apparently the walls are out of alignment because the front and back walls are tied together with steel cables.

In other cases, masonry or wood walls are tied together with steel tie rods. Often the turnbuckle in the center of the tie rod is concealed with a lathe-turned wooden block. The failure of such a tie rod caused a wooden church wall to collapse, and cost the life of a New Jersey fire officer.

In Southern California, there is a timber truss warehouse building in which steel rods parallel the bottom chord. The rods might have been required to tie the walls together to increase earthquake resistance, because the walls were eccentrically loaded by the trusses, or to provide greater tensile strength in the bottom chord of the truss. These situations may be very significant in a fire.

Other structural elements may be tied to the trusses. In a Washington, D.C. theatre, the tie rods holding the canopy were tied to the trusses. The trusses were heavily involved in fire. The loosened tie rods allowed the canopy to become an **undesigned cantilever** which pulled the top 20 feet of the masonry front wall out, almost to the point of overturning. Alert junior officers, who were studying building construction, recognized the hazard and cleared the area.

Hazard Control

Some measures have been taken to mitigate such truss hazards.

Automatic Sprinklers

Some truss-floor multiple dwellings have been protected with residential sprinkler systems, meeting NFPA 13R, *Standard for the Installation of Sprinkler Systems in Residential Occupancies up to Four Stories in Height*. The purpose of these systems is to prevent flashover, save lives, and permit the occupants to escape. There have been a significant number of successful operations. However, these are partial systems and should not be expected to provide the fire suppression or control that full sprinkler systems have provided. The sprinkler systems cover selected portions of occupied spaces. It is very likely that the sprinkler system will control any fire originating in the protected areas of an apartment. However, if the fire originates in or penetrates the void, the sprinklers will not be in a position to control the fire.

A fire in a fully-sprinklered college dormitory destroyed the building. The fire originated in or reached the unsprinklered voids.[23]

This author examined a 13R sprinklered, combustible dwelling in Texas. The floor trusses were cantilevered out to form the exit balcony—the hazard described earlier in the "Catastrophic Potential" section. The balcony was one entire non-firestopped void. The firestopping at the front wall included pieces of gypsum board buttered into place with cement. Very likely it was penetrated in one or more places. A failure of one of these barriers would admit fire to the entire balcony truss void. It is not hard to envision a serious loss of life in such a structure. Unfortunately, such a loss of life might misguidedly deal a fatal blow to the concept of residential sprinkler protection.

In De Kalb County, Georgia, fire resulting from a plumber's propane torch developed in the truss void of an almost completed three-story apartment house. There was no contents load. The sprinkler system was operational, but the sprinklers did not fuse. A fire fighter fell through the second floor into the fire when the truss collapsed. Fortunately, his partner used a hose line which protected him from the fire until he could be pulled out.[24]

Chief Kelly of New Smyrna Beach, Florida related an incident in Orlando where a fire started at a defective ceiling light fixture in a first floor apartment closet of a three-story, brick-veneered, plywood-sheathed, truss-floor apartment house. The fire extended to upper floors on the outside of the wood sheathing, concealed by the brick veneer.

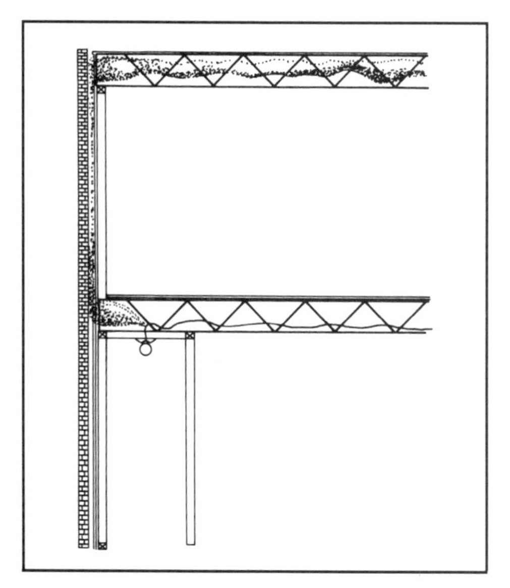

Fig. 12-19. Fire in a closet light fixture spread through the truss void, extending to the outside wood sheathing. Because the fire was concealed by the brick veneer, it worked its way back into the upper floor truss, collapsing the floor. This fire would not have been controlled by a 13R sprinkler system.

The truss floor of the third floor sagged over a foot and was too dangerous to walk on. Had this building been protected with a sprinkler system, it would not have affected the outcome.

One method of combating truss void fires might be to install standard high-temperature sprinklers in the truss void. It is conceivable that the sprinkler would delay operation until considerable heat had built up in the void. This would cause a steam cloud to generate, possibly sufficient to temporarily suppress the fire in the void. This would be similar to the indirect method of attack on fires in ship holds, developed by LLoyd Layman during WWII.[25]

Fire-Retardant Treated Wood

The Walt Disney Corporation built apartment houses at Disneyworld to house employees working at Epcot Center. The buildings are sprinklered according to NFPA 13R and the trusses are made of fire-retardant treated wood. This solution is resisted by the industry because most codes apply a 25 percent strength penalty for the use of treated wood. The recent failures of roofs made of FRT plywood tend to cause treated wood of any type to be dismissed.

Technical Studies

While the attitude of some in the industry has been that fire fighters are misinformed about truss-related hazards, there are those who have recognized the problem. This writer was invited by the Truss Joist Company to address the annual meeting of the Forest Products Research Association in 1990 on "Fire Performance of Light Wood-Frame Structures." From floor comments, I noted that some in the audience realized for the first time that improved design does not eliminate the problem of existing structures.

Whatever improvements are made, if any, would be many years down the road. Fires are fought in existing buildings, not in future concepts. A truss manufacturer's letter to a fire publication protested its attention to the fatal Hackensack collapse, complaining that "Those trusses were 40 years old." Because of terrible multiple-death fires in apartment houses, New York City passed the Tenement House Act of 1903 which mandated massive changes in the design of apartment houses to provide life safety for the tenants. Unfortunately, the effective date was delayed a year. Plans were filed for thousands of old-law tenements, known to be unsafe, during the delay. Many people died in such buildings, and an uncounted number are still occupied. Some have been remodeled and improved.

Kirk Grundahl, P.E., of Qualtim Technologies International, had formed a working group of interested parties in the industry and the fire service to determine what improvements might be made to truss

Fig. 12-20. These flat roof apartments in Somerville, Massachusetts are being "improved" by the addition of triangular trusses. This leaves the old smoky tar-and-gravel roof hidden in the attic. Know your buildings! (Capt. D. Mager, Boston Fire Department)

voids. The effort has been taken over by the NFPA Research Foundation.

The NFPA Research Foundation has invited interested persons and organizations to participate in planning the test phase of its national fire research project on engineered light-weight construction.

In addition, the U.S. Forest Products Laboratory at Madison, Wisconsin has received a grant from the U.S. Department of Agriculture to address critical barriers to advances in structural wood engineering associated with concerns for the fire safety of new structural members such as trusses. The Project Summary notes that "The concern of fire fighters for the safety of trusses has been most visibly expressed by Francis L. Brannigan in his National Fire Protection Association book, *Building Construction For The Fire Service*."

My own objection is that even if the structural stability of the trusses is improved, the hazard of the accumulation of carbon monoxide in the truss loft will still remain.

Wooden I-Beams

The wooden beam sawn out of a tree trunk is inefficient in terms of weight and cost. Recently the sawn beam has been replaced by wooden I-beams (composite wood joists).

Look at the end of a steel I-beam. Since the steel is extruded through a die, the designer can choose the most efficient shape. Both the top flange and the bottom flange are wide to cope with compressive and tensile loads. The web, which separates the top from the bottom, is thin, just sufficient to keep the top and bottom apart.

A sawn wooden beam can be thought of as "containing" an I-beam with "surplus" wood along the sides. The surplus wood is what makes it possible for fire fighters to stand and operate on a burning structure. This "fat" has been the basis of interior fire fighter tactics. As long as only the "fat" or surplus wood is burning, a fire fighter is relatively safe. Until it burns away, the beam is structurally sound. The gradual sagging of sawn beams due to the loss of the exterior wood often gave warning of impending collapse.

Wooden I-beams are another method of making buildings lighter. The beams consist of solid or laminated 2 × 4 for the top and bottom chords, and a plywood of oriented strand board (OSB) web between them. They are manufactured in lengths up to 60 feet. Contrasted with the sawn beam, the "fat" or surplus (structurally unnecessary wood along the sides) is no longer present.

The web of the I-beam is often penetrated by utilities, and in some cases sizable holes are cut into it so that any fire gets a grip on both sides of the I-beam at the same time, guaranteeing early failure.[26]

Often, cut-off ends of I-beams are used as "firestopping." Be alert to them.

Automatic Sprinkler Problems

Recent unsatisfactory fire experience in combination with fire tests results in buildings with exposed composite wood joists (wooden I-beams) up to 30 inches in depth points out basic problems of protection which will need to be addressed if sprinklers are going to be relied on for the protection of wooden I-beam structures.

The current NFPA 13 Standard addresses composite joists up to 16 inches deep. Tests of different designs of sprinklers for deep composite joists have produced mixed results.[27]

Collapses

The following is in a letter from Deputy Chief Frank Cruthers of the New York Fire Department. Note the competent size up and action of the first officer to arrive, and that the action was not the result of experience but of training. Training and knowledge are far less costly than experience.

> "A New York fire officer operating at an odor-of-gas emergency in a five-story tenement building under renovation, noted that wooden I-beams had been substituted for many of the sawn beams. He informed all first alarm units and forwarded information for inclusion in the Critical Information Dispatch System. However, this information was not provided with the dispatch. It is keyed to the address. The fire was reported "in the vicinity of . . . ".
>
> Fortunately, the unit which had discovered and reported the wooden I-beams was the first-arriving engine at the fire. On arrival heavy fire was coming from a second floor window. The officer, observing the wooden I-beams from the street, realized that this was the building previously reported. He ordered his unit to operate from a safe exterior position and notified the Battalion Chief in charge of the fire, who immediately ordered an outside operation and verified that all his forces were withdrawn from the building.
>
> Within eight minutes after the arrival of the first unit, the front of the second and third floors collapsed. Heavy concentrated loads of gypsum board stacked on the floors, awaiting installation, were a contributing factor.
>
> In the opinion of officers in command, there would have been casualties, except for the training the officer had received and his alertness."

A Fairfax County, Virginia fire fighter went through a wooden I-beam floor. The house was under construction, and the floor was the only fuel for the fire.

In another case, a residence in Clayton County, Georgia suffered a severe loss when wooden I-beam floors collapsed.

Inconsistent Construction

It is not possible to determine the construction of a building based upon what is seen at the first place opened for examination.

In a Ronald McDonald House in Gainesville, Florida, wooden I-beams span from the exterior walls to the interior corridor walls. The corridor is about six feet wide. The corridor is spanned with 2 × 6 sawn beams.

The flat roof of a restaurant consists of wood and steel tubular trusses from the exterior walls to the interior corridor walls. In this case, too, the corridor is spanned with sawn 2 × 6 beams.

A steak restaurant has a hip roof. The sloping sections have sawn rafters. The flat top is made of deep parallel-chord gusset-plate trusses. The restaurant interior—and all the voids—are sprinklered.

Many older buildings are being rehabilitated. Extensions and additional stories may be built of trusses or wooden I-beams. Older buildings often had high ceilings for ventilation. Mezzanine floors, supported on trusses, may have been inserted. It is unsafe to assume that a building is consistent in its elements.

Fig. 12-21. In Colorado Springs, Colorado, this church has plank and beam, sawn joists and wood and tubular steel trusses in the roof, all supported on an unprotected steel column.

Second Roofs

Truss roofs are sometimes added to flat-roof buildings to improve their appearance or to deal with leaky roofs. In *Firefighting in Canada*, August 1991, Lorne Ulley describes a million dollar fire in such a newly renovated motel in St. Laurent, Quebec.

The fire started in a guest room and was extending upon arrival of the fire department. When the fire reached the new undivided roof, it was soon fully involved. The principal damage was to the roof. The fire fighters kept the roof fire from dropping down.

Case Studies

The following cases illustrate numerous characteristics of trusses in structures.

Phoenix, Arizona fire fighters were ventilating the roof of a large one-story residence. The roof was of masonry tiles on chipboard on wood trusses. At 17 minutes into the fire the roof failed. Four fire fighters fell into the attic. The failure was due to the fallout of the metal gusset plates. The chipboard remained intact. The fire fighters were saved from serious injury or death because they were wearing full protective equipment. An excellent dramatic video showing members of the truck company falling through the truss roof was made by the Phoenix, Arizona Fire Department. Copies of this excellent tape are available at a nominal cost from the Phoenix Fire Department.

On February 19, 1990 Tualatin, Washington fire fighters were ordered out of a bowstring-truss paper warehouse only minutes before the roof collapsed. Sprinklers coated with paper dust were ineffective. The dramatic collapse can be seen on video.[28]

Captain Randy Mathis describes a fire where North Spartanburg, South Carolina fire fighters had initiated an interior attack on a truss-roofed commercial building. The rear sector officer noticed that the roof had dropped four to six inches. The building was immediately evacuated and a defensive attack mounted. It was found that 8 of 30 trusses had lost their top gusset plates, and total collapse was averted only by the fact that gusset plates in the bottom chord had been protected by gypsum board.

According to Battalion Chief Martinez, fire fighters in Ontario, California noted very little smoke as they approached a supermarket. The front section had a roof of steel bar joists. Its dimensions were 80 feet × 100 feet. The attached rear section had open wood and steel tubing trusses. The attack teams were pulled out just before the collapse of the roof, about an hour after arrival.

Paterson, New Jersey units avoided entering a blazing battery factory with a bowstring truss roof. The roof collapsed minutes after arrival.

Faulty wood trusses may have caused several roof collapses in the Pacific Northwest. The suspected trusses were wooden, open-web joists with glued finger joints. About 200 buildings contain the trusses which may be faulty.[29]

New York fire fighters were withdrawn from a one-story garage with a bowstring truss roof, in anticipation of collapse. When it occurred, fire fighters at the doorway were knocked down by the blast of air generated as the roof fell. It is not sufficient to be barely clear of the building.[30]

Skokie, Illinois fire fighters were advised en route of a bowstring truss building, from computerized preplans. Department procedures call for withdrawal if progress is not made quickly by the initial attack. Shortly after personnel withdrew, and a roll call was held, the roof collapsed.[31] The article refers to bowstring trusses. The SOP should apply to all trusses.

A western New York State fire company hooked up to a bad hydrant as they prepared to enter a free-standing fast-food restaurant. The delay in getting water providentially slowed the attack team's entry. The light-weight parallel-chord truss roof collapsed before they could enter.

Deputy Chief Fekete, of the Louisville, Kentucky Fire Department, reports on a truss roof collapse 25 minutes after the alarm. Air conditioners, supported by the roof trusses, fell in five minutes. Several fire fighters were injured. Later there was a ceiling collapse in an area where there was only light smoke.[32]

A Chesterfield, Virginia church had light-weight roof trusses. The roof was struck by lightning. Fire was showing from the roof on arrival but there was only light smoke in the church. Five fire fighters pulling the ceiling inside the church were temporarily trapped when the roof collapsed an estimated 30 minutes after lightning struck.

Orlando, Florida fire fighters were operating at a fire in a one-story concrete block building with a steel truss roof covered with corrugated sheet metal. The contents were waste paper. The temperature of structural steel is much higher in an unvented fire than a vented fire. Moving steel pushed down the block end gable wall onto the loading platform. Fortunately fire fighters had moved off the platform moments earlier.

NFPA Journal (formerly *Fire Command*) performs and excellent service by publishing an annual summary of fire fighter deaths in the July/August issue. Some years ago, a truss roof collapsed, fatally trapping an Irving, Texas fire fighter. His death was classified under the category "Exposure to Fire Products." This obscures the fact that the collapse of the building actually caused his fatal burns. Since then the classification of fatalities has been fine tuned.

The roof of a Nagadoches, Texas plant was supported on 330 foot laminated bowstring trusses. The top and bottom chords were formed from nine laminated plants 1.5 inches × 7 inches making the member about 15 inches × 7 inches. The trusses collapsed about 20 minutes after the fire was first noticed.[33]

A Grand Prairie, Texas industrial building was involved in fire. There was a ceiling (probably combustible) 10 feet below the trusses. No fire was visible inside the building upon arrival. There is an excellent picture of the collapsed steel bowstring trusses.[34]

The Farmington Hills, Michigan Fire Department responded to a fire in a private dwelling, a three-level home with 6000 square feet of space, a five-car garage and a converted four-car garage. The walkout basement included a built-in swimming pool. Fire conditions prevented access to the seat of the fire, despite an aggressive interior attack. Approximately 15 minutes into the fire the second floor collapsed. The floors were light-weight tubular steel and wood trusses. Fortunately, the intensity of the fire kept fire fighters off the second floor and out from under it.[35]

The Struthers, Ohio Fire Department had just upgraded their pre-fire plan for the Lyons Medical Supply warehouse when the fire struck. The high fire load, lack of built-in fire protection and the presence of steel bowstring trusses, forced the chief to abandon offensive operations and concentrate on protecting an exposure.[36]

Los Angeles, California fire fighters operating a line into the windows barely escaped when the collapse of an arched rib roof on a body shop brought down the brick walls. There are excellent dramatic pictures accompanying the story.[37]

New York fire fighters were operating at a fire in an abandoned commercial garage. The only fuel was the wooden-bowstring truss-supported roof. The roof was preventing the overturning of the wall from a projecting sign. When the roof trusses failed, the sign brought down the wall. There was no overt warning. The warning is in understanding buildings and the consequences of individual component failures.

A Gillette, Wyoming fire fighter died when the light-weight truss roof of a church collapsed as he attempted to ventilate. In a *Fire Engineering* article an investigator is quoted, "Usually before a roof collapses, it will sag and you can feel it. Everyone said this one went from a good roof to no roof almost instantly." A chief hazard of truss roofs is that instant collapse is the rule not the exception. The article also continues "... convinced that their fire ground procedures were correct, officers have not drastically changed procedures for roof operations."[38] A number of letters critical of the above quote were received and two were published.[39]

Paul McFadden, a New York City fire officer, properly recommends a defensive attack on cellars in which wooden I-beams support the first floor. This good advice should be extended also to wood or steel truss floors. Both are very liable to early collapse.[40]

Orange County, Florida fire fighters responded to a one-story concrete block building with a wood truss roof. The roof collapsed about 30 minutes after arrival and five minutes after crews were ordered off.[41]

Chief Sanders related an incident in April 1989 where Louisville, Kentucky fire fighters were ordered out of a two-story commercial building just prior to the complete collapse of the light-weight truss roof.

In late 1989, according to Deputy Chief Gary Skan and Captain Peter Scale, the Clinton Connecticut Volunteer Fire Department had a fire in a four-unit truss-floored, truss-roofed town house. The occupant was at home. The fire started at a kitchen light fixture and apparently smoldered in the truss void. The occupant, thinking he heard "mice" in the ceiling, investigated and found nothing. The fire in the meantime had reached the attic by a typical interconnecting vertical void. When the occupant opened the attic pull-down stairs, he found the attic fully involved. As fire fighters prepared to vent the roof, it self-vented, with two CO ignitions. Air horns and radios were used to evacuate the building. Note the similarity to an Orlando, Florida fire cited earlier. In both cases, the fire started in the void. Sprinklers would not have caught this fire either.

There have been many collapses of wood-truss roofs under construction. These roofs are unstable until the roof sheeting is nailed. Some collapses were caused by concentrated loads of roofing, others by the wind.

Chief Mittendorf of the Los Angeles City Fire Department discusses the relative hazards of sawn joist rafters and light-weight roof trusses in a 1988 *Fire Engineering* article.[42]

Note: Despite this discussion, by no means should anyone assume that the attention given to truss and composite wood joist floors in this chapter represents unqualified approval of sawn wood joists. While they have stood up to many fires, a number of fire fighters have lost their lives on or under joisted floors.

Fig. 12-22. Just because trusses and wooden I-beams are dangerous, don't ignore the hazard of sawn joists. They have cost many fire fighters their lives.

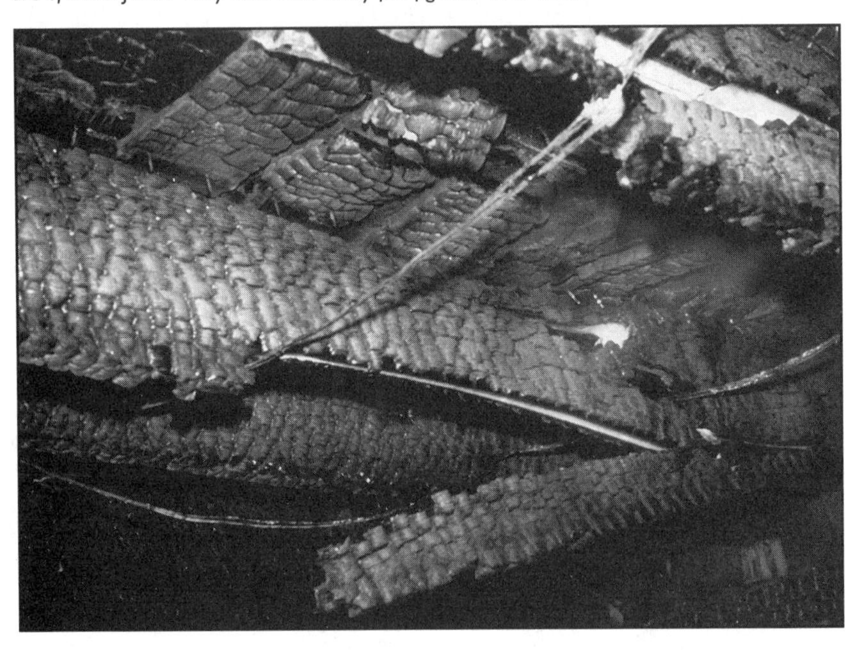

Tactical Considerations

These construction practices and materials demand major changes in fire fighting tactics. Fire fighters can no longer automatically "get the roof," or go charging pell mell into building fires. Fire departments must know more about the buildings in which they fight fires. Catastrophes and lawsuits will force change on the stubborn.

In the case of trusses or wooden I-beams, department SOPs should provide for evacuation as soon as trusses are involved in fire, or in the case of steel trusses, when they are being substantially heated.

I am struck with the apparent nonchalance with which fire fighters open up potentially lethal voids. It is very possible to have a raging fire overhead with only light smoke showing. It is a meaningless term when the source of the smoke cannot be seen.

Be extra careful when responding at night to a fire in a fast-food restaurant. Light smoke may be showing in the brightly lighted interior. After a line is stretched to the interior, the ceiling is pulled or tiles are raised. In most cases, minor fire is found and extinguished. In the worst case, an attic fire is accelerated and the roof falls.

All personnel at a fire scene should be alert to exterior conditions. Fire or heavy smoke from the roof indicates that the trusses are on fire. As with wood truss floors, truss voids in ceilings may harbor a well-concealed fire that is ready to burst out with almost unbelievable fury when oxygen is admitted to a void containing heated carbon monoxide. The reaction can range from a deflagration to a detonation, from a backdraft or flashover to an explosion capable of blowing a building apart. It is not an exaggeration to say that the opening of a void is the equivalent of disturbing a suspicious package. Most often nothing happens, but once in awhile—disaster.

Any ceiling below a truss void should be pulled and examined by disciplined fire fighters under control, standing near a doorway for rapid escape. If there is a fire barrier in the attic, the same procedure should be repeated on the other side of the barrier.

Do not misunderstand the term "fire-rated" (meets flame-spread requirements) suspended panel ceiling. It does not provide any fire resistance, but does provide a void to hide the fire and store carbon monoxide. If the steel grid of a suspended ceiling comes down, fire fighters could become entangled in the grid and be confused or trapped in a type of steel net. The area can be pitch black. A fire fighter, with no idea of what has happened, could thrash around until the SCBA air gives out.

Train personnel to stay close to walls or other internal barriers to total collapse such as shelving, or to be below the net if the ceiling

Tactical Considerations, continued

falls. If caught the best procedure might be to roll onto the back and wriggle out. There is less potential for the breathing apparatus to be snagged in the grid.

Consider the use of a piercing nozzle from an aerial platform bucket to pierce the void horizontally. It is possible that the generated steam would accomplish instant suppression.

Fire inside may destroy the interior connection of wooden I-beams or trusses cantilevered out to form a balcony. The balcony will collapse when loaded with unwary fire fighters. Never forget that a cantilever is truly a seesaw. When one end goes up, the other end goes down.

Light-weight steel trusses can reach elongation and failure temperatures very rapidly, possibly even before the arrival of fire fighters. Steel trusses should be cooled with hose streams from a safe location. Do not worry about wetting contents or "throwing water through smoke." If sprinklers had been installed they would be operating through smoke. A necessary part of sprinkler function is prewetting exposed combustibles.

If fire is in the truss void, heat will move through it, elongating bar joists some distance from the fire and endangering fire fighters in apparently safe locations.

Do not rely on fire-rated combustible construction or good codes. Do not rely on firestopping or fire barriers. Examine and protect them as necessary. The many deficiencies cited in the text can be found across the country.

After occupants have been removed from a structure, the guiding principle should be, "No building is worth a fire fighter's life."

Do not concentrate solely on bowstring trusses. All trusses are dangerous.

Be aware of concentrated loads such as equipment or storage.

Snow or ice on a roof has already used up some or all of any reserve strength the roof has available.

Building hazards and proper actions must be known to all fire fighters, not just commanders. Recall the rapid fatal truss failure in Orange County, Florida at a time when only the first-due engine was on the scene.

Examine multiple dwellings in your area for the serious defect of allowing the truss or I-beam floors to extend out to form the stairway platform. Make every effort to stop this dangerous practice.

Arrange for first-alarm assignments sufficient to evacuate the building over ladders. Notify the tenants of the necessity for immediate movement on the sounding of any fire alarm. Make it clear that the fire department should be called if smoke is even

Tactical Considerations, continued

smelled. This might indicate a hidden fire. Check the quality of the firestopping. Make local officials aware of your firesafety concerns to avoid comebacks like, "You never told us."

In other words, make a huge fuss. The emphasis on the safety of occupants may be more effective than the safety of fire fighters. The official attitude, "We pay them to take those risks" is embedded in the law in the form of the fire fighter's rule, which states that fire fighters assume the risks of their profession. Some judges, upholding the rights of a property owner, have even applied this dreadful rule to unpaid volunteers.

Be well aware of the deficiencies of partial sprinkler systems. Do not place absolute confidence in standard sprinkler protection for commercial buildings with deep wooden I-beams.

Be sure fire fighters operating on roofs are wearing full protective gear. Note that full gear has saved the lives of fire fighters who went through a roof.

Particularly note the use of masonry tiles on roofs. They provide a heavy dead weight.

Be aware of roofs sheathed in foamed plastic sheets in energy-efficient structures. Even if the plastic is inhibited to prevent flaming, it may melt and break down.

When going into a defensive mode because of potential roof collapse, keep personnel well clear of doors and windows, due to the blast of fire which may come out when the roof collapses.

If ventilation on a dangerous roof is necessary, fire fighters should be fully supported on an aerial apparatus. Even this is risky because the collapse of the roof may result in an upward burst of fire which might engulf the bucket.

Recognize that the old indicators of collapse—softening, strange noises and cracking—may or may not be available. Sudden catastrophic collapse is the rule, not the exception.

Above all, know your buildings. Make all information and knowledge accessible of building conditions which might be hazardous during a fire. Develop a record and retrieval system which will assure that the necessary information is made known to responding units immediately on dispatch.

New Jersey has passed a law to assist fire fighters. It requires that all buildings with trusses must have a prominent sign outside the building which indicates truss construction. Such a sign would be a useful supplement to information permanently recorded in the computer data bank.[43]

Wooden I-beams and trusses of all types will force a fundamental change in fire fighting tactics, or lives will be lost. It is possible that

Fig. 12-23. Working off an aerial apparatus is much safer than being supported on the roof truss or sawn joist. In addition, fire fighters should be fully protected and wear their SCBA. (Bill Noonan, Boston Fire Department)

Tactical Considerations, continued

someday a fire officer will be found personally negligent in a civil suit, and an ambitious prosecutor may well file criminal charges for deliberately placing employees in extreme danger. Changes in building codes or better enforcement of existing codes will do little to reduce the fire fighting hazard presented by these construction techniques. There is little in the building code specifically concerned with fire fighter safety. This is the responsibility of the fire department.

Remember: BEWARE THE TRUSS!

References

The addresses of all referenced publications are listed in the Appendix.

1. Brannigan, F. "Surveys Aid in Preparation for Handling Large Fires." *Fire Engineering*, January 1948.
2. Ellison, Huntington, Mickadeit. *Building Construction, Materials and Types of Construction*, Sixth Edition. John Wiley, New York, N.Y., Figures 1-16 and 1-17.
3. Swensson, K. "Opryland's Cascades." *Modern Steel Construction*, March/April, 1989, p. 9.
4. "The Case of the Rising Roofs." *New Shelter*, January 1986, p. 11.
5. Senft, J., Bendtson, B., and Galligan, W. "Weak Wood, Fast Grown Trees Make Problem Lumber." *Journal of Forestry*, August 1985.
6. "Failures, Wood Truss Roof." *Progressive Architecture*, October 1986, p. 57.
7. Isner, M. NFPA Fire Investigation Report, Nursing Home Fire, Dardanelle Arkansas, March 13, 1990.
8. "Long Spans Need Care, Panel Says." *Engineering News Record*, May 28, 1981.
9. Orange County Fire and Rescue Investigation Team. "Orange County Fatal Fire." *Fire Engineering*, July 1989, p. 31.
10. Dunn, V. "Firefighting on Sloped Peaked Roofs." *Firehouse*, June 1985, p. 48.
11. Dunn, V. "Truss Collapse, Final Report." *Firehouse*, June 1985, p. 36.
12. Eisner, H. "Trusses Can Kill." *Firehouse*, September 1988, p. 55. Also Eisner, H. "Hackensack NJ, One Year Later." *Firehouse*, August 1989, p. 67. Also Corbett, G. "Five Fall in Hackensack." *Fire Engineering*, October 1988, p. 22.
13. Mittendorf, John W. and Brannigan, Francis W. "The Timber Truss: Two Points of View." *Fire Engineering*, Vol. 144, No. 5, May, 1991, p. 51.
14. "Cabled Truss Curves As It Rises." *Engineering News Record*, October 13, 1988; "Shortcuts to Long Spans." *Engineering News Record*, September 8, 1988.
15. Ryan, V. and Bender, E. "Fire Endurance of Open Web Steel Joist Floors with Concrete Slabs and Gypsum Ceilings." NBS Report No. 141, Baltimore, Maryland, August 21, 1954.
16. Corbett, G. "Lightweight Wood Truss Floor Construction." *Fire Engineering*, July 1988, p. 41.
17. Schaffer, E. "How Well Do Wood Trusses Really Perform during A Fire?" *Fire Journal*, March/April 1988, p. 57.
18. Hughes, S. "A Rare Look at Flashover." *Fire Command*, August 1986, p. 21-23.
19. Sylvia R. "New Types of Construction Require Rapid Opening Up, Larger Streams." *Fire Engineering*, March 1980, p. 28. See also *Fire Engineering*, Letters to the Editor, June 1980, p. 10.
20. Mittendorf, J. "Lightweight Construction Tests Open Fire Service Eyes to Special Hazards." *Western Fire Journal* (now American Fire Journal), January 1982, p. 23.
21. Straeske J. and Weber, C. "Testing Floor Systems." *Fire Command*, June 1988, p. 47.
22. Steckler, K., Quintiere, J., Klote, J. "Johnson City Fire—Key Factors." Center for Fire Research NIST (formerly NBS), 1990.
23. Bell, J. "184 Evacuated in Dormitory Fire." *Fire Journal*, January 1984, p. 50.

24. Turner, P., De Kalb County, GA, Fire Department Forum, *Firehouse*, December 1989 p. 8 and correspondence.

25. Layman, L. "*Attacking And Extinguishing Interior Fires*." Fourth edition. National Fire Protection Association, 1960

26. Report. Truss Joist Corp., Boise, Idaho, 1989.

27. Report. Forest Products Research Society Meeting. Salt Lake City, Utah, June 1990.

28. American Heat Video "Truss Roof Collapses and Construction." 1990.

29. Cohn, L. "Trusses Suspect in Collapse." *Engineering News Record*, September 10, 1987, p. 12.

30. O'Hagen, J. "Truss Fire and Collapse." *WNYF* (With New York Firefighters), 4th issue 1984.

31. Wilms, P. "SOPs to Live By." *Fire Command*, August 1990, p. 42. See also Letter to Editor, October 1990.

32. Fekete, J. "We Have a Roof Cave In." *Fire Command*, January 1984.

33. Timoney, T. "Texas Plywood Manufacturing Plant Destroyed." *Fire Journal*, March 1985.

34. Courtney, N. "Fire Record: Rope Manufacturer." *Fire Journal*, January 1986, p. 15.

35. Kurzeja, W. "Light-Weight Truss Construction Gives Up More Lessons." *Fire Engineering*, March 1989, p. 32.

36. Ryan, T. "Preplanning Power." *Firehouse*, December 1988, p. 32.

37. Simmons, A. "Falling Wall." *Fire Command*, September 1984, p. 26.

38. Vosnik, B., Ernst, R., Ernst, L. "Coping With A Brother's Death." *Fire Engineering*, September 1990, p. 62.

39. Austin, H. and Sepanski, D. Letters to The Editor, *Fire Engineering*, December 1990, p. 17.

40. McFadden, P. "Cellar Fire Wrap Up." *Fire Engineering*, September 1987, p. 8.

41. *Firefighter's News*, August/September 1990, p. 38.

42. Mittendorf, J. "Joist Rafter vs. Lightweight Truss." *Fire Engineering*, July 1988, p. 41.

43. Dyson, J. "Truss Roof Construction Readily Known." *ISFSI Voice*, October 1990, p. 25.

Author's Notes

My reply to reference 17 was published in June, 1988 in *Fire Engineering*, "Are Wood Trusses Good For Your Health?" This article fully demonstrates that ASTM E 119 tests do not assure safety to fire fighters. This article was also published in the *ISFSI Voice*, March 1989, p. 18. "Amend the Code," a sidebar article by this author, was also published in June, 1988 in *Fire Engineering*. It expresses my opinion on why an attempt to eliminate the trusses by code will probably fail.

Fire Journal asked Gordon Routley, P.E., SFPE, then Assistant to the Chief, Phoenix, Arizona, to write an article on the subject of wood trusses. His excellent article, "How Wood Trusses Perform During a

Real Fire," appeared in *Fire Journal,* January/February 1989.

The Schaffer article will be quoted by truss proponents for years to come, particularly since it appeared in an NFPA publication (NFPA does not necessarily agree with articles published). All fire departments should have a copy of Routley's article readily at hand.

Several letters to the editor supporting the fire service position were published. Lieutenant C. Varone, Providence, R.I. Fire Department, *Fire Journal,* July-August 1988, p. 6.; Brannigan, F. *Fire Journal,* September/October 1988, p. 94. Letters objecting strongly to light-weight wood trusses from Clark, A., and Deputy Chief Dunn, FDNY, *Fire Journal,* March 1989.

There were a number of letters to the editors from manufacturers and sellers of wood trusses.

Truss Plate Institute, *Fire Engineering,* September 1988, p. 26. Lumbermate Company, *Fire Engineering,* November 1988, p. 20.

Brannigan, F. "A Truss Manufacturer is Angry." *Firehouse,* June 1989. This article is accompanied by a set of pictures showing the extremely dangerous practice of extending the trusses out of the structure to form the stairway platform.

It should be noted that Mr. Shaffer did not write any objections either to this author's or Routley's articles.

13

Automatic Sprinklers

When this writer started his career in fire protection, automatic sprinklers were familiar only to fire protection personnel. Most citizens, including those who reported on fires, were completely unfamiliar with them. Times have changed. One of the first questions many reporters now ask at serious fires is "Is there a sprinkler system in the building?"

Automatic sprinklers are fast becoming a fact of life in buildings. They are being installed in new buildings in compliance with codes to reap the substantial benefits granted by such protection, and in recognition of the potentially catastrophic financial consequences, particularly liability, presented by serious fires.

Slowly the need to retrofit sprinklers into buildings which should have been sprinklered when they were built also is being recognized.

Sprinklers were once almost exclusively installed in factory and mercantile buildings. For decades, the only sprinklered high-rise office building in the country was the headquarters of the National Board of Fire Underwriters.

In recent years, special sprinkler designs have been developed specifically for multiple dwellings and single-family homes.

*Note: This chapter should be studied in conjunction with Chapter 14, "Rack Storage." The essential element of fire safety in rack storage is adequate sprinkler protection. Much pertinent material is found in only one chapter.

Types of Sprinkler Systems

There are six major classifications of automatic sprinkler systems.[1,2] Each type of system includes piping for carrying water from a source of supply to the sprinklers in the area under protection. The six major classifications of systems are:

Wet Pipe Systems

These systems employ automatic sprinklers attached to a piping system containing water under pressure at all times. When a fire occurs, individual sprinklers are actuated by the heat, and water flows from the sprinklers immediately.

Dry Pipe Systems

These systems have automatic sprinklers attached to piping which contains air or nitrogen under pressure. When a sprinkler is opened by heat from a fire, the pressure is reduced to the point where water pressure on the supply side of the dry pipe valve can force open the valve. Then water flows into the system and out any opened sprinklers. Dry pipe systems are typically installed in areas subject to freezing.

Preaction Systems

These are systems in which there is air in the piping that may or may not be under pressure. When a fire occurs, a supplementary fire detecting device in the protected area is actuated. This opens a water control valve which permits water to flow into the piping system before a sprinkler is activated. When sprinklers are subsequently opened by the heat of the fire, water flows through the sprinklers immediately —the same as in a wet pipe system.

Deluge Systems

These systems have all sprinklers open at all times. When heat from a fire actuates the fire detecting device, the deluge valve opens and water flows to, and is discharged from, all sprinklers on the piping system, thus deluging the protected areas, usually flammable liquid operations.

Combined Dry Pipe and Preaction Systems

These include the essential features of both types of systems. The piping system contains air under pressure. A supplementary heat detecting device opens the water control valve and an air exhauster at the end of the unheated feed main. The system then fills with water and operates as a wet pipe system. If the supplementary heat detecting system should fail, the system will still operate as a conventional dry pipe system.

Special Types

Special Types of systems depart from requirements of the NFPA Sprinkler Standard in such areas as special water supplies and reduced pipe sizes. They are installed according to the instructions that accompany their listing by a testing laboratory.

Factory Mutual Engineering Corp. has prepared an excellent "Pocket Guide to Automatic Sprinklers." The National Fire Sprinkler Association also publishes extensive material available to all concerned with sprinkler protection.

Space will not permit the duplication of other information about sprinklers obtainable from other available sources. This chapter covers material not generally presented elsewhere.

At the end of the chapter recent sprinkler developments are presented. These include:

* The use of sprinklers to protect glass fire barriers, as in atriums (the interior vertical opening in many recent mid- and high-rise buildings)
* Special designs of sprinklers for flammable liquid storage
* Early Suppression Fast Response (ESFR) sprinkler, which promises improvement in the protection of rack storage.

Sprinkler Installation Incentives

The various model codes provide different incentives for the installation of sprinklers. Unfortunately, the codes do not provide any penalties for sprinklers that are out of service.[3]

Some who are opposed to these incentives or concessions argue that "they are giving away the store" to get sprinklers installed. Indeed, some fire protection professionals believe that a lot of fire protection eggs are being put into the one basket. The question is, "who, if anyone, is watching the basket." This question is discussed in detail later in this chapter. First, you will learn the code-related, site-development and tax or insurance incentives that make sprinkler installation attractive.

Code Incentives

Typical code incentives or concessions[4] generally are available in the following areas.

* Heights and areas
* Construction of corridors and tenant separations
* Interior wall, ceiling and floor finishes
* Travel distances to exits

Fig. 13-1. Montgomery County, Maryland fire officials closed this department store until the sprinklers were back in service. Because the weather was cold, Kensington fire fighters laid frost-protected hose lines into the fire department connection.

* Exit widths
* Standpipe requirements, hose station, and water flow
* Fire detection systems
* Draftstopping in attic spaces

Detailed information and provisions of specific codes are available also from the National Fire Sprinkler Association.

Site Development Incentives

* Fewer fire hydrants with greater spacing
* Reduced fire flow, small supply pipe
* Increased fire department response time
* Increased allowable distance from public access way
* Street width reduction
* Cul-de sacs allowances

Tax or Insurance Incentives

* Elimination of value of sprinkler system from assessed valuation
* Property tax rebates
* Elimination of water department fees
* Insurance premium reductions

Opposition to Sprinklers

Not everyone is enthusiastic about the spread of requirements for automatic sprinklers.[5,6,7] The concrete and gypsum industries and the manufacturers of spray-on fire proofing or other form of protective encasement of steel beams object to trading off passive fire protection for sprinklers, citing the importance of protecting the structure of the building, the imperfection of mechanical systems, and the possibility of explosion, arson, or earthquake. In their view, sprinklers are a secondary defense.

Another chief argument involves loss of compartmentation. As has been shown in the chapters on Garden Apartments and Concrete Construction, the compartmentation concept is seriously flawed in many cases.

Furthermore, there is still much opposition to any requirement for retroactive installation of sprinklers. While much of the opposition is financial, the specious argument that such requirements are unconstitutional has found some favor. This argument is without merit with respect to United States law.[8] Montreal, Canada has recently passed a very strong retroactive sprinkler law.[9] It was passed in the aftermath of a high-rise fire in 1986.

Many opponents claim that sprinklers are ugly. However, there is no evidence that anyone ever refused to enter a building because it was unattractively sprinklered. Much of the cost, particularly of a retroactive installation, is caused by hiding the sprinkler system. If the argument of overall sprinkler cost is an issue, the fire department should be prepared with the cost of a bare bones system, and point out that aesthetic costs are the option of the owner, not a fire protection requirement.

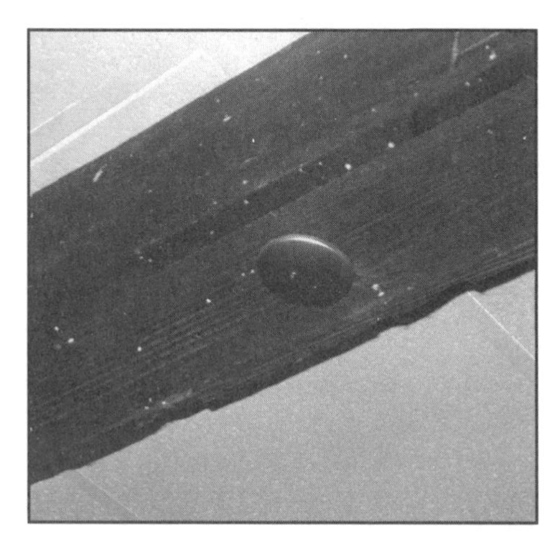

Fig. 13-2. The round metal disk conceals a sprinkler. Concealing the sprinkler system increases the cost somewhat but gives protection without the sprinkler showing.

Fig. 13-3. A large part of the cost of retrofitting the sprinklers into this Los Angeles hotel is required to hide the system. When retrofitted sprinklers are resisted because of cost, the fire department should distinguish between the cost of the basic needed protection, and the optional cost of concealing the system.

The probability of a serious fire in any given office building or other building with many occupants is extremely low. It is also a fact, however, that in the typical unsprinklered glass-enclosed office building with interior stairways and a substantial fire load, the consequences of a serious fire during working hours could be very severe — with multiple fatalities.

Those who do not wish to provide the sprinkler protection often argue from "good experiences." In effect they "need" fatalities to be convinced. They are unconvinced by rational arguments. Unfortunately this is the case often accepted by public officials.

The charge is sometimes made that there is little experience with sprinklers in high-rise buildings. After the terrible Triangle Shirtwaist fire in New York City in 1911, the law was changed to require that all factories over six stories in height be sprinklered. In the greatest migration of an industry in history, the entire garment industry moved into high-rise-sprinklered loft buildings (rented factory space) in mid-town Manhattan in the 1920s and 1930s. In these buildings, a large body of experience on successful sprinkler operations has been accumulated.[10]

After the spectacular First Interstate Bank fire in Los Angeles in 1988, passage of a sprinkler law for all new and existing high-rise buildings in California seemed assured. However, this is not the case. All buildings, built after 1974 and over 75 feet high throughout the state, must be sprinklered. In Los Angeles all pre 1974 high rises must be sprinklered. A law that would have required statewide retrofit failed – according to one source, due to the great number of state-owned buildings which would have required sprinklers.

Architects who know the codes very well can often circumvent sprinkler requirements. Because of the loss of fire fighter's lives in cellar fires, some cities have required sprinklers for basements in excess of a certain size, usually 2500 square feet. One of the solutions is to cut the basement into sections with firewalls, so that no section is larger than the legal limit. It is likely that the fire doors will not function, so the result can be a much more difficult and dangerous fire.

Another of the solutions is to keep the building barely below the height at which sprinklers are required.

Tactical Considerations

If a building owner has demonstrated a lack of interest in adequate protection and fire fighter safety, such situations may require a creative response. Fire fighters should not use phrases such as "surround and drown" and "big water show" for heavy caliber defensive operations. Substitue "improvised sprinkler system." Explain to reporters inquiring about sprinklers that it is not the sprinkler pipes but water that puts out the fire. The owner failed to supply the pipes so the fire fighters are doing their best under the circumstances.

Much opposition to sprinklers develops from a lack of understanding of the threat of fire and the operation of sprinklers. In Evanston, Illinois, the Byer Museum was destroyed by fire in 1985. Despite this fact, the curator of another Evanston museum was quoted as saying "It is a questionable practice to have sprinklers in a museum. A sprinkler system could do more harm to an art collection than a fire, particularly if the fire were contained or the sprinkler system malfunctioned."[11,12]

The National Gallery of Art in Washington, D.C. had a policy of NOT lending their art to sprinklered museums. The Smithsonian Institution was actually asked to return some works on permanent loan when National Gallery officials learned that the Smithsonian was retrofitting sprinklers. The Chief of the Fire Protection Division of the Smithsonian and a museum director met with National Gallery curators. As a result, the National Gallery now will lend art to sprinklered museums. There are a few museums, however, which will not.

In another case, a senior NIST engineer pointed out the hazards of the large accumulation of acrylic plastic in a museum workshop to a museum curator. The engineer's visit was abruptly terminated.

In 1990 in Chicago, a $75 million loss occurred in two buildings being renovated that housed nine art galleries. The loss of artwork amounted to $50 million. Fire doors had been removed from connecting openings. The building in which the fire started had been sprinklered but the system was out of service pending the installation of a new system.

The valuable exhibits in one Canadian museum are protected by sprinklers, but for some reason the lounge and gift shop are not. Fire in the contents of these areas could provide enough smoke to seriously damage the exhibit area. Heat might operate sprinklers that would not be hitting any fire. Such "economy" or "selective placement" is penny-wise and pound-foolish.

Popular Misconceptions

Fears arise out of some of the many misconceptions about sprinklers.

⋆ "The sprinkler system will discharge on even a trifling fire." This is not true. In fact, it takes a sizable fire to activate a conventional sprinkler. In some cases, the sprinklers should be supplemented with other fire protection measures. If the fuel for the triggering fire happens to be valuable art work, a serious loss may take place in the first few minutes.

For example, very valuable original prints of birds by John James Audubon are displayed in a sprinklered wooden house in Key West, Florida. The amount of fire needed to trip the sprinklers would have already caused damage to the collection. Valuable items should be kept in a non-combustible environment.

⋆ "The entire building will be drowned when the sprinklers go off." In ordinary systems, the sprinklers go off one at a time, not all at once. Only in deluge systems do all the sprinklers discharge at once.

⋆ "Water does more damage than fire." This is not true. There is no "water damage" at a total loss. Wet material from books to computers has been salvaged. Heat-damaged or burned material is destroyed. In 1960 the fleet aircraft carrier USS Constellation suffered a devastating fire. Many computers that had been soaked with salt water and subjected to freezing for months were recovered. Burned and heat-damaged computers were not. The major damage in many of today's fires is caused by greasy, often corrosive smoke from burning plastic.[13]

This writer has often recommended sprinklers for laboratories. Some laboratory scientists have a great fear of water damage and endorse CO_2 extinguishing systems as a panacea. This author once successfully overturned opposition to sprinklers by pointing out that sprinklers provided academic freedom — the opportunity to attempt something dangerous without causing a disaster.

A prominent fire protection engineering firm designed a sprinkler system for a nuclear laboratory with 1/4- and 1/8-inch sprinklers, (strainers had to be provided in the lines) to satisfy the fears of scientists about water damage. The water discharge was calculated to be just enough for the existing fire load. No controls existed to prevent the typical build up of fire load. This writer characterized the system as a "Tinkler" system, designed only to annoy the fire.

The various national authorities who govern airlines are studying the possibility of sprinklers for aircraft, since so many lives of persons who survived crashes have been lost in post crash fires. An English physician is opposed because he fears that people would breath in droplets containing dissolved toxic chemicals. He totally misses the point that the purpose of the sprinklers is to drastically reduce the production of toxic products of combustion. The same sort of objection has been heard from hospital and nursing home administrators.

★"The pulling of a manual fire alarm box will set off all the sprinklers." This is not the case.

At times the efficiency of sprinklers is deceiving. Consider the operation of one sprinkler which snuffs out a fire. The amount of burned material is relatively small. If the sprinkler flowed 15 minutes, a reasonable time, about 225 gallons of water would be dumped. The observer concerned about water damage has no idea of what might have happened had the fire continued to burn.

All fire department personnel should be alerted to the potentially serious damage caused by comments to the media made by uninformed occupants or careless remarks by fire fighters. The fire ground commander should make it a point to explain to reporters the success of the sprinkler operation.

★ "The pipes might leak." This is unlikely. Sprinkler piping is tested after installation to 200 psi. Domestic piping is not tested to this extent. Sprinkler leakage losses are minor, and are often due to careless use of lift trucks, or result from freezing. If sprinklers are undesirable due to fear of possible leakage, then logically, all plumbing should also be eliminated. A properly-installed system is supervised. An alarm will sound and the fire department will respond when water flows. No such alarm is provided on domestic piping.

★"Smoke is the big killer so smoke detectors are better than sprinklers." A smoke detector does nothing to suppress the fire which is generating the toxic smoke. The sprinklers suppress the fire and the

Fig. 13-4. Glen Echo, Maryland fire fighters move rapidly to clear out water discharged from sprinklers that saved the building. Quickly clearing out the water prevents unnecessary damage.

production of smoke. However, sprinklers are slow to operate on a smoldering fire which is generating toxic gas. Detectors are necessary for life safety, especially in sleeping occupancies, even though the sprinklers will control the fire as soon as it breaks out.[14,15]

There have been tests of smoldering fires in which temperatures generated by the fire were too low for sprinkler activation, yet lethal concentrations of carbon monoxide gas were produced. The concentrations developed over several hours, but potential victims in such situations may be disabled or disoriented before a fatal dose is received.

The NFPA has no record of a multiple-death fire (a fire which kills three or more people) in a completely sprinklered building where the system was properly operating. The only exception to this record exists in an explosion or flash fire, or where fire fighters or plant employees were killed during fire suppression operations.[16]

*"We have smoke detectors and the fire department is right down the block." The fire department may be at another incident. Patrick E. Phillips, whose credentials as SFPE Fellow and chairperson of more than one NFPA fire alarm committee are more than adequate, has

Fig. 13-5. In this modern hotel room the smoke detector will alert the occupants to a smoldering fire in the room. The sprinkler might not operate on a smoldering fire before toxic levels of carbon monoxide are generated.

repeatedly warned that fire protection and fire detection are not synonymous. Only if the fire alarm is received where there is someone ready, willing and able to control the fire could detection be considered to be protection. Many fires spread so fast that even immediate detection and response are ineffective. The myth that there are 15 minutes available to control a fire should be laid to rest.

A full box alarm assignment of the Boston Fire Department which had responded to a minor fire was only a short distance away from the Cocoanut Grove Night Club where 492 people died in 1942.

The previously cited NFPA films "Fire, Countdown to Disaster" and "Fire Power" both clearly and dramatically demonstrate the speed of fire development.

⋆ "Sprinklers cause damage to libraries." Some librarians are opposed to sprinklers. A typical argument states that, "the water will wash the classification labels off the backs of the books" or "books are hard to burn, we've never had a fire."

The major concern is not the local branch library but the main library with its extensive and generally irreplaceable collections.

The Los Angeles City Library fire is a case in point. This massive concrete structure was built in 1926. A concrete "fort" enclosed book stacks and racks for book storage. Upward ventilation was provided throughout the stacks to prevent mildew. It would be hard to imagine a better built-to-burn design, but such is typical of libraries. Not only did the design extend the fire, but the open construction permitted water to flow down to books below. The arsonist ignited the fire at the top of the stacks. Using as little water as possible, the manual fire fighting effort required 1,250,000 gallons of water. The cost of the fire loss was about $20 million. By contrast, the initial fire could have been controlled or extinguished by operating no more than two or three sprinklers.[17, 18, 19, 20]

The solution is full automatic sprinkler protection. Librarians should be informed that all the stack areas of the Library of Congress are now protected with cycling on-off sprinklers. Even so, because of narrow passages and stairways, and difficulty in venting, a fire in the library would be a tough fight.

Note also that older libraries often used marble for the flooring. If unsupported, as in the Library of Congress, such floors can look good after heat exposure, but fail under a fire fighter's weight.

Unfortunately, library stacks are perfect for spreading fire. They are simply racks storing combustible material without fire separation between levels. Ventilation is almost impossible—spaces are very constricted and passages are tight and tortuous. The heat produced in such a fire is extreme. At the Yeshiva University Library fire in New York City, 40 engine companies were used in rotation to operate two lines so as to reduce injuries.

Fig. 13-6. The cycling on/off sprinkler opens when the set temperature is exceeded, closes when the temperature drops, and opens again if the temperature rises. (Courtesy Factory Mutual)

A number of municipal libraries are now sprinklered. An architect planning a new library asked the then Asst. Director of the Salt Lake City Library, "What do you like most about your library?" Her reply shocked him, "It's sprinklered."* She went on to explain that libraries are public buildings. It is almost impossible to exclude anyone. With the sprinklers in place, a disturbed person with a grievance can't destroy the library. Libraries are very vulnerable to arson. As many as 85% of recent library fires were due to arson.

⋆ "Smoke detectors set off all the sprinklers." A television serial showed a cigarette setting off a smoke alarm which in turn set off all the sprinklers in a building. The network has reportedly defended such misrepresentations by claiming "creative license." If someone advances this as a fact, it might be useful to mention that it was probably seen on a fictional TV show.

Fire Service Activities to Correct Erroneous Opinions

In years past, the sprinkler industry did little or nothing to overcome negative public perceptions. Promotion of the use of sprinklers was left to the public fire protection forces and the insurance industry. More recently the sprinkler industry has supported Operation Life Safety (OLS), a consortium of the International Association of Fire Chiefs, the United States Fire Administration and the private sector. The purpose of OLS is to champion the increased use of sprinkler systems in all types of residential properties.

Sprinkler Demonstrations

Some simple demonstrations have been useful in correcting erroneous opinions. The U.S. Fire Administration sponsors demonstrations of the life saving efficiency of quick-response sprinklers. Conversely, some technicians in laboratory and computer occupancies are concerned that sprinklers will go off in a minor fire. To demonstrate the amount of fire it takes for a standard sprinkler to operate, use the following.

> Attach a sprinkler to a length of steel pipe fed by a garden hose. Build a sizeable bonfire of fast-burning fuel. Hold the sprinkler over the fire. Because there is no heat reflection, the sprinkler will be extremely slow to activate. Many who feared that the sprinkler would operate on a trifling fire will be amazed and converted.
>
> With a little more effort, a more elaborate demonstration is possible.

* Eileen Brannigan Longsworth, Director of Library Services, Salt Lake County, Utah.

Install piping with one or two sprinklers in an available shack or even a large packing box. Set up a fast-burning light fuel fire. The entire mass will be burning before the sprinkler activates. Point out the impossibility of controlling the fire with extinguishers.

Sprinklers and Flammable Liquids

Many believe that sprinklers will be totally ineffective on flammable liquids. At the Norfolk Naval Base the following demonstration was used to show that sprinklers would control the fire and make it possible for employees to get close enough for an extinguisher attack.

For You to Do

Cut a 55-gallon steel drum in half. Mount it on four steel legs. Insert a 1/4-turn valve in the bottom equipped with a long handle. Set the entire assembly in a steel pan about eight inches high. Put the pan under sprinklers in a demonstration shed. Place gasoline in the drum. Using the long handle open the valve so that gasoline is flowing. Ignite the gasoline in the drum and pan. The sprinklers will operate quite rapidly, suppressing but not extinguishing the fire. Carbon dioxide fire extinguishers can readily complete the job.

Fire Service Misconceptions

⋆ "The building is sprinklered, there is no problem." Sprinklers are not a universal remedy for all fire problems. In Chapter 9, Fire Growth, this author has noted ceilings of burlap and other high-flame spread materials. This hazard should not be dismissed because "it's sprinklered." It is very doubtful that the sprinklers would operate in time to control the flame spread.

It is worth repeating that if a life hazard is so severe that it cannot be controlled by sprinklers, then the hazard should not be permitted to exist.

A number of deficiencies of sprinkler systems that might contribute to failure are covered in this chapter. Others are discussed in Chapter 14, Rack Storage. Some members of the fire service seem to believe that it is better that such problems not be discussed.[21]

⋆ "Supplying the Fire Department Connection (FDC) is a secondary operation." The sprinkler system should be backed up as soon as possible after arrival. The more water a sprinkler discharges, the more heat that is absorbed. The supply pressure should be the maximum safe pressure because the increased density of the water pattern may be crucial. Some sprinkler systems have a higher working pressure than

Fig. 13-7. It is questionable if sprinklers would control a fast-moving fire in a rattan ceiling unless the rattan had been adequately flame-retarded.

others. Preplans should include a statement as to the pressure to be supplied and maintained. The 1991 fire in the Meridian Building in Philadelphia (See Chapter 11) was stopped by 10 sprinklers on the thirtieth floor. It was estimated that the discharge from these sprinklers was three to four times normal due to the pressure supplied by the fire department. It is very doubtful that the sprinklers could have stopped the fierce fire attack if they were operating at only normal pressure.

⋆ "Sprinklers should be shut down as soon as possible to prevent excessive water damage or to clear the air." Factory Mutual statistics show 23 fires in 10 years with losses of $43 million (1986 dollars) in which premature closing of valves by fire fighters was a major factor.[22] When a sprinkler system is controlling a fire, cooled gases are often driven downward and obscure vision. Sprinklers should not be shut off as long as hot water is falling down. All visible fire should be extinguished. The fire fighter who shuts a valve should be in full fire equipment with a radio, and remain at the valve so it can be opened instantly if fire breaks out again.

* "Residential Sprinkler Systems (NFPA 13R, *Standard for the Installation of Sprinkler Systems in Residential Occupancies up to Four Stories in Height*, and NFPA 13D, *Standard for the Installation of Sprinkler Systems in One- and Two-Family Dwellings and Mobile Homes*) are the same as other sprinkler systems." They are not. Residential sprinkler systems are intended to limit flashover and to hold a contents fire until the occupants can escape. There have been a significant number of successes with these systems.[23,24]

Residential structures, however, are only partially sprinklered. To keep costs down, thus making the systems cost acceptable, sprinklers are omitted from certain areas in residences. This is particularly significant when a residential sprinkler system is used in a multiple dwelling with truss floors, or in a building with interconnected voids such as in rehabilitated buildings with lowered ceilings.

Some object to the term "partial sprinkler system." It is argued that this term *should be applied* to systems where sprinklers are omitted from areas vulnerable to fire. NFPA 13R and 13D systems are based on data which indicate the omitted areas are *not* vulnerable to fire.[25]

It is possible that there was no experience in the data base relative to the combustible trussvoid which exists in truss floor buildings. It is a fact that fires originate in void spaces which would be unprotected by sprinklers. Electrical fires and failures of metal fireplace flues are but two possible causes of fires.

This writer served for several years as a consultant to the National Bureau of Standards to determine why fires in combustible multiple dwellings extended beyond the area of origin. A number of the fires investigated started in or very early extended to void spaces, which would be unsprinklered under Standards NFPA 13D or NFPA 13R.

The following few cases are representative of other fires in "sprinklered buildings" which started or developed in areas that were unsprinklered or would be unsprinklered — and thus unaffected by sprinklers — under the standards cited.

* Williamsburg, Virginia: A "sprinklered" renovated dormitory at the College of William and Mary was destroyed because of a void fire.[26]

* Syracuse, New York: Four fire fighters lost their lives in a fire in a fraternity house which had been equipped with a corridor sprinkler system for life safety.[27]

* De Kalb County, Georgia: Fire resulting from a plumber's propane torch developed in the truss void of a nearly completed three-story apartment house with no contents load. The sprinkler system was operational, but the sprinklers didn't fuse. A fire fighter fell through the second floor into the fire when the truss collapsed. Fortunately, his partner used a hose line to protect him from the fire until he could be pulled out.[28]

★ Orlando, Florida: Fire started in a defective ceiling light fixture in a first-floor apartment of an unsprinklered three-story, brick-veneered, plywood-sheathed, truss-floor apartment house. The fire extended to upper floors on the outside of the wood sheathing, concealed by the brick veneer. The truss floor of the third floor sagged over a foot.

★ San Antonio, Texas: A spilled flammable liquid ran down into the truss void of an unsprinklered apartment house and was accidentally ignited. Fire fighters had barely complied with orders to get out when the floors collapsed.[29]

★ Lake Benbrook, Texas: In a fire witnessed and photographed by this writer, an outside rubbish fire extended through the exterior wall into the truss void of an unsprinklered townhouse.

★ In plain language, the sprinkler systems in these and many other buildings are partial sprinkler systems. This is not to denigrate the usefulness of sprinklers in saving the lives of occupants. As stated, there have been many successful sprinkler operations. Balancing the potential structural failures with the positive life safety benefits, this writer is strongly in favor of the partial systems — in fact, this author's house was partially sprinklered in 1962.

However, fire departments should not expect to receive from residential sprinklers the level of structural protection performance they are accustomed to receiving from standard sprinklers.

Fire Department and Sprinklers

In 1898, the International Association of Fire Chiefs resolved to have nothing to do with the practice of pumping into the new fangled invention called the automatic sprinkler.[30] Today, fire chiefs are enthusiastic supporters of automatic sprinklers. However, some fire departments seem to act as if "sprinklers are none of their business."

Some insurance companies have seemed, in the past, to have had the opposite side of the same attitude. Until high rack warehouses presented a new problem, insurance companies, relying heavily on sprinkler protection, seemed to operate as if the fire department did not exist. For instance, Norman Thompson (then Director of Research for Factory Mutual) wrote a very important book, *Automatic Sprinklers*, published in 1951 by NFPA, which led to the adoption of the then new spray-type "standard sprinkler." The book makes no reference to the fire department supporting the sprinkler system. This writer can recall a list of 20 things to do when fire occurs in the plant — the last item was "call the fire department."

There is no doubt that a properly designed, adequately supplied sprinkler system can almost guarantee that a fire will not destroy a building. In order to "sell" sprinkler protection, concessions are made with respect to other fire protection features, such as exit travel distances, number of exits, flammability of surface finishes, size of fire areas, protection of void spaces, and other time-proven fire protection and life safety features.

It was noted earlier that almost all fire protection emphasis is being placed on sprinklers, with the responsibility for them left with fire departments. The building department, with no personnel or system to make ongoing inspections, is out of the picture after the Certificate of Occupancy is issued. The insurance company often compensates for its risks by raising the premiums or refusing to renew the risk (see the discussion of insurance in Chapter 14).

The Sawgrass Mall is a huge two million square-foot covered mall with nine anchor stores in Sunrise, Florida. The fire protection is almost completely active, consisting of full sprinkler protection with multiple water supplies, in a structure designed for automatic mechanical smoke removal.[31]

Because sprinkler protection is so important, the fire department's responsibility cannot be permitted to end when the sprinklers are installed and approved.

Fire fighters making fire suppression, preplanning or fire prevention inspections should be acutely aware of the many conditions which can cause sprinkler failure.

The fire department should take action when notified of a sprinkler shut off or other disabling functions. Keep up to date on systems being installed or temporarily out of service.

When a sprinkler system is pronounced "out of service" the degree of impairment should be given. For example:

- "System can be supplied through fire department connections."
- "Mechanics on scene can restore system in ten minutes."
- "System is completely unavailable."

In Cartaret, New Jersey, a fire in the S&A Plastics plant severely exposed a Unocal storage facility containing hundreds of drums of flammable solvents. This building in turn exposed a number of tanks. Fire Superintendent Richard Greenburg had reviewed the plans for a sprinkler system being installed in the drum storage building. He had the system supplied through the fire department connections (FDC) and, assisted by a Unocal employee, opened the valves. Fourteen sprinklers checked the spread of the fire. Five sprinklers had opened before the water was turned on, demonstrating that the potential for a disastrous extension was very real.[32]

Fig. 13-8. Sprinklers are provided **(A)** *to protect columns and* **(B)** *to protect beams in this Florida Mall. (Chief H. Howard, Sunrise, Florida)*

Why Were the Sprinklers Installed?

As you have seen, there are several reasons for installing sprinklers. The nature of the reason may help to determine fire department action concerning impairments.

NFPA 101®, the Life Safety Code®, permits a number of cost reductions if sprinklers are installed. For instance, there can be a greater travel distance to exits and fewer exits than required in an unsprinklered building. This provides substantial savings in the building of a shopping mall.

In the case of a sprinkler system installed in accordance with the Life Safety Code, the following proposition seems valid.

"If the exits were located as permitted in a sprinklered building, and the water was turned off, the building would be unsprinklered, the exits would be inadequate and the occupancy would be illegal."

In this scenario what action should the fire department take when it learns that the sprinkler system is out of service in a department store, a shopping mall or any structure with a population of thousands? The lifesafety of all the occupants depends almost totally on adequate sprinkler protection.

There have been several fires in unsprinklered department stores with a heavy loss of life. Over 100 Christmas shoppers and employees died in the Taiyo Department Store in Kumamoto, Japan in 1973. Sprinklers were being installed in the new addition of the store that was under construction, but the law did not apply to the existing structure.[33]

In this writer's opinion, the fire department action should be the same as if a large proportion of the exits were blocked. In such a case, the fire department would have no hesitation in shutting the store until the condition was corrected. This doesn't mean that many stores would have to be shut down. When the alternatives are properly explained to management before the crisis arises, the work is done when the store is closed.

A fire department which does not accept the scenario, or doesn't plan to take any action when sprinklers installed for life safety are shut down, would be well advised to send the following inquiry to the city attorney:

> Counselor:
>
> Consider a building in which Life Safety Code modifications, particularly reduction in exits, were permitted because the building was sprinklered. What is our legal position if:
>
> we were notified that the sprinklers were turned off while the building was occupied,
>
> no special action was taken in such a case,
>
> a loss of life occurred during a fire, or
>
> a key factor in loss of life was that the victims could not get out in time?

In addition to legal action against the city, is it reasonable to expect that senior officers could be sued successfully?

s/ Fire Chief

P.S. Is our situation any different if the department store owner is a good friend of the mayor?

Since the legal situation may vary depending upon why the system was installed, the fire department should identify the reason why each sprinkler system was installed.

If the sprinkler system was installed for life safety or in accordance with another code provision, there still may be a legal responsibility to take action if the sprinklers are inadequate or out of service.

Even if there were no legal requirement for the installation of sprinklers, the loss of the sprinklers may create such a serious situation that the fire department can act under general legal provisions authorizing the abatement of serious hazards.

All too often an old brick and timber building with sprinklers shut down or removed could produce a disastrous fire which destroys or threatens to destroy adjacent structures. For example, in 1983, a spectacular fire in Donaldson's department store did $90 million damage to the adjacent fire resistive high rise of the Northwestern State Bank in Minneapolis, Minnesota.[34] A similar fire occurred in Indianapolis in 1973.[35]

Fire Department Policy

The fire department should have a formal policy on the subject of sprinkler system disabilities. The policy should cover such items as:

* Fire department notification. It may be necessary to seek the cooperation of other licensing authorities to place the burden of notifying the fire department directly on the sprinkler contractor. This does not exclude required notification by owner or tenant.
* Formal legal action. Situations may arise where the fire department should take legal action outside of the department operations, such as shutting down a department store.
* Informal action. Situations may arise where informal actions are taken such as modifying pre-fire plans, or providing emergency water supply, or working with management to establish fire watches or shutting down hazardous processes.
* Authority to modify requirements.
* Formal personnel instruction.

Fire Department Instruction

A program of fire department education regarding impairment of sprinkler systems should include training on:

* A basic knowledge of sprinkler protection, including types of systems and the adequacy and reliability of the water supply.

* An understanding of the different reasons why sprinklers are installed.

* Fire department policy.

* Detailed knowledge of situations which decrease or destroy the efficiency of sprinklers.[36]

Company Level Inspections

Sprinkler impairment can be divided into water supply and water distribution problems.

The design and adequacy of the water supply system is beyond the scope of this chapter. It is assumed here that the designed supply is adequate.

At times, however, this assumption is not valid. This author was pleased to see that an old wooden hotel in Alaska had been sprinklered, down to a sprinklered shower stall. The unconnected sprinkler riser, however, brought the revelation that, "the contractor never finished the job." In addition, the siamese connection was so located that a supply line would make it impossible to open the exit door.

Furthermore, many sprinkler systems in low-rise buildings depend solely on the city water main pressure to provide adequate sprinkler flow. In many areas of the country, continuing drought has forced water supply utilities to reduce pressure to conserve water. The pressure at upper stories of low-rise buildings may be inadequate for sprinkler discharge. In this case the owner should be required to install a pump to provide adequate pressure.

The fundamental purpose of a sprinkler system is to hit the incipient fire with enough water to suppress it. There are too many ways for failure to occur to cover all of them here, but a few examples should get fire fighters looking more critically at sprinklered buildings, and make routine inspections more productive.

Making the following determinations are all within the capability of adequately trained fire suppression personnel making a company level inspection. Some can be determined visually. Others require answers from management or maintenance personnel. Any questions should be referred to the designated fire department authority for resolution. All deficiencies should be formally documented, but the system should provide for immediate verbal reporting of serious problems. Fire department policy should direct the action to be taken in various circumstances.

Tactical Considerations

The fire department should not manipulate any valves or any other equipment during an inspection or test. Such action would leave the department exposed to being accused of error or negligence if there is a future system failure.

Some fire departments reset dry valves or replace sprinklers and restore systems. This practice may open the department to liability. If this is done, the owner and the insurer should be notified immediately, by receipted document, that temporary repairs have been made to the system and the owner has the responsibility of having the work checked or redone by a qualified sprinkler contractor. Permanent files should be kept of such notification. It may be years before a problem arises.

Outside the Building: Check for several items.

* Are flow tests made to assure that the designed supply is actually available?

* Are static supply elements, such as overhead or surface tanks, or pond intake in good condition and protected from freezing where necessary?

* If the system depends on pumps, is there a check on maintenance and test procedures? Is the power supply vulnerable to a fire? (At one naval base the cables passed over a barracks which burned.)

* If the power company cuts power to the building, will the pumps fail? Are emergency power supplies adequate? Is the emergency system tested regularly. Where a pump is installed, a manifold for testing the fire pump is usually provided. Each 2½-inch outlet indicates 250 gpm of pump capacity.

Fig. 13-9. The sprinkler system in the Florida strip mall is totally dependent upon adequate power to the water pump. The fire department should be sure that it can be depended upon if a fire occurs.

* Are fire department siamese connections (FDC) available? In earlier years they were optional. In many industrial plants, fire department connections were omitted from individual buildings. As a result, when a fire main is placed out of service, the sprinklers served by that main are out of service. If there is no FDC, it is not possible to use hose lines to supply the system. Major plumbing efforts are required to keep the system supplied. All systems should have FDC connections whether or not required by code.

If there is a possibility of pumping contaminated water into the sprinkler system, the check valves which prevent backflow into the domestic system should be checked regularly.

* Are FDC properly protected from damage? All openings should be capped. When caps are missing, open pipe is a target for debris which will cut off the flow of water. The heavy brass caps popular years ago frequently were removed. Use cast iron or plastic caps.[37]

* Are all FDC labeled to indicate the type of service and area covered?

The piping from the FDC to the system is dry and thus unsupervised. Nothing is apparent if it is broken. In modern construction, connections are often quite distant from the building, under driveways and parking lots and subject to traffic pressure and corrosion. The system should be subject to hydraulic pressure tests from the FDC to the check valve, as is required by at least one city.

Outside valves are most often post indicator valves (PIV) which indicate open or closed. The valves should be locked as a minimum. The best system includes an electrical central tampering alarm.

Some PIVs are located horizontally in the wall of the building. Sometimes the valve is located in a curb box on the sidewalk or in the street. Often this box is difficult to locate. It may even be paved over. A piece

Fig. 13-10. Missing caps on a fire department connection invite acts of vandalism such as this.

Fig. 13-11. *It is obviously quite some distance from the FDC to the building. The dry unsupervised main is laid in filled ground. It could very well be broken. The time to find out is before a fire.*

Fig. 13-12. *The outside indicator valves are chained to discourage tampering, and alarmed to detect as little as a quarter turn of the valve. But who is attending the tampering alarm? And what action will take place if the alarm sounds?*

of apparatus may be parked on it. A sign should be posted on the building giving distances and directions to help locate the valve box.

Inside the Building: The designers of sprinkler systems are aware of the possibility of closed valves. Unlike other valves, sprinkler valves

Fig. 13-13. The OS&Y valve is open, but is is not locked and there is no tamper alarm.

are made to indicate their position. If the stem is protruding, the valve is open. This is called an OS&Y valve (outside, screw and yoke). In the best system, electrical alarms transmit a signal if the valve is turned. At the very least, valves should be chained and locked. The sprinkler valves are often hidden in locked closets. In some malls the sprinkler valve for one store is located in another store. Signs should indicate the location, and it should be entered on the preplan.

⋆ Is there a system for reporting any valves closed for maintenance? The system should assure that all closed valves are reopened. In a huge department store 29 sprinkler valves were closed and 28 were opened. A fire fighter fatality resulted when fire occurred in the area covered by the closed valve.

⋆ The details of the system for controlling shut valves and the special precautions taken when valves are shut are very much the fire department's business. This is especially the case if the sprinklers were installed by law, or if exemptions were granted with respect to other code requirements.

⋆ Is there a potential for freezing of all or portions of the sprinkler system? Are adequate precautions taken?[38]

⋆ Are old sprinkler systems or inadequately designed systems in use? Dry pipe systems which have "gone wet," either for fires or by accident, are often found to be plugged by scale carried by the inrushing water. A record should be kept of these occurrences and they should be reported to a designated authority for investigation.

Fig. 13-14. A closet such as this one might be difficult to find in a hurry. Its location and the area that it covers should be entered on the preplan.

Often industrial buildings and warehouses are built on speculation. The builder sometimes puts in a minimum sprinkler system to make the building attractive to tenants. The fire load introduced by the tenant may be far beyond the capacity of the system. While specific analysis is beyond the company inspection level, some fire loads, such as boxed records, foamed plastics, aerosols, or rubber tires, are easily recognizable as being beyond the capacity of the typical sprinkler system.

Many shopping malls are only partially sprinklered. Often a structure was built as an open mall and the stores were unsprinklered. Then when the mall was enclosed, sprinklers were provided in the open area but not in the individual stores. In other cases, major department stores which anchor the mall are sprinklered while other stores are not. In one case a fire in an unsprinklered central mall area broke windows in a department store and supermarket. Over two hundred sprinklers operated in those two stores, which had to be closed for two months. Damage was $7 million.[39]

Mall investors often want to keep down their costs and sometimes provide only the sprinkler main, leaving each store owner to install sprinklers. A very serious hazard exists when a store under construction is left unsprinklered. All vacant stores in a mall should be required

Fig. 13-15. Although several shops are still under construction, this mall is in full operation. Because these stores haven't been sprinklered, there is an open invitation to a life-threatening uncontrolled fire.

to be sprinklered because they are often rented temporarily to other occupants for storage.

The construction of a high-rise building represents a tremendous investment. The owner wants to start using or renting available space as soon as possible. This leads to a very hazardous situation of construction work, with all its hazards, being carried on in an occupied building. Even if the building is sprinklered, very often the most hazardous floors, those under construction, are not because sprinklers go in after the ceiling is in place.

The building department should be alerted to this hazard, and insist that the sprinklers be in service throughout the building before any partial occupancy is permitted. If the building department is complacent, the fire department should take every means possible to mitigate the hazard. Prospective tenants, who are relying on the sprinkler system, should be put on notice as to the hazard.

This author first noted this hazard in the headquarters of a major bank which had sprinklered its new headquarters. A lower floor, loaded with construction shanties and other ready fuel, was unsprinklered. The elevator doors were not yet installed because they might be damaged by the workers. (Serious high-rise fires in Boston and New York were described in Chapter 11 under partial occupancy of buildings under construction.)

Fig. 13-16. The Fairbanks, Alaska Fire Marshal required full sprinkler protection throughout this mall before it opened. This store is properly sprinklered.

There seems to be great interest in preserving historic buildings. Often these buildings are just minimally restored. Sometimes the building department requires sprinklers in these structures. Often the sprinklers just cover the accessible areas and do not cover the voids. This may be deliberate. The sprinkler system is needed for life safety and the sprinklering of the voids will make the cost prohibitive. If this is the case and the fire starts in, or penetrates the void, the building may be destroyed. You learned earlier that a "fully sprinklered" renovated dormitory at the College of William and Mary was destroyed because of a void fire.

In some historic buildings, sprinklers are installed in the non-public areas and omitted in the public areas in deference to "historic authenticity." This is called selective placement of sprinklers. While doing restoration work, an architect recently recommended sprinklers be installed in the attic and the basement of the Old State House in Boston but not on the intervening floors because the sprinklers were not "architecturally compatible." However, providing adequate fire protection without compromising historic values has often been accomplished.[40]

The historic 187-year old John C. Calhoun Mansion, called Fort Hill, is maintained on the campus of Clemson University in South Carolina. A museum of valuable 18th and 19th century art, furnishings and historic memorabilia, it has been sprinklered since 1966.

In May 1988 an arsonist set a fire inside Fort Hill, but the sprinklers extinguished it with moderate damage. The historic structure still stands. Two weeks before the fire, a visiting historic preservation conservator referred to the visible, non-authentic sprinkler system as an embarrassment. He recommended a fixed halon deluge system. The University wisely disregarded his recommendation.[41]

McGill University in Montreal housed two important research units in an old mansion called the Lady Meredith House. There were no sprinklers. A basement fire spread up through the unfirestopped walls and open staircases, necessitating five alarms. Many research papers were lost.[42]

Fire officers trying to convince the director of such an enterprise to spend scarce research money on fire protection might point out that computers often are a target of thieves, and fires are often set to conceal the theft.

Are all areas of the building "seen" by sprinklers? Unsprinklered voids often exist. An old office building in Memphis, Tennessee has been extensively modified over the years, including the installation of a light-hazard sprinkler system. The sprinklers were obstructed by air ducts, and the system was not properly maintained. A fire originated in a void space. It took scores of fire fighters three hours to literally "dig" the fire out of the walls, cocklofts and other concealed spaces.

Fig. 13-17. This structure within a mall was required to have an open roof so the sprinklers could "see" an incipient fire. An alternative would be to put sprinklers into the structure.

In another case, a combustible building was sprinklered except for a three-foot high cockloft. Fire originated in the cockloft and burned for 11 hours. The loss was $600,000.

The sprinkler protection of a dinner theater was impaired by the construction of a mezzanine without sprinklers underneath. A fire occurred underneath the mezzanine and gained headway until ceiling sprinklers operated and sounded the alarm. The loss was much greater than it would have been had proper sprinkler protection been maintained.

In some cases, unsprinklered voids are created by a building within the building. Offices on factory floors, special storage rooms and mezzanines, are typical of the structures which can provide a place for the fire to grow unchecked, burst out, and overpower the sprinkler system.

These areas should be sprinklered or their ceilings should be of lightweight plastic panels approved for such use, which melt out when exposed to fire. Plywood will not burn through in time.

In some cases, large ceiling expanses of department stores were fitted with panels to conceal the sprinklers. This might be dangerous. Melting and falling panels might be taken by occupants to mean the structure is failing and cause panic.

NFPA 13 permits omission of sprinklers where the ceiling is within six inches (152 mm) of the joists. Tests conducted at the National Research Council of Canada showed that fire can spread readily in these voids.[43]

A two million dollar loss occurred in a midwest shopping mall when fire extended through a combustible void above the ceiling of the sprinklered building. The roof was a conventional metal deck.[44]

Another major loss was suffered at Tinker Air Force Base in Oklahoma when a metal deck roof fire, started by roofers, could not be controlled by sprinklers installed only below the ceiling line. (See Chapter 7 for a discussion of Metal Deck Roof Fires.)

Are sprinklers uncompromised? Sprinklers should remain unpainted and untainted by other materials. Painted sprinklers should be replaced. This writer has seen at least two cases where spray fireproofing was applied to sprinklers. To counter the belief that sprinklers are ugly, concealed sprinklers have been developed. A flat dome, usually enameled to match the ceiling, conceals a sprinkler. It will drop out when heated. Concealed sprinkler covers should not be painted when the ceiling is painted. Paint might interfere with the cover dropping. One fire department was embarrassed when an area with concealed sprinklers was reported to be unsprinklered.

★ Some systems are equipped with cycling (on-off) sprinklers. These sprinklers will open when the temperature increases, and close again when the temperature drops. This reduces water damage and permits lower water requirements. These sprinklers must be installed on wet pipe systems. The Census Bureau suffered severe water damage when,

Fig. 13-18. Sprinklers loaded with paper dust, such as this one, failed to function in an Oregon paper warehouse fire. Good housekeeping is essential if sprinklers are to be exptected to operate. (L. Loar, Tualatin Valley Fire and Rescue)

on their own initiative, without competent advice, they attempted to further reduce the potential for accidental water damage by installing cycling sprinklers on a preaction system. When water was admitted it flowed from many sprinklers. Furthermore, a rumor was widespread among computer system managers that the sprinklers had been responsible for the damage.

The Atomic Energy Commission, a predecessor of the Department of Energy, was a pioneer in complete sprinkler protection for some of the most senstive computers in the country.

Exposure protection is sometimes provided by sprinklers outside the building. In cold climates, these are subject to freezing. Protecting them from freezing should not require shutting down the entire system.

In other cases, open outside sprinklers all discharge when a valve is opened manually. Such systems are very useful and can release fire fighting units for other tasks. The fire department should know the location of the valve and how to get to it when the building is closed. This information should be incorporated in the pre-fire plan.

Sprinkler Fraud

In 1984, the fire protection community was shocked and dismayed to learn of a major case of sprinkler fraud. As part of an investigation

Fig. 13-19. The exterior water spray system protects the fire-resistive high rise from a combustible exposure. The fire department should be aware of how to turn on such a system.

of an arson fire at a sprinkler installation company, a sprinkler contractor in California was discovered to have installed unconnected sprinklers which were simply glued to the ceiling. This author has been told there were a number of other similar situations. While many new trends start in California and work eastward, let us hope this does not become a trend with sprinkler fraud. In any case, those inspecting and checking sprinkler systems should look beyond the obvious.[45]

Management

Management is receiving lots of attention in the fire service today. Unfortunately, much fire service management effort is concerned with managing the fire department, rather than managing the fire problem. The integrity of sprinkler protection is a big fire management problem, and fire departments would do well to get started on managing it!

Special Situations

Flammable-liquid fires can be controlled with sprinklers but they present special problems.

* Most flammable liquids float on water. There is the potential for flowing flaming liquid, which can spread fire and injure fire fighters.

* Flammable liquids have a high Btu content per pound and a high rate of heat release.

* Flammable liquid containers can result in a BLEVE (Boiling Liquid Expanding Vapor Explosion).

* Aerosol cans can act as flaming rockets.

* Runoff water may create a significant contamination problem.

The best control method for flammable liquid fires is to keep the containers from overheating and control the flowing fire on the floor.

The containers can be kept from overheating if the sprinklers can wet every container exposed to heat. The use of conventional sprinklers on flammable liquid fires creates the problem of controlling run off to prevent environmental problems. It appears that foam water sprinklers would substantially reduce the problem. The National Fire Protection Association Research Foundation is currently funding a research project to measure the effeceiveness of foam water sprinklers.

* The floor fire also can be contained by the use of trench floor drains that are designed to smother the fire.[46]

Protection of Glass Fire Barriers

The use of glass, both on the exterior and interior of buildings for separations which may be designed to resist fire, or which may accidentally serve this purpose, is increasing. In some cases, lines of sprinklers have been used to protect the glass. It is important that all the glass be wet. In the case of hot fires the wetting must occur very early to avoid thermal shock to hot glass.[47,48] Window treatments, such as draperies, should not be installed between the glass and the sprinklers.

Early Suppression Fast Response Sprinklers (ESFR)

There are a number of failures of sprinklers to perform properly in high rack storage warehouses (see Chapter 14, Rack Storage for details).

This has led to new research to develop sprinklers which would not be designed just to control the fire but to suppress it.[49] This requires early discharge of a larger quantity of water — as much as two or three times that of standard sprinklers. In fact, ESFR sprinklers are designed to approach the old fire fighters slogan "put the wet stuff on the red stuff." They also parallel fire departments' general abandonment of booster lines for interior attack in favor of higher capacity hose lines, and where necessary, a blitz attack (see Chapter 3). Conventional sprinklers cool the ceiling and pre-wet surrounding storage. ESFR sprinklers dump most of the water directly on the fire.

The sprinkler orifice is 0.7 inch; conventional sprinklers are 0.5 inch. When systems are converted to ESFR, in most cases the existing piping is not adequate in size.

Tests indicate that ESFR sprinklers can protect rack storage of high-density plastics up to 25 ft (7.6M) in height without the use of in-rack sprinklers.

As of this writing, plans are underway for protection of other high-challenge situations such as rolled tissue paper, expanded plastics, and rubber tires.

The use of ESFR sprinklers permits greater flexibility in warehouse layout because the need for in-rack sprinklers is eliminated for many commodities up to a specified height.

Prewetting

Prewetting of exposed contents is an essential part of sprinkler operation. Many palletized materials are shrink-wrapped in an envelope of tight plastic. Sprinklers cannot prewet the material. In some cases this has resulted in a failed sprinkler attack.

Fig. 13-20. Because the stock is shrink-wrapped in plastic, the sprinklers cannot prewet it if a fire occurs. Also note that the stock is piled directly on the floor rather than being on skids. This would cause severe damage from water.

Case Studies

The following case studies cited in the literature contain detailed information regarding sprinklers and provide insight into many of the concepts presented in this chapter.

Birr, T. "On The Job, Eugene Oregon," *Firehouse*, September 1986. The fire occurred in a four-story wood frame commercial building. The basement and corridors were sprinklered. Lazy smoke was reported on arrival. Fire spread through non-firestopped voids. After three hours, a defensive mode was ordered. The fire damage was confined to the lower part. Upper floor offices received only smoke damage. The city will require firestopping if the building is rebuilt.

Though not mentioned in the article, the building was almost undoubtedly of balloon frame construction. Firestopping was practically unknown when this building was constructed. The cost of retrofitting firestopping can be very high.

Biswanger, W. "$8 Million Blaze Guts Fitchburg Warehouse." *Firehouse*, January 1987. An old mill was being converted into apartments. A welder was cutting away pipes. A multi-alarm spectacular fire resulted. The article notes that nothing is known about the sprinkler system in the building. A large empty combustible building is as bad at least as a leaking gas truck.

Biswanger, William. "Nine Alarm Fire Ravages Lowell (Mass.) Mill Complex." *Firehouse*, September 1987. Huge daytime fire in an old mill. The sprinklers in the original fire building had been cut off after a problem about two months earlier. Apparently this was unknown to the fire department. The $30 million loss was a substantial blow to the city's tax income, over 400 persons lost jobs.

The only defense against a raging fire in an old mill is a sprinkler system in full service. This fire proves again, if proof is needed, that reliance on manual fire fighting is inadequate. The fire department must know the status of every such critical sprinkler system, and maximum pressure must be exerted to get the system back in service. In the interim, extraordinary measures should be taken to provide some special protection.

Budnick, E. and Fleming, R. *Fire Journal*, September/October and November/December 1989. Report on the National Fire Protection Research Foundation's Quick Response Sprinkler (QRS) Research Project. Update on the efforts to develop an engineering procedure for the use of QRS technology in NFPA 13 Occupancies, and an engineering basis for QRS system design for NFPA 13 applications.

Byrne, R. "Miracle at Philadelphia." *Fire Journal*, November/December 1987. The article describes a state-of-the-art fire alarm system at the Bank of the United States building (circa 1820) which houses a number of historic documents. There is no recognition in the article that fire detection is not fire protection. A sidebar article takes up the point of sprinkler damage vs. hose streams.

Carter, H. "Cultural Fire Protection, Correcting Some Misconceptions." *Fire Journal*, January 1987. Why institutions such as libraries and museums often resist sprinkler protection.

Eisner, H. "Raging Hamtramck Fire Prohibits Defensive Attack." *Firehouse*, November 1984. The sprinkler system in this heavy timber building was overwhelmed by a flammable liquids fire. Poor water supply left a huge mutual aid force almost helpless.

Eisner, H. "10 Alarm Warehouse Inferno Wipes out Firehouse." *Firehouse*, April 1985. A spectacular fire in a brick and heavy timber building, in which the sprinkler system was out of service, threatened a conflagration on New York's West Side. Engine 54 was ordered to evacuate immediately but they took time to pull their hose line out. Company members were hospitalized. When the order is given to evacuate, the response should be immediate. It would be well for department procedures to specifically order that equipment be left, when it delays getting the fire fighters to a place of safety.

Eisner, H. "Three Alarmer Strikes Jersey City Warehouse." *Firehouse*, November 1987. A typical industrial area heavy timber warehouse fire at which the Btu's greatly outnumber the water. The fire department was not aware that the sprinkler system was out of service and that there were no stairways in the building.

Keith, G., and Garner, D. "Automatic Sprinkler Systems: What the Fire Department Needs to Know." *Fire Engineering*, April 1991, p. 71. FM also has prepared an excellent "train the trainer" program entitled "Fighting Fires in Sprinklered Buildings."

Klein, M. and Fleming, R. *Fire Journal*, March/April 1989. Opposing views on whether NFPA 13R sprinkler systems are partial sprinkler systems.

Lougheed, G. and Richardson, J. "Sprinklers in Combustible Concealed Spaces. *Journal of Fire Protection Engineering*, Volume 1, Number 2, 1989. NFPA 13 permits omission of sprinklers for voids below combustible wood joists where the ceiling membrane is within six inches of the joists. Tests indicated that such voids represent a major path for fire extension.

Morehart, Jonas. "Sprinklers in the NIH Atrium." *Fire Journal*, January/February 1989. A display in an atrium ignited and the atrium sprinklers were slow to operate.

Neilson, R. "Standing Firm on Sprinkler Retrofit," *Fire Engineering*, June 1987. Oakbrook, Illinois passed and has enforced retrofitting of sprinkler protection in commercial occupancies. Notable is the fact that the city retrofitted its own buildings.

Porter, William and Schofield, Barbara. "Making a Landmark Wooden Victorian College Building Firesafe." *Fire Journal*, July 1985. The historic 90 year-old wooden structure is the pride of Nyack College. A costly effort is required to make a balloon-frame building fire safe. The installation of automatic sprinklers was simply not enough. The building was gutted to the frame and rebuilt. The final cost was about two thirds the estimated cost of a new structure. Many rehabilitation efforts are principally cosmetic. The figures given above give some indication of the cost of making "a silk purse out of a sow's ear."

"Residential Sprinklers Put to the Test." *Fire Command*, March 1984. A fire occurred in a Cobb County, Georgia apartment building equipped with a residential sprinkler system. The fire was extinguished.

Sachs, G. "Wolf Trap Reprise." *Fire Engineering*, September 1987. Wolf Trap Park, a large wooden open air theatre, had suffered several major fires. Finally, the U.S. Park Service was convinced that built-in fire protection was in order. This fire damage was limited to $20,000 by excellent manual fire fighting by the Fairfax County Virginia Fire Department.

The article does not comment on the fact that apparently neither sprinklers nor fire alarm were provided in the void space where the fire occurred. The fire alarm sounded when heavy smoke broke down through the suspended ceiling. It is unclear what the effect of this sudden appearance of heavy smoke might have had on an audience. There is no mention of whether the alarm or sprinkler systems will be extended to the unprotected areas.

Snadekie, James and Loveman, Laurie. "Stable Fires, Sad but Preventable". *Fire Engineering*, April 1986.

Sprinklers had been shut down in the Belmont Park horse barn, but no special watch precautions were taken. In the fire, 45 horses died. Despite water problems the fire department saved part of the wooden horse barn. Practical guidance for fire safety in the construction and operation of stables is provided.

Wentworth, P., and O'Reilly, J. "Backdraft and Blaze Gut Landmark (Providence R.I.) Building." *Firehouse*, March 1987.

Fire occurs in a huge vacant combustible building in a congested area. Because the building was unheated, the sprinklers were shut off. First alarm at 6:31; all fire fighters evacuated at 7:18 due to intense heat and heavy smoke; at 7:43 a tremendous backdraft occurred on the fifth floor. There were multiple layers of ceiling. There was no fire department siamese connection on the sprinkler system. Fire fighters succeeded in opening a valve in the basement. Note the backdraft over an hour after the first alarm which refutes the concept that a backdraft will occur only early in the fire.

Yao, Cheng. "The ESFR Sprinkler System, a New Approach." *Fire Journal*, March/April 1985. The development of new Early Suppression Fast Response sprinklers are designed for high-challenge fires which cannot ordinarily be controlled by ceiling sprinklers.

References

The addresses of all referenced publications are contained in the Appendix.

1. Cote, A., *ed. Fire Protection Handbook*. 17th ed, National Fire Protection Association, 1991.

2. Bryan, J. *Automatic Sprinkler and Standpipe Systems*. 2nd ed, National Fire Protection Association, 1990.

3. Isman, K. "Sprinklers and the Model Codes." *Society of Fire Protection Engineer's Bulletin*, May/June 1989.

4. Paige, P. "Sprinkler Incentives Make Believers out of Owners and Builders." *American Fire Journal*, June, 1990, p. 43.

5. Coon, J. Walter. "Trading Off Life Safety." (Position Paper and Information Sheet on NFPA 13D) *Gypsumation*, July 1985, Gypsum Association. (Also in *Building Standards*, January/February 1986, International Conference of Building Officials.)

6. Progressive Architecture. "No High Ground." October 1980, p. 88.

7. Sperck, J. "Don't Allow Sprinkler Tradeoffs." *Engineering News Record*. Letter to Editor, April 1984, p. 9.

8. Brannigan, V., J.D. "Applying New Laws To Existing Buildings: Retrospective Fire Safety Codes." *Journal of Urban Law*, Vol. 60, Issue 3, Spring 1983, p. 447.

9. Stronach, I. "City Brings In Retroactive High-Rise Sprinkler Law." *Fire International*, February/March 1990, p. 23 (Also in *Operation Life Safety Newsletter*, July 1990).

10. Powers, R. "Sprinkler Experience in High-Rise Buildings." Technology Report 79-1, Society for Fire Protection Engineers, 1979.

11. Enstad, R. "Blaze at Evanston Museum Not Arson, Fire Chief Says." *Chicago Tribune*, January 3, 1985.

12. Kirsch, A. "Preserving Today's Treasure for Tomorrow." *Factory Mutual Record*, January/February 1990, p. 3.

13. Courtney, N. "Sprinkler Extinguishes Computer Fire." *Illinois Fire Journal*, March/April 1989, p. 19.

14. Bill, R. Jr., and Hsiang, Cheng Kung. "Evaluation of An Extended Coverage Sidewall Sprinkler and Smoke Detectors In a Hotel Occupancy." *Journal of Fire Protection Engineering*, Vol. 1, No. 3, July/September 1989.

15. Bill, R. Jr., "A Life Safety Team: Smoke Detectors and Sprinklers in Hotels." *Fire Journal*, May/June 1990, p. 28.

16. Letters, A Reply from John R. Hall, Jr., Director, NFPA Fire Analysis & Research Division. *Fire Journal*, January/February 1989, pps. 6 and 8.

17. Isner, M. "Fire in Los Angeles Central Library." *Fire Journal*, March/April 1987, p. 56.

18. Simmons, A. "On The Job, Los Angeles." *Firehouse*, August 1986, p. 33.

19. Morris, J. "The Los Angeles Central Library Fire." *Library Association Record*, September 1986, p. 441.

20. Morris, J. *The Library Disaster Preparedness Handbook*. American Library Association, Chicago, 1986.

21. *American Fire Journal*, "Remarks and Sparks." March 1987, May 1987.

22. *Factory Mutual Record*. "Bergenfield Burn." January/February 1988.

23. Teague, P. "Residential Sprinklers." *Fire Journal*, September/October 1988, p. 46.

24. *Fire Journal*. "Reported Residential Sprinkler Successes." September/October 1988, p. 59.

25. Fleming, Russ. "A Closer Look at the NFPA Residential Sprinkler Standards." *Fire Journal*, March/April 1988, p. 51. (Also "Letters." *Fire Journal*, March/April 1989.

26. Bell, J. "184 Evacuated in Dormitory Fire." *Fire Journal*, January 1984, p. 50.

27. Demers, D. "Four Firefighters Die in Syracuse Fire." *Fire Command*, October 1978, p. 20.

28. Turner, P. "De Kalb County, GA, Fire Department Forum," *Firehouse*, December 1989, p. 8.

29. Corbett, G. "Lightweight Wood Floor Construction, A fire Lesson." *Fire Engineering*, July 1988, p. 41.

30. O'Brien, D. "A Centennial History of the International Association of Fire Chiefs." IAFC, Washington, D.C., 1973.

31. Howard, H. "Megamalls, A Fire Protection and Suppression Hazard." *Fire Engineering*, March 1990.

32. Solomon, R. and Lemoff, T. "It Pays To Know The Territory." *Fire Command*, March 1990, p. 23.

33. Lathrop, J. "Taiyo Department Store Fire." *Fire Journal*, May 1974, p. 42.

34. Best, R. *Fire Journal*, "Demolition Exposure Fire Causes Over $90 Million Loss to Minneapolis Bank," July 1983, p. 60.

35. Sharry, J. "Group Fire in Indianapolis, Indiana." *Fire Journal*, Vol. 71, No. 4, July 1974, p. 38.

36. Moore, J. "Why Sprinkler Systems Fail." *Industrial Fire World*, December 1988.

37. Corbett, G. "Fire Department Connections." *Fire Engineering*, February 1988, p. 40.

38. Courtney, N. "Frozen Pipes Defeat Fire Protection System, Michigan." *Fire Journal*, January/February 1989, p. 15.

39. *Factory Mutual Record*, "Spotlight on Business." November/December 1987, p. 17.

40. "Fire Safety Retrofitting In Historic Buildings." Advisory Council on Historic Preservation and the U.S. General Services Administration, August 1989.

41. Abraham, J. "Brothers Charged in Arson Fires." *Fire Command*, May 1989, p. 25.

42. Ulley, L. "Montreal Blaze Claims Research Documents." *Firefighting in Canada*, April 1990, p. 27.

43. Lougheed, G.D. and Richardson, J.K. "Sprinklers in Combustible Concealed Spaces." *Jounral of Fire Protection Engineering*, Vol. 2, April-June 1989, p. 49.

44. Herzog, C. and Broehm, K. "Million Dollar Loss Occurs In Shipping Mall Fire." *Fire Command*, August 1988, p. 18.

45. Hart, S. "Sprinkler Fraud in California." *Fire Journal*, May/June 89, p. 37.

46. Davenport, J. "Protection Guidelines Revised For Storage of Flammable and Combustible Liquids." *The Sentinel (Industrial Risk Insurers)*, Vol. XLV, No. 3, (Third Quarter) 1988, p. 3.

47. Kim, A. "Sprinkler Protection For Window Assemblies." *SFPE Bulletin*, November/December 1989, p. 13.

48. Kim, A. and Lougheed, G. "The Protection of Glazing Systems With Dedicated Sprinklers." *Journal of Fire Protection Engineering*, Vol. 2, No. 2, April/May/June 1990, p. 49.

49. *Factory Mutual Record*, "Sprinkler Technology." May/June 1989, p. 9.

14 Rack Storage

Once there was a simpler time when small fire departments had small fires, and big fire departments had big fires. Those days are gone forever! The advent of rack storage has brought potential major fire problems to Anyplace, USA.

Warehouse Fire Problems

Many serious fires have occurred in the long history of warehouse occupancies. The principal contributing factors have remained the same.

- Huge concentrations of fuel
- Tremendous dollar values
- Few employees per unit of area
- Failure to segregate extra hazardous materials such as flammable liquids
- Failure to raise bottom layer of stock above floor, thereby preventing water damage and possible collapse
- Vulnerability to arson
- Failure of management to give serious attention to the potential fire problem.
- Inadequate fire protection, either in initial design or in maintenance.

The potential major fire problem of large storage is not limited to huge warehouse complexes. Storage areas of manufacturing facilities and retail establishments, shipping and receiving areas, general, rental storage warehouses, and outside storage adjacent to buildings present similar problems.

Pallets

The problem of fire in warehouses was already significant when material was being stacked only by hand. About 40 years ago the lift truck became available. With this device, stock could be stacked on pallets, and pallets could be stacked one atop the other. This development provided a massive new dimension to the fire protection problem of warehouses. The pallet storage system provided as much as 36 times the surface area as boxes stacked solid. Initially, this problem was not universally recognized. Automatic sprinkler systems had such an excellent record that it was taken for granted that a sprinklered warehouse was safe.

Sprinklers

In a meeting with fire chiefs held at NFPA after the 1987 Sherwin Williams warehouse fire in Dayton, Ohio, Paul Fitzgerald, Chief Operating Officer of Factory Mutual, is quoted, "The chiefs were concerned that sprinklers were overtaxed so quickly. We were surprised that they lacked access to data that would have predicted that sprinklers couldn't have handled such a fire. In short, we both learned to appreciate the other's position and ways in which we could work better."[1]

After World War II the U.S. Navy had $70 million in radar equipment stored in a wooden warehouse at Norfolk Naval Air Station (NAS). The stock was palletized, and stacked 20 feet high. An ordinary hazard sprinkler system was supplied by NAS high-pressure mains. This writer ordered a preplan developed for a massive heavy caliber defensive attack to attempt to limit a loss. Navy fire protection engineers, steeped in 99% sprinkler efficiency tradition, derided the plan. One important feature of the plan was to ensure that pumpers were not hooked to high-pressure hydrants, but to domestic hydrants or overboard suction from the bay. This would prevent a post fire conclusion that the disaster was due to the fire department depriving the sprinkler system of water. Serious loss experience with such occupancies led to the concept of locating sprinklers within the stack.

A Serious Risk

This belief of sprinklered warehouse "invincibility" is still apparent in some cases. This author recently had the opportunity to look at a Southern California warehouse that stores records.

The records, stored in cardboard boxes, are stacked 15 or more feet high. The masonry-wall building has a plank-on-wood bowstring truss roof. The building department required steel tie rods to be installed along the bottom chord of the trusses to handle the tensile load. The work was done either because of weaknesses in the trusses or to increase earthquake resistance. The tie rods, heated in a fire to 1000°F, will elongate and slack off. The trusses will then revert to their previous weak condition at the same time they are being attacked by the fire.

The building is sprinklered. However, there are no sprinklers in the stacks. There are several bays to the warehouse. Fire doors in the masonry walls do not appear to be in operating condition. The boxes have corrugated lids which fit over their tops. The stacked boxes with the protruding lids as well as the narrow aisles are ideal for very rapid flame spread.

A potential scenario includes the following:

A fire starts in the rear of the warehouse. Ignition is caused by a lift truck, somebody smoking, or arson. The sprinklers operate, but without adequate effect.

Arriving fire fighters stretch a hose line to the rear of the warehouse, through the narrow aisle. Suddenly the fire flashes down the aisle on the surface of the boxes. The fire fighters are in very serious trouble.

Fig. 14-1. There are no sprinklers in the heavily loaded stacks of this Southern California warehouse. The only sprinklers are high in the overhead. Fire fighters stretching a line through this narrow aisle could be in serious danger.

Shelving

While pallets are still used for storing some commodities, many other items require shelving. Lift trucks also make it possible to place stock on shelves. These layers, in effect, miniature floors, created a whole new fire problem.

An estimated $14 million loss occurred in a rack storage warehouse in Kernersville, North Carolina in March 1981. Sprinklers were not provided in the racks in a section almost 16 feet high. Though the fire had been burning only 10 minutes when the fire department arrived, an interior attack was impossible. About 15 minutes later, the roof started to collapse. There was no fire department connection to the sprinkler system.

In the past the fire department connection (FDC) was optional. NFPA 13, "Standard for the Installation of Sprinkler Systems," now requires the connection except for systems of 20 sprinklers or less, or when permission has been obtained from the "authority having jurisdiction" (AHJ). Since the sprinkler system is a vital element of the fire department's suppression, the "AHJ" should be the fire department and no such exception should be granted.

Modern Rack Storage Warehouses

A new type of building, the huge one-story warehouse in which the contents are stored on unprotected steel racks, has made its appearance across the country. Typically, such buildings are of noncombustible construction. The size can be almost unlimited. The K-Mart warehouse which was destroyed by fire was as large as 33 football fields. Of the destroyed warehouses listed below, the smallest was the Sherwin Williams facility, which was as large as five football fields.

In these warehouses, merchandise is handled by mechanical equipment. In the most advanced warehouses the operation is fully or partially automated. Computers keep stock records, and a signal from a computerized system operator can send the traveling elevator to the correct location. The elevator platform then ascends to the proper level, and the desired item, for example, a refrigerator, pallets of radios in polystyrene packaging, or automobile parts, is loaded onto the elevator. In some cases, the stock moves automatically all the way to the truck which will carry it to its destination.

A rack storage warehouse, depending on its height, is like a multistoried building without the fire resistance provided by even the poorest floor. Ordinary sprinklers located at the roof line are completely inadequate to control such a fire hazard. Sprinklers must be provided at intermediate levels as well. Within certain limits, Early Suppression Fast Response (ESFR) sprinklers, (See Chapter 13, Sprinklers) may be able to suppress a fire without in-rack sprinklers.

Fig. 14-2. The railroad car is about to be loaded with the discarded cardboard boxes for recycling. The care is in a multimillion dollar warehouse, vital to the operation of a grocery chain. Is this warehouse the place to store such low value, easily ignited, difficult to suppress fuel? A fire in a loaded railroad car would be beyond the reach of the sprinklers. (Fire Department, Halfway, Maryland)

Some warehouses are provided with interior rail tracks so cars can be unloaded within the warehouse. A railroad car represents a major unsprinklered area within a warehouse. A fire in a loaded railroad car within the building would be a major problem.

Tactical Considerations

Preplan to use deluge guns immediately on such a fire. Anticipate the opening of sprinklers over a wide area. Discuss ahead of time the advisability and possibility of moving the car out of the building. Is a tractor or other unit which can move a car available?

Dry Storage of Boats

A special type of rack storage warehouse stacks boats several levels high in open or partially enclosed rack structures. These warehouses are usually unsprinklered. The boats are of combustible fiberglass, and many contain fuel. A hot, fast-growing fire with early structural collapse can be anticipated. Plans should include immediate use of heavy-caliber streams in a defensive attack for a fire in such a facility.

Library Stacks

Libraries are the original stack storage buildings; not the familiar branch library but the large main libraries which have multi-level stack areas where the books are stored. Patrons do not usually go to the stacks. Stack attendants retrieve the requested books. Most stack areas are built to a common pattern. Stack levels are about six feet high so attendants can reach the shelves without a ladder. The floors are open to air circulation to prevent mildew. A better system to guarantee the spread of fire and destruction of the books in case of fire can hardly be imagined.

The stacks are often structurally part of the building, so that fire damage to unprotected steel may also have structural consequences.

Libraries are often targets of arsonists. The trend is to open stacks to the public, thus reducing the security which existed when only employees had access.

Still some librarians cannot appreciate the risk. In 1984 Baltimore's Enoch Pratt Free Library embarked on a renovation program. The management refused to include sprinklers.

A major HPR insurer conducted actual fire tests to show how fire could involve stacks of books.

As was discussed in Chapter 13, many librarians have been reluctant to install sprinklers, fearing water damage.

Fig. 14-3. A test fire demonstrates the vulnerability of library books in stacks. (Industrial Risk Insurers)

Some Major Losses

In 1985 there were 159 warehouse fires, each with a loss of at least $1 million.

Some of the largest warehouse fires in recent years include:

- Supermarket General — Edison, N.J., 1979, $50,000,000
- Montgomery Ward — Bensonville, Ill., 1978, $30,000,000
- K-Mart Distribution Center — Falls Township, Penn., 1981, $100,000,000+
- MTM Partners — Elizabeth, N.J., 1985, $150,000,000
- Service Merchandise — Garland, Tex., 1987, $30,000,000
- Sherwin Williams — Dayton, Oh., 1987, $49,000,000

In current dollars these losses approximate $600,000,000. All these buildings were sprinklered. In all cases the sprinklers were overcome by the fire.[2]

The Problem

There are several contributing elements in the potential for such huge loss fires:

- Modern contents of warehouses are increasingly higher-hazard materials such as plastics, flammable liquids and gases, and chemicals. Even non-combustible materials are packaged in combustible containers and packing.
- Huge quantities of material in one fire area provide fire loads which can easily overcome the best defenses.
- Higher and deeper aisles limit access, and encroach on the space over the tops of piles required for effective sprinkler or hose stream operation.
- The size and construction of the buildings often make even the heaviest caliber fire streams ineffective.
- Automatic sprinkler systems as installed are often inadequate for the potential fire problem.
- Automatic sprinkler systems that are adequate for the job as installed, can be defeated by changes in the operation and storage patterns of the warehouse.

Fire Defenses

The elements of fire defenses for these structures and occupancies are multifaceted. They include:

- Application of Fire Protection Standards
- The Building
- Static Defenses-Structural
- Dynamic Defenses-fire alarm and suppression
- Management Attitude
- Public Officials Attitude
- Insurance Companies
- Employees
- Fire Department Operations

Standards

There are five NFPA Standards applicable to these occupancies.

NFPA 231 Standard for General Storage

NFPA 231C Standard for Rack Storage of Materials

NFPA 231D Standard for Storage of Rubber Tires

NFPA 231E Recommended Practices for the Storage of Baled Cotton

NFPA 231F Standard for the Storage of Roll Paper

For a complete discussion of these standards see the NFPA *Fire Protection Handbook*, Seventeenth Edition, Section 8, Chapter 5.

Other standards or recommended practices are available from Factory Mutual (*Factory Mutual Record*), and Industrial Risk Insurers (*Sentinel*), two major industrial insurance corporations.

The Building

Typically, these types of buildings are code classified non-combustible. The term is a code classification, not necessarily an accurate description. If the building is concrete, remember that concrete is inherently non-combustible, that is, it will not burn. It is not inherently fire resistive, however. Unless it is fire resistive it may disintegrate and collapse in a fire. It is important to remember that if the code says "non-combustible" as is most often the case, the building is not fire resistive.

Tilt-slab concrete buildings with precast wall panels are very common structures. They are very hazardous. The roof is necessary to stabilize the walls. Concrete T-beam roofs are subject to spalling or disintegration of the concrete below the tendons. Tendons exposed to

800°F (less than that produced by a self-cleaning oven) lose their prestress capacity.

Wooden beams can burn away or fail at the steel splices or connections. Steel beams first elongate, and can push the walls down. At higher temperatures they fail and can pull the walls in. I recently witnessed a fire in a warehouse under construction. The only fuel was foamed plastic insulation. Elongating steel bar joists pushed the walls out of alignment. The contractor immediately replaced the tormentors which braced the walls when the building was under construction. The roof was no longer trustworthy.

The wall panels in these structures are often sealed with a sealant. In a fire the sealant burns out. Fire can then extend through the joint into or from any exposure area such as a shed erected against the wall.

If the roof is a conventional metal deck built-up roof of layers of asphalt and roofing paper, it can burn independently of the original fire, elongating the bar joists to push down the walls, possibly quite some distance from the original fire. At a later stage the bar joists will collapse. (See the discussion of metal deck roofs in Chapter 7, Steel Construction.)

The storage racks are structures in their own right. They are subjected to the same force of gravity as any building. Their gravity resistance system consists of very light-weight steel framing. It is extremely vulnerable to fire damage and failure.

Fig. 14-4. The stacks are light-weight structures. If the sprinklers do not control the fire, collapse is probable.

This hazard exists in many lumber yards or similar occupancies. Metal shelves supported on columns of light-weight steel, with many punched-out holes for flexibility and reduced weight are used for storage of stock. These are structures within the building and extremely vulnerable to collapse.

Passive Defenses

Fire Walls and Fire Doors

Subdividing fire areas is accomplished by the use of fire walls. Fire walls in steel structures probably are not free standing. (See the discussion of fire walls in Chapter 7, Steel Construction.)

Fire walls are considered to be **passive** fire protection, but there are very few fire walls without openings. The integrity of the fire wall depends on the active closing of fire doors when a fire occurs. As many of 50 percent of fire doors in supposedly well-protected properties have been found to be inoperative. Overhead rolling doors are particularly susceptible to failure. In one case where 12 doors were trip tested, only three operated.[3]

Water Spray

In the K-Mart warehouse, water spray had been substituted for fire doors. Flaming aerosol cans shot like rockets through the spray to greatly extend the fire.

Active Defenses

Automatic Sprinklers

The principal **active** defense is automatic sprinkler protection. For almost one hundred years sprinklers have had an unparalleled record of suppression or control of incipient fires. Unfortunately, this superb record cannot be taken as an indication of what can happen in a high or dense storage warehouse.

Every one of the huge losses cited earlier and many other large losses occurred in sprinklered buildings.

Fig. 14-5. Note the sprinklers installed in the stacks of this newly opened warehouse. (Fire Department, Halfway, Maryland)

> "The bottom line in warehouse properties is this: the fixed protection [i.e., sprinklers] either has — by specific design — the ability to control a fire in the structure,
>
> OR
>
> the property must be viewed as unprotected. There is no in between!"[4]

In too many cases the sprinkler protection is inadequate and fails very early, sometimes before the fire department arrives. See Chapter 13 for a discussion of possible modes of failure of sprinkler systems.

One of the causes of failure is the early distortion and collapse of the steel roof from which the sprinkler system is suspended. This can be caused by the exposure presented by a fire in stored pallets. Factory Mutual has developed recommendations to prevent the special problems presented from idle pallets from occurring.[5, 6]

The development of Early Suppression, Fast Response (ESFR) sprinklers gives promise that even high-challenge fires can be suppressed, rather than just controlled by the sprinklers. Only time and actual fires will tell.

Recently, a warehouse storing rolled paper was partially destroyed. Rolled paper is a severe challenge to a sprinkler system. The system will be inadequate if it is not designed for the hazard. The first sprinkler system using ESFR sprinklers on rolled paper was installed in Sweden in 1989.

As these fire hazards grow in size, the concept of sprinkler protection as we have known it may be superseded by or coordinated with fixed oscillating nozzles, similar to those used in the Indianapolis Coliseum. Such systems are used to protect huge outdoor hazards such as large lumber piles. The contents of a warehouse is similar to a huge pile of fuel. See Chapter 7 for a picture of such a fixed nozzle.

Foam System Protection

Some sprinkler systems are equipped to deliver Low Expansion Foam. These systems are used where there is a potential for flammable liquid fire.

There is also some use of high-expansion foam. The *Chicago Tribune's* rolled-paper warehouse is protected with a massive high expansion foam system. Twenty-two massive foam generators are mounted on the ceiling. In six minutes the entire warehouse could be covered or filled with foam, 50 feet deep.[7, 8]

Attitudes

Management Attitude

A distinction in attitude can be made between the management of a major national corporation operating a distribution warehouse, which probably has a loss control department, and entrepreneurial, currently fashionable bulk distribution businesses.[9]

The fact that such tremendous risks for potential losses are developing seems to have caught American management unaware. In the minds of many managers, the possibility of a disastrous fire is the reason why insurance is purchased. The annual cost of insurance makes it easy to factor it into the cost of the operation. Thus the problem is disposed of by financial problem solvers.

Management attitudes can range from "There is a serious fire potential here, and it's my job to see that it doesn't happen," to "These fire types are petty bureaucrats who are warts on the backside of progress."

It is unlikely that management is fully familiar with the details of serious fires. The fire department should be as familiar as possible with the circumstances of such disasters. The closer the pre-disaster situation at a facility matches the fire experience, the more credibility it has for the warehouse manager. It takes a lot to overcome "It can't happen here."

Fig. 14-6. Tests like this one demonstrated the necessity of in-rack sprinklers. The tests showed that such fires could not be handled by sprinklers alone and that fire department assistance is necessary. (Factory Mutual Research Corp.)

At the 1991 Industrial Fire Protection Conference of the International Society of Fire Service Instructors, a manager of the Levitz Furniture Co., which has a strong fire protection program, discussed some of the problems of maintaining fire protection in warehouses.

According to a research psychologists, three psychological perceptions, however irrational, govern risk assessment.[2]

- The individual often thinks he or she controls the risk. Temporary aisle storage or extra high racks may be allowed without fear in a warehouse because the managers believe it will be or could be removed promptly. The same person may fear flying because total control rests with somebody else.
- The hazards usually show up as "one big event." Statistically, most warehouse fires are caught while still small and easily controlled — on the order of a trash can fire. Many are never reported. Although we have seen how quickly a small fire can become very big, the perception is that future small fires will similarly be controlled.
- The risk is either familiar or unfamiliar. Familiar dangers are harder to fear over a long time while unfamiliar risks are almost impossible not to fear. A warehouse manager surrounded by familiar storage has a tendency to ignore possible dangers because they are so familiar. The unfamiliar working of a distant nuclear power plant, however, might cause considerable fear.

It is not sufficient to simply lay out the risk facts for the manager. Fire professionals must be armed with sufficient detailed factual experience to convince the manager that the fire official's perception of the situation should become the manager's. The break comes when the attitude changes from a hostile "What's your problem?" question to a businesslike one, such as, "We have a problem all right, what's the solution?"

There was a major fire in the Smithsonian Institution. At a meeting afterwards, I was seated next to an official whom I knew had bitterly opposed the installation of a fire door recommended by Harold "Bud" Nelson. The fire door had saved from damage or destruction the flag which inspired the Star Spangled Banner and the national stamp and coin collection. Not admitting that I knew of his opposition, I remarked on the fact that the unwanted fire door had performed admirably. He said, "I'm the one who opposed that door. I've learned one thing here. If you design a major building, assume there will be a major fire." When we can get management to operate from the premise that there will be a major fire sooner or later, the rest is easy.

Attitudes of Public Officials

Public officials may not be too helpful. Warehouses generate taxes and employment. Officials may pressure the fire department to "go easy" on potentially lucrative facilities. They might show more interest if a case can be made for potential serious air pollution or contamination of groundwater.

Insurance Company Assistance

The basic business of insurance companies is finance. Insurance companies are known as huge cash cows because of the tremendous revenues generated by premiums, and the cash or equivalent reserves which must be kept available to meet claims.

An insurance company can and often does lose money in the underwriting side of the business. Typically, losses are equal to about half the premium income, and expenses represent the other half. If these two total more that 100% of premium income, the company suffers an underwriting loss. This loss may be made up by investment income.

There is constant tension between the underwriting and financial sides of the company. The underwriters are pressured to push out the limits of risks to increase premium income. Some insurance companies, such as Factory Mutual and Industrial Risk Insurers, specialize in Highly Protected Risks (HPR).

A highly protected risk occupancy is defined as one which has a loss-prevention-responsive management, and the following performance specifications, as related to storage occupancies.[2]

- HPR are protected by properly designed automatic sprinklers with adequate water supply.
- The sprinkler system is capable of controlling a fire incident without human intervention.
- HPR sprinkler system alarms are supervised to give early notification.
- The early detection and alarm system also gives the responding fire department an opportunity to extinguish the fire.

A number of the huge fires cited earlier were insured by HPR companies. They obviously did not meet the HPR standard because in those cases the sprinkler protection was inadequate. These huge losses were a learning experience for the insurance companies. Sometimes a very poor risk is insured by an HPR company. Often this occurs as a part of a package of good risks. Just as a bookmaker lays off bets when overloaded with too many bets on one horse, insurance companies **reinsure** part or even all of a risk they do not wish to keep. Today's insurance market is international.

The insurer can make suggestions to the insured but cannot require anything. If the risk is deemed unsatisfactory, there are three recourses: raise the premium; reinsure all or a part of the risk; cancel the policy.

The usual solution is one of the first two. In other words, the solution to many problems which decrease fire safety is often financial.

The relationship between the insurer and the insured is confidential. The insurer cannot disclose potential problem areas to anyone without the consent of the insured. This points up the desirability of maintaining good personal relations with insurance inspectors. Valuable information may be conveyed indirectly, by such means as body language, for example.

If the insurance inspector recommends any procedure which is counter to the practices of the fire department, it is wise to get the recommendation in writing. Long after the fire is out, some insurance executive might judge the fire department incompetent and a serious lawsuit could result.

Since the insurance system is a privately managed taxing system that apportions the cost of fire mismanagement out to all, the public may someday awaken to such risks. The extent to which all of us contribute to make good on huge losses, which are totally unnecessary in the light of available technology, may someday be limited.

Employees

The fire department should be aware of the attitude of warehouse employees. These attitudes can range from open hostility to management, which indicates the potential for arson, to a real gung-ho attitude on the part of a fire brigade who believe that they can handle fires without any outside help.

There should be strong emphasis, backed up by warehouse management, on the immediate reporting of even any possible fire event, such as odors or smoke haze. Many serious fires have gained headway while employees, fearing to give a false alarm, searched for the fire.[10]

Employees should be forbidden by their management to shut off any sprinkler valves before the fire department arrives. Supervisors, faced with an emergency, often extend their authority into areas in which they are not qualified. The fire department should participate in employee training in the handling of fire extinguishers. Emphasis should be placed on the limitations of fire extinguishers. Of particular note is the futility of bringing extinguishers from all over the plant, and their inability to deal with a rapidly growing fire or one on the tops of stacks. Employees should be well trained in the use of hose lines provided in the building. They provide a much greater fire killing capacity. Be sure the pressure is adequate to have a substantial delivery to the top of the piles.[11]

A million dollar warehouse fire was fought initially by employees with hand extinguishers. They never thought to use the hose lines.[12]

It is particularly beneficial to have the hose lines fed from the sprinkler system. This makes it less likely that there will be an unreported fire, as the water flow will trigger the sprinkler alarm.

Fire department contacts with employees should be maintained discreetly. This has resulted in anonymous tips which indicate the need for a "routine inspection" which "happens" to discover a serious problem.

Fire Department Actions

With regard to these warehouse structures the fire department has several distinct areas of interest.

- Initial planning and plan review prior to construction.
- Inspection of construction and witnessing of fire protection equipment tests.
- Routine and special inspections during operations.
- Regular liaison with the warehouse manager.

- Adequate planning for fire suppression, "if and when."

These activities can be greatly complicated if the warehouse property is located in two or more governmental jurisdictions.

Initial Planning and Plan Review

As soon as the municipality is aware that such a structure is planned, a task force should be assembled from all concerned agencies to address the many considerations.

Water supply should be adequate for sprinklers and manual fire fighting. Water supply for manual fire fighting should not decrease sprinkler supply below that necessary for effective performance. Particularly in the case of tilt-slab buildings, hydrants should be located well out of the collapse zone to avoid loss of apparatus, or subjecting fire fighters to a hazardous removal operation.

Electrical service to fire pumps should be unaffected by a fire in the warehouse. If power must be cut to the warehouse, the pumps should continue to operate. Standby diesel power should be adequate enough to replace the entire electrical pump supply.

In one recent fire the loss was increased by an inadequate water supply. One half hour after the alarm, the prestressed concrete roof was starting to fail. In error, an engine supplying sprinklers was hooked up to a yard hydrant instead of a city hydrant. The first due officer noted rapid fire growth. The power company cut the power to the building, turning off the fire pumps. The diesel back up was not operational. After the fire it was determined that it was unnecessary to have cut the power.[13]

Often the only access to such a facility is through a single guarded gate. Additional road access and/or protected gates at distant points on the perimeter should be available for emergencies.

The potential for air contamination from fire smoke or other air toxics, and the potential for ground water contamination from fire suppression water should be planned for. This became a serious problem in the Sherwin Williams Paint Warehouse fire in Dayton Ohio.[14]

Evacuation routes must be established in advance.

Who is responsible for warehouse fire protection design? Will the sprinkler design be adequate?

For many years sprinkler systems were designed by using standard pipe schedules. Computers have made hydraulic design possible, in which calculations match the water flow to the fire load as closely as possible. The lower the water flow the less costly the system. It is customary for the same company to design and install the sprinkler system. There is a temptation to design the system to the minimum to keep the cost down.

Fig. 14-7. Is it an exit or isn't it? The sign on the door would cause great confusion in a fire. It should be changed to read "Emergency Exit Only." (Fire Department, Halfway, Maryland)

> Factory Mutual counsels ". . . .. Base fire protection plans not only on what is being stored today, but on what will be in the warehouse tomorrow. Protect for the commodity that represents the greatest fire challenge that may be stored there, and with storage at the greatest height the building will allow."[15]

Higher stacking means that a minimally designed system will prove inadequate.

Insist on fire department siamese connections (FDC) being included into the sprinkler system. While the on-site pumping system and mains may be adequate, the siamese connection is the only practical method for the fire department to supplement the sprinklers. It is also necessary to supply the system and keep up the water flow and alarm feature if the pump or main is out of service.

Hose lines fed from the sprinkler system should be required for employee use. The flow of water will trigger an alarm. This reduces the possibility of unreported fires.

Consultants: Consulting assistance is often important for these types of large warehouses. As stated, the adequate design of the sprinkler system is absolutely vital, and many municipalities lack the necessary expertise to achieve this. The municipality should consider engaging an experienced consulting engineer to review the plans.

Collapse Potential

These structures have a high collapse potential. Most rack storage warehouses are tilt slab concrete. When the roof fails the walls may fall out.

There should be enough room between the building and security fences to allow for a safe area for fire department operations, considering outward collapse.

Fire Doors

The benefits of fire doors should not be underestimated. The use of water spray in lieu of fire doors in fire walls has not always been effective. Plans for fire doors should provide sturdy physical protection for the doors to prevent damage, and proper housekeeping should ensure that stock is not being placed against them.

Clearance

Sprinkler clearance is a critical factor. Make it clear to management that the fire department insists on three feet of clearance between tops of stacks and the sprinklers. Clearance of 18 inches is often cited as adequate for sprinklers, but it is not adequate for hose streams. It is a good practice to paint a line around the walls at that height and paint "Fire Zone Keep Clear" on the wall above. The space above that line should be regarded as the property of the fire department, not the owner. The useful storage volume of the building should be calculated without the area above the line. In fires that have destroyed these types of warehouses, combustibles were piled too close to, and in some cases, above, the sprinklers. In one case, cardboard cartons were stored a few inches from a 400-watt, 808-volt ceiling-mounted light which ignited the top of the stack 22 feet above the floor.

Inspections During Construction

Whenever tests are conducted on the sprinkler systems, fire doors, particularly overhead rolling doors, or other fire protection features, fire department personnel should be present.

Routine and Special Inspections

Every effort should be made to assure management that the fire department's efforts are directed toward maintaining the warehouse as a vital contribution to the economy of the community. If this attitude is well received, a genuine cooperative effort may be achieved. On the other hand, management may be hostile, and perhaps require that search warrants be obtained before any inspection can be made. Maybe fire departments should adopt a policy that fire fighters will go no farther during a fire than is permitted during an inspection.

Fig. 14-8. Fire fighters from local volunteer companies give up a Saturday morning to become familiar with a new warehouse. Few citizens are aware of the hours spent in training for emergencies. It might be worthwhile to invite a reporter along from the local newspaper. (Fire Department, Halfway, Maryland)

Fire prevention education commends neatness. However, neat, orderly storage on racks may create the impression that there is no hazard, when in fact the warehouse may be a potential disaster.

During inspections make a particular effort to note when there is an increase in the hazard classification of the commodities stored.[16, 17, 18]

Storage in the aisles can extend the fire and overwhelm the sprinkler system. The usual excuse, "It's only temporary," can be countered with, "Right, the building may not be here in the morning."

Few people are aware of how fast a fire can develop. Make sure that warehouse personnel on all shifts, including contract cleaners and guards, are instructed by management that the fire department is to be called even if a fire is only suspected. First aid or brigade fire fighting must not delay the alarm. Be familiar with the stories of major fires and use them in classes and casual conversation.

As part of their evening rounds, shift commanders should swing by such warehouses and note any special problems which appear.

Get specific information about the warehouse sprinklers and water supply. Set forth the special precautions to be taken if sprinklers are out of service. How are fire pumps started if they fail to start automatically?

Suggest that the insurance company inspector visit the fire department when an inspection is made.

Industrial Risk Insurers has produced and excellent movie, "High Piled Stock." It is well worth seeing, particularly to see the effects of the "flues" between stacks, and to gain an appreciation of the hazards of collapse. The film was made, however, with all the roof vents in the test building fully open. Had the vents not been open it would have been impossible to make the movie due to total smoke obscuration. Contact the IRI Loss Prevention Training Department for this film and other IRI materials.)

Warehouse Loss Examples

Three Potlach Corporation fire fighters and two other employees were killed, while seven other employees and two Lewiston, Idaho fire fighters were injured in the warehouse collapse of water-soaked bales of tissue papers. These tissues were discards. They were in 500-700 pound bales. The sprinkler-water-soaked bales collapsed on fire fighters early in the fire. A second collapse trapped rescuers.[19]

In Garland, Texas two fire fighters were trapped by falling debris in a 200,000 square-foot Service Merchandise warehouse.

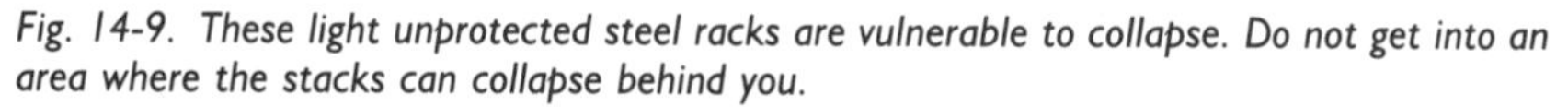

Fig. 14-9. These light unprotected steel racks are vulnerable to collapse. Do not get into an area where the stacks can collapse behind you.

An Indiana grocery warehouse had concrete block walls, wood bowstring trusses and a dry pipe sprinkler system that were inadequate for the fire load. This warehouse was "supervised" from Baltimore, Maryland. The roof trusses collapsed about ten minutes after personnel were ordered off the roof.[20]

Preplanning

A specific member of the fire suppression forces should be assigned as liaison officer to be completely informed on all factors which affect fire fighting operations in a building. It would be the Liaison Officer's concern to see to it that all vital information is disseminated to all those who should have it and that pre-fire plans are kept updated.

The warehouse manager should designate a specific senior subordinate to maintain liaison with the fire department liaison officer.

Get all the information you can, but take all assurances with a large dose of salt. The Fire Chief is the one most concerned with the safety of fire fighters. Others have additional, "important" considerations — financial, operating, or turf-protecting.

Tactical Considerations

Rack storage warehouses are being erected in many locations protected by fire departments that have never faced such a fire problem. These warehouses present an extreme hazard to the life of a fire fighter. Officers and fire fighters alike must recognize that a completely different tactical situation will exist. If the fire is attacked with standard tactics learned and practiced on ordinary dwellings and mercantile buildings, a terrible loss of fire fighter's lives may result.

The most important consideration in preplanning is the safety of fire fighters. Operations at a fire in a rack storage warehouse require detailed advance study and planning. This must start when the structure is proposed. You can't make it up as you go along.

Fire departments whose experience is principally in residential fires may have a "dwelling fire mindset" which can result in disaster in a warehouse fire. Two recent fire service press articles by Captain Robert Obermayer[21] and Bill Gustin[22] provide some excellent guidance on this subject.

Intelligence Acquisition

The following intelligence or informational needs regarding dense warehouse storage facilities are by no means an exhaustive list. Each property is a unique problem.

Contents: Are the contents flammable, or packed in flammable plastic? Will they absorb water? We noted above the loss of five lives in a warehouse collapse from bales of soaked tissues.

Some materials, such as plastic trays for meat markets, are packed with combustible strapping so the bundles will collapse in a fire. The hope is that this will aid extinguishment. Perhaps so, but this is small comfort to the fire fighter buried in the debris.[23]

A Chicopee, Massachusetts warehouse stored acrylics banded in metal and nylon. The nylon bands failed and accelerated the fire. The building was sprinklered but there were no in-rack sprinklers, and no supervision of the sprinklers. The roof failed early in the fire and dropped the sprinkler system. The loss was $5 million.[24]

Aerosol cans pressurized with flammable gases act like self-propelled rockets. They have been a major contributing factor in several disasters, such as the K-Mart, Supermarkets General, and Sherwin Williams fires. Try to isolate aerosol cans and any other extra hazardous materials.

Rolled paper presents a difficult fire fighting problem. The paper rolls exfoliate, i.e., peel off layers like an onion, and flying paper extends the fire.

Storage arrangements are usually patterned without fire hazards taken into consideration. In one case, Navy Supply was adamant about storing vital airplane propellers in an unsprinklered wooden warehouse, and toilet paper in a fire-resistive sprinklered building for inventory purposes. When the situation was shown to a senior operations Admiral, he became irate, and the propellers and toilet paper storages were exchanged that day.

Assuming the sprinkler system was originally adequate for the prospective occupancy, storage contents changes may have made it unlikely or impossible that a fire would be controlled. Some examples are:

- Storage Commodity changes
- Flammable liquids introduced
- Storage quantity increases
- Storage height increases
- Storage arrangement changes
- Temporary storage in aisles
- Use of solid rack shelves

Tactical Considerations

If the changes are such that the sprinklers cannot control a fire, the building should be considered as unsprinklered, and tactics should be adjusted accordingly.

On the Fireground

All personnel should be aware of the known weak points of the warehouse structure and construction. Particular emphasis should be laid on the potential for collapses or failures discussed in earlier chapters:

- A combustible metal deck roof fire
- Failure of pre-tensioned concrete T-beams
- Failure of truss roof
- Failure of connections or splices of heavy timber roof
- Potential fire collapse of tilt-slab walls outward

Tactical Considerations

The four corners of the building are the least dangerous areas with respect to wall collapse. In any case, apparatus should be located beyond the collapse zone. If hydrants are located in the collapse zone, consider using large diameter hose to feed pumpers located outside the zone.

Racks are lightweight steel structures in themselves. They are subject to collapse due to the heating of steel. Fire fighters should never advance beyond the point where racks might collapse behind them.

Racks may be erected across the openings at the far end of aisles, making dead end aisles. Furthermore, some warehouses have movable racks to create more storage room. When stock is to be handled, an aisle is made by moving racks, thus closing other aisles. It is conceivable that an electrical malfunction during a fire could cause racks to shift. The operation should be carefully studied to determine any hazards that might occur during a fire.

Solid Rack Shelves

After the Triangle Shirtwaist fire (New York, 1912) in which 142 died in an unsprinklered high-rise garment factory, the garment industry moved into sprinklered high-rise buildings in midtown New York. Garment making generates huge amounts of combustible scraps which can

Fig. 14-10. Is this scene a recipe for disaster? The sprinklers are high in the overhead. There are no sprinklers in the stacks because the local code did not require them. The shelves are four feet deep. A raging fire can roll through the lower shelves undeterred by the sprinklers. Note the light-weight steel shelves. Collapse would be inevitable. Don't risk lives in a warehouse like this.

accumulate under cutting tables. Fires can rage under these tables, unaffected by the sprinklers. The width of garment cutting tables is permitted to be 40 inches. Some authorities having jurisdiction have interpreted the 40-inch concept to permit several levels of unsprinklered shelving no more than 40 inches wide. In this author's opinion this interpretation is disastrously mistaken.

Ventilation

An argument rages as to whether it is better to close up the building and let the sprinklers do the job, or to vent it and attempt a combined manual/automatic attack. The jury is still out on that one. Partisans of each approach present convincing arguments.

In an excellent article in *Fire Journal*, J. Talbert summarizes both sides of the argument:[25]

> "Smoke and heat venting of one story industrial warehouses has been a controversial subject for many years. Of the improved risk insurance carriers, FM recommends against vents. IRI recommends vents strongly. Kemper does not strongly recommend or reject vents but has asked that automatic vents be converted to manual (FD) operation. It generally recommends manual vents for new construction."

The article gives a good summary of the pros and cons. The conclusions are that the resources spent on vents would be better applied to insuring adequate sprinkler design, locked or supervised valves, and an ongoing inspection system to ensure against sprinkler impairment.

In a letter to *Fire Journal*, John Edwards of Hampshire, United Kingdom, takes issue with the FM position and states that British research shows that ventilation delays the onset of poor visibility.[26]

The most comprehensive tests ever carried out on the operation of ventilators and sprinklers together were completed recently in Ghent, Belgium, by Colt International and the U.K. Fire Research Station. Tests were conducted in a 12,000 cubic-meter building. Data was collected every second for up to ten minutes at 120 points for each of 38 tests.[27] As of late 1991 the conclusions of the tests were not available.

Tactical Considerations

Venting warehouse roofs can be inadequate and dangerous. It should not be undertaken just as a routine, "get the roof." Only after preplan consideration has determined that ventilation can be carried out safely should it be attempted. Experience in ventilating ordinary roofs is not adequate.[28]

Handline Operations: In a number of warehouse fires smoke has suddenly banked down to the floor leaving fire fighters unsure of the way out. The hoseline is the fire fighter's lifeline. "Follow the hose back to safety if lost," is one of the oldest instructions to new fire fighters. Remember, however, that a hose line fed from an interior hose outlet is not a lifeline. If interior outlets are used, lifelines should be strung to the exterior from the outlet.

Acting Captain Frank Cheatham of Phoenix, Arizona ordered a second alarm on arrival at a 77,000 square-foot carpet warehouse. His crew advanced a line through heavy smoke with zero visibility. Cheatham got to the seat of the fire and then moved beyond the hose line. His air gave out. When found he had a weak pulse and no respiration. Fortunately, he was revived.

Tactical Considerations

No fire fighters should be permitted to move beyond the point where a defined path to safety exists. Those who think this unnecessary should be taken to a warehouse in breathing apparatus with the face piece blanked out, spun around a few times and told to find the way out.

Tactical Considerations, continued

In some high-rise procedures, engine companies are paired up in order to stretch lines and provide adequate relief. This practice might well be adopted for huge warehouses, particularly considering the minimum staffing of fire companies in many cities.

Be prepared to use heavy stream appliances inside the building. The Wilmington, Delaware Fire Department develops 300 GPM streams from lightweight portable deckpipes fed by 1¾-inch hose lines. A hand cart, or better still a tractor with several hundred feet of dual 2½-inch lines and a portable monitor, would be a useful piece of equipment to be provided by and located at the warehouse. Many fire fighters have been in ordinary residences when their air supply gave out. The relatively short distance to safety has been lifesaving. This experience is not sufficient to impress on personnel the hazards of running out of air in a large warehouse. It may be that the time delay between the warning bell and the loss of air is not long enough.

Fire Doors:

Fire doors have a poor operating record for many reasons. Fire door deficiencies are usually considered minor violations because it often takes only effort, not big money to correct them. It generally is a truck company function to check all the fire doors and close those which should be closed. In any case the function should be specifically assigned.

A fire occurred in an old St. Louis brick and timber warehouse with a deteriorated sprinkler system. Early in the fire, before it appeared necessary, personnel were assigned to close fire doors between the fire section and other sections of the warehouse. Suddenly the fire accelerated and the interior was untenable. The closed fire doors stopped the fire.[29] *In another case in a warehouse loaded with urethane furniture, fire spread through partially open fire doors.*[30]

Fire fighters should not pass through a fire door without blocking it. Overhead rolling doors are particularly dangerous. In one case the strength of all six trapped fire fighters was necessary to open a sliding fire door which had jammed into a bent retainer. The block should be removed when all fire fighters have returned through the opening.

Automatic Sprinklers:

Be sure the sprinklers are operating. The sound of falling water, water on the floor, and noises from sprinkler risers are all indications of water flow. If the fire is not being suppressed, check that the flow is adequate. Previously-obtained information from gauge readings may be useful.

Fig. 14-11. This Mobile, Alabama fire door failed to operate, permitting the fire to extend.

Tactical Considerations, continued

Back up the sprinklers to the maximum, operating pumpers at a pre-determined pressure setting. This is particularly important for lightweight steel roofs. If the roof fails, the sprinklers are lost.

The Fire Ground Commander should make sure that the fire pumps are working properly, and get them started if necessary. This also should be done in accordance with previously determined procedures.

Sprinklers should not be shut down hastily, or to limit water damage when there is any potential for the fire increasing in intensity. Sprinklers are not an auxiliary appliance. They are the principal fire suppression method. If possible, sprinklers should be shut down from outside valves, which are likely to be more accessible if needed than inside valves.

Factory Mutual is on record, "Extinguishing the fire is more important than water damage." This position should be taken seriously because, when a fire occurs, the insurance company, up to

Tactical Considerations, continued

the limit of the policy, is, in effect, the owner. If in doubt about the application of this statement to a particular property, check with the insurance company.

In any case, the fire fighter who shuts a valve must be fully equipped with radio, fire clothes and SCBA to be able to open the valve instantly if so ordered by the Fire Ground Commander.

If a warehouse is heavily involved on arrival, it is most likely lost. Use extreme caution. Probably the most productive effort would be to back up the sprinklers to the maximum. Note from Chapter 11 that the sprinklers which stopped the Meridian Plaza, Philadelphia high-rise fire were probably delivering four times the ordinary water flow. Any systems which have collapsed should be shut off, if possible, to conserve water.

Cooperation:

The best hope for successfully combatting fire in a rack storage building rests in the cooperation among the owner, the operator (often different from the owner), the insurer, and the fire department, not only of the city involved but also of those who might be called for mutual aid. It is not unlikely that a building is situated in more than one jurisdiction. The problem of who is in command should be settled before the fire.

Protecting the Department and the City Treasury

A fire department's planning should look beyond the fire. Before the ashes have cooled, the attorneys and financial experts of the insurance company may be searching for some way to get someone else to pay the loss. The amounts of money involved can be staggering.

The 1985 fire in a warehouse in Elizabeth, New Jersey, cited earlier, resulted in a record out of court settlement of over $80 million. There were over 200 litigants and 60 lawyers involved in the case.[31]

The plaintiffs claimed the defendants were negligent in storing aerosol paint cans in a building that lacked internal fire walls, and that the sprinkler system in the paint storage section had broken at least a week before the fire.

Fire departments have no control over tenants, owners and insurance companies suing one another. It is the fire department's responsibility to be sure that it and the taxpayers are protected against raids on the public purse.

"Hired Gun" experts are readily available to point out the errors in fire suppression, which in their view, caused or contributed to the loss.

One easy way to generate a lawsuit is to interfere with the flow of water to the sprinklers. In a likely scenario, the engine company rolls in. There is little or nothing showing. It's a long lay to the city main. The yard hydrant is close by. The pumper is hooked up to the yard hydrant to supply a small line. The fire blooms. The pumper then supplies a ladder pipe.

The expert might contend the sprinklers were containing the fire, but only until they were robbed of water by the pumper feeding the ladder pipe. The expert might say the sprinklers were containing the fire until they were shut off by the fire department to get a look at the fire or to limit water damage. The expert might contend the sprinklers could have controlled the fire with a little help, but the fire department failed to supply the siameses.

The best way to defend a lawsuit is to prepare in advance. A good preparation may convince the plaintiff that the chances of prevailing are so low as to make a suit impractical. A good preparation may strengthen the position of the city attorney, making the argument for a disgraceful settlement invalid.

In self defense, the fire department should request, in writing, a statement from the management of any potential high loss facility, as to their degree of compliance with the HPR definition given earlier in this chapter. The request may very well go unanswered but should be repeated at regular intervals.

A solid, well documented file of deficiencies observed — and, without exception, reported to the management in writing — could be of great help to the defense of the taxpayers against an unwarranted suit. This is a hardball situation, however. It is not the best assignment for an inspector who believes that "you catch more flies with sugar than with vinegar." A reasonably competent attorney can make sincere off the record cooperation sound like negligence.

Personal Safety

The art of warehouse protection is inexact, to say the least. Impressive looking systems have failed totally. Even if the management's planning is adequate, the execution is often faulty. All planning should place the safety of fire fighters first. This is the fire department's responsibility, no one else is taking care of it.

References

The addresses of all referenced publications are contained in the Appendix.

1. Teague, P. "Member Profile: Paul M. Fitzgerald." *Fire Journal*, NFPA, January/February 1989, p. 9.
2. "Before The Fire—Fire Prevention Strategies for Storage Occupancies." NFPA White Paper, Ad Hoc Committee, 1988.
3. *Factory Mutual Record*. "Plan/Review, A Blueprint for Success." January/February 1989, p. 15.
4. Lambert, J. A. *The Sentinel*, Industrial Risk Insurers, First Quarter 1988, p. 3.
5. Cassacio, E. "Problems Stack Up; Storage of Idle Pallets." *Factory Mutual Record*, May/June 1988.
6. *Fire Engineering*, "Dispatches." May 1984, Vol. 137, No. 5, p. 23.
7. *Industrial Fire World*, "Chicago Tribune Protects Paper Warehouse From Fire." August 1989, p. 6.
8. Kirsch, A. "Reducing The Risks of Roll Paper Storage." *Factory Mutual Record*, September/October 1989, p. 3.
9. Snyder, J. "Buying Into Bulk Suggested for the '90s." *Orlando (FL) Sentinel*, January 29, 1990, Business Section, p. 6.
10. Brannigan, F. "Delayed Alarms-A Focus for Public Fire Educators." *Fire Engineering*, October 1984.
11. Cook, B. "People," *Factory Mutual Record*, May/June 1989, Vol. 66, No. 3.
12. Courtney, N. "Glass Bottle Storage." Fire Record, *Fire Journal*, September/October 1989, p. 21.
13. Keith, W. "On The Job, Missouri." *Firehouse*, March 1988, p. 41.
14. Isner, M. "Sherwin Williams Fire." *Fire Journal*, March/April 1988, p. 65.
15. "Protection." *Factory Mutual Record*, May/June 1989, p. 6.
16. Anderson, D. "Advances in Commodity Classification." *Factory Mutual Record*, March/April 1990, p. 8.
17. Barritt, J. "Commodity Classification." *The IRI Sentinel*, First Quarter 1988, p. 7.
18. Sotis, L. "Commodities and Storage Arrangements." *Factory Mutual Record*, May/June 1989, p. 13.
19. Best, R. "Storage Collapse Kills Five." *Fire Command*, Vol. 47, No. 11, November 1980, p. 20-24.
20. O'Brien, A., and Redding, D. "Large Loss Fires in the United States During 1983." *Fire Journal*, November 1984, p. 37.
21. Obermayer, R. "The Dwelling Fire Mindset." *Fire Engineering*, March 1991, p. 154.
22. Gustin, B. "Modern Warehouses." *Firehouse*, June 1990, p. 38.
23. Lenton, T. "SPI Fire Tests of Stored Plastics in Warehouses." *Fire Journal*, Vol. 73, No. 3, May 1979, p. 30.
24. Denault, V., and Beauchemin, R. "Delayed Alarm Results in $5,000,000 Loss at Chicopee Warehouse." *Firehouse*, November 1984, p. 42.
25. Talbert, J. "Smoke and Heat Venting for Sprinklered Buildings." *Fire Journal*, May 1987, p. 26.

26. "Letters: Feels Sprinklers and Vents Are Both Necessary." *Fire Journal*, May/June 1988, p. 6.

27. "Sprinklers and Vents, New Research," *Fire International*, February/March 1990, p. 10.

28. "Insulated Metal Deck Roof Fire Tests." Report, Factory Mutual Engineering, May 1955.

29. Sachen, J. "Multiple Alarm Blaze Races Through Warehouse." *Firehouse*, May 1984.

30. Hall, G. "Smoky Five Alarm Blaze Consumes Warehouse." *Firehouse*, July 1985, p. 39.

31. "$80.5 Million Settlement Reached in 1985 New Jersey Warehouse Fire." *New York Times*, (AP) June 10, 1990, p. 10.

Additional Readings

Berkol, A. "Sprinklers Control Arson Fires in Rack Storage Warehouse in Mt. Prospect Illinois (Oct. 16, 1988)," Report 030 of the Major Fire Investigations Project of the US Fire Administration.

Fire Protection Handbook, 17 Edition, 1991, Section 8, Systems Approach To Property Classes. Chapters 4, 5, 6, and 7, and Section 5, Suppression, Chapters 9 through 14.

Isner, M. "Summary Fire Investigation Report: Cold Storage Warehouse fire, Madison, Wisconsin, May 3, 1991," National Fire Protection Association, 1991.

Appendix A
Last Minute Updates

The writing and publication of a book takes place over an extended time. One by one the chapters are "locked up." Very often significant information becomes available shortly thereafter. Included in these updates are a number of such items.

Chapter 2

In September, 1991, a New York City fire fighter, searching the top floor of an abandoned building for homeless occupants, died when the ceiling collapsed and raging fire burst out on him.

Chapter 3

In October, 1991, a spectacular multi-alarm fire occurred in the wooden cooling tower atop the 24-story John F. Kennedy Federal Office Building in Boston. The fire sent debris crashing to the street. It was reported that debris fell among some of the 1000 employees who were evacuated.

In October, 1991, Oakland, California suffered more than a $1 billion loss in brush fires that destroyed over 3000 homes and killed 24 people. This writer was told that while some houses with tile roofs were destroyed, the only houses that did survive had tile roofs.

Chapter 4

A fire started in antique knob and tube wiring in a basement and extended through the floor joists to non-firestopped walls and to the attic. The fire destroyed a historic 80-year-old church in San Jose, California in March, 1991. The fire had spread extensively before an occupant smelled smoke. The smoke developed so fast that a call to 911 could not be completed. The Captain of the first-due engine transmitted second and third alarms, and realized that the fire couldn't be stopped; all that could be done was to protect exposures. Interior fire fighters were driven out when fire burst from the voids. (Larson, R. "On The Job California — General Alarm Fire Engulfs Historic San Jose Church." *Firehouse*, October 1991, p. 70.)

Further information has become available on the collapse of the tile covered facade which killed a Los Angeles County fire fighter. Failure of gusset plate roof trusses precipitated the collapse. Another similar collapse has occurred. Alerted by warning sounds similar to those heard in the previous collapse, fire fighters evacuated in time.

A fire seriously damaged a huge 19th century State Hospital In Worcester, Massachusetts. Construction workers had cut off a water main and replaced it with a much smaller temporary main. At the fire the water supply was totally inadequate until big supply lines were laid by mutual aid companies. (Hayes, B. "On The Job: Massachusetts." *Firehouse*, November 1991.)

Tactical Considerations

Red paint and chrome do not put out fires; water does. A fire department without water is like a battleship without ammunition, an impressive looking bluff; fire can't be bluffed. A fire department should be constantly vigilant to protect its ammunition. This often requires battles. Water companies and departments often resent any interference in "their business."

A January 1992 fire totally destroyed a gymnasium at Rhode Island College in Providence, Rhode Island. The bearing walls were block with brick veneer. The roof was supported on steel trusses. The floor was made of several layers of wood supported on steel beams, designed to be springy for basketball. There was a three-foot void below the floor down to the basement ceiling. The fire was in the void. In about 40 minutes elongating steel was causing cracks in the wall. The FGC ordered a defensive mode a few minutes later. Shortly thereafter wall collapse commenced. There were a number of other contributing factors to the total loss, estimated at over $3,000,000.

Chapter 5

An advertisement for US Gypsum in BOCA (The Building Official and Code Administrator), for January/February 1991, p. 5, describes a combustible floor/ceiling assembly which achieves a two-hour rating. The ceiling consists of two layers of specified ⅝" gypsum board. On top of the joists there is 1 inch of gypsum panel. Three inches of "thermal fire blankets" in the void provide an acoustical barrier, according to the ad.

"Preplanning for Apartment Complex Response" by J. Burkush and J. Forrest in *Fire Engineering* for January 1992, p. 67, is an excellent discussion of the absolute necessity of preplanning for fires in apartment complexes.

Chapter 7

An article by Jane Carminati in *Firehouse* for December 1991, "Long Island Shopping Mall Fire Claims the Lives of Two," appears to describe a metal deck roof fire, though the term is not used. The dense clouds of heavy black smoke which greeted the first-due engine company and the description of the roof lend credence to the belief. The fire started in artificial flowers, often very combustible, and capable of delivering sufficient heat to the metal deck to initiate a self-sustaining metal deck roof fire.

Gary Long, Fire Marshal of East Whiteland Township, Pennsylvania, provided an interesting clipping from the Winter 1991 Issue of *Building System Sentinel*, published by the Metal Building Manufacturers Association. The article describes a new "standing seam" metal roof installed atop a leaking "bar joist roof" (possibly a convention metal deck roof), of a 125,000 sq-ft building in Little Chute, Wisconsin. The roof is raised six inches to provide room for 6 inches of glass fiber insulation. The pitch of the roof was also changed to provide improved water run off.

Tactical Considerations

If such a roof is installed over a conventional metal deck roof, the insulation above the roof would likely cause heat to be retained, thus possibly increasing the rate of flame spread of a metal deck roof fire. In addition, flammable gases given off by the decomposing roll of roofing material might accumulate in the void and ignite violently if the roof is vented in the conventional manner. I believe ventilation of a metal deck roof fire is dangerous and unproductive; in my opinion the safest attack is with hose lines from below to cool the steel and thus stop the production of gases.

In December 1991, four fire fighters died in a fire in Brackenridge, Pennsylvania. According to preliminary information, the building had concrete slab floors supported on unprotected steel. The basement had a large fuel load of flammable liquids. About 40 minutes after fire department response, a portion of the first floor collapsed. A huge fireball erupted from the basement through the floor opening and the fire fighters died. Further collapse of the first floor ensued. As of this writing the NFPA is preparing a detailed report on this tragedy.

Tactical Considerations

Stay alert. Be aware of the hazards to fire fighters of a fire in an unprotected steel building.

Chapter 8

A reinforced concrete high rise 757 feet tall was built in New York on a lot 50 feet wide, sandwiched between New York's musical jewel, Carnegie Hall, and a famous restaurant. Further information can be found in *Engineering News Record*, Jan 14 1991, p. 30.

Chapter 9

A spectacular fire swept up 11 stores in a reinforced concrete apartment building in Toronto, Ontario. A roofing contractor had installed a plastic trash chute from the roof to a refuse container on the ground. There was fire in apartments on 11 floors. A reinspection disclosed that the contractor had reinstalled an identical plastic chute. A Stop Work Order was issued. (Speed, A. "Apartment Building Suffers Unusual Fire Spread," *Firefighting In Canada*, September 1991, p. 6.)

Factory Mutual Research has begun testing wall and ceiling/roof assemblies in a 50-foot (15m) Corner Test structure. Assemblies which pass the test can be used at increased heights without automatic sprinkler protection being required for this structure. (The contents, however, may require sprinklers.) The test uses 750 pounds of oak pallets stacked 5 feet high as fuel. The test is run for 15 minutes. ("For the Record." *FM Record*, September/October 1991, p. 23.)

The May/June, 1991 issue of *NFPA Journal*, p. 77, provides a picture of the ruins of Boston's tragic Cocoanut Grove fire in November, 1942 in which 492 people died. Shown are the tell-tale round globs which indicate acoustical tile (at that time surely combustible) glued to the ceiling.

Chapter 10

Gary Kraus, Administrator of the Sutter County, California Fire Department, related a 1988 incident which occurred when he was Chief of

Suisun City, California, in which positive pressure ventilation caused hidden fire to break out.

The building was an old commercial building, occupied by the U.S. Post Office. Due to alterations there were several ceilings. The basic fire, apparently incendiary, was extinguished with one blast from a small line. It appeared to be unnecessary to open up the structure; in addition there was the need not to compromise the arson investigation. A fan was set up to provide positive pressure ventilation. This continued for about a half hour while the arson investigation was made.

At that point fire fighters commenced to pull the ceiling. Simultaneously, smoke was seen from the outside front of the building. After a three-hour fight the loss of the roof and ceiling was held to about one fourth of the building.

Chief Kraus notes that PPV may accelerate or even drive fire into voids in this type of construction, and that the needs of overhaul and investigation should be balanced.

Chapter 11

The October, 1991 issue of *Fire Engineering* contains an excellent article, "High Rise Flows with PRV's: The Boston Tests," p. 64, by W. A. Damon, PE, a project director for Schirmer Engineering. The article describes tests made on Boston's Prudential Complex. The entire article is must reading for any fire department concerned with pumping into standpipes.

The widely accepted fire department practice of pumping into standpipe hose outlets to replace a defective siamese connection or supplement the supply is not possible if the standpipe is equipped with pressure regulating valves, which must be distinguished from pressure restricting valves.

Fire department pumpers must pump into the standpipe system at pressure equal to those generated by the building pumps. Unless special high-pressure pumpers are available. "Tandem pumping" (described in a sidebar article by D. Fornell, p. 66) is necessary. Special hose should be set aside to handle the pressures generated. All connections should be tied. The pumping area should be kept clear of personnel.

Tactical Considerations

Fire departments should check pumping pressures with respect to local buildings.

Chapter 12

The Salt Lake County, Utah, Fire Department's *Firewatch* reported that fire fighters critiquing a half million dollar fire they had fought in a truss roof factory, discovered that nine wooden trusses had been burnt through and there was a 400 sq-ft sag in the roof. As a result "a more thorough inspection of trusses and roof will be made before committing members regarding an interior attack."

In September, 1991, three Santa Ana, California fire fighters were injured in a collapse of a roof and second floor. The floor and roof were supported on sawn wooden joists, hung on hangars from laminated wood girders. It appears that a long preburn and air conditioning units on the roof contributed to the collapse.

Repeat: Do not let concern about trusses blind you to the hazard of wooden beams. Note that beams hung on brackets are less secure from fire than beams set atop the girder. See Fig. 2-45 in Chapter 2 of this text.

An engineer making an inspection for another purpose discovered a visible tensile failure crack in one of two 6" × 14" timbers forming the bottom chord of one of the 90-foot long, 130+ year-old wood trusses of the Philadelphia Academy of Music. The failure constituted a condition of imminent danger.

When the trusses were examined, two other failed trusses were discovered. One was so severe as to require immediate evacuation. The other was delayed until the close of the concert season.

The trusses were repaired with steel rods running end to end and with steel splice plates. The repair was extremely delicate because of the age and the beauty of the building. The consultants agreed that the probable cause was high winds channeled by nearby high-rise buildings. Winds of 69 mph were recorded, just 13 days before the failures were discovered. The channeling is believed to have raised the effective speed to 140 mph. (Levenson N. "A Sudden Unusual Force." *Architectural Record,* February 1990, p. 146.

Tactical Considerations

Many very old spectacular buildings have heavy timber trusses. Even weather-protected wood is subject to many forms of failure.

A prefire evaluation should include at the least a visual examination of the trusses for obvious defects. It would, of course, be better to have the building department engineers make a professional evaluation.

Chapter 13

The Washington Post of September 15, 1991 (p. A 23) reported that the U.S. Office of Management and Budget opposed legislation to require sprinklers in all new or remodeled Federal Office Buildings because of cost.

A Letter to the Editor on p. 30 of *Fire Engineering* for October 1991, by Deputy Chief (Ret.) Elmer (Bud) Chapman of FDNY, a high-rise expert, points out the significant fact that the same type of pressure reducing valves (PRV) which provided inadequate pressure to Philadelphia fire fighters at the One Meridian Plaza fire are also used on sprinkler systems in high-rise buildings. Since these valves can remain in the closed position for years, he makes an important recommendation that both legislation and NFPA Standard 13 require inspection and servicing annually.

Chapter 14

The San Antonio, Texas Fire Department had developed a plan for warehouse fire protection which puts the burden on the occupant to disclose detailed information as to contents, by answering nine questions in what is called a "Commodity Letter." Based on the information provided, the plans are approved or appropriate fire protection requirements are given to the contractor/tenant.

I consider the program to be an excellent example of modern fire loss management. Every fire department with a warehouse problem should read the referenced articles. (Corbett, G. "Avoiding Warehouse Problems, The San Antonio Approach," *Fire Engineering*, November 1991, p. 38, and the "Commodity Letter," p. 42.

In May, 1991, a loss of over $100 million occurred in a cold storage warehouse complex in Madison, Wisconsin. The sprinkler system was inadequate for the risk.

The referenced article describes a roof fire separate from the fire at the floor level. From the pictures it would appear that this was a combustible metal deck roof fire as described in Chapter 7. (Isner, M. "$100 Million Fire Destroys Warehouses." *NFPA Journal*, November/December 1991, p. 37.

Appendix B
Addresses of References

American Iron and Steel Institute
1133 15 St., NW
Washington, DC 20005

American Fire Journal
19072 E Artesia Blvd. #7
Bellflower, CA 90706-6299

American Heat
240 Sovereign Court
St. Louis, MO 63011

Concrete Masonry Industry
5420 Old Orchard Rd.
Skokie, IL 60077-4321

Consumer Reports
Consumers Union
101 Truman Ave.
Yonkers, NY 10703-1054

Fire Chief Magazine
307 N. Michigan Ave.
Chicago, IL 60601

Gypsum Association
810 First St., NE
Washington, DC 289 5540

International Conference of Building Officials (ICBO)
5360 South Workman Mill Rd.
Whittier, CA 90601

Speaking of Fire
International Fire Service Training Association
IFSTA Oklahoma State University
Stillwater, OK 74078-00118

Sentinel IRI (Industrial Risk Insurers)
85 Woodland St.
Hartford, CT 06102

Record
Factory Mutual
1151 Boston-Providence Turnpike
P.O. Box 9102
Norwood, MA 02062

Fire Engineering
Park 80 West Plaza 2
Saddle Brook, NJ 07662

Nat'l. Automatic Sprinkler Assn.
Robin Hill Corporate Park
Route 22
P.O. Box 1000
Patterson, NY 12563

International Association of Fire Chiefs (IAFC)
4025 Fair Ridge Drive
Fairfax, VA 22030

Advisory Council on Historic Preservation
Old PO Building, Suite 809
1100 Penn Ave., NW
Washington, DC 20004
(202)786-0543

WNYF New York Fire Department
Fire Academy
Randalls Island
New York, NY 10035

Farolito, also "804" Newsletter
Walter W. Maybee
2541 Camino Alfredo
Santa Fe, NM 87505

Progressive Architecture
600 Summer St.
Stamford, CT 06904

Firehouse
445 Broad Hollow Rd.
Melville, NY 11747

Firefighting in Canada
222 Argyle Ave.
Delhi, Ontario N4B 2Y2

Bulletin of Society of Fire Protection Engineers and
Journal of Fire Protection Engineering
Society of Fire Protection Engineers (SFPE)
1 Liberty Square
Boston, MA 02109

Firefighter's News
248 S. Rehoboth Blvd.
P.O. Box 165
Milford, DE 19963

Fire Engineer's Journal
Institute of Fire Engineers
148 New Walk
Leicester, LE1 7QB
England

NFPA Journal, (also *Fire Journal* and *Fire Command*)
Fire Protection Handbook
Fire Technology
National Fire Protection Association
One Batterymarch Park
Quincy, MA 02269-9101

The American Society of Mechanical Engineers
345 East 47th St.
New York, NY 10017

Viking Penguin
40 West 23rd St.
New York, NY 10010

Underwriters Laboratories
333 Pfingsten Rd.
Northbrook, IL 60062-2096

Industrial Fire World
P.O. Box 9161
College Station, TX 77840

Journal of Applied Fire Science
Baywood Publishing Company
26 Austin Ave.
P.O. Box 337
Amityville, NY 11701

Francis L. Brannigan, SPFE
2041 Daylily Rd.
Scientists Cliffs
Port Republic, MD 20676

John Mittendorf
Fire Technology Services
22422 Sunlight Creek
El Toro, CA 93630

Elmer Chapman
15 Guild Lane
Levittown, NY 11756

Phoenix Fire Department
520 W. Van Buren St.
Phoenix, AZ 85003

National Forest Products Assn.
1250 Connecticut Ave., NW
Washington, DC 20036

Index

A

Accountability, 10, 633-634
Acoustical tiles, *See* Tiles
Active fire protection, 614
Adhesives, 302, 303, 392-393
Adobe, 145
Aerial apparatus, 39, 183, 559
Aggregates, 327
Air conditioners, roof, 17, 182, 218
Air conditioning
  in high-rise buildings, 467, 485-487
  individual room units, 485
  multi-floor systems, 486
  single-floor systems, 485-486
  and smoke movement, 485
Air ram, 19
Aircraft, 398, 573
Alarms
  classification of, 220
  unwarranted, 441-442
Alleys, 197
Alterations, 170, 178, 362
  added dead load and, 17
  to ceilings, 389-391
  fire resistance and, 105
  hazards of, 389-391, 399, 403
  to high-rise buildings, 456, 457, 498
  to occupied buildings, 498, 592
  truss hazard and, 525-526
Aluminum, 328
  fire characteristics of, 35-36, 268
American Medical Association (AMA), 37
Ammonium nitrate, 7
Anchors, 123
Angles, 260
Arched trusses, 519
Arches, 74-76
  braced or tied, 75
  brick segmental, 26, 160
  cast-iron, 148
  failure of, 76, 160
  flat, 76, 334
  flat concrete, 334
  Gothic, 75
  hollow-tile, 76
  laminated wood, 85, 130
  masonry, 76, 160
  Roman, 74, 75
  segmental, 26, 75, 116, 334
  steel, 75
  stone, 46
  tied, 75, 519
  trefoil, 75
  trussed, 520
  wood, 76, 116-117
Arson, 133, 179
Asbestos
  fire characteristics of, 37
  fireproofing, 291, 292
  hazards of, 24, 37, 268, 292
Asbestos cement shingles, 120
Asbestos felt siding, 120
Ashlar masonry, 145
Asphalt asbestos protected metal (AAPM), 269
Assemblies
  fire rated, 200, 223-224, 252, 294-297, 318, 536
  fire-resistive, 169, 200, 246, 251, 283
  wood, non-standard tests of, 541-542
ASTM E84, Tunnel Test, 410-411
ASTM E119, *See* Fire-Resistance Test
Atmospheric conditions, 480-485
Atriums, 467-468
Attics, 217, 230
  fires in, 48, 59, 91, 232
  vents in, 229
Axial loads, 27, 29, 47, 81

B

Backdraft, 13, 14, 24, 39, 422
Backup, 15
Balconies, 57, 163, 176, 217, 320, 544, 548
Balloon frame buildings, 96-100, 107
Ballrooms, 179
Bar joists, 56, 167, 200, 526, 535, 536
Bars, steel, 261
Basements, 191, 571
Bays, 61
Beams, 52-61
  aluminum, 35, 36
  bending moment of, 60-61
  cantilever, 56, 57, 163, 334
  carrying capacity and depth of, 53-55
  connections, 173-174, 181
  continuous, 56, 334, 335

drop-in, 85, 181
false, 135, 364
fixed, 56
girders (*see* Girders)
grillage, 56, 166, 182
I-, 3, 63, 118, 166, 261, 541, 546-547, 551-553
joists, 56, 83, 92, 167
lintels, 56, 148, 158, 162, 172
loading of, 48, 60, 333-334
needle, 59
overhanging, 56, 85, 181
penetrations, 132
reaction of, 60
simple, 56
soldier, 279
span distance of, 55
spliced, 85-86, 115-117, 181
suspended, 59, 60, 163
T-, 334
transfer, 59, 174, 520
types of, 56-60
Belted, 270
Bending moment, 60-61
Bents, 61
wind—, 61
Board and batten siding, 120
Boards, 92
Boats, dry storage of, 609
Bombline, 128
Bowstring truss, 10, 519-520, 534
Box columns, 261
Box girders, 261
Braced or tied arches, 75, 519
Braced walls, 162-163
Bracing
of arches, 75
compressive load on, 62
diagonal, 25
excavation, 278-281
K-, 25
loss of, 65
portal, 25
stability, indication of, 162
temporary, 66-67, 270, 271
of walls, 71, 72-73
Bracket, 56
Breen, David, 412
Brick nogging, 126
Brick veneer siding, 122-124
on non-wooden building, 124
Brick veneer walls, 51, 68, 99, 123, 147, 269
archors, 123
on concrete, 326
Brick walls, 49, 68, 73
effect of fire streams on, 167
Brick- and block-composite walls, 51, 147
Bricks
fire characteristics of, 33
imitation, 148
poor quality, 157
Bridges
collapse of, 347
concrete, 340, 347, 373-374
over tanks, 39
post-tensioning, 340, 347
railroad, 75
steel, 276-278
suspension, 45, 53
Brunacini, Alan, 203
Btu (British thermal unit), 30, 31
Building alterations. *See* Alterations
Building codes, 18, 41, 42, 248-249
atriums and, 468
improving, 415, 503
inadequate, 282-286
steel structures and, 319-320
void spaces and, 191
Building Construction (Ellison), 99
Building construction. *See* Construction principles
Building contents. *See* Contents, building
Building Failures (McKaig), 47, 154
"Building fire," dispatch, 2, 11, 89
Building materials
combustible, 33, 41
deficiencies of, 170-173
fire characteristics of, 32-40
fire-resistance ratings of, 249
imitation, 368
weight saving, 16, 462-463, 517
Building Officials Conference of America (BOCA) Code, 18
Buildings
classification fallacies, 32
under construction (*see* Construction, buildings under)
see also types of buildings such as Fire-resistive
Bulkheads, 273
Buttresses, 72, 75, 146

C

Cables
flammability classification, 386

steel (*see* Steel cables)
suspension, 45, 49, 53
with ties, 132
Calculated risks, 287-289
California bungalows, 90
Caloric value, 30
Canopies, 163, 177, 193, 194
Cantilevered, 49, 60, 194
balconies, 57
beams, 56, 57, 163, 334
fire escapes, 58
platforms, 357
trusses, 521
walls, 70-71, 145
Carbon monoxide
explosion of, 382, 421-424, 543
flammability range of, 543
ignition of, 14, 23, 24, 43, 147, 164, 187, 189, 229
physiological effects of, 420, 543
see also Toxic gases
Carcinogens, 37
Carpeting, 399
fire tests on, 414
Cast iron, 33
Casting, 327
continuous, 328
Cast-in-place concrete, 327, 372
Cast-iron beam box, 166
Cast-iron buildings, 62, 144, 148
Cast-iron columns, 63, 82, 84, 148, 174-176, 244
Cavity walls, 73, 145, 146-147
Ceiling spaces, 186, 363
Ceilings
failure of, 37, 557-558
false, 186
insulated, 36-37, 296, 384
plank and beam, 103
plaster, 386
suspended, 297, 318, 363, 382, 456, 532
tile, 153, 185, 250, 294-296, 363, 382, 387, 392, 413, 456
tin, 150, 272
wooden suspended, 136
Cement, 325
Cement-asbestos board, 24, 267
Centroid, 27
Chairs, 327
Chamfer, 92
Channels, 261
Chapman, Elmer, 489
Chases, 164-165
Chemicals, fire-retardant, 115
Chief Officers, 4
Chipboard, 117, 119
Chords, 519
Churches, 191, 283, 314
Chutes, 497
Cinder blocks, 326, 330, 342
Civil Engineering Handbook, 22
Coatings
fire-retardant, 408
intumescent, 112
tar, 269
Cockloft, 15, 104, 149, 164, 189, 296
Collapse
of arenas, 528
of balconies, 558
of bridges, 347
of ceilings, 557-558
during construction, 342-349
of floors, 23, 178-179, 345-346
hazard, 12
imminent, 2, 44
indicators, 154-165
lift-slab, 348-349
of loar-bearing walls, 68
during overhaul, 3
partial, 79
of post-tensioned concrete, 354-356, 377-378
of precast concrete, 348, 357, 372-373
progressive, 23-24, 347
of reinforced masonry, 348
Ronan Point—, 23
of roofs, 10, 19, 180, 181, 530, 559
of trusses, 552-553
of walls, 168
see also Loads
Columns, 61-67
bays, 61
bents, 61
box, 261
bracing of, 65-67
connections, 173-174
Euler's law, 64-66
fire testing, 244, 245, 246, 249
hollow, 62, 63
H-shaped, 63, 261
interior, 48
intermediate, 64
Lally, 291, 329
load-carrying capacity, 61
long, slender, 64, 65
paper, 47
piers, 64, 146, 332
pilasters, 72, 146

pillars, 61
protection on, 298-299
removal of, 205
shapes of, 61-62
struts or rakers, 61
studs, 93, 97, 103
temporary bracing of, 66-67
types of, 64-67
wall, 26, 72
Combined dry-pipe and preaction systems, 566
Combustible, 33, 41
meanings of, 40-45
protected, 44
versus fire-resistive, 251
Combustible gases
detonation of, 23
in void spaces, 189-190
see also Carbon monoxide
Combustible liquids, 30
Combustible multiple dwellings, 215
See also Garden apartments
Combustible trim, 364
Combustions, products of, 419, 428
Communication, dispatch, 1-3
Communication systems, 463
Compartmentation, 232, 235, 365, 442, 491, 569
Composite and combination columns, 328
Composite concrete-steel floors, 50
Composite construction, 51, 328
Composite floors, 50, 328
Composite materials, 46, 50
Composite structural elements, 50-52
Composite walls, 51, 68, 145, 147
Composites, 50-52
Compression, 29
Compressive forces, 45, 46-47
Concrete, 325
cast-in-place, 327
cored, 336
curing of, 325, 345
deteriorated, 360-361
failure of, 350
falsework, 36, 343-345, 347, 350, 358
fiberglass reinforced, 331
fire characteristics of, 34, 369, 373
fire load, 362
fireproofing, 359
fire-resistance, 162, 359-360, 362
heat absorption, 362
non-combustibility of, 162
over steel, 290-291
parging, 146
plain, 330, 333
precast, 8-, 83, 326, 327, 330, 348, 357, 372-373
prestressed, 338-340
spalling, 33, 331
strength of, 325, 330
tensioned, 369-371
testing, 46
waffle, 335, 363
see also Concrete buildings; Concrete construction
Concrete beams and girders, 333-334, 360, 364
Concrete blocks, 326, 330, 342
Concrete buildings, 325-326
collapse of, 342-349, 356
exterior extensions, 367-368
fire problems in, 359-378
precast, 357
progressive collapse in, 24
steel in, 264-265, 361-362
under construction, 350-359
see also Concrete; Concrete construction
Concrete columns, 328, 329, 332, 333
Concrete construction, 325-378
bridges, 373-374
combustible trim, 364
composite, 328
definitions, 327
fire department problems with, 327
fire resistance, 359, 364
framing, 326
hazard of collapse, 342-349, 356-357, 365-367, 372, 377
high-rise buildings, 463
interior finish, 363-364
lift slab, 329, 348-349
monolithic, 329
pretensioning and post-tensioning, 339-340, 347, 354-357, 370-371, 377-378
structural elements, 332-342
tilt slab, 167, 332, 373
types of, 327
voids, 363-364
see also Concrete; Concrete buildings
Concrete floors, 229, 334-338, 345-346, 365-367, 369-371, 463
Concrete masonry unit, 146
Concrete panels, 268, 284, 328, 367, 368
Concrete partitions, 369
Concrete roofs, 47, 361
Concrete walls
blocks, 26
precast, 71, 145

Condominiums, 215
Connections, 48, 80-86
  beam to column, 173-174
  beam to girder, 173
  concealed, 82
  in concrete, 372
  failure of, 81-84
  finger joints, 117
  fire department (FDC), 578, 588, 606, 622
  gravity, 82
  metal, 116
  in monolithic concrete structures, 80
  in ordinary construction, 173-174
  pinned, 80
  plastic design, 81
  poor, 33
  in precast concrete buildings, 80
  in rigid-framed buildings, 80
  steel, 83, 117, 263
  stress and, 11
  truss, 521
  types of, 80
  weakest, 48
  wet joints, 80
Construction, buildings under, 60
  collapse of, 342-349
  concrete, 350-359
  fire potential, 351-353
  inspections of, 623
  partial occupancy of, 499-501, 592
  steel-framed, 270-271, 281-282
Construction Failures, (Feld), 154
Construction principles, 9-86
  application of force to materials, 45-50
  arches, 74-76
  beams, 52-61, 85-86
  building materials, 32-40
  building terminology, 9-10
  columns, 61-67
  combustible and noncombustible, 40-45
  composites, 50-52
  connections, 80-86
  fire load, 29-32
  foundations, 79
  gravity resistance systems, 10-11
  loads, 16-29, 48, 78-79, 81
  recycled buildings, 19-20
  rigid frames, 77-78
  roofs, 74
  shells and domes, 78
  structural elements, 52-86
  structural stability, 21-13
  walls, 67-74
Consultants, 622
Containment, 252
  of fire, 431-437
  see also Smoke
Contamination, smoke, 425-428
Contents, building, 30, 539
  and fire growth, 383-388, 400-402, 404, 493-498
  fire-resistant tests on, 226
  warehoused, 627
  water-soaked, 167, 179
Continuous beams, 56, 335
Continuous casting, 328
Continuous slipforming, 328
Converted buildings, *See* Recycled buildings
Cooling towers, wooden, 129
Core construction, 25, 464, 470
Cored concrete, 336
Cork, 394, 494
Corner Test (FM), 413
Cornices, 193-194
Corrosion, of impregnated wood, 111
Corrosive fumes, 39
Corrosive gases, 428
Corrugated metal siding, 121
Corrugations, 47
Course, 146
  header or bond, 146
  stretcher, 146
Cranes, tower, 279-281, 358-359
Critical Radiant Flux (CRF), 414
Croker, Edward F., 8
Cross walls, 68, 146
Curtain walls, 67, 69, 92, 166, 199, 268

D

Dalby, Calvin, 8
Dampers, tuned-mass, 481
Dead loads, 16-18
  added, 17-18
  concentrated, 26
Deckguns, 129
Decks, 113
Decorations, 400-402
Deflection, 47, 53
Deformation, 45
Deluge guns, 105, 128, 134
Deluge systems, 300, 566
Demising walls, 69
Demolition, 60, 81
  hazard of, 132, 207
  power to order, 343

Deputy Chiefs, 4
Diaphragm construction, 114
Dimension lumber, 92
Dip tanks, 408
Disneyworld, 145, 549
Dog iron, 174
Domes, 78, 322
  geodesic, 78
Dougherty, Thomas, 442
Downspouts, 614, 165
Draftstopping, 106, 107, 109
Drain times, 22
Dripping-pool fire, 38
Drop panels, 328
Dry-pipe systems, 80, 566, 590
Ducts
  aluminum, 36
  grease, 273
  insulation in, 386
Dunn, Vincent, 207
DuPont Plaza Hotel fire, 37, 485, 498, 508
Dwellings
  converted, 19, 20
  multiple, 109
Dynamic fire protection, 316, 319

E

Early suppression fast response (ESFR)
  sprinklers, 598-599, 608, 615
Earthquakes, 25, 146
Eccentric loads, 27-29, 81, 163
Elevated water tanks, 18, 19
Elevator shafts, 69, 429, 479
Elevators
  for evacuation, 477
  in high-rise buildings, 464, 477-479
  termination of, 25
Emergency command, 1
Emergency force, 5
Encasement method, fireproofing, 290-291
End matched, 92
Energy conservation, 42-43
Engineered wood, 92, 114-119
Escalators, 437-438
Euler's law columns, 64-66
Evacuation
  command, 1, 2, 12, 90, 168, 202
  elevators for, 477
  of high-rise buildings, 473
Excess flow valve, 352
Exhibition halls, 320-321
Exits, 474, 584
Explosions
  backdraft, 24
  cooking-gas, 23
  load and, 23
  see also Carbon monoxide
Explosives, storage of, 24
Exposure hazard, 90, 105, 202, 238, 459-460
Exposure protection, 121, 197, 596
Exterior extensions, 367-368
Exterior Insulation and Finish Systems (EIFS), 368
Extinguishers, portable fire, 219, 376, 620

F

Fabrics, 393-394
Factory Mutual, 309, 386, 413
Falsework, 36
  failure of, 347, 350
  problems of, 343-345, 358
Fatalities, 13-15
Ferries, 288
Fiberboard
  combustible, 40
  high-density, 393
  low-density, 119, 387
  regulating, 388-389
  sheathing, 119, 385
Fiberglass reinforced concrete, 331
Filing cabinets, 20
Films, 12, 38, 219, 226, 404-405, 439, 538, 575, 625
Filters, 426
Finger joints, 117
Fire barriers
  attic, 217
  defects in, 234-235
  glass, 598
  horizontal, 436-437
  interior structural walls as, 192
  masonry bearing walls as, 195-200
  row frame buildings, 126
Fire Chiefs, 5, 6
"Fire: Countdown to Disaster" (NFPA film), 12, 219, 226, 404-405, 439, 538, 575
Fire department connections (FDCs), 578, 588, 606, 622
Fire departments
  decision making, 6-7
  notification of, 220
  problems at garden apartments, 221-223

problems with concrete construction, 327
responsibilities, 151, 633-634
and sprinklers, 581-598
Fire doors, 195-196, 245, 311, 614, 623, 631
closure devices, 431, 433-434
smoke-sensitive releases, 434-436
Fire education, 219, 438-439, 585-586
Fire escapes, 58, 263, 455
Fire growth, 381-416
alterations effect on, 399, 403
building elements and, 383-388
contents and, 383-388, 400-402
decorations and, 400-402
eliminating hazards to, 392
examples of, 382-383
fire loads and, 402-403, 492-498
furnishings and, 403
in high-rise buildings, 460-461, 465-466
inadequate mitigation of hazards, 392-399
interior finish and, 386-399, 406, 493-494
rapid, control of, 406-409
void spaces and, 389
Fire limits, 145
Fire loads, 29-32, 318
concentrated, hazards of, 278
in fire-resistive buildings, 31, 250, 362
and flame spread, 492-498
in high-rise buildings, 492-498
and rubbish handling systems, 496-397
of today, 402-403, 457
"Fire Power" (NFPA film), 38, 226, 404, 575
Fire resistance, 152
of assemblies, 169, 200, 246, 251, 283
combustibility and, 251
of concrete, 259-260
fire load and, 362
high-rise buildings and, 452
inherent, 43-45
legal issues in, 248-250
and mass, 17, 330
minimum requirements, 248
of partitions, 34, 369
principles of, 243, 253
standards for, 244-245
testing (*see* Fire-Resistance Test) versus protected combustible, 231
Fire severity, estimating, 250-251
Fire streams
effect on brick walls, 167
weight of, 21
Fire tests
on assemblies, 541-542
on building materials, 409-415
early, 243-245
faked, 415
residential, 404
on trusses, 538-541
see also Fire-Resistance Test
Fire towers, 464
Fire walls, 69, 192, 195-200, 233-234, 252, 311, 614
penetrations in, 69, 80, 233
tied, 309-310
Fire-cut, 83, 84, 165
Firefighting, 6
high-rise fires, 471, 478-479, 480-481, 490, 631
and increased load, 21
metal deck roof fires, 306-307
safety, 239
warehouse, 15, 612-616, 631-633
Fireground safety, 203
Fireproof, 152, 252, 336, 453
Fireproofing
concrete, 359
directly applied, 285, 286, 291-292, 317, 462
encasement method, 290-291
individual, 289
membrane, 289, 292-293
steel, 289, 359, 462
tendons, 355
types of, 289-298
with water, 300
Fire-rated, 296, 413, 557
assemblies, 200, 223-224, 252, 294-297, 318, 536
Fire-Resistance Test (NFPA 251/ASTM E119), 223-227, 231, 224-252, 274, 293, 336, 413, 538-541
Fire-resistive buildings, 11, 12, 21, 104-105, 250
early, defects of, 454-458
exposure to, 105-106
fire load in, 31
wood in, 104-105
see also Fire resistance
Fire-retardant chemicals, 115
Fire-retardant coatings, 408-409
Fire-retardant treated wood (FRT), 111-112, 183, 549
Firestopping, 106-107
in balloon-frame buildings, 97

codes, 296
in garden apartments, 218, 232
inherent, 107
legal, 107, 184
perimeter, 252
and truss floors, 109
types of, 107-108
wood, 107
First Interstate Bank fire, 402, 466, 503, 506, 571
Fixed beams, 56
Fixture mounting, 407
Flame spread. *See* Fire growth
Flame spread ratings, 249, 409-411
Flameover, 13
Flameproofing, 408-409
Flammable, 41
Flammable gases, 421
Flammable liquids, 323, 496, 578, 597-598
Flash point, 41
Flashover, 13, 14, 39, 405, 421
Flat plate, 335
Flitch plate girders, 50, 118
Floors
collapse of, 23, 178-179, 345-346
composite, 50, 328
composite concrete-steel, 50
concrete (*see* Concrete floors)
concrete topped, 152, 336, 337
cutting through, 369-371, 458
diaphragm, 25, 50
fire tests on, 243-244, 245
fire cut, 83, 84
fireproof, 26
hollow tile, 244
penetrations, 178-179, 365, 366
self-releasing, 83, 84, 174
tile arch, 455, 458
truss (*see* Truss floors)
wooden, 26
Flying buttresses, 146
Foam core panels, 39, 96
Foam core roofs, 39, 96, 181
Foam systems, 300, 322-323, 616
Folded plate, 47
Footings, 328
Forces
applying to materials, 45-50
flexural or bending, 67
horizontal, 309
Forcible entry, 476
Foundations, 79
anchoring to, 83
FPETOOL, 32
Friable, 24, 267
Fume hoods, 18, 176
Fumes, from decay-treated wood, 80
See also Gases; Toxic gases
Funishings, 403, 539
Fusible links, 431

G

Gang nails, 83
Garages, parking, 288, 361, 469, 472
Garden apartments, 215-240
case studies, 237-240
characteristics of, 216-220
fire department problems with, 221-223
fire walls, barriers, and draft stops, 233-235
firestopping in, 218, 232
interior construction, 217-218
management of, 219-220
number of stories, 216
protected combustible construction, 223-232
sprinklers in, 216, 236
tenants of, 219-220, 237
walls in, 216
water supplies, 222-223, 239
Gas fires, 303
Gas laws, 107
Gas leaks, 147
Gas service/lines, 221-222, 372
Gases
combustible, 23, 189-190
cooking, 23
corrosive, 428
fire (*see* Toxic gases)
flammable, 421
liquefied petroleum (LPG), 352
movement of, 298-299
natural, 147
toxic (*see* Toxic gases)
Gasoline fires, 277
Geometry, 53
Girders, 56, 310
bridge, 340
box, 261
built-up, 56
concrete, 333-334, 360, 364
connections, 173
flitch plate, 50, 118
heavy timber, 84
spandrel, 56, 262

Glass
  fire barriers, 598
  fire characteristics of, 36
  exteriors, 466-467
  falling, 36, 466
  wired, 36, 197
  see also Windows
Glass Fiber
  formwork, 36
  insulation, 36-37
  in panels, 39
Glass fiber-reinforced polyester resin plastic, 37
Glued laminated timbers, 92
Glulam, 116
Gore, Robert B., 360
Gravity, smoke and, 429
Gravity connections, 82
Gravity resistance systems, 10-11, 21
Grillage, 56, 166, 182
Gypsum board
  brick veneer on, 148
  classifying, 41
  deficiencies of, 224, 234-235
  for enclosure of steel, 314
  fire characteristics of, 34, 223
  fire rating of, 223-224, 365
  fire-resistance testing, 224-227
  firestopping with, 108
  installation of, 228-230
  penetrations of, 228-230
  shaft liners, 16
  sheathing, 119, 227-228
  wall and roof panels, 39

H

Hall, J.R., 276
Handicapped persons, 477
Handline operations, 630
Hazardous processes, 24
Header or bond course, 146, 147
Heavy timber construction, 129-130, 166, 176, 206
  case studies, 211-213
  protection of, 206
Heavy timbers, 92
High Efficiency Fire Resistive (HEFR) filters, 426
Highly Protected Risk (HPR) insurance, 313, 434, 618-619
High-rise buildings, 451-452
  alterations to occupied buildings, 498
  evacuation of, 473, 477
  exits in, 474-475
  exposure problems, 105-106, 459-460
  fire load and flame spread in, 492-498
  fires in, 105, 424, 452-453, 460, 465-466, 478-479, 489-490, 504, 506
  forcible entry and, 476-477
  general classifications, 454-469
  life safety and, 451, 455
  maintenance operations, 496
  occupants of, 475, 498-501
  rubbish handling systems, 496-497
  smoke movement in, 479-485
  wind load and, 25, 480-481
  see also High-rise construction
High-rise construction, 451-511
  air conditioning, 467, 485-486
  aluminum railings in, 35-36
  atriums, 467-468
  automatic sprinklers, 468, 501-505, 571, 592
  building inventory items, 505
  building materials, 462-463
  case studies, 506-508
  communications systems, 463
  compartmentation, 491
  elevators, 464, 477-479
  exteriors, 464, 472
  floors, 463
  fire resistance and, 452, 454-458
  framing, 270
  historical notes on, 452-461
  interior core structure, 464, 470
  modern, 461-469
  parking garages, 469, 472
  partial occupancy during, 499-501
  roofs on, 79
  smoke removal systems, 487-490
  stairways, 459, 464, 470, 475, 491-492, 500
  suspended, 49
  windows and glass, 36, 466-467
  see also High-rise buildings
Highway structures, 276-278
Hollow tiles, 51, 160-161, 335
Hollow masonry walls, 75, 145, 146
Hollow tile walls, 28, 51, 160-161, 335
Hollow walls. *See* Cavity walls
Hoods. *See* Fume hoods
Hoops, 332-333
Hose, 105
House trailers. *See* Mobile homes
H-shaped columns, 63, 261
Hurricanes, 25, 83
Hydrants, 222, 284

I

I-beams, 3, 63, 118, 166, 261, 541, 546-547, 551-553
Ice, 10, 19, 163, 182
Impact loads, 16, 22-24
  lateral, 23-24
Impregnated wood, 110-111
  corrosion of, 111
  delamination of, 114
Individual fireproofing, 289
Inflammable, 42
Inspections, 5-6, 434, 586
  during construction, 623
  prefire planning, 5-6
  routine and special, 623-625
  sprinkler systems, 586, 587-591
  unannounced, 5
Insulation
  air-duct, 386
  altered fire characteristics with, 42
  cable, 386
  ceiling, 36-37, 296, 384
  electrical, 386
  foamed plastic, 147, 385-386
  glass fiber, 36-37
  roofing, 34-35
  wall, 147
Insurance
  HPR, 313, 434, 618-619
  renters', 237
Insurance companies, 568, 581-582, 618-619
Interior designers, 403-404
Interior finish
  in concrete buildings, 363
  fire growth and, 386-399, 406, 493-494
  wooden, 105
Interior floor finish, 414
Intermediate columns, 64
Interstitial space, 520
Inversion layer, 480
Inverted king post truss, 521

J

Joist spaces, 189
Joists, 56, 92
  composite, 552
  steel or bar, 56, 167, 200, 526, 535, 536
  wood, fire cut, 83
Juvenile wood, 524

K

Kilojoule, 30
KIP, 45
Knickerbocker Theater collapse, 28, 160

L

Laboratories, 573
Ladder pipe, 307-308
Ladder tower, 307-308
Ladders, 27
Lally columns, 291, 329
Laminated timbers, 92, 115-117
Language, dispatch, 1-3
Lapse conditions, 480
Lattice trusses, 524
Lawsuits, 19, 634
Ledger board, 82
Left-in-place form, 335
Library stacks/libraries, 33, 575-576, 577, 610
Life safety, 451, 455
Life Safety Code. *See* NFPA 101
Lift slab construction, 329
  collapse of, 348-349
Light smoke showing, 185-186, 440, 557
Lightwells, 178
Limited combustiblity, 41
Lintels, 56, 148, 158
  failure of, 172
  reinforced concrete, 162
  steel, 162
Liquefied petroleum gas (LPG), 352
Liquids, combustible, 30
Live loads, 10, 16, 18-19, 457
  added, 18-19
  concentrated, 26
Load-bearing walls, 67, 68
Loads, 16-29
  axial, 27, 29, 47, 81
  changes in, 22, 29, 457-458
  classification of, 26-27
  compressive, 62, 64
  concentrated, 26
  dead, 16-18
  definitions of, 16-29
  design, 18, 49, 50
  eccentric, 27-29, 81, 163
  fire (*see* Fire Load)
  impact, 16, 22-24
  imposition of, 26-29

lateral, 81, 164
live, 10, 16, 18-19, 457
overspray, 19
repeated, 24
static, 24
stress of, 11
superimposed, 52, 55
suspended, 48, 176-177
tensile, 47, 529
torsional, 29
transmission of, 78-79
twisting, 29
undesigned shifting of, 29, 81
uniformly distributed, 55
water, 18, 19, 21-22, 182
wind, 25-26, 62, 67, 71
Log cabin, 93-94
Long, slender columns, 64, 65
Lumber, 92
boards, 92
definitions and types of, 92, 117
dimension, 92, 137
matched, 92
treated, 111

M

McCormick Place fire, 283-285, 320, 386
Magnesium fires, 277
Maintenance operations, 496
Malls, 437, 582, 591-592
Management
attitudes on sprinklers, 616-618
in garden apartments, 219
industrial, 288
Mansards, 104, 233
fake, 194
Marble, 33, 152
Marquees, 22, 59, 163
Masonry
construction terms, 145-146
decorative, 68
safety factor, 49
Masonry columns, 146
Masonry walls, 23, 25, 43, 62, 70, 74, 76, 123, 144
as fire walls, 233
hollow, 75, 145, 146
materials used in, 145-146, 147, 268-269
rubble, 146
solid, 146
structural stability of, 155-168
see also Ordinary construction
Masonry wire trusses, 147
Mass, 53, 330
Matched lumber, 92
May Day, 3
Membrane fireproofing, 289, 292-293
Membrane roofs, 35
Metal deck roofs, 153, 166, 274, 536
fire problem, 302-309
fires, 303-305, 306-307
on non-combustible buildings, 305-306, 308-309
Metal siding, 120
corrugated, 121
Mezzanines, 176, 177, 179, 278
Mill construction, 130, 204-207
conversions, 205-206
see also Heavy timber construction
Mittendorf, John, 99, 308, 445, 535, 541
Mobile homes, 35, 117, 118
Monoammonium phosphate, 115
Monolithic construction, 329, 372
connections in, 80
Mortar
chemical reaction of, 157
fire characteristics of, 33
sand-lime, 33, 82, 157-158
Mortise, 173
Motels, 192, 194, 215
Mudsills, 345
Multiple dwellings, 109
Museums, 571-572
Mushroom caps, 329
Mutual aid, 221

N

Nail heads, 227, 235
National Institute of Standards and Technology (NIST), 274
Natural gas, loss of odorant, 147
Natural stone, fire characteristics of, 33
Needle beams, 59
Nelson, Harold (Bud), 32
Neoprene, 424
Neutral axis or plane, 53
New York City Tenement House Act of 1903, 218, 550
NFPA 70, *National Electrical Code*, 407
NFPA 101, *Life Safety Code*, 411-412, 414, 584-585
NFPA 251 (ASTM E119). *See* Fire-Resistance Test

NFPA 253, *Radiant Flux Test*, 414-415
NFPA 255, *Surface Burning Characteristics of Building Materials*, 410, 411, 412
NFPA 256, *Standard Methods of Fire Test of Roof Coverings*, 125
NFPA *Fire Protection Handbook*, 41
NFPA standards on sprinklers, 580, 595, 612
NFPA standards on storage, 612
Nitrocellulose, 420
Non-combustible, 40, 41
Non-combustible buildings, 42, 104-105
Non-load-bearing walls, 67, 68, 69
Novelty siding, 120
Nozzles
  monitor, 471
  piercing, 558
  swivel tip, 100

O

Obermayer, Robert, 92
Occupants
  of garden apartments, 219-220, 237
  of high-rise buildings, 475, 498-501, 592
  prefire planning and, 45, 219-220
Offices, open-plan, 398-399, 494
O'Hagan, John, 16, 489
One-high story, 278
Open-air structures, 134
Operation Life Safety (OLS), 577
Ordinary construction, 89, 130, 143-207
  case studies, 207-211
  characteristics of, 144-153, 340-341
  connections, 150, 173-184
  fire barriers, 192, 195-200
  floors, 178-179
  masonry construction terms, 145-146
  problems of, 153-155
  recent, 152-153
  renovation and restoration, 151-152
  risk analysis, 200-203
  roofs, 180-184
  structural stability of interiors, 169-184
  structural stability of masonry walls, 155-168, 195-200
  unprotected steel in, 314
  void spaces, 184-195
  walls, 146-148, 149, 150, 155-168
  see also Heavy timber construction; Mill construction
Oriented strand board (OSB), 39, 92, 115, 551
Overhanging beams, 56, 85, 181
Overhaul safety, 3, 14, 39, 202
Owners, rights of, 154-155

P

Pallets, 606, 615
Panel points, 519
Panel walls, 67, 69, 269
Panels
  aluminum sandwich, 269
  automobile, 39
  concrete, 268, 284, 328, 332, 367, 368
  corrugated, 39
  drop, 328
  fire-retardant, 39
  foam core, 39, 96
  glass fiber reinforced, 37
  gypsum, 39
  imitation concrete, 368
  insulating, 269
  metal, 269, 270
  plastic, 39
  wood and plastic roof, 119
Paper, 30, 393
  shape of, and compressive loads, 47
Paper wrapping, 117
Parallel-chord truss, 10, 520
  steel, 535, 547
  wooden, 118, 135
Parapets, 163, 183, 200
Parging (or pargetting), 146, 157
Parking, fire department, 221
Parking garages, 288, 361, 469, 472
Partition walls, 67, 69, 199-200, 310
Partitions
  fire-resistive, 34, 369
  head-high, 102
Party walls, 69, 197-199, 233-234, 235
Passive fire protection, 316-318, 319, 614
Pause conditions, 480
Phillips, Patrick E., 574
Piers, 26, 64, 146, 332
Pilasters, 72, 146
Pillars, 61
Pinned connections, 80
Pintle, 174
Piping
  as columns and beams, 170
  gas, 221-222
  sprinkler, 134, 573, 621-622
  in vertical voids, 218
Plain concrete, 330, 333

Plane of weakness, 164-165, 337
Plank and beam construction, 101-102, 117
Planks, 117
  tongue and groove, 117
Plaster, 386
Plastic design, 81, 271
Plastics, 37-40
  corrosive acids from, 428
  fire characteristics of, 30, 35, 37-40, 119, 147, 395-398, 401
  foamed, 39, 120
  glass fiber-reinforced polyester resin, 37, 39, 268
  insulation, 35
  roofing panels, 119
  thermoplastics, 38
Plates, steel, 261
Platform frame buildings, 99, 100-101
Platforms
  cantilevered, 357
  temporary construction, 57
Platoon system, 7
Plenums, 296
Plugging effect, 424
Plywood, 92, 114-115
  compressive strength of, 46
  delamination of, 114
  fire characteristics, 30, 31, 395
  FRT, 111-112, 183, 282
  impregnated, 114
  as interior finish, 114
Plywood roofing, 111-112, 114, 119, 153, 183-184
  treated, 111, 182-184, 282
Plywood sheathing, 119
Plywood siding, 120
Polycholorinated biphenyls (PCBs), 427
Polyisocyanate, 335
Polyurethane, 395, 401, 422, 424
Polyvinyl chloride (PVC), 420
Positive pressure ventilation (PPV), 445-447
Post and frame construction, 94-96
Post indicator valves (PIV), 588
Post-tensioning, 330, 339
  hazards of, 347, 354-357, 369-371, 377-378
Preaction systems, 566
Precast concrete, 326, 327, 330, 367
  collapse of, 348, 357, 372-373
Precast concrete tilt slab walls, 71
Prefire planning, 1, 3-8
  exposure to high-rise buildings, 106
  fire load, 31
  and fire resistance, 252-253
  highway structures, 277
  inspections, 5
  loads, 20, 29
  marking dangerous buildings, 167-168
  metal deck roofs, 307-308
  occupants, 45
  rack-storage warehouses, 621-622, 626
  rate of heat release, 31, 32
  surveys, 6
  unprotected steel buildings, 320-322
Preservation, 152, 205-206
Pressure-treated wood, 111
  loss of strength in, 111
  toxic fumes from, 113
Prestressed concrete, 338-340
Pre-tensioning, 330, 339
Protected combustible construction, 223-232
  versus fire-resistive construction, 231
Protected structures, 215-216. *See also* Garden apartments
Protective clothing and equipment, 14. *See also* SCBA
Public fire education, 219, 438-439
Purlins, 262

Q

Queen post truss, 522

R

Rack-storage warehouses, 601-634
  collapse potential, 622-623, 628
  employees in, 620
  fire defenses, 612-616
  fire department actions, 620-633
  fire loads, 32
  fire problems, 15, 605-610, 611
  fire walls and fire doors, 614, 623
  major losses, 611, 625-626
  management in, 616-618
  modern construction of, 608-609, 612-617
  pallets in, 606, 615
  preplanning, 621-622, 626
  shelving in, 608, 614, 628-629
  sprinklers in, 598, 599, 606, 608, 614-616, 621, 631-632
  see also Library stacks/libraries

Radiant Flux Test, 414-415
Radiant heat, 320-321
Radiant Panel Test, 413
Radio language. *See* Communication
Radioactive materials, 425
Rafters, 52. *See also* Beams
Railings, aluminum, 35
Rain, 73
Rakers, 61, 279
Ranch houses, 90-91, 218
Rate of Heat Release (RHR), 30, 31
Rattan, 394
Recycled buildings, 18, 19-20, 59, 172, 176, 179, 189, 205, 314. *See also* Alterations
Refrigerator plants, 484
Reinforced concrete, 34, 46, 70, 83, 330
  fallacies of, 346-347
Reinforced masonry, 146, 340-342
  collapse of, 348
Reinforcing bars or rods, 331, 332
Renovations, 132, 151. *See also* Alterations; Recycled buildings
Repeated loads, 24
Reshoring, 345
Residential sprinkler systems, 580
Residential structures, design load, 50
Restorations, 151, 593
Ribbon board, 97
Rigid frames, 77-78
  connections with, 80
  precast concrete, 77
  steel, 77, 266
  wooden, 77
Rising roofs, 137, 524
Risk analysis, 200-203, 617-618
Risks, calculated, 287-289
Rolled paper storage, 615, 616, 627
Roana Point Collapse, 23
Roof trusses, 520, 528, 530-531, 536, 546
Roofing materials, classification of, 125
Roofs
  air conditioners on, 17, 182, 218
  "approved," 303
  architectural types of, 181
  asphalt, 120
  ballasted, 35
  bowstring-truss, 75
  built-up, 302
  cable-supported, 53
  Class I, 304
  collapse of, 10, 18, 19, 180, 530, 559
  concrete, 47, 361
  dead loads on, 17-18, 28, 79, 182
  fabric, 322
  flat, 18, 19, 553-555
  foam core, 39, 96, 181
  hazards, 181-183
  hip, 553
  membrane, 35
  metal deck (*see* Metal deck roofs)
  metal-plate truss, 137
  peaked, 217, 265
  plastic, 39
  plywood (*see* Plywood roofing)
  rising, 137, 524
  sawn joist, 74, 181
  second, 554
  single-ply, 34-38
  sprinkler protection for, 303
  tile, 91-92
  truss (*see* Truss roofs)
  ventilating, 35, 180, 182, 183, 308, 630
  wood, fire characteristics of, 180-181
  wood and plastic panel, 119, 181
  wood shingled, 120, 124-126
Rope, manila hemp, 46
Rough lumber, 92
Row frame buildings, 126-128, 186, 190, 198, 199, 202, 215. *See also* Garden apartments
Rubbish handling systems, 496
Rubble masonry, 146
Running bond, 69
Rural areas, 81

S

Safe working stress, 49
Safes, 19-20, 24, 26
Safety factor, 49, 50
Sand-lime mortar, 33, 82
Scaffolding, 60, 62, 65
SCBA (self-contained breathing apparatus), 14, 35, 39, 113, 421, 424
Screwbacks, 290
Scuppers, 204
Self-closing doors, 431, 433-434
Self-releasing floors, 83, 84, 174
Self-storage facilities, 273
Self-weight, 16
Serpentine walls, 72
Shakes, 120
Shear forces, 45
Shear walls, 25, 69, 70
Sheathing
  combustible fiberboard, 40

exterior, 119-120
foamed plastic, 120, 385-386
gypsum, 119
interior, 185
low-density fiberboard, 119, 385
plywood, 119
Shells, 78
Shingles, 120
asbestos cement, 120
asphalt, 126
roofing, 120, 124-126
siding, 120
wood, 90, 120
Shipment of dangerous goods, 7
Siding, 120-124
electrical hazard, 121
exposure protection, 121
types of, 35, 120-124, 136, 268
Signs, hanging, 163
Simple beams, 56
Six Flags Haunted House fire, 38
Slate, 33
Slipforming, 331
continuous, 328
Smoke
cold, 484
contaminated, 425-428
light, 185-186, 440, 557
measuring the obscuration of, 412
physiological effects of, 419, 424
wind and, 26, 480-481
see also Smoke movement
Smoke blowers, 445
Smoke control, 245
Smoke damage, 425, 437
Smoke detectors, 90, 91, 434, 435, 440-441, 573, 574, 577
types of, 441
Smoke movement, 298-299, 429-430
air conditioning and, 485-486
atmospheric conditions and, 480-485
in high-rise buildings, 479-485
stack effect, 481-485
Smoke removal systems, 26, 487-490
in unsprinklered buildings, 489, 490
Smoldering fires, 574
Snow, 28, 182
Soffits, 101
Soldier beams, 279
Solid masonry walls, 146
Space frame, 520
Spalling, 33, 331, 373
Spandrel girders, 262
Spandrel space, 270, 459
Spliced timbers, 85-86, 115-117
Splines, 92
Split-level houses, 101
Sprinkler Experience in High-Rise Buildings (Powers), 502
Sprinklers, automatic, 565-601
adequacy of, 32, 151, 316, 615
for atriums, 468
case studies, 600-601
clearances, 623
combined dry-pipe and preaction systems, 566
for cooling towers, 129
cycling (on-off), 595
deficiencies of, 578, 591
deluge systems, 300, 566
demonstrations, 577-578
dry-pipe systems, 80, 566, 590
early suppression fast response (ESFR), 598-599, 608, 615
fire department and, 580-598
fire service misconceptions about, 578-581
and flammable liquids, 578
foam systems, 300, 322-323, 616
fraud, 596-597
in garden apartments, 216, 236
in high-rise buildings, 468, 501-505, 571, 592
impairment or shutdown of, 582, 584-586
inspections, 586, 587-591
installation incentives, 567-568
NFPA standards on, 580, 595
opposition to, 134, 502, 569-572, 616-620
partial systems, 580-581, 591
piping for, 134, 573
popular misconceptions about, 572-577
prewetting, 599
preaction systems, 566
for protection of steel, 299-300, 316, 322
quick-response, 577
for rack storage warehouses, 598, 599, 606, 608, 614-616
recent developments in, 567
residential systems, 580
retroactive installation of, 569
special types of systems, 567
for truss hazards, 548-549
wet-pipe systems, 566, 595
Stack effect, 481-482
summer, 484
winter, 483

Stair shafts, 69
Stairways/stairs
 accommodations or access, 475
 combustible, 218
 enclosures, 431-432
 exterior fire tower, 459
 and fire growth, 431
 in high-rise buildings, 459, 464, 470, 475, 491-492
 pressurized, 491-492
 relocated, 178
 scissor, 464, 470, 500
Standpipes, 457
Static loads, 24
Steel
 cold-drawn, 33, 256, 279, 301
 combustiblity, 257-258
 connectors, 83, 117, 263
 as a construction material, 262
 corrugated, 47, 264, 328, 335
 elongation of, 166, 256, 259, 273-274, 373
 failure of, 29, 33, 167, 259, 274
 fire characteristics, 33, 34, 83, 255-262, 271-275
 fire resistance of, 255
 fireproofing, 289-298, 359
 hot, water on, 259
 in non-steel buildings, 262
 safety factor, 49
 sheets of, 273
 strength of, 255
 structural, 256, 261
 testing, 46
 thermal conductivity, 257
 unprotected, 131-132, 153, 216, 265-266, 279, 287, 308, 312-314, 361-362
 weight of, 261
 see also Steel buildings; Steel construction
Steel beams, 17, 153, 166-167, 273, 546-547, 551
Steel buildings, 265-271
 high rise, 270
 metal deck roof problems, 302-309
 prefab, 266-267, 278
 protection of, 311-322
 test experience, 285-286
 see also Steel; Steel construction
Steel cables, 45, 49, 53, 162, 278, 279, 301, 338, 339, 370, 535
Steel columns, 166, 244, 283, 314, 317, 332
Steel construction, 255-323
 building code problems, 319-320
 definitions, 260-262
 fire doors, 311
 fire walls, 309-311
 framing, 266, 270, 271, 326
 highway structures and bridges, 276-278
 plastic design in, 271
 spans, 266, 267
 sprinkler protection, 299-300, 322
 walls, 267-270
 see also Steel; Steel buildings
Steel expansion joints, 272
Steel gusset plates, 83
Steel roofs, 35, 282. *See also* Metal deck roofs
Steel tendons, 256, 264
Steel tension rods, 132
Steel trusses, 29, 160, 282, 312, 314, 320, 522-523, 535, 558
Steel walls, 269
Steiner Tunnel Test, 410-412
Stone veneer siding, 124
Strain, 45. *See also* Stresses
Strands, 339
Streams. *See* Fire streams
Stressed skin construction, 114
Stresses, 10, 45
 impact, 23
 see also Loads
Stretcher course, 146
Structure design, elements of, 47, 48
Structural elements, 52-86
 arches, 74-76
 beams, 52-61
 columns, 61-67
 composite, 50-52
 connections, 80-86
 foundations, 79
 interior, effect on walls, 165-167
 rigid frames, 77-78
 roofs, 74
 shells and domes, 78
 stress and, 11, 23
 transmission of loads, 78-79
"Structural fire," diapatch, 3, 11, 89
Structural stability
 loss of, 12
 of masonry walls, 155-168
Struts, 61, 519
Stucco, 121
Stud channel, 107
Studs, 93, 97, 103
Superimposed loads, 52, 55
"Surveys Aid in Preparation For Handling Large Fires," 4

Suspended beams, 59, 60, 163
Suspended high-rise buildings, 49
Suspended loads, 48, 176-177
Suspension cables, 45, 49, 53

T

Talbert, J., 629
Tar, 307, 309
T-beams, 334
Tees, 261
Telephone facilities, 427
Telephone poles, 167
Temperature rods, 331
Tenants. *See* Occupants
Tendons,339
  cutting, 371
  protection of, 355
  whipping, 347
Tennessee retirement home fire, 40
Tenon, 96, 173
Tensile forces, 45, 46-47, 147
Tensile impact, 22-23
Tension, 29. *See also* Post-tensioning
Terra cotta tiles, 146, 147, 290
Thermal energy, 479-480
Thermoplastics, 38
Tied arches, 75, 519
Tiebacks, 279
Ties, 332-333, 519
Tile roofs, 91-92
Tiles
  acoustical, 185, 250, 294, 318, 387, 392, 398, 460-461
  hollow, 28, 51, 160-161, 335
  terra cotta, 146, 147, 290
Tilt slab construction, 167, 332, 357
  hazards of, 373
Timbers, 93
  heavy, 92
  imitation, 135-136
  spliced and laminated, 92, 115-117
Tin ceilings, 150, 272
Tornadoes, 29
Torsional loads, 29
Tower cranes, 279-281, 358-359
Town houses, 126, 215. *See also* Garden apartments
Toxic gases
  decay-treated wood and, 80
  Habel's rule and, 420
  membrane roofs and, 35
  physiological effects of, 419-421, 424
  plastics and, 38, 39
  pressure-treated wood and, 80
  transformers and, 427
  see also Carbon monoxide
Traffic controls, 106, 278
Training schools, 12
Transfer beams, 59, 60, 174, 520
  undesigned, 60
Transformers, electrical, 24, 427
Triaged, 259
Triangle Shirtwaist Factory fire, 451, 453, 570
Trim
  combustible, 364
  interior, 493-494
Trunnels, 94, 173
Truss floors, 109, 520, 537-538, 541, 546, 548-549
Truss frame construction, 103-104
Truss voids, 536, 542-543, 548-549, 557
Trussed arches, 520
Trusses, 55, 517-560
  aluminum, 36
  arched, 519
  bowstring, 10, 519-520, 534
  cantilevered, 521
  case studies, 554-560
  cast-iron, 519
  collapse of, 552-553
  components of, 519
  compressive load on, 66, 521
  concrete, 519
  connectors, 521
  defects, 524
  defined, 518
  excessive loads and, 527, 528
  exterior, 521
  failure of, 3, 61, 83, 104, 522, 524, 528-533, 541
  fire-resistance ratings, 538
  floors, 109, 520, 537-538, 541, 546, 548-549
  gusset-plate, 524, 529, 533
  hammerbeam, 518
  hazard control, 548-551
  hazards of, 103, 517, 525, 544-546
  heavy, 533-535
  inverted king post, 521
  lattice, 534
  masonry wire, 147
  parallel-chord, 10, 118, 135, 520, 535, 547
  principle types of, 519-521
  principles, 518, 521-527

problems with, 522-525
queen post, 522
resistance to collapse, 18
rising, 137, 524
roof, 74, 240, 520, 524, 528, 530-531, 536, 546, 554
steel, 29, 160, 282, 312, 314, 320, 522-523, 535, 558
timber, 524, 547
triangular, 518, 529-530
Vierendeel, 25, 520-521
wooden, 83, 118, 135, 153, 529, 534
Trussloft, 104, 109, 542-543
Tube construction, 25
Tubes, steel, 261
Tunnel Test (ASTM E84), 410-414
Tunnels, 443

U

Ultimate strength, 49
Underwriters blocks, 326
Underwriters Laboratories, 293
Unite area measurement, 45
Underground buildings, 44
Unreinforced masonry, 146

V

Vacant buildings, 133
Valves, sprinkler, 352, 588-590
Vapor seals
bituminous, 153
paper, 42, 384-385
Veneer walls, 67, 68, 146, 147. *See also* Siding
Ventilation
dangers, 15, 35, 446
necessity of, 442-447
positive pressure, 445-447
principles of, 489-490, 629-630
on roofs, 35, 180, 183, 308, 630
techniques, 444
Vents
attic, 229
automatic, 442-443
cockloft, 149
floor line, 429
manual, 629
Vertical openings, 454, 459, 497. *See also* Elevator shafts; Stairways
Vierendeel truss, 520-521
Vinyl siding, 120
Void spaces, 15, 100, 150, 580
in concrete buildings, 363-364
and fire growth, 389
hazards of, 184-195, 594-595
large, 190-191, 467-468
in mixed construction, 193
Voussoirs, 75, 160

W

Waffle concrete, 335, 363
Walkways, 79
Wall columns, 26, 72
Walls, 67-74
basement, 112-113
bearing, 23, 92, 268
"blowout," 24
bracing of, 71, 72-73, 146, 162
breaching, 73, 168, 342
brick (*see* Brick walls)
brick veneered (*see* Brick veneered walls)
cantilever, 70-71, 145
cavity, 73, 145, 146-147
collapse of, 68, 168
combination, 310
composite, 51, 68, 145, 147
cracks in, 160
cross, 68, 146
curtain, 67, 69, 92, 166, 199, 268
demising, 69
fire (*see* Fire walls)
fire testing, 245
holes in, 161
hollow, 75, 145, 146-147
hollow tile, 28, 160-161
impact loads and, 23
load-bearing, 67, 68
masonry (*see* Masonry walls)
in masonry buildings, 99
non-load-bearing, 67, 68, 69
panel, 67, 69, 269
parapet, 157
partition, 67, 69, 199-200, 310
party, 69, 197-199, 233-234, 235
serpentine, 72
shear, 25, 69, 70
stabilizing, 161
unstable, 161
veneer, 67, 68, 146, 147
wind load and, 67, 71
wood-treated, 112-113
Warehouses. *See* Rack storage warehouses

Water
  fireproofing with, 300
  on hot steel, 259-260
  loads, 18, 19, 21-22, 182
  materials soaked by, 167, 179
  ponding, 19
Water curtains, 106
Water damage, 313, 425, 572, 573, 575, 579, 610
Water distribution, 586
Water pressure, 586
Water spray systems, 311, 614
Water supplies
  garden apartments, 222-223, 239
  for spray systems, 312
  for sprinklers, 502, 586, 621
Water tanks, elevated, 18, 19
Web, 519, 551
Wet joints, 80
Wet-pipe systems, 566, 595
Wide-flange shapes, 261
Wilshire fire, 105
Wind, and smoke movement, 26, 480-481
Wind bent, 61
Wind loads, 25-26, 62
  on walls, 67, 71
Windows
  concealed, 191, 197
  double or triple glazed, 36, 42, 467
  energy efficient (EEWs), 467
  in high-rise buildings, 36
  unbreakable, 36, 467
  see also Glass
Wood, 93
  as a building material, 100
  compressive strength, 46
  dangerous treated, 112-113
  engineered, 92, 114-119
  fire characteristics of, 33, 180-181
  fire-retardant treated (FRT), 111, 549
  firestopping, 107
  grain of, 114
  impregnated, 110-111
  laminated, 92, 115-117
  limitations of, 114
  lumber, 92, 137
  in non-combustible and fire resistive buildings, 104-105
  pressure-treated, 111, 113
  protection from ignition, 110-112
  strength of, 136-137
  surface coated, 112
  uses of, in construction, 93
  see also Wood construction
Wood arches, 76, 116-117
Wood construction, 25, 89-138
  advantages and disadvantages, 130-131, 137-138
  cooling towers, 129
  definitions, 92-93
  fire-resistive buildings, 104-106
  firestopping, 106-109
  heavy timber, 129-130
  non-combustible buildings, 104-105
  open area structures, 134
  rehabilitation and demolition hazard, 132
  roofing, 124-126
  row frame buildings, 126-128
  sheathing, 119-120
  siding, 120-124
  types of, 90-92, 93-104
  unprotected steel in, 314
  vacant buildings, 133
  wood in, 110-115, 135-137
  see also Wood
Wood lath, 93
Wooden beams, 17, 54, 85, 158-160, 172, 551
Wooden columns, 62
Wooden cooling towers. *See* Cooling towers
Wooden floors, 26
Wooden roofs, 112-113
  fire characteristics of, 180-181
  shingled, 120, 124-126
  see also Plywood roofing
Wooden shingles, 90
Wooden suspended ceilings, 136
Wooden trusses, 83, 118, 135, 153, 529, 534
Woodway Square Apartments fire, 125
Wythe, 146

Z

Zees, 261
Zinc chloride, 115